Reliability Analysis of Modern Power Systems

About the Authors

Professor (Dr.) R. K. Saket (Member'12–Senior Member'18, IEEE) is a professor with the Department of Electrical Engineering, Indian Institute of Technology (Banaras Hindu University), Varanasi, Uttar Pradesh, India. Previously, he was a faculty member at Government Engineering College, Rewa (Madhya Pradesh), India; Birla Institute of Technology and Science, Pilani (Rajasthan), India; Sam Higginbottom University of Agriculture, Technology and Sciences, Allahabad (Uttar Pradesh), India; and University Institute of Technology, Rajiv Gandhi University of Technology, Bhopal (Madhya Pradesh), India. He has provided his industrial services to Electrical Safety Division of the Department of Energy, Government of Madhya Pradesh, Ujjain Division (Madhya Pradesh), India, as an electrical safety engineer. He has more than 25 years of academic, industrial, and research experience. He is the author/co-author of six national/international patents, approximately 200 scientific articles, book chapters, and research papers in indexed international journals and prestigious conference proceedings. He received the Best Paper Award(s) at Thammasat University, Bangkok, Thailand on 23 January 2017 and The Maldives National University, Republic of Maldives on March 11, 2023. He is the author/co-author of the prestigious textbook *Modern Power System Analysis*, 5th edition, 2022, published by McGraw Hill, New Delhi, India. He has supervised 21 PhD research scholars and 55 MTech students. He has delivered many technical talks and is honored as a resource person in power system reliability engineering. He has delivered his academic and research lecture(s) at World Academy of Science, Engineering & Technology, Paris (France) on 21–23 September 2014; International Theravada University, Yangon (Myanmar) on 2–4 December 2014; Thammasat University, Bangkok (Thailand) on 21–23 January 2017; Tokyo City University, Tokyo (Japan) on 9–11 March 2018; Brunel University, London (UK) on 20–23 February 2023; and The Maldives National University, Male City (Maldives) on 11–12 March 2023. His research interests include reliability engineering, electrical machines and drives, modern power system reliability, reliability enhancement of industrial components and systems, and reliability aspects in renewable energy systems.

Professor Saket is a fellow of the Institution of Engineers (India), fellow of the Institution of Electronics and Telecommunication Engineers (India), senior member of IEEE (USA), member of IET (UK), and a life member of the Indian Society for Technical Education, New Delhi (India). He is an associate editor of the *IET Renewable Power Generation* (UK), *IET Electrical Systems in Transportation* (USA), and *IEEE Access* (USA). He is the Managing Guest Editor of the *IEEE Journal of the Electron Devices Society* (USA), *Computers & Electrical Engineering*, Elsevier, and *Electrical Engineering*,

Springer Nature. He is an Editorial Board Member of the *Journal of Electrical Systems* (Algeria), and *Engineering, Technology and Applied Science Research* (Greece). He has received many awards, honors, and recognitions for his excellent academic and research contributions including the prestigious Gandhian Young Technological Innovation Award (2018) appreciated by the hon'ble president of India at Rashtrapati Bhavan, New Delhi, India; Design Impact Award (2018) appreciated by Padma Vibhushan Ratan Tata at Mumbai, India; Nehru Encouragement Award (1988 and 1990) awarded by the hon'ble chief minister of Madhya Pradesh state government, Bhopal, India; and IEEE IAS Global Distinguished Educator Award 2023 (USA) at the Maldives National University, Republic of Maldives.

Professor (Dr.) P. Sanjeevikumar (Member'12–Senior Member'15, IEEE) received the PhD degree in electrical engineering from the University of Bologna, Bologna, Italy, in 2012. He is a full professor in electrical power engineering at the Department of Electrical Engineering, Information Technology, and Cybernetics, University of South-Eastern Norway, Norway. He has authored over 750 scientific papers and received the Best Paper cum Most Excellence Research Paper Awards from IET-SEISCON'13, IET-CEAT'16, IEEE-EECSI'19, IEEE-CENCON'19, and five best paper awards from ETAEERE'16-sponsored Lecture Notes in Electrical Engineering, Springer book. He is a fellow of the Institution of Engineers (India), the Institution of Electronics and Telecommunication Engineers (India), and the Institution of Engineering and Technology (UK). He received a lifetime achievement award from Marquis Who's Who in USA in 2017 for contributing to power electronics and renewable energy research. He has been listed among the world's top 2% of scientists since 2019 by Stanford University, USA.

He is an editor/associate editor/editorial board for refereed journals, in particular the *IEEE Systems Journal (USA), IEEE Transaction on Industry Applications (USA), IEEE Access (USA), IET Power Electronics (UK), IET Electronics Letters (UK),* and *International Transactions on Electrical Energy Systems, Wiley;* subject editorial board member—*Energy Sources—Energies Journal*, MDPI; and the subject editor for the *IET Renewable Power Generation (UK), IET Generation, Transmission and Distribution (UK)*, and *FACETS* Journal (Canada).

List of Contributors

Almoataz Y. Abdelaziz
Faculty of Engineering & Technology
Future University in Egypt
Cairo
Egypt

Rahul Agrawal
Department of Electrical Engineering
Indian Institute of Technology (BHU)
Varanasi, Uttar Pradesh
India

K. S. Anand Kumar
Electrical Division
CSIR National Aerospace Laboratory
Bangalore, Karnataka
India

Liladhar Arya
Department of Electrical Engineering
Shri G.S. Institute of Technology & Science
Indore, Madhya Pradesh
India

Rajesh Arya
Department of Information Technology
Acropolis Institute of Technology and Research
Indore, Madhya Pradesh
India

Vanitha Bagana
Department of Electronics and
Communication Engineering
National Institute of Technology
Jamshedpur, Jharkhand
India

Ramesh C. Bansal
Department of Electrical Engineering
University of Sharjah
Sharjah
United Arab Emirates

Om P. Bharti
Department of Electrical Engineering
Government Polytechnic
Ghazipur, Uttar Pradesh
India

Udaya M. Bhaskara Rao
Engineering Division
CSIR National Metallurgical Laboratory
Jamshedpur, Jharkhand
India

Sanjay Kumar Chaturvedi
Subir Chowdhury School of Quality and
Reliability
Indian Institute of Technology Kharagpur
Kharagpur, West Bengal
India

Laxman Chaudhary
Department of Electrical Engineering
Indian Institute of Technology (BHU)
Varanasi, Uttar Pradesh
India

Sharat Chandra Choube
Department of Electrical & Electronics Engineering
University Institute of Technology, RGPV
Bhopal, Madhya Pradesh
India

Gautam Singh Dohare
Department of Electrical, Electronics & Communication Engineering
Galgotias University
Greater Noida, Uttar Pradesh
India

Dhawal Dwivedi
Department of Electrical Engineering
Indian Institute of Technology (BHU)
Varanasi, India

Stuart Galloway
University of Strathclyde
Glasgow
UK

Heeralal Gargama
Subir Chowdhury School of Quality and Reliability
Indian Institute of Technology Kharagpur
Kharagpur, West Bengal
India

Suprakash Gupta
Department of Mining Engineering
Indian Institute of Technology (BHU)
Varanasi (UP)
India

Debasis Jana
Department of Mining Engineering
Indian Institute of Technology (BHU)
Varanasi (UP)
India

Kalpesh B. Kela
Department of Electrical Engineering
L.D. College of Engineering
Ahmedabad, Gujarat
India

Baseem Khan
Department of Electrical and Computer Engineering
Hawassa University
Hawassa
Ethiopia

Mohd. Tauseef Khan
Department of Electrical Engineering
Rajkiya Engineering College
Banda, Uttar Pradesh
India

Atul Koshti
Department of Electrical Engineering
GES'S R.H. Sapat College of Engineering
Management Studies & Research
Nashik, Maharashtra
India

Dwarkadas P. Kothari
Department of Electrical Engineering
Indian Institute of Technology
Delhi
India

Amit Kumar
Flameproof and Equipment Safety Department
CSIR-Central Institute of Mining and Fuel Research
Dhanbad, Jharkhand
India

Deepak Kumar
Department of Mining Engineering
Indian Institute of Technology (BHU)
Varanasi (UP)
India

Sachin Kumar
Department of Electrical Engineering
G.B. Pant Institute of Engineering and Technology
Pauri-Garhwal, Uttarakhand
India

Sandeep Kumar
Department of Electronics and Communication Engineering
G.B. Pant Institute of Engineering and Technology
Pauri-Garhwal, Uttarakhand
India

Kanhaiya Kumar
Electrical & Electronics Engineering Department
Galgotias College of Engineering & Technology
Greater Noida, Uttar Pradesh
India

Jyoti Maurya
Department of Electrical Engineering
Indian Institute of Technology (BHU)
Varanasi, Uttar Pradesh
India

Saad Mekhilef
Department of Electrical Engineering
Swinburne University of Technology
Melbourne
Australia

Joydeep Mitra
Department of Electrical Engineering
Michigan State University
Michigan
USA

Nga T. T. Nguyen
Department of Electrical Engineering and Computer Science
University of Wyoming
Wyoming
USA

Jyotsna Ogale
Department of Electronics Engineering
Samrat Ashok Technological Institute
Vidisha, Madhya Pradesh
India

Pradeep Purey
Department of Electronics Engineering
Maharaja Ranjit Singh College of Professional Sciences
Indore, Madhya Pradesh
India

Santosh S. Raghuwanshi
Department of Electrical Engineering
Medi-Caps University
Indore, Madhya Pradesh
India

Rajiv Nandan Rai
Subir Chowdhury School of Quality and Reliability
Indian Institute of Technology Kharagpur
Kharagpur, West Bengal
India

Chandrima Roy
Department of Electronics & Communication Engineering
Heritage Institute of Technology
Kolkata, West Bengal
India

Bharat B. Sagar
Department of Computer Science and Engineering
School of Engineering, Harcourt Butler Technical University
Kanpur, Uttar Pradesh
India

R. K. Saket
Department of Electrical Engineering
Indian Institute of Technology
(Banaras Hindu University)
Varanasi (UP)
India

Devkaran Sakravdia
Department of Electrical Engineering
Government Engineering College
Ujjain, Madhya Pradesh
India

P. Sanjeevikumar
Department of Electrical Engineering
IT and Cybernetics
University of South-Eastern Norway
Porsgrunn, Norway

Kumari Sarita
Department of Electrical Engineering
Government Engineering College
Aurangabad, Aurangabad, Bihar
India

Jitendra Singh Shakya
Department of Electrical Engineering
Samrat Ashok Technical Institute
Vidisha, Madhya Pradesh
India

Pushpendra Singh
Department of Electrical Engineering
Rajkiya Engineering College
Banda, Uttar Pradesh
India

Saumya Singh
Department of Electrical Engineering
Indian Institute of Technology (BHU)
Varanasi, Uttar Pradesh
India

Smriti Singh
Department of Electrical Engineering
Indian Institute of Technology (BHU)
Varanasi, Uttar Pradesh
India

Sunil K. Singh
Department of Electrical Engineering
Shri Ramswaroop Memorial University
Lucknow, Uttar Pradesh
India

Sandeep K. Soni
Department of Electrical Engineering
Shiv Nadar Institution of Eminence
Delhi, NCR
India

Bhavik N. Suthar
Department of Electrical Engineering
Government Engineering College
Bhuj, Gujarat
India

Eram Taslima
Department of Electrical Engineering
Indian Institute of Technology (BHU)
Varanasi, Uttar Pradesh
India

Lakhan Singh Titare
Department of Electrical Engineering
Government Engineering College
Ujjain, Madhya Pradesh
India

Aanchal Singh S. Vardhan
Department of Electrical Engineering
National Institute of Technology
Jamshedpur, Jharkhand
India

Akanksha Singh S. Vardhan
Department of Electrical Engineering
Shri GS Institute of Technology & Science
Indore, Madhya Pradesh
India

Lokesh Varshney
Department of Electrical, Electronics & Communication Engineering
Galgotias University
Greater Noida, Uttar Pradesh
India

Aanchal Verma
Department of Electrical Engineering
Indian Institute of Technology (BHU)
Varanasi, Uttar Pradesh
India

Foreword

Writing a foreword is both a pleasure and a challenge. Still, I always enjoy writing the foreword to recommend bookwork in my fields of power, energy, drives, reliability, renewables, and power systems technologies. This application benefits society and solves power demand, reliability, power quality, generation, transmission, and distribution crises through modern power systems. I read this book titled *Reliability Analysis of Modern Power Systems*, Wiley-IEEE Publications, authored/edited by learned colleagues Prof. R. K. Saket, Department of Electrical Engineering, Indian Institute of Technology (Banaras Hindu University), Varanasi, Uttar Pradesh, India, and Prof. P. Sanjeevikumar, Department of Electrical Engineering, IT and Cybernetics, University of South-Eastern Norway, Porsgrunn, Norway, which has proven exciting and innovative research findings for the readers. In particular, power generation, transmission, and distribution reliability are continuously gaining attention as crucial, empowering technology with a smart control approach to enhance the system's reliability and efficiency. These challenges are solvable through modern power electronics, with higher flexibility, adjustable electronics loads, energy storage systems, and renewable energy systems, including windmills, solar PV, batteries, supercapacitors, hydropower, and tidal power.

On reading, the topics that draw strong reader attention are the basic principles and scientific importance of reliability theory, the Bayesian approach and useful life prediction, genetic algorithms, artificial neural networks in reliability-based optimization, and reactive power control variables rescheduling; and also reliability evaluation of power electronics converters and devices in modern power system applications, reliability assessment of multilevel inverters, design and development of microgrids, and reliability aspects in snubber circuits for industrial power applications. I hope the readers will enjoy reading this book and be able to grasp the research findings as a textbook for bachelor's, master's, and PhD programs in universities and further future enhancement in modern science, engineering, technology, and skills!

Finally, I congratulate the editors, authors, Wiley-IEEE Publishers, reviewers and editorial assistants, and the IEEE and Wiley press production team; without their support, it would not be a successful outcome. Credit goes to their effort in making flyers and editing, proof correction, etc.

With Best Compliments

Prof. Frede Blaabjerg, Fellow IEEE
Villum Investigator Professor,
Center of Reliable Power Electronics (CoRPE),
Department of Energy Engineering,
Aalborg University, Aalborg, Denmark

It is a pleasure to write the foreword to this important book, *Reliability Analysis of Modern Power Systems,* edited by Dr. R. K. Saket and Dr. P. Sanjeevikumar. The book has been contributed by many authors who are university professors specialized in respective topics and are editors. The book is hybrid, heterogeneous, and interdisciplinary in nature. All the contributions appear authoritative in nature. The book contains 26 chapters.

Reliability is an extremely important topic in modern power systems. It is particularly important for tomorrow's large and complex smart grid power system, where many generators (fossil, nuclear, and renewables with storage elements) are interconnected through long transmission, distribution, and utilization systems to supply reliable electricity to consumers at economical prices. Smart grid uses state-of-the-art technologies in power systems, power electronics, computers, communications, information, cyber, and AI to optimize system availability, reliability, power quality, security, and energy efficiency with optimum resource utilization.

The book deals with reliability theories and their applications in power systems very comprehensively. The topics include reliability studies in the distribution system, photovoltaic energy system, hydro-power system, power electronic converters (including devices and snubbers), electric vehicles, and microgrids. It also covers studies on genetic algorithms and artificial neural networks in reliability-based optimization, parametric estimation models, security of transmission systems, probabilistic voltage security assessment, reactive power control rescheduling, etc.

The book can be considered an important reference book for engineers involved in designing reliable power systems. It can also be considered a graduate-level text. Selected topics from the book can also be taught in undergraduate courses.

It should be mentioned here that Dr. R. K. Saket and Dr. P. Sanjeevikumar are well known internationally for their research contributions. While Dr. R. K. Saket specializes in reliability theory and applications, Dr. P. Sanjeevikumar is more renowned in the area of power electronics. The publication of this book is very timely and will be accepted with enthusiasm.

Best Compliments

Prof. Bimal K. Bose, PhD, DSc (Honoris Causa), IEEE Life Fellow
Emeritus Chair Professor of Electrical Engineering
(Formerly Condra Chair of Excellence in Power Electronics),
Department of Electrical Engineering and Computer Science,
The University of Tennessee, Knoxville, USA
http://en.wikipedia.org/wiki/Bimal_Kumar_Bose

Preface

Reliability is an old concept and a new discipline in engineering and technology. *Reliability Analysis of Modern Power Systems* (RAMPS) has evolved from our deep interest in academics and innovative research and our long-standing involvement in quantitative reliability evaluation and application of probabilistic techniques to modern power system problems. Modern power systems have grown larger and spread over larger geographical area with many interconnections between neighboring systems and integration of renewable energy sources. The applications benefit society and solve power demand, power quality, reliability, generation, transmission, and distribution crises through modern power systems. The systems that involve high-risk and high-cost factors need to be reliable enough. This book is basically intended to present important aspects of reliability and its analysis in the field of modern power systems. The primary objective of RAMPS is to present the historical importance of reliability engineering, dependability assessment of power components, design enhancement for reliability, impact of renewable energy systems, and innovative research findings for engineering students, PhD research scholars, practicing power engineers, and technical managers. The emphasis throughout is on practical applications, and the mathematical concepts described are accordingly suitable for solving the types of reliability problems covered in RAMPS. The references given in the bibliographies are limited to those considered to provide a direct continuation of the chapter material, with an emphasis on experimental research and practical applications. Tables, figures, flowcharts, and graphs are provided to complement the analytical methods described, and numerous worked examples are included. Power generation, transmission, and distribution are continuously gaining attention as crucial, empowering technology with a smart control approach to enhance the system's reliability and efficiency. The challenges are solvable through modern power electronics with high flexibility, adjustable electronic loads, energy storage systems, and renewable energy systems including wind energy conversion systems, solar photovoltaic, electric vehicles, hydropower, and tidal power.

The text of RAMPS is divided into six sections as shown in Figure P.1. The first section describes the historical background, reliability principles, and comprehensive importance of reliability engineering; the second section addresses the design for reliability and design enhancement for dependability of modern power components and systems; the third section provides information about reliability assessment aspects for transmission system of modern power systems; the fourth section describes the reliability issues associated with conventional power distribution system; the fifth section of RAMPS illustrates the reliability analysis methodologies of modern distribution systems

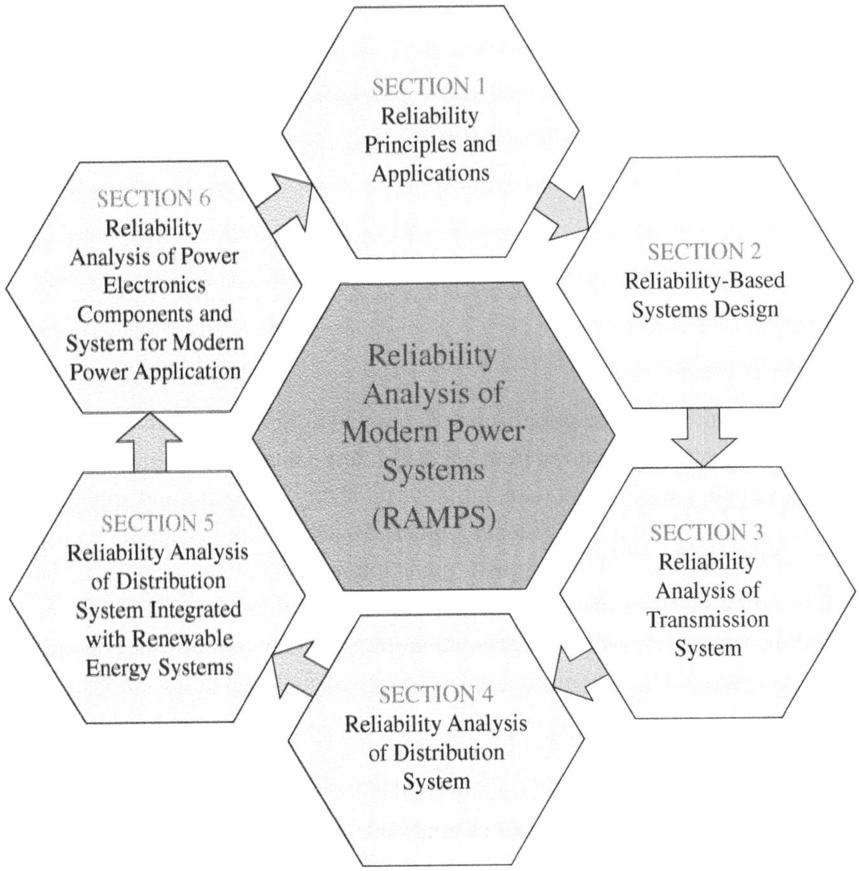

Figure P.1 Organization of *Reliability Analysis of Modern Power Systems.*

integrated with renewable energy systems; and the reliability analysis of power electronics components, integrated circuits and systems for modern power system applications are described in the sixth section.

The sections of the RAMPS textbook are subdivided into 26 innovative chapters as shown in Figure P.2. *Section 1: Reliability Principles and Applications* provides an overview of reliability principles and comprehensive applications in the fields of modern engineering and technology, where the basic concepts of reliability, life distribution of power components, reliability analysis of composite power system structure, failure distribution functions, and useful life predictions are described in detail. In Chapter 1, the basic reliability principles, applicability of bathtub curve, and scientific importance of reliability theory with numerical illustrations are described in detail. The Bayesian approach for reliability evaluation and remaining useful life prediction is presented in Chapter 2. The evaluation of basic reliability indices using state enumeration method with industrial numerical problems is described in Chapter 3. In Chapter 4, the methodologies for reliability evaluation of network with industrial problems are discussed. The probabilistic approach for standby and load-sharing system reliability evaluation is illustrated in Chapter 5.

Section 2: Reliability-Based Power Systems Design contains four innovative chapters. The physical reliability methods and design for system reliability including experimental questions and industrial real problems are presented in Chapter 6. Design for maintainability and availability analysis

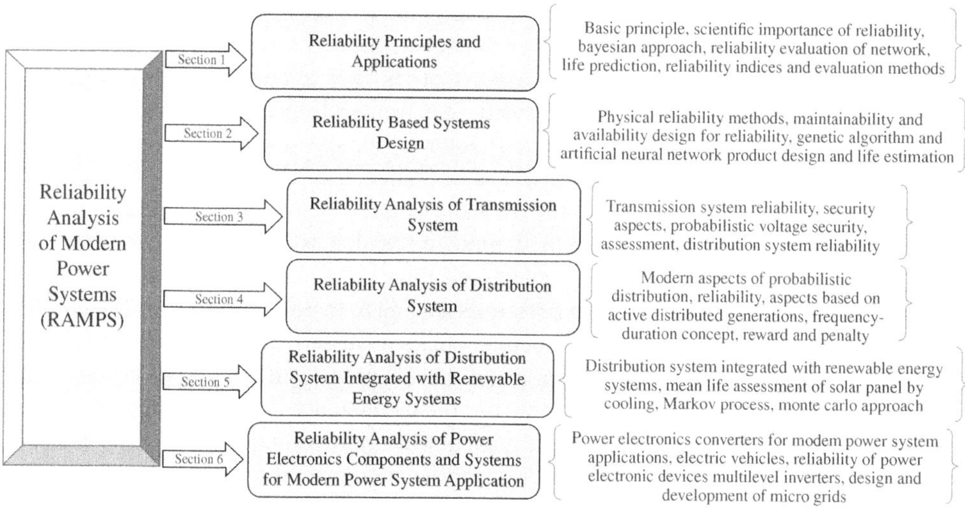

Figure P.2 Section classification of RAMPS and organization of the chapters.

for system design are described in Chapter 7. Chapter 8 presents the genetic algorithm and artificial neural networks in reliability-based design optimization. The parametric estimation models for minimal and imperfect maintenance are illustrated in Chapter 9.

Section 3: Reliability Analysis of Transmission Systems contains only two innovative chapters. The transmission system reliability evaluation including security aspects and dependability illustrations is illustrated in Chapter 10. In Chapter 11, the probabilistic voltage security assessment and stability enhancement using rescheduling of reactive power control variables of modern power systems are presented.

Section 4: Reliability Analysis of Distribution Systems is incorporated with four comprehensive chapters. Chapter 12 presents the modern aspects of probabilistic distributions with industrial numerical problems for reliability evaluation of engineering systems. Reliability enhancement of electrical distribution systems considering active distributed generations is discussed in Chapter 13. In Chapter 14, the reliability enhancement strategy for electrical distribution system considering reward and penalty is presented and supported by industrial data. The reliability analysis of composite distribution system using frequency–duration concept is incorporated in Chapter 15.

Section 5: Reliability Analysis of Distribution Systems Integrated with Renewable Energy Systems is decorated with five innovative chapters. The reliability assessment of distribution systems integrated with renewable energy systems is presented in Chapter 16. In Chapter 17, the reliability evaluation and performance analysis of hybrid photovoltaic energy systems for rural electrification using Markov process are described successfully. The probabilistic distributions and Monte Carlo approach for reliability evaluation of SEIG-based modern power systems are presented in Chapter 18. In Chapter 19, the reliability and mean life assessment of solar panels by cooling are discussed. The reliability assessment of different topologies in photovoltaic systems is incorporated into Chapter 20.

Section 6: Reliability Analysis of Power Electronics Components and Systems for Modern Power System Applications is incorporated with six innovative chapters. The reliability evaluation of power electronics converters for modern power system applications is covered in Chapter 21. In Chapter 22, the reliability assessment of sub-components of electric vehicles for performance enhancement of grid-integrated power systems is incorporated excellently. The reliability assessment of multilevel inverters for modern power system applications is illustrated successfully in Chapter 23. In Chapter 24, the reliability aspects of snubber circuits for industrial power applications are discussed. The reliability assessment of power electronics devices and systems for modern power applications is incorporated into Chapter 25. Finally, the reliability aspects of the design and development of microgrids are illustrated in Chapter 26.

The requirements of the course syllabi of the All India Council for Technical Education (India), University Grants Commission (India), Indian Institute of Technology (IITs), American Society for Quality Control (USA), and the Institute of Quality Assurance (UK) in reliability engineering are covered, so the book will be suitable for use in courses leading to these organizations. The emphasis on practical approaches and experimental methodologies in modern power system engineering, comprehensive coverage of reliability standards and specifications, and the overall layout of the book should make it equally suitable as a general, up-to-date reference for use in academia, industry, and government agencies. We hope that the readers will enjoy reading this book of reasonable length and comprehensive coverage. This book is timely and addresses the needs of engineers, professionals, and researchers. Readers will be able to grasp the research findings as a textbook for graduation, postgraduation, and PhD programs in universities and will refer it for future enhancement in modern science, engineering, technology, and skills!

Finally, we congratulate the authors, co-authors, expert reviewers, editorial assistants, Wiley-IEEE Publishers, and the IEEE and Wiley press production team; without their valuable support, it would not be a successful outcome.

18 January 2024
Varanasi (UP) India

R. K. Saket
P. Sanjeevikumar

Acknowledgments

While writing this textbook, we have had the benefit of valuable advice and suggestions from many academicians, senior faculty colleagues, practicing engineers, and other innovators who used the earlier editions of the book entitled *Modern Power System Analysis*, authored by D.P. Kothari, I.J. Nagrath, R. K. Saket, and our previous research publications. I want to thank the expert reviewers of *Reliability Analysis of Modern Power Systems* (RAMPS) for their many helpful suggestions. To the extent that I have been able to respond to their criticism, the text has been improved immeasurably. All these individuals have influenced this edition. We express our thanks and appreciation to them. The publishers and authors gratefully acknowledge the suggestions and also express their appreciation for all those reviewers who took the time to review the book. We hope this support and response will continue in future.

We acknowledge our sincere thanks and regards to Prof. L.D. Arya, Professor of Power System Reliability Engineering, Shri G.S. Institute of Technology and Science, Indore, Madhya Pradesh, India, for his wholehearted encouragement, excellent contributions, and cooperation to complete this prestigious book. I owe this moment of satisfaction with deep gratitude for his valuable time, exhorting guidance, and valuable suggestions during the various stages of this book. His lifelong experience in the field of power system reliability engineering and awareness of recent developments in these areas have been very useful to us during the writing of this book. No amount of words can really suffice to express our sincere thanks to him for discussing the problems critically and providing concrete suggestions.

We are heartily thankful to Prof. D.P. Kothari, Former Director and Professor of Power System Engineering, Indian Institute of Technology, New Delhi, India, for his valuable guidance, dedicated efforts, keen interest in developing the writing skills, mental satisfaction, encouragement, and helpful criticism at various stages of the book chapter preparation. His positive attitude, unassailable optimism, constant encouragement, and invaluable assistance made it possible to bring out this book in its present shape.

Prof. R. K. Saket would like to acknowledge sincere thanks and kind regards to his great Guru, Prof. I.J. Nagrath, Former Deputy Director and Professor of Power System Engineering, Birla Institute of Technology and Science, Pilani (Rajasthan), India, who provided the historical opportunity to join the PhD program under his valuable supervision and helped directly or indirectly wherever needed during his research work and BITSian life.

Prof. R. K. Saket would like to express his heartily regards and dedication to his great father Parinivutta Ram Das Saket Vardhan and beloved mother Parinivutta Terasita Devi who gave him inspiration and perennial encouragement adventurously during his engineering education and academic career. He will be failing his basic duties if he does not mention the golden soil of his small village Dhaurahara, located in the Vindhya–Rewanchal region of Madhya Pradesh, India, which was far away from the city and was lacking the basic amenities such as electricity, quality education, drinking water, and road during his engineering education, even then he could rise to this level, could publish his innovative research work in SCIE journals of all the global publishers, and could write many books including prestigious RAMPS, Wiley-IEEE (USA).

Prof. P. Sanjeevikumar would like to express his heartily regards and dedication to his parents, the late Mr. Padmanaban Sanjeevi and the late Mrs. Kamala Padmanaban, for their inspiration and encouragement adventurously during his engineering education, academic engagements, and research activities.

The editors would also like to thank the higher authorities, learned faculty colleagues, and research scholars of the Indian Institute of Technology (Banaras Hindu University), Varanasi, Uttar Pradesh, India, and the University of South - Eastern Norway, Norway, who have been very generous and helpful toward us for maintaining a suitable atmosphere, providing manuscript preparation support, and extending all facilities required for the completion of this prestigious book. The editors, and authors acknowledge the financial support extended by Science and Engineering Research Board (SERB); a statutory body of the Department of Science & Technology (DST), Government of India; New Delhi, India under the sponsored research project sanction order No. EEQ/2021/000177, and project code No.: R&D/IIT(BHU) Varanasi (SERB)/EE/21-22/09/348/ 02.05.2022.

The editors are thankful to Ms. Victoria Bradshaw and Ms. Storry Aileen, Senior Editorial Assistant for Electrical Engineering, Technology & Engineering Careers; Wiley – IEEE publishers (USA) for handling peer-review process of the RAMPS proposal and for providing appreciable improvement suggestions to enhance the book proposal contents. We would also like to acknowledge Ms. Kavipriya Ramachandran, Managing Editor; Mr. Deenadayalu Govindanagaraj, Content Refinement Specialist; Mr. Vasanth Koilraj, Support Service Administrator; and production staff of Wiley – IEEE publishers (USA) for completing publication requirements and copyright agreement formalities for contributors.

The editors are thankful to Prof. Bimal Kumar Bose, Department of Electrical Engineering and Computer Science, the University of Tennessee, Knoxville, USA, and Prof. Frede Blaabjerg, Department of Energy Engineering, Aalborg University, Aalborg, Denmark, for writing the foreword notes on this book.

We want to express our affection to the future talents of next-generation Aanchal, Aakanksha, Siddharth, Siddhant, Kishankumar, and Yomeka for providing emotional support and natural happiness directly or indirectly throughout the publication process of this textbook.

18 January 2024

R. K. Saket
P. Sanjeevikumar

Section 1

Reliability Principles and Applications

1

Basic Principles and Scientific Importance of Reliability Theory

Aanchal Verma[1], Akanksha Singh S. Vardhan[2], Vanitha Bagana[3], R. K. Saket[1], and P. Sanjeevikumar[4]*

[1] *Department of Electrical Engineering, Indian Institute of Technology (BHU), Varanasi, Uttar Pradesh, India*
[2] *Department of Electrical Engineering, Shri GS Institute of Technology & Science, Indore, Madhya Pradesh, India*
[3] *Department of Electronics and Communication Engineering, National Institute of Technology, Jamshedpur, Jharkhand, India*
[4] *Department of Electrical Engineering, IT and Cybernetic, University of South-Eastern Norway, Porsgrunn, Norway*

1.1 Introduction

Reliability, an age-old concept, has evolved into a contemporary discipline within modern engineering and technology. Originating as an extension of probability theory, reliability theory found its early applications in industries such as nuclear, electronics, and aerospace after World War II. As engineering systems become more complex, the demand for elevated reliability has emerged as an essential requirement. Reliability as a concept was primarily formulated for nonrepairable technical systems due to the critical importance of initial failure in such contexts.

Over time, reliability studies expanded to encompass diverse engineering fields, both for systems that are repairable and those that are not. The evolution of reliability theory traces back to World War II, with formal evaluations emerging as a response to issues with German missiles. However, the challenges of data scarcity, computational limitations, and a reluctance to employ probabilistic techniques have progressively diminished.

In this chapter, our objective is to outline contemporary reliability evaluation techniques suitable for modern power systems while elucidating the significance of various reliability indices. As reliability often intertwines with safety concerns, particularly in scenarios involving human lives, it becomes integral to broader safety considerations (Saket et al. 2022).

Reliability is a pivotal factor in engineering design, influencing system effectiveness and accounting for constraints such as cost. The analysis also weighs factors like operating environment, operational time, and service quality. To quantitatively assess system utility, reliability indices rooted in probability theory are essential. Qualitative judgments are insufficient for engineers planning and designing systems.

Operating environment plays a substantial role in reliability modeling, affecting failure and repair rates. Consideration of time in operation is imperative, as the probability of successful system operation diminishes over prolonged periods. Equally crucial is the quality of service, as the system's availability does not necessarily guarantee consumer satisfaction.

In essence, reliability can be described as the likelihood of a component or system fulfilling its intended function adequately throughout the designated period of operation, contingent upon the given operating environment.

Reliability Analysis of Modern Power Systems, First Edition. Edited by R. K. Saket and P. Sanjeevikumar.

1.2 Basic Concept of Reliability Engineering

Reliability is described as the probability that a component or system effectively carries out its intended function within the designated timeframe and under the specified operating conditions. The classical definition of reliability underscores four fundamental attributes: the likelihood of success, adequate functionality, duration of operation, and prevailing operating conditions.

Reliability engineering delves into the exploration of equipment lifespan and failure. By applying scientific principles and mathematical methodologies, it investigates the aging and failure patterns of devices. The goal is to gain insights into failure mechanisms, leading to enhanced product designs that promote longevity and mitigate the consequences of failure. The emphasis here is on design improvement.

A pivotal aspect is that new system and product designs must demonstrate safety and reliability before being manufactured and utilized industrially. This practice is widespread across various engineering domains, encompassing both nonrepairable and repairable systems.

Electric power systems serve as a prime example of environments where an exceedingly high level of reliability is anticipated (Bansal et al. 2002; Arya et al. 2000, 2001). The reliability concept is often divided into two facets: security and adequacy. Adequacy concerns the presence of ample facilities within the system to meet customer load demands. This encompasses energy generation and the requisite transmission and distribution infrastructure. Probabilistic models are commonly employed for adequacy assessment across different segments of the power system, like generation, and load. In contrast, security pertains to the system's capacity to respond effectively to internal disruptions.

In engineering applications, concepts necessitate quantification, translating into procedures that establish correlations between various attributes and numerical scales for measurement. This necessitates the conversion of reliability into measurable quantities through suitable definitions. Reliability is rooted in the mathematical realm of probability, and its practical implications can vary significantly across different contexts.

Electric power systems exemplify environments where exceptional reliability is imperative. For numerous power systems, customers typically experience an average of 2–3 hours of interruptions per year, with even shorter durations in certain regions (Bharti et al. 2019; Kumar et al. 2020; Saket et al. 2006, 2007, 2009). Industries with stringent requirements also demand a high degree of reliability. Consequently, reliability has consistently been a pivotal factor in the design, arrangement, maintenance, and operation of electrical power systems. Reliability theory's formal concepts and methodologies have been extensively employed to evaluate power system reliability from various angles.

1.2.1 The Role of Reliability Engineering

The chapter also delves into the primary challenges and potential solutions presented in this book. Additionally, it explores the contemporary context, including the implications of digital transformation and the influence of the Industrial Internet of Things (IIoT) on the functions of reliability engineers. The central role of the power reliability engineer (PRE) is to elevate the reliability of essential assets, which encompass fixed power plants, earth-moving equipment, power-related components, and the associated utilities and facilities. Reliability is established through the integration of reliability engineering principles into the asset management (AM) system, alongside a rigorous and organized assessment of physical failures, asset conditions, and component reliability data. Reliability engineering is a cornerstone discipline that ensures products, systems, and services operate as intended for their intended lifespan. This comprehensive field plays a pivotal role in numerous industries, contributing to better safety, cost-effectiveness, enhanced productivity, customer satisfaction, and adherence to regulations. The importance of reliability engineering is described as follows:

1. **Improving Safety:**
   Reliability engineering serves as a safeguard against catastrophic failures. In industries where human lives are at stake, like modern power systems, aviation, and medical devices, reliable products are essential. Consider an aircraft: A reliable one is less likely to crash, instilling confidence in passengers and airline companies alike. Similarly, medical devices, such as life-support systems, must function without fail to prevent harm to patients.
2. **Reducing Costs:**
   Reliable products and systems require fewer maintenance and repair interventions, translating into substantial cost savings for businesses. Think about a reliable car; it demands fewer repairs compared to an unreliable counterpart. The same principle applies to industries like energy, where a reliable power plant requires less maintenance than an unreliable one, minimizing downtime and costs.
3. **Increasing Productivity:**
   Operational efficiency is a hallmark of reliability engineering. When products and systems operate reliably, they contribute to enhanced productivity. A reliable factory machine, for instance, can churn out more products per hour than an unreliable one. This efficiency boost directly impacts the bottom line and overall business success.
4. **Enhancing Customer Satisfaction:**
   Customers are naturally drawn to products and services they can rely on. Reliability fosters customer satisfaction, which, in turn, leads to increased sales and higher profits. Imagine a customer purchasing a new car and experiencing minimal issues over time; they are more likely to become brand advocates and recommend the product to others.
5. **Regulatory Compliance:**
   Numerous industries are subject to stringent regulations that demand adherence to specific reliability standards. For instance, the aviation industry operates under strict regulations governing aircraft reliability. Compliance is crucial not only for ensuring safety but also for avoiding legal repercussions and maintaining the integrity of the industry.

1.2.2 Key Features of Reliability Engineering

1. **Failure Analysis:**
   Failure analysis is at the core of reliability engineering. By identifying the root causes of failures, engineers can develop strategies to prevent similar issues in the future. This proactive approach not only enhances product performance but also contributes to long-term cost savings.
2. **Risk Assessment:**
   Reliability engineering involves quantifying risks associated with potential failures. This information helps prioritize preventive measures, ensuring that resources are directed where they are most needed. Effective risk assessment minimizes surprises and contributes to a well-structured risk management strategy.
3. **Probabilistic Modeling:**
   Probabilistic modeling leverages mathematical models to predict the likelihood of failure. These predictions inform decision-making processes regarding design, maintenance, and

testing strategies. By quantifying potential failures, businesses can allocate resources more effectively and make informed choices.

4. **Life Cycle Management (LCM):**
   Reliability engineering extends throughout the entire life cycle of a system or product. From design to disposal, engineers engage in activities such as reliability testing, maintenance planning, and failure data analysis. This comprehensive approach ensures optimal performance and longevity.

1.3 Scientific Importance of Reliability in Modern Technology

The electric power supply system's reliability can be described as its ability to consistently offer users uninterrupted and satisfactory service. This criterion of quality encompasses the maintenance of power supply frequency and voltage within defined tolerances. Certainly, the level of reliability customers encounter will naturally vary depending on their particular geographic locations. Furthermore, different segments of the power network, including distribution, generation and transmission systems, will demonstrate varying levels of reliability.

It is obvious that in order to meet the expected customer-level reliability, each of these individual components and systems must maintain an elevated level of reliability. The concept of adequacy revolves around ensuring that the system possesses ample facilities to meet consumer demands. This encompasses the generation of adequate energy and the interconnected transmission and distribution networks responsible for efficiently conveying this energy to end consumers. Adequacy predominantly pertains to static conditions and does not encompass system disturbances.

In contrast, security concentrates on the system's capacity to effectively respond to disturbances originating within itself. Security is intricately linked to the system's ability to counteract diverse disruptions it may encounter, spanning from localized to widespread effects and even encompassing the loss of significant generation and transmission facilities.

Dividing the system into its primary functional zones provides a pragmatic approach. These zones encompass the composite generation, generation systems and transmission (bulk power) systems, and distribution systems. Each major functional zone can be further subdivided, facilitating the focused analysis of specific subsets of the issue. Notable subzones include individual generating stations, substations, and protection systems.

Establishing the required level of system-generating capacity to guarantee a reliable power supply is a crucial component of power system planning and operation (Huang et al. 2019; Gandini 1990; Freixas and Pons 2008). This intricate issue can be divided into two separate domains: static and operational capacity demands. In power system planning, the static capacity domain focuses on the long-term assessment of the system's overarching need, while the operating capacity domain deals with the short-term evaluation of the actual capacity required to meet a specific load demand. Both of these considerations must be taken into account during the planning phase, especially when exploring alternative facilities. After decisions are made, the short-term requirement becomes a matter of operational concern.

In the context of static capacity evaluation, a fundamental parameter for generating units is the probability of encountering a forced outage at a future point in time. This probability, referred to as unit unavailability in engineering systems and historically referred as the unit-forced outage rate in power system applications, serves as a fundamental element for evaluating the system's static capacity.

1.4 Basic Concept of Probability Theory

The Bayesian approach has an important aspect related to conditional probability for assessing the reliability of power system components. The Bayes theorem is introduced, along with a brief explanation. A sample space denoted as S, let A_i be a series of exhaustive and mutually exclusive events that $P(A_i) > 0, \forall_i$. Within "S" sample space, consider an event "B" where the probability of event "B" occurring, $P(B) > 0$. Then,

$$P(A_i/B) = \frac{P(A_i)P(B/A_i)}{\sum_i P(A_i)P(B/A_i)} \tag{1.1}$$

Proof: Consider A_i as an event within the set S, and B is event where $P(B) > 0$. Subsequently, in accordance with the definition of conditional probability,

$$P(A_i/B) = \frac{P(A_i \cap B)}{P(B)}$$

and

$$P(B)P(A_i/B) = P(A_i \cap B) \tag{1.2}$$

$$P(B/A_i) = \frac{P(A_i \cap B)}{P(A_i)}$$

$$P(A_i)P(B/A_i) = P(A_i \cap B) \tag{1.3}$$

From (1.1),

$$P(A_i/B)P(B) = P(A_i)P(B/A_i)$$

$$P(A_i/B) = \frac{P(A_i)P(B/A_i)}{P(B)}$$

$$P(B) = \sum_i P(A_i)P(B/A_i) \tag{1.4}$$

We now assert that, given the event (A_i) being mutually exclusive and collectively exhaustive within the set S, it can be concluded that $\cup_i A_i = S$ and that the individual A_i events are nonoverlapping or disjoint.

$$B = B \cap S = B(\cup_i A_i) = \cup_i (B \cap A_i)$$

Thus,

$$P(B) = P[\cup_i (B_i \cap A_i)] = \sum_i P(B \cap A_i)$$

Thus, by axiom of probability

$$P(B) = \sum_i P(A_i)P(B/A_i) \tag{1.5}$$

Exercise 1.1: Among a collection of circuit breakers, Factory F_1 produces 1000 of them, with 20 being faulty. Factory F_2 manufactures 4000 circuit breakers, and 40 of them are defective. Factory F_3 contributes 5000 circuit breakers, with 50 being faulty. All these circuit breakers are combined into a single stockpile. What is the probability that a defective circuit breaker selected at random being produced in Factory F_1?

Solution:
Consider A_1, A_2, and A_3 as events representing the selection of circuit breakers produced by F_1, F_2, and F_3, respectively. The overall number of products is given by:

$$N(S) = 5000 + 4000 + 1000 = 10{,}000$$

$$p(A_1) = \frac{1000}{10{,}000} = \frac{1}{10},$$

$$P(A_2) = \frac{4000}{10{,}000} = \frac{2}{5}$$

$$p(A_3) = \frac{5000}{10{,}000} = \frac{1}{2}$$

Let B represent the event where the selected circuit breakers are found to be faulty.
Then

$$p(B/A_1) = \frac{20}{1000} = \frac{1}{50}$$

$$p(B/A_2) = \frac{40}{4000} = \frac{1}{100}$$

$$p(B/A_3) = \frac{50}{5000} = \frac{1}{100}$$

To calculate $P(A_1/B)$ using Bayes' theorem:

$$P(A_1/B) = \frac{P(A_1)P(B/A_1)}{\sum_{i=1}^{3} P(A_i)P(B/A_i)}$$

$$= \frac{\left(\frac{1}{10}\frac{1}{50}\right)}{\left(\frac{1}{10}\frac{1}{50}\right) + \left(\frac{2}{5}\frac{1}{100}\right) + \left(\frac{1}{2}\frac{1}{100}\right)} = \frac{2}{11}$$

Exercise 1.2: For a power transformer, rejection can occur due to either a flawed design or other reasons. The chance of having a flawed design is 0.1, and the probability of rejection specifically due to this design issue is 0.95, while it is 0.45 for rejections arising from different causes. In the event of a power transformer's rejection, what is the probability that this rejection is specifically due to a design issue?

Solution:
Consider events F_1 and F_2 to represent whether the design is flawed or not, and let event A represent the rejection of the power transformer.
Then, $P(F_1) = 0.1$, $P(F_2) = 0.9$ and $P(A/F_1) = 0.95$, $P(A/F_2) = 0.45$,

$$P(F_1/A) = P[\text{rejection due to faulty design}]$$

$$\frac{P(F_1)P(A/F_1)}{P(F_1)P(A/F_1) + P(F_2)P(A/F_2)} = \frac{0.1 \times 0.95}{0.1 \times 0.95 + 0.9 \times 0.45} = 0.19$$

Exercise 1.3: Three machines are involved in manufacturing wattmeters for power system applications, with 10,000, 20,000, and 30,000 watts, respectively. Defectives are produced by these machines at rates of 1%, 2%, and 1%, respectively. What is the probability of a randomly chosen defective wattmeter is manufactured by the third machine?

Solution:

Assume that A_1 represents a wattmeter manufactured by the first machine, A_2 by the second machine, A_3 by the third machine, and B denotes the case where the selected wattmeter is identified to be faulty.

Hence $n(S) = 30{,}000 + 10{,}000 + 20{,}000 = 60{,}000$

$$P(A_1) = \frac{10{,}000}{60{,}000} = \frac{1}{6}$$

$$P(A_2) = \frac{20{,}000}{60{,}000} = \frac{1}{3};$$

$$P(A_3) = \frac{30{,}000}{60{,}000} = \frac{1}{2}$$

$$P\left(\frac{B}{A_1}\right) = \frac{1}{100}; P\left(\frac{B}{A_2}\right) = \frac{2}{100}; P\left(\frac{B}{A_3}\right) = \frac{1}{100}$$

To find the $P\left(\frac{A_3}{B}\right)$ by Bayes theorem:

$$P\left(\frac{A_3}{B}\right) = \frac{P(A_3)\,P\left(\frac{B}{A_3}\right)}{\sum_{i=1}^{5} P(A_i)\,P\left(\frac{B}{A_3}\right)}$$

$$= \frac{\left(\frac{1}{2} \times \frac{1}{100}\right)}{\left(\frac{1}{6} \times \frac{1}{100}\right) + \left(\frac{1}{3} \times \frac{2}{100}\right) + \left(\frac{1}{2} \times \frac{1}{100}\right)}$$

$$= \frac{1}{\left(\frac{1}{600}\right) + \left(\frac{2}{300}\right) + \left(\frac{1}{200}\right)} = \frac{1}{200} \times \frac{600}{(1+4+2)} = \frac{3}{8}$$

1.5 Basic Concepts of System Reliability

Based on the fundamental concept of reliability, the key characteristics encompass probability, proper functioning, duration, and operational context of the system. The reliability of both components and systems frequently hinges on the duration they have spent in service. Hence, in the context of reliability research, an essential element to consider is the failure-time distribution, which portrays how the time until component failure is distributed under defined environmental circumstances. Components within a power system might be categorized as nonrepairable or repairable. A component earns the designation of "repairable" if it can be restored to its initial state post-failure without disrupting the system's operation.

Key concepts in power system reliability include availability, maintainability, mean up time (MUT), loss of load probability (LOLP), mean time to repair (MTTR), mean down time (MDT), mean time to failure (MTTF), and mean time between failures (MTBF). A practical approach to describing this distribution is through its corresponding instantaneous failure rate.

1.5.1 Bathtub Curve

The bathtub curve is a graphical representation in Figure 1.1 that illustrates how the rate of failure of an item changes over time. This curve demonstrates the initial period of an item's life, known as the "infant-mortality period," during which the failure rate goes down. Afterward, it reaches a stable intrinsic failure rate level, maintaining this rate during its useful life. As the item continues to age and surpasses its designed lifespan, the curve then shows an increase in the failure rate, known as wear-out failures. This curve is constructed by mapping three distinct phases: (i) The initial infant-mortality failure rate upon introduction, (ii) the ongoing random failure rate during its operational lifespan, and (iii) the wear-out failure rate as the product exceeds its designated lifespan.

The hazard function represents the pace at which units still in operation experience failure. Reliability analysts frequently employ this function to elucidate the failure patterns of devices and systems. Instantaneous rate of failure i.e. hazard function, may be decreasing, increasing, or constant. Usually in the life of a component/device/system all three types of failure rates are predominant in three stages of the system. These three failure rate are described in three different regions called infant-mortality period, random failures, and wear-out failures, as shown in Figure 1.1. This curve also shows the combination of three regions known as λ-curve/bathtub curve. Curves (a), (b), and (c) represent endogenous failure rates, exogenous failure rate, and wear-out failure rates, respectively. Curve (d) represents total failure rate. Even in the early stages, wear-out failure could be minimal, but it emerges as the dominant factor in curve (c) as it enters region III. Similarly, even in debugging period random failures may be present. The incorporation of the hazard function is derived from the observation that devices collectively display a hazard pattern that mirrors a bathtub over the course of their lives. The shape of the curve is meant to demonstrate the concept that the aging process in a population of devices goes throughout distinct phases. During initial phases of the devices' lifespans, failures happen at a relatively elevated rate. This initial phase of high failure rates is often attributed to weaknesses in some devices within the population, resulting from manufacturing defects, material flaws, or other physical irregularities. Following the initial phase, referred to as the infant-mortality period, the device population transitions into a stage known as the functional life period, characterized by a relatively low and stable hazard function. Ultimately, when individuals in the population approach the later stages of their lives, survivors encounter a rising failure rate as a result of accumulated wear and tear. The hazard model, or the bathtub curve, can be dissected into three distinct sections:

(i) Initial period,
(ii) Constant hazard model, and
(iii) Wear-out region.

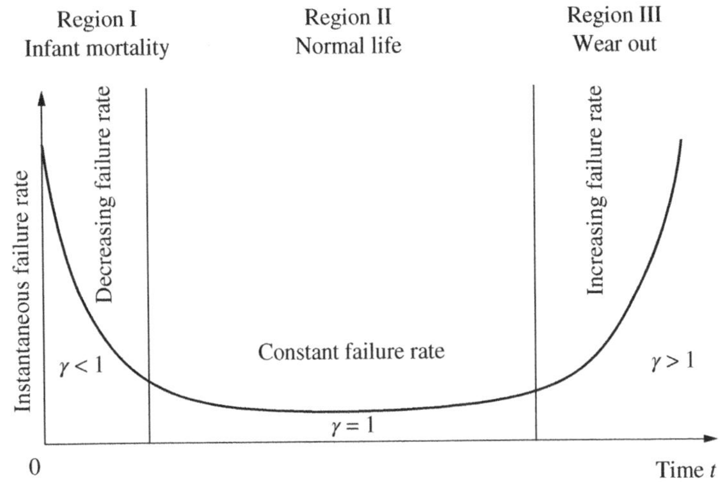

Figure 1.1 Bathtub curve.

1.5.1.1 Initial Period (debugging/Burn-In/Infant-Monolith Period)

Initial period where the failure rate is decreasing is known as debugging period or infant-mortality period. Failure rate is decreasing in this period because as time passes it is design/manufacturing defects are removed. Failure in this period is due to endogenous reasons. That is internal regions are mainly responsible for the system's failure rate. During the burn-in phase, a diminishing hazard function implies a decreased likelihood of an impending failure as time progresses. In general, this is best suited model for the initial phase of a component's operation, characterized by a hazard function that consistently diminishes over its entire lifespan.

1.5.1.2 Constant Hazard Model

The second stage represents the device's useful period, during which the failure rate remains relatively stable. Failures in this region are also known as random failures. Here in this useful period, failures are by chance and reasons are exogenous. Reasons for failures are external, e.g. overstressing, accident, etc. The constant hazard model is relevant to components in which the probability of failure within a given interval remains consistent throughout their entire lifespan. During the middle phase of their life, there is no anticipation of unusual clusters of failures, and thus, the so-called chance failures with relatively low and unchanging hazard rates take precedence.

1.5.1.3 Wear-Out Region

In the third region which is known as wear-out region, aging effect dominates, and increasing failure rate is observed due to wear-out. The rising hazard function basically demonstrates that these components are gradually becoming more susceptible to failure as they age, particularly within a group of the same size that has endured for the same duration, is expressed as:

$$h(t) = f(t)/R(t)$$

where $R(t)$ = reliability function;
$f(t)$ = probability density function;
$h(t)$ = hazard function of the system or device.

1.5.1.4 Burn-In Screening

This is a design and test procedure to filter out the defective, weak or low quality product from the population, so that they do not turn-out at a later stage during application as shown in Figure 1.2. This reduces the risk at higher level and minimizes the possibility of rework. Ideally, only those

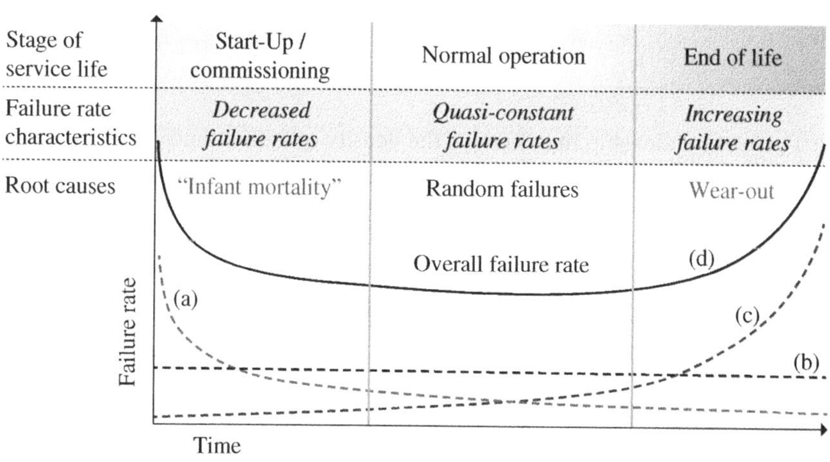

Figure 1.2 Representation of hazard function.

components must be used in the assembly of the system, which have undergone burn-in screening test already, because it considerably enhances the system reliability.

Burn-in is an essential step if the output of the production is not matching the reliability specifications. It must be ensured that the burn-in stress or temperature is not so high that the units under test are damaged. Burn-in screening is an expensive process and thus the time period for which the screening is done is limited. How long a burn-in screening is run depends on the desired level of reliability. Some of the related references are (Saket 2011) and (Saket 2013).

1.5.2 Reliability Functions

To delve into this concept further, a component taken into consideration that commences its operation at a specific time, for example, $t = 0$. Here $f(t)$ represents the probability density associated with the time until the component fails. This probability density (t) indicates the likelihood of the component failing in the range of time t and $t + \Delta t$, denoted by the function $f(t)\Delta t$.

Consequently, the probability of the component failing within the range from 0 to time t can be calculated using the cumulative distribution function $t, F(t) = \int_0^t f(x)\mathrm{d}x$. The reliability function, which denotes the probability of the component enduring until time t, is articulated as $R(t) = 1 - F(t)$. This function complements the cumulative distribution by indicating the likelihood of survival up to a certain time t.

The probability of a successful outcome, which is equivalent to the probability of the component failing in the range of t and $t + \Delta t$, is given by $F(t + \Delta t) - F(t)$. The probability of failure in this interval, considering the component's survival until time t is articulated as:

$$\frac{F(t + \Delta t) - F(t)}{R(t)} \tag{1.6}$$

Dividing by Δt, we determine the mean failure rate within the time interval from t to $t + \Delta t$, given that the component has survived up to time t:

$$\frac{F(t + \Delta t) - F(t)}{\Delta t} \cdot \frac{1}{R(t)} \tag{1.7}$$

By approaching the limit with Δt converging to zero, we obtain the instantaneous failure rate, commonly well known as the failure rate:

$$Z(t) = \frac{F'(t)}{R(t)} = \frac{1}{R(t)} \frac{\mathrm{d}F}{\mathrm{d}t}(t) \tag{1.8}$$

Finally, observing that the PDF $=f(t) = \frac{\mathrm{d}F(t)}{\mathrm{d}t}$,

$$\text{we get the relation } Z(t) = \frac{f(t)}{1 - F(t)} \tag{1.9}$$

Now, let us derive a pivotal relationship that conveys the density of failure times using the failure rate function. By utilizing the relationship between the reliability function $R(t) = 1 - F(t)$ and consequently, $F'(t) = -R'(t)$, the expression for $Z(t)$ is formulated as follows:

$$Z(t) = -\frac{R'(t)}{R(t)} \tag{1.10}$$

After successfully solving this differential equation for $R(t)$, we acquire

$$\log R(t) = -\int_0^t Z(x)ax$$

$$R(t) = \mathrm{e}^{\int_0^t Z(x)ax} \tag{1.11}$$

Table 1.1 Relation between probabilistic functions.

–	F(t)	f(t)	R(t)	h(t)
$F(t)$	—	$\int_0^t f(t)\mathrm{d}t$	$1 - R(t)$	$1 - \mathrm{e}^{-\int_0^t h(t)\mathrm{d}t}$
$f(t)$	$\frac{\mathrm{d}F(t)}{\mathrm{d}t}$	—	$-\frac{\mathrm{d}R(t)}{dt}$	$h(t)\mathrm{e}^{-\int_0^t h(t)\mathrm{d}t}$
$R(t)$	$1 - F(t)$	$\int_0^\infty f(t)\mathrm{d}t$	—	$\mathrm{e}^{-\int_0^t h(t)\mathrm{d}t}$
$h(t)$	$\frac{\mathrm{d}F(t)}{\mathrm{d}t}/[1 - F(t)]$	$\frac{f(t)}{\int_0^\infty f(t)\mathrm{d}t}$	$-\frac{\mathrm{d}}{\mathrm{d}t}\log R(t)$	—

Furthermore, by utilizing the relationship $f(t) = Z(t)R(t)$, we obtain $f(t) = Z(t)\exp\left[-\int_0^1 Z(x)ax\right]$. This formula is referred to as the general equation for the distribution of failure times. Table 1.1 displays the relationships between reliability functions.

For constant failure rate, the following relations are given for reliability assessment of the power components.

(1) Reliability function

$$R(t) = \mathrm{e}^{-\int_o^t h(t)\mathrm{d}t}$$

put $h(t) = \lambda$ (constant failure rate of the components)

$$R(t) = \mathrm{e}^{-\int_o^\infty \lambda \mathrm{d}t}$$

$$\Rightarrow \quad R(t) = \mathrm{e}^{-\lambda t}$$

$$R(t) = (1 - P_f)$$

The failure probability of power component

$$P_f = [1 - R(t)]$$

$$\Rightarrow \quad P_f = (1 - \mathrm{e}^{-\lambda t})$$

(2) The probability density function (PDF) of power component

$$f(t) = \frac{-\mathrm{d}R(t)}{\mathrm{d}t}$$

$$\Rightarrow \quad f(t) = \lambda \mathrm{e}^{-\lambda t}$$

(3)

$$\mathrm{MTTF} = \int_0^\infty R(t)\mathrm{d}t$$

$$\mathrm{MTTF} = \int_0^\infty \mathrm{e}^{-\lambda t}\mathrm{d}t$$

$$\mathrm{MTTF} = \frac{1}{\lambda}$$

$$\Rightarrow \quad R(t) = \mathrm{e}^{-\lambda t}$$

1.5.3 Mean Time to Failure of Component

Our focus frequently lies in determining the average time until a component fails (MTTF), rather than focusing on the specifics of each failure. This parameter is assumed to remain consistent for all components that share identical designs and function under same operating conditions. With life test data available for a group of N items and their corresponding failure times as $t_1, t_2, \ldots t_n$, the MTTF is articulated as follows:

$$\text{MTTF} = \frac{1}{N}\sum_{i=1}^{n} t_i \tag{1.12}$$

Although when a component's characteristics are captured through its reliability function, the MTTF is calculated as the arithmetic mean of the random variable T, which signifies the duration until the component experiences a failure.

Henceforth,

$$\text{MTTF} = E[T] = \int_0^{\infty} tf(t)\,dt \tag{1.13}$$

But $f(t) = \frac{dF(t)}{dt} = -\frac{dR(t)}{dt}$.

Hence,

$$\text{MTTF} = -\int_0^{\infty} t\,dR(t)\,dt = -[tR(t)]_0^{\infty} + \int_0^{\infty} R(t)\,dt$$

$$= \int_0^{\infty} R(t)\,dt \qquad (\because R(\infty) = 0)$$

$$\text{MTTF} = \int_0^{\infty} R(t)\,dt \tag{1.14}$$

The MTTF is determined by applying the Laplace transform to $R(t)$,

$$\text{MTTF} = \int_0^{\infty} R(t)\,dt \underset{t\to\infty}{Lt} \int_0^{\infty} R(t)\,dt \tag{1.15}$$

However, $\underset{t\to\infty}{Lt} \int_0^t R(x)\,dx = \underset{s\to 0}{Lt}\, R(s)$,

Thus, $\text{MTTF} = \underset{s\to 0}{Lt}\, R(s) \int_0^t t^2 f(t)\,dt - (\text{MTTF})^2$

Remark: $\text{Var}(T) = E[T - E(T)]^2 = E[T^2] - [E(T)]^2$.

1.5.4 Additional Examples

Exercise 1.4: In a composite power system, where $R(t)$ represents the reliability function:

$$R(t) = 1/e^{\sqrt{0.001t}}, t \geq 0$$

a) Calculate the reliability for a mission lasting 50 hours.
b) Demonstrate that the hazard rate exhibits a decreasing trend.
c) Determine the reliability for a 50-hour mission, considering a wear-in period of 10 hours.
d) Find the design life required to achieve a reliability of 0.95, considering 10 hours of wear-in period.

Solution:

The system's reliability function is: $R(t) = 1/e^{\sqrt{0.001t}}, t \geq 0$

a)

$$R(50) = 1/e^{\sqrt{0.001 \times 50}} = 0.9512$$

b)

$$\lambda(t) = \frac{-R'(t)}{R(t)} \Rightarrow -e^{\sqrt{0.001t}} \times \frac{1}{e^{\sqrt{0.001t}}} \times \left(-\sqrt{\frac{0.001}{t}}\right)$$

$$\Rightarrow \sqrt{\frac{0.001}{t}}, \text{ which is a decreasing function of } t.$$

c)

$$R\left(t/T_0\right) = \frac{R\left(T_0 + t\right)}{R\left(T_0\right)}$$

$$R(50/10) = \frac{R(60)}{R(10)} = \frac{e^{\sqrt{0.001 \times 10}}}{e^{\sqrt{0.001 \times 60}}} = 0.8651$$

d)

$$\frac{R\left(t_D + 10\right)}{R(10)} = 0.95$$

$$R\left(t_D + 10\right) = R(10) \times 0.95 = 1/e^{\sqrt{0.001 \times 10}} \times 0.95$$

$$\Rightarrow \quad \sqrt{\left(t_D + 10\right) \times 0.001} = 0.15129$$

$$t_D = 12.89 \text{ hours.}$$

Exercise 1.5: In the context of a protective relay circuit with an MTBF of 0.8 years and under the assumption of random failures, address the following inquiries:

i) Determine the likelihood that the circuit will remain operational without experiencing failure over the course of 1 year.
ii) Calculate the probability of encountering over two failures within the initial year.
iii) Ascertain the anticipated count of failures anticipated annually.

Solution:
Given the random occurrence of failures, the count of failures within a time interval of "t" is modeled by a Poisson distribution, given by $P[N(t) = n] = \frac{e^{-\lambda t}(\lambda t)n}{n!}, n \geq 0$, where λ signifies the rate of failure. Consequently, the time gaps between failures are distributed exponentially with an average of $1/\lambda$.

Now MTTF $= 1/\lambda = 0.8$ year

$\therefore \quad \lambda = 1.25$ per year

i)

$$P[N(1) = 0] = \frac{e^{-\lambda} \hat{\lambda}^0}{0!} - e^{-125} = 0.2865$$

ii)

$$P[N(1) > 2] = 1 - e^{-\lambda} \left[\frac{\hat{\lambda}^0}{0!} + \frac{\hat{\lambda}^1}{1!} + \frac{\hat{\lambda}^2}{2!}\right]$$

$$\Rightarrow 1 - 0.2865\left[1 + 1.25 + \frac{(1.25)^2}{2}\right] = 0.1315.$$

iii)

$$E[N(t)] = \lambda t$$

$$E[\text{Annual failure count}] = \lambda = 1.25.$$

Exercise 1.6: In a redundant power system with "n" independent, identical components, each having a constant failure rate λ, the formula to compute the MTTF is expressed as: MTTF = $\frac{1}{\lambda}\sum_{i=1}^{n} {}^{n}C_{i}\frac{(-1)^{i-1}}{i}$. What is the minimum system reliability when the failure rate is $\lambda = 0.02$ per hour?

Solution:
If $R_s(t)$ is the reliability of system expressed as $R_s(t) = 1 - (1 - e^{-\lambda t})^n$ where the reliability of individual components is represented as $e^{-\lambda t} = 1 - \sum_{i=0}^{n}(-1)^i\ {}^{n}C_i e^{-i\lambda t} = -\sum_{i=0}^{n}(-1)^{i-1}\ {}^{n}C_i e^{-i\lambda t}$

$$\therefore \quad \text{MTTF} = \int_0^\infty R_s(t)\,dt$$

$$= \sum_{i=1}^{n}(-1)^{i-1}\ {}^{n}C_1 \int_0^\infty e^{-i\lambda t}dt = \frac{1}{\lambda}\sum_{i=1}^{n} {}^{n}C_i\frac{(-1)^{i-1}}{i}$$

When $n = 2$, $\text{MTTF} = 200 \times \sum_{i=1}^{2} {}^{2}C_i\frac{(-1)^{i-1}}{i} = 200\left[2 - \frac{1}{2}\right] = 300$

When $n = 3$, $\text{MTTF} = 200 \times \sum_{i=1}^{3} {}^{3}C_i\frac{(-1)^{i-1}}{i} = 200\left(3 - \frac{3}{2} + \frac{1}{3}\right) = 200 \times \frac{11}{6} = 366$

Hence the required minimum value of $n = 4$.

Exercise 1.7: The power transmission system comprises five consecutive segments within its protection scheme. Each individual segment has a consistent failure rate of 0.23 failures per year, with the failures of segments being independent from each other. Compute the system's MTTF and its reliability function.

Solution:
The failure rate remains constant, and the failure times are distributed exponentially with a parameter value of $\lambda = 0.23$.
For a single segment, the reliability function is represented as $R_1(t) = e^{-(0.23)t}$.
The system comprises five consecutive segments. As these segments experience independent failures, The system's overall reliability is represented as follows:

$$R(t) = R_1(t)R_2(t)R_3(t)R_4(t)R_5(t)$$

$$= e^{-(0.23)t}.e^{-(0.23)t}.e^{-(0.23)t}.e^{-(0.23)t}.e^{-(0.23)t} = e^{-(1.15)t}$$

$$\text{MTTF} = \int_0^\infty R(t)dt = \int_0^\infty e^{-(1.15)t}dt = \left[\frac{e^{-(1.15)t}}{-1.15}\right] = 0.8696$$

Exercise 1.8: A newly acquired distribution transformer exhibits a constant daily failure rate of 0.03 during continuous operation, along with a fixed daily repair rate of 0.2.

i) Calculate the interval availability for the initial 30 days and determine the steady-state availability.

ii) Establish the steady-state availability when introducing a stand-by unit, assuming zero failures during stand-by mode.
iii) In the scenario where both units are operational, ascertain the steady-state availability.

Solution:

$\mu = 0.2$ per day, $\lambda = 0.03$ per day

a)

$$A_1(T) + \frac{\mu}{\lambda + \mu} + \frac{\mu}{(\lambda + \mu)^2 T}[1 - e^{-(\lambda+\mu)T}]$$

$$A_1(30) + \frac{0.2}{0.03 + 0.2} + \frac{0.03}{(0.03 + 0.2)^2 X 30}[1 - e^{-(0.03+0.2)30}]$$

$$= 0.8696 + 0.0031 + 0.8727$$

$$A(\infty) = \frac{\mu}{\lambda + \mu} = 0.8696$$

b) In the stand-by redundant system,

$$A_i(\infty) = \frac{\lambda\mu + \mu^2}{\lambda^2 + \lambda\mu + \mu^2}$$

$$= \frac{0.03 \times 0.2 + 0.2^2}{0.2^2 + 0.3 \times 0.2 + 0.03^2}$$

$$= \frac{0.046}{0.0469} = 0.9808$$

c) In the setup of an active redundant system

$$A_i(\infty) = 1 - [1 - A(\infty)]^2$$

$$\Rightarrow 1 - [1 - 0.8696]^2$$

$$\Rightarrow 0.9830$$

1.6 Conclusion

The analysis provided in this chapter creates a framework for correlating the reliability of power systems to the reliability of their individual components across various equipment designs. The utilization of system configurations based on binary component states and independent components allows for an approach to reliability analysis that simplifies the complex system. It is imperative to focus on component-level reliability due to the clear and well-defined impact of component's reliability on the reliability of overall system. Particularly in the context of composite power system designs, the ability to independently assess the reliability of individual components is crucial for achieving elevated levels of overall reliability.

The chapter offers practical examples to showcase the application of these methods, illustrating both the advantages and limitations of various techniques. These practical cases are supported by real-life examples taken from the author's substantial experience in academia, research, and consulting. The chapter provides a thorough exploration of fundamental concepts in probability theory, the use of redundant systems to enhance reliability, evaluations of power component structures with a focus on reliability, application of the Hazard model for failure analysis of components, and the presentation of various probability distributions related to reliability. Solved and unsolved numerical examples are included in each section and subsection, offering a comprehensive and illustrative view of the concepts discussed in the chapter.

References

Arya, L.D., Choube, S.C., and Saket, R.K. (2000). *Composite System Reliability Evaluation Based on Static Voltage Stability Limit*, 133–139. Journal-Institution of Engineers India Part EL Electrical Engineering Division.

Arya, L.D., Choube, S.C., and Saket, R.K. (2001). *Generation System Adequacy Evaluation Using Probability Theory*, 170–174. Journal-Institution of Engineers India Part EL Electrical Engineering Division.

Bansal, R.C., Bhatti, T.S., and Kothari, D.P. (2002). Discussion and closure of "bibliography on the application of probability methods in power system reliability evaluation". *IEEE Transactions on Power Systems* 17 (3): 924.

Bharti, O.P., Saket, R.K., and Nagar, S.K. (2019). Reliability assessment and performance analysis of DFIG-based WT for wind energy conversion system. *International Journal of Reliability and Safety* 13 (4): 235–266.

Freixas, J. and Pons, M. (2008). Identifying optimal components in a reliability system. *IEEE Transactions on Reliability* 57 (1): 163–170.

Gandini, A. (1990). Importance and sensitivity analysis in assessing system reliability. *IEEE Transactions on Reliability* 39 (1): 61–70.

Huang, X., Coolen, F.P.A., Coolen-Maturi, T., and Zhang, Y. (2019). A new study on reliability importance analysis of phased mission systems. *IEEE Transactions on Reliability* 69 (2): 522–532.

Kumar, S., Saket, R.K., Dheer, D.K. et al. (2020). Reliability enhancement of electrical power system including impacts of renewable energy sources: a comprehensive review. *IET Generation, Transmission & Distribution* 14 (10): 1799–1815.

Saket, R.K. (2011). Reliability evaluation of defence support systems. In: *Innovations in Defence Support Systems-2: Socio-Technical Systems*, vol. 338 (ed. L.C. Jain, E.V. Aidman, and C. Abeynayake), 241–286. Springer-Verlag.

Saket, R.K. (2013). Design aspects and probabilistic approach for generation reliability evaluation of MWW based micro-hydro power plant. *Renewable and Sustainable Energy Reviews* 28: 917–929.

Saket, R.K., Bansal, R.C., and Singh, G. (2006). Generation capacity adequacy evaluation based on peak load consideration. *The South Pacific Journal of Natural and Applied Sciences* 24 (1): 38–44.

Saket, R.K., Bansal, R.C., and Singh, G. (2007). Reliability evaluation of power system considering voltage stability and continuation power flow. *Journal of Electrical Systems* 3 (2): 48–60.

Saket, R.K., Bansal, R.C., and Singh, G. (2009). Power system components modelling and reliability evaluation of generation capacity. *International Journal of Reliability and Safety* 3 (4): 427–441.

Saket, R.K., Kothari, D.P., and Nagrath, I.J. (2022). *Modern Power System Analysis*. Tata McGraw-Hill Publishing Company.

2

Bayesian Approach for Reliability Evaluation and Remaining Useful Life Prediction

Debasis Jana, Suprakash Gupta, and Deepak Kumar

Department of Mining Engineering, Indian Institute of Technology (BHU) Varanasi (UP), India

2.1 Introduction

The quality of trustworthiness of any system, product, or service over a specified period and conditions is termed reliability. This quality of any system, product, or service is inherently uncertain and stochastic. Engineers and researchers are facing difficult challenges with accessing the evaluation of the reliability of any complex system. There are relevant statistical inference methods are available for reliability analysis. For example, the evaluation of hardware reliability can be approached through the reliability block diagrams (RBD), fault tree analysis (FTA), Markov and semi-Markov models (MKV), and Petri Nets models. The models like Gaussian processes, vector regression, and the Markov-modulated Poisson process are used in the software reliability study. ATHEANA, CREAM, and SPAR-H are utilized for the human reliability analysis. Every approach to assessing reliability has a unique set of benefits and drawbacks that are inherently associated with it. Due to the dependencies between interconnected components or factors and environmental effects, evaluation of reliability using RBD, and FTA is difficult because of the restriction of binary variables. Markov and Petri net models have problems like state space explosion.

Power system reliability is the ability of an electrical power system to deliver electricity consistently and without interruptions to its customers. It is a measure of how dependable and resilient the power system is in providing electricity to meet the demands of consumers. Reliability in power systems is crucial because interruptions in the power supply can have significant consequences for various sectors, including residential, commercial, industrial, and critical infrastructure. Power outages can result in economic losses, inconvenience, compromised safety, and disruptions in essential services. Ensuring high reliability in the power sector is very important. For ensuring power system reliability, several key factors like adequate generation capacity, robust transmission, and distribution infrastructure, effective system planning, fault detection and rapid response, redundancy and backup systems, maintenance and asset management are considered. Power system reliability is typically measured using metrics such as the Customer Average Interruption Duration Index (CAIDI), System Average Interruption Duration Index (SAIDI), and System Average Interruption Frequency Index (SAIFI). These metrics quantify the customers' average duration and frequency of power outages.

The reliability estimation of a complex system is challenging and often requires the incorporation of multiple techniques and multiple sources of data for building a credible statistical model. Reliability evaluation using a probabilistic graphical model like Bayesian network (BN) is a research

Reliability Analysis of Modern Power Systems, First Edition. Edited by R. K. Saket and P. Sanjeevikumar.

interest in recent days that is attracting considerable attention from researchers and reliability engineers. BN is an acyclic-directed graphical model that is capable of characterizing and analyzing uncertainty effectively, making it a significant problem-solving model in real-world encountered problems. Within the literature, the application of BNs extends to many fields including risk analysis, fault diagnosis, safety engineering, and uncertainty prediction.

2.2 Bayesian Network

BNs are graphical models to describe probabilistic relationships within a group of random variables. BNs are a directed acyclic graph, consisting of points (represent variables) and directed edges connecting these points. In BN, the arc-starting nodes are called the parent or cause nodes, and the arc-ending node is the child or effect node. For example, if an arc starts from node A and ends in node B, then node A is the parent of node B and B is the child of A. A root node is a node without a parent, and a node without any child node is considered a leaf node. A typical BN is shown in Figure 2.1. In Figure 2.1, the "parent or cause nodes" is the node at the tail (nodes E1, E2, E3, and E4), and the "child or effect node" is the node at the head (node IE1 and IE2).

The conditional probability is the main-stay of a BN and is defined as the likelihood of occurrence of an event depending on a previous event occurrence. Each node in a BN has a CPT that illustrates how the state of the node is characterized by every state of its parents' conditional probability. The joint probability distribution is also calculated from the CPT. The conditional probability is mathematically expressed as:

$$P(A|B) = \frac{P(A).P(B|A)}{P(B)} = \frac{P(A,B)}{P(B)} \tag{2.1}$$

In the equation, the hypothetical event is denoted by A and B is the observed evidence; PA prior probability before observing B; $P(B|A)$ is the conditional probability that B is present in every state of A; $P(B)$ is the marginal probability; $P(A|B)$ is the posterior probability after observing B, and $P(A, B)$ is the joint probability.

The joint probability distribution function is estimated using the conditional independence and the probability theory chain rule. The joint probability distribution of a set of variables $U = \{X_1, \ldots, X_n\}$ is expressed as follows:

$$P(U) = P(X_1, \ldots, X_n) = \prod_{l=1}^{n} P(X_i|P_a(X_i)) \tag{2.2}$$

where $P(X_i|P_a(X_i))$ denotes the conditional probability of X_i, and $P_a(X_i)$ is the parent node of X_i.

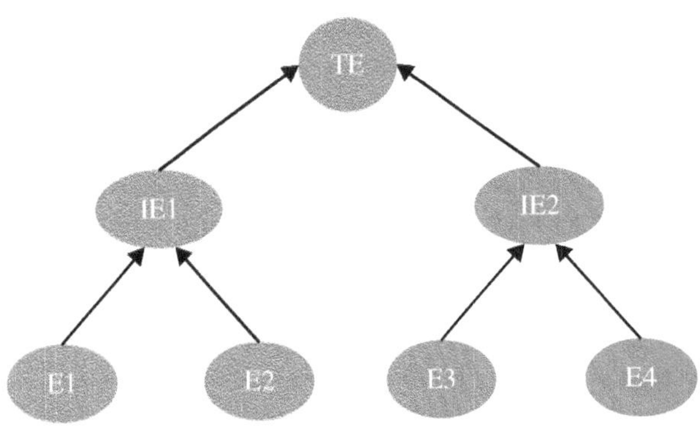

Figure 2.1 Typical BN with two parent nodes and one child node.

Through the marginalization of the joint probability distribution, system reliability can be obtained. During the BN reliability analysis, each event's prior probability is updated by new information or posterior probability. The posterior probability will be updated by conditional probability function or Bayesian inference as soon as a new observation or evidence E is given, as follows:

$$P(U|E) = \frac{P(U,E)}{P(E)} = \frac{P(E|U) \cdot P(U)}{P(E)} = \frac{P(E|U) \cdot P(U)}{\sum_i P(E|U) \cdot P(U)} \tag{2.3}$$

Example 2.1 Let us attempt to comprehend the workings of BNs with a simple illustration. For this illustration, let us pretend that we are modeling a student's marks for an examination. The marks depend on two additional factors: examination difficulty (ed) and intelligence level (IQ) of the student.

Exam difficulty (ed) is a discrete variable and measures the difficulty of the exam with two possible values, 0 for easy and 1 for difficult. The easier the exams, the lower the value.

Intelligence Level (*iq*) represents the student's IQ level. It is also discrete with two values 0 for low and 1 for high. It is a measure of the student's overall intelligence.

Additionally, the student's intelligence level leads us to another component, which is the students' aptitude score (as), which can be further broken down. With the marks that the student has earned up to this point, he is now able to get admission to a specific university.

Therefore, we have five variables: student marks (m), exam difficulty (ed), intelligence (iq), aptitude score (as), and university admission (ad). The probability distribution values of the given five variables in the form of CPT and the BN are presented in Figure 2.2.

The joint probability distribution is given by

$$P(m, \text{ed}, \text{iq}, \text{as}, \text{ad}) = P(\text{ed}).P(\text{iq}).P(m|\text{iq}, \text{ed}).P(\text{ad}|m).P(\text{as}|\text{iq})$$

where the probability of exam difficulty is denoted by P(ed), the probability of intelligence is denoted by P(iq), $P(m|\text{iq}, \text{ed})$ is the conditional probability of marks obtained given his/her

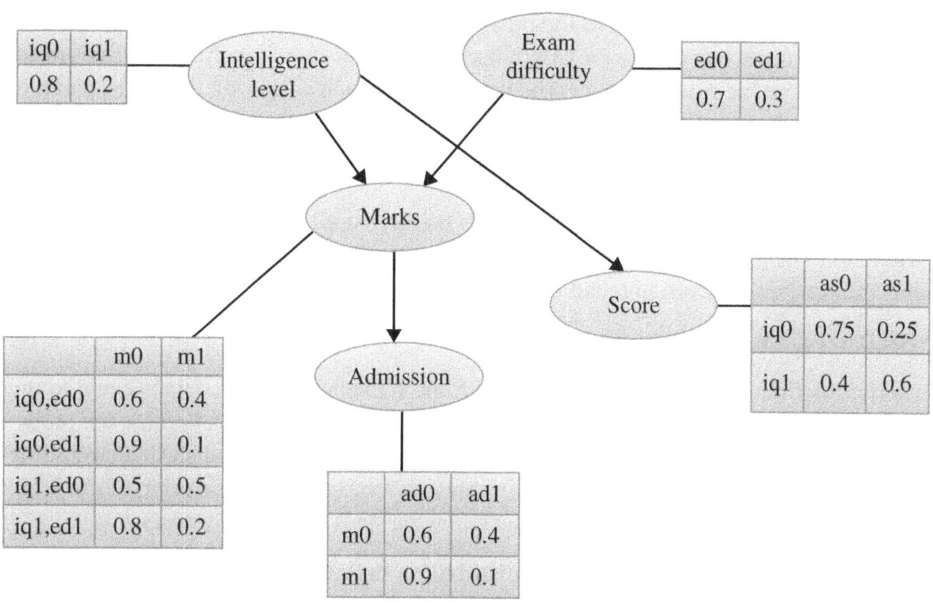

Figure 2.2 BN example.

intelligence level and exam difficulty, $P(\text{ad}|m)$ is the conditional probability of getting admission given students' marks and $P(\text{as}|\text{iq})$ is the conditional probability of students' aptitude score given his her intelligence.

Now, find out what is the probability that a student with a low aptitude score and a low intelligence level will still be able to pass the exam and get into the university even if the difficulty of the exam is hard.

For the given scenario, the joint probability distribution can be written as

$$P(m = 1; \text{iq} = 0; \text{as} = 0; \text{ad} = 1) = ?$$

From the CPT and BN presented in Figure 2.2, the probability for the given conditions is calculated as

$$P(m = 1; \text{iq} = 0; \text{as} = 0; \text{ad} = 1) = 0.1 \times 0.1 \times 0.75 \times 0.3 \times 0.8 = 0.0018$$

Practice 2.1 Determine the likelihood that a student will have a high IQ and a high aptitude score, the test will be easy, and yet the student will still not pass the test and will not be admitted to the institution.

2.3 Bayesian Reliability

The Bayesian statistics make it possible to conduct simultaneous evaluations of the reliability of the complex system in addition to its subsystems and its individual components' reliability. The purpose of the Bayesian reliability evaluation is to get a posterior distribution that is as accurate as feasible, and then to utilize this distribution to compute failure rate estimates with confidence intervals, which are referred to as credibility intervals. A credibility interval is used to represent an unknown parameter. Several statistical distributions and parameters that define the models are utilized for reliability estimates. The exponential and the Weibull are frequently used distributions to estimate the mean of failures data. On the other hand, the binomial is frequently used distribution to predict the chance of failure for pass/fail systems. To be more precise, we cannot make probability claims about the real parameter. The Bayesian approach deals with the model parameters even though they are random values. There is no presumption that the model parameters are random; rather, we utilize the tools of probability to convey our uncertainty about the values of the parameters. Mean time between failures (MTBF) is a good example of such parameters. Creating a prior distribution for the model parameters is the first step, and then we use that revised prior distribution to help us derive the updated assessment or posterior distribution for the model parameters. The concept of Bayesian reliability can be expressed using the following formula:

$$f_{\text{pos}}(R|\text{Data}) \propto \text{Lik}\,(\text{Data}|R) f_{\text{por}}(R) \tag{2.4}$$

The posterior distribution, denoted by the f_{pos} with the reliability R, given data is equal to the product of the prior knowledge, denoted by f_{por} with likelihood function (Lik). Because the probability function of the data is the only thing that is considered in a standard statistical analysis, the basic shift in mathematics that occurs in a Bayesian analysis is the addition of the prior information.

To perform a Bayesian reliability evaluation, the following steps are typically necessary:

1. **Specify Prior Distribution:** Define a prior distribution that represents the initial knowledge or beliefs about the reliability parameters. This distribution can be based on expert opinions, historical data, or other relevant information.

2. **Collect Data:** Gather data on system performance, failures, or any other relevant information related to reliability.
3. **Likelihood Function:** Specify a likelihood function that represents the statistical relationship between the data and the reliability parameters. This function describes how likely the observed data is under different reliability scenarios.
4. **Bayesian Inference:** Combine the prior distribution and the likelihood function to obtain the posterior distribution using Bayes' theorem. The posterior distribution represents the updated belief about the reliability parameters given the prior knowledge and the observed data.
5. **Posterior Analysis:** Perform various analyses on the posterior distribution to obtain reliability metrics of interest, such as failure rates, MTBF, or system availability. Uncertainty intervals can also be calculated to quantify the uncertainty associated with these metrics.
6. **Sensitivity Analysis:** Assess the sensitivity of the results to different assumptions or inputs. This helps identify influential factors and provides insights into the reliability behavior of the system.

2.4 Application of BN in Reliability and Remaining Useful Life

BN provides an effective framework in the context of probabilistic events and is recognized as a robust formalism to represent random variables with complex dependencies between them. The main utility of a BN is to estimate the posterior probabilities of random variables set by combining the prior knowledge with the real-time observation of some other variables. BN framework provides a compact representation of the event-dependent failure behaviors and common-cause failures in today's complex systems. In addition, BN is also a solution to the state space explosion problem associated with Markov chain-based modeling approaches (Liu and Singh 2010).

The BN-based dependability evaluation primarily consists of four processes, which include structure modeling, parameter modeling, inference, and verification and validation. For evaluating the reliability of any field of system, like hardware, structures, software, and human reliability, the aforesaid four steps are necessary to perform. Figure 2.3 provides a flowchart that outlines this process in detail.

A typical hardware system comprised of two subsystems is presented in Figure 2.4. The parallel subsystem comprises three components connected in parallel, whereas the series subsystem consists of three components connected in series. In BN, the status of each component should be represented by the root node, and the reliability of the entire system should be represented by the final leaf node.

2.4.1 BN Structure Modeling

The first step of the BN model deals with qualitative aspects corresponding to the directed acyclic graph. Determining nodes and identifying connection arcs are the two main steps involved in this step of modeling. The techniques utilized to accomplish BN structure modeling include knowledge representation for reliability estimation of structures, hardware, and humans; mapping technique for reliability analysis of hardware, and humans; and structure learning.

The structure of the BN is developed in such a way that it considers the impact linkages between the system's components responsible for monitoring, data analysis, and the distribution of information. The minimal cut-set matrix approach is used to extract the structure of the BN. The BN structure is developed by making use of the logical relationships existing between the components.

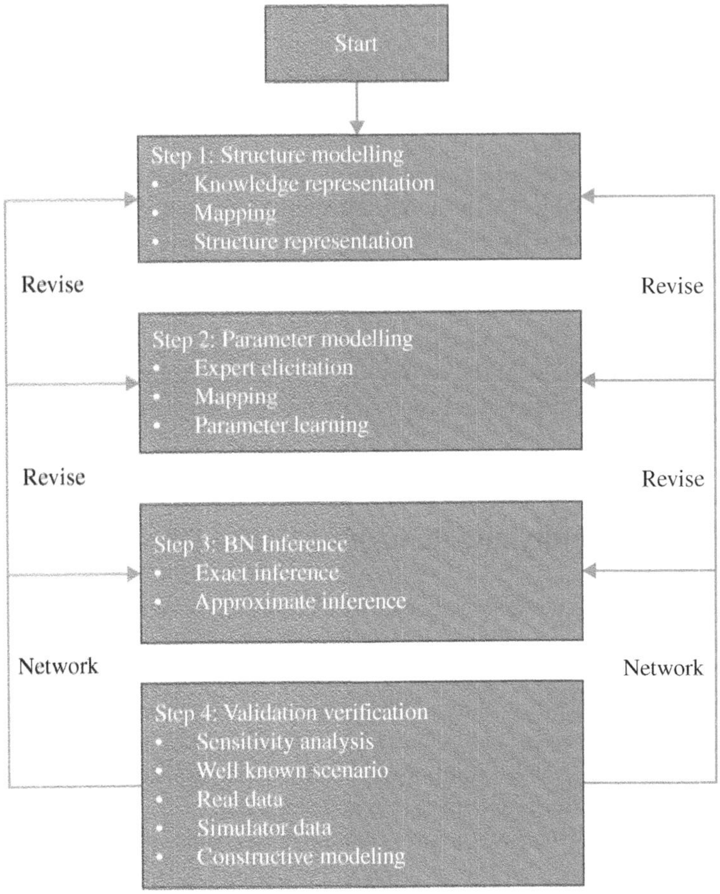

Figure 2.3 BN method flowchart.

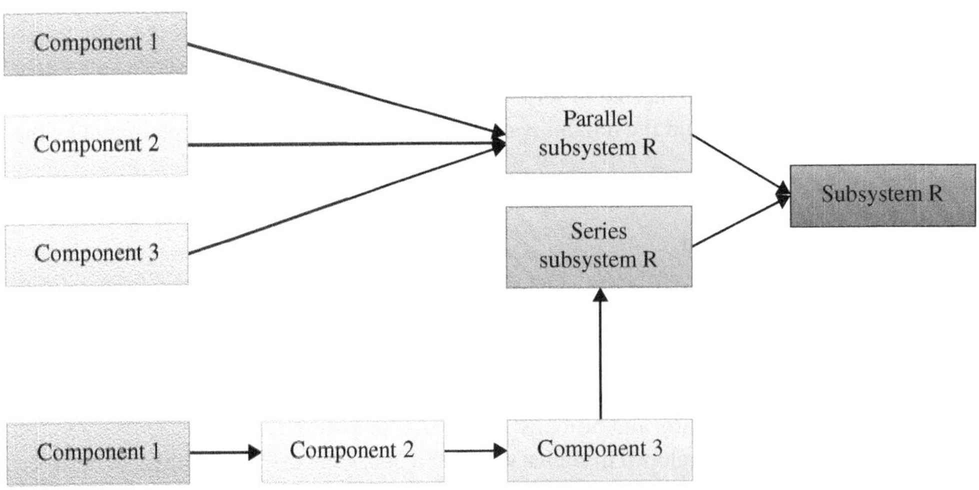

Figure 2.4 Typical BN of a series-parallel system for reliability study.

These methods of representing information are extremely subjective and have the potential to produce dependability evaluation models. The developed network is often too large in techniques like mapping and knowledge-based representation and takes considerable computational time. This makes the structure learning technique preferable during BN structure modeling.

The algorithm for structure learning is typically the machine learning algorithm. The component failure data along with the complete system failure data, are used to learn the structure of BN. The reliability evaluation using this method is more accurate than those obtained using methods that involve knowledge representation and mapping. This is one of the advantages of using this structured learning method. The apparent downside of using structure learning is that it can be difficult or even impossible to obtain adequate data for exact structure learning.

The BN reliability evaluation method has already been incorporated into power system reliability modeling. For example, Eliassi et al. (2015) used BN for the evaluation of composite power systems by giving importance to components. In their work, the training data has been generated by utilizing the Monte Carlo simulation with state sampling. They established the BN structure with the data by using common structure learning algorithms.

2.4.2 BN Parameter Learning

The process of BN parameters learning involves assigning a prior probability as well as a conditional probability. When evidence is gathered, the posterior probability of variables is referred to as the conditional probability of those variables. Typically, a CPT is utilized in the case of discrete variables, while with continuous variables, a conditional probability distribution is utilized. The techniques that are used to accomplish BN parameter modeling include expert opinion and parameter learning for hardware, structures, and software reliability analysis and mapping, particularly for hardware reliability analysis.

BNs' capacity to learn from data is a particularly useful trait. Only the CPT parameters are learned from historical data during parameter learning, even though the structure of the BN may be easily perceptible. The BN's structure must be learned, nevertheless, when using structure learning. This implies that historical data must be used to realize the interdependence between the domain variables.

Every unknown parameter in a CPT must be learned from historical data. A beta distribution defined in Eq. (2.5) is used for estimating the parameter of CPT:

$$P(\theta) = \text{Beta}(\theta|\alpha, \beta) = \frac{\neg(\alpha+\beta)}{\neg(\alpha).\neg(\beta)}.\theta^{\alpha-1}.(1-\theta)^{\beta-1} \tag{2.5}$$

According to the researchers, The Beta distribution is effective for parameter estimation of a CPT. This is because it has a convenient property that, over the interval [0, 1] they are nonzero and integrate to 1. The distribution parameters α and β are the number of events + 1. Also, the distribution mean value is $\frac{\alpha}{\alpha+\beta}$. This implies that a proportionally larger α value gives θ gets closer to unity, signifying a larger data set gives more certainty about θ.

2.4.3 BN Inference

Interface is the most important characteristic of a BN. In this stage of network evaluation, the evaluated probability of networks is updated when new observations are given. For hardware and software reliability evaluation, exact inference, and mapping inference are used methods for performing BN inference.

A variable that was observed and instantiated in the BN serves as the evidence. By giving some evidence or observations, the probability of one or more variables can be estimated in the BN inference. This indicates that BN generates the posterior probability for a target node of interest given a collection of input values, or evidence. The probability distribution in the BN is calculated by combining the data using various inference techniques.

There are both precise and approximate methods for carrying out Bayesian inferences. Variable elimination algorithms and junction tree algorithms are two commonly used precise inference techniques. In the variable elimination methodology, the variables are eliminated in a particular order. In addition to improving computing efficiency, this provides a better approach to rewrite the inference computations. On the other hand, the junction tree methods transform the multiple connected BNs into junction trees. Then, inferences are carried out using the Lauritzen-Spiegelhalter algorithm (Lauritzen and Spiegelhalter 1988) or variable elimination (Shafer-Shenoy algorithm).

When the model is large, with hundreds of variables, the precise inference approach is slow. According to the literature, approximation techniques frequently perform well in such situations. Variational approaches (mean-field approximation), parametric approximation, sampling techniques, loopy belief propagation (Boyen-Koller algorithm, Boyen and Koller 1998), etc., are a few examples of approximation techniques.

2.4.4 Validation and Verification

BN model validation is an extremely important facet of the reliability evaluation process. This is because these aspects instill a level of acceptable confidence in the calculated results. How accurately the BN model reflects the reality of the evaluated reliability is the validation. In BN analysis, for verification and validation, sensitivity analysis is the most used technique for hardware, structures, software, and human reliability. This can be done by utilizing contrastive modeling, simulated data, and real data.

2.5 Dynamic Bayesian Networks

The dynamic modeling of system behavior is the focus of the most recent work in the field of real-time reliability evaluation. DBN is used to represent temporal aspects of a system's behavior, which makes it easier to evaluate reliability (Murphy 2002). By including pertinent temporal dependencies inside the random variables, which will represent the variables' dynamic behavior over time, the Bayesian network model is expanded to a DBN model. To formalize complicated dynamic system models using prior knowledge of the system's functional and malfunctioning states, Weber and Jouffe (2006) introduced an approach based on Dynamic Object-Oriented Bayesian Networks (DOOBNs).

DBNs are a visual model for performing causal analysis and time-varying probabilistic inference in the presence of uncertainty in the system. Contemporaneous dependencies and noncontemporaneous dependencies are the two categories of dependencies found in a DBN. Contemporaneous dependencies are said to be when the arcs or connections between nodes represent variables within the same period. Arcs between nodes representing variables at various times are referred to as non-contemporaneous dependencies.

A typical DBN with two time slices t and $t+1$ is presented in Figure 2.5. The two-slice temporal Bayesian network (2TBN) that defines the transition model $P(X_t \mid X_{t-1})$ can be defined as given in Eq. (2.6):

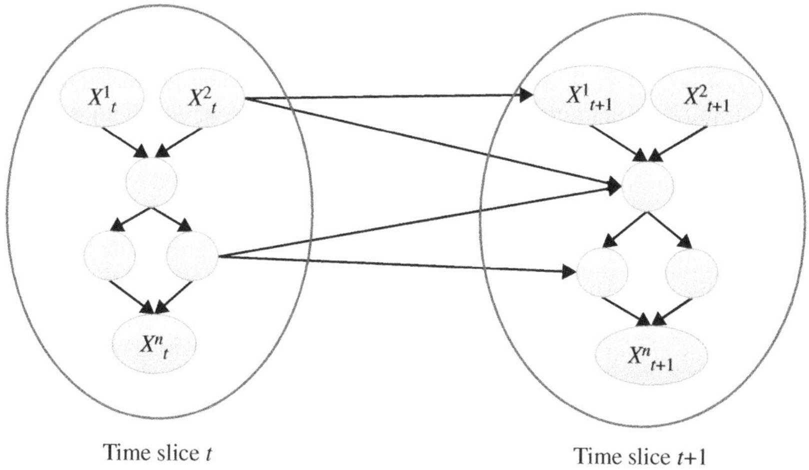

Figure 2.5 DBN consisting of two-time slices.

$$P(X_t \mid X_{t-1}) = \prod_{i=1}^{N} P\left(X_t^i \mid P_a\left(X_t^i\right)\right) \tag{2.6}$$

where $P_a\left(X_t^i\right)$ are the parents of X_t^i and X_t^i is the ith node at time t.

In a 2TBN, the node in the second time slice has its associated CPT or conditional probability data (CPD) depending on the variable type defined by $P\left(X_t^i \mid P_a\left(X_t^i\right)\right)$ for all $t > 1$. For a sequence of length T, the joint probability distribution of the 2TBN network is given in Eq. (2.7):

$$P(X_{1:T}) = \prod_{t=1}^{T}\prod_{i=1}^{N} P\left(X_t^i \mid P_a\left(X_t^i\right)\right) \tag{2.7}$$

2.6 Advantages and Limitations of BN and DBN

The Bayesian approach for reliability evaluation is a powerful tool that integrates prior knowledge, data, and statistical models to obtain reliable and up-to-date assessments. It provides a robust framework for decision-making and can be applied in various industries, including engineering, manufacturing, and infrastructure systems.

In the context of reliability evaluation, the Bayesian approach offers several advantages:

1. **Incorporation of Prior Information:** Bayesian analysis allows the integration of prior knowledge or beliefs about system reliability into the evaluation process. This prior information can be based on expert opinions, historical data, or previous studies. By combining prior knowledge with new data, the Bayesian approach provides a more comprehensive and informed analysis.
2. **Flexibility in Modeling:** Bayesian methods allow for flexible modeling of the underlying reliability distributions. Different distribution models can be specified based on the specific characteristics of the system under evaluation. This flexibility enables the incorporation of complex and realistic models that better represent the reliability behavior of the system.
3. **Sequential Updating:** One of the key features of Bayesian analysis is the ability to update estimates as new data becomes available. This is particularly useful in reliability evaluation, where data on failures or system performance may accumulate over time. The Bayesian approach

allows for sequential updating of reliability estimates, resulting in more accurate and up-to-date assessments.

4. **Uncertainty Quantification:** Bayesian inference provides a natural framework for quantifying uncertainty in reliability evaluations. Through the posterior distribution, which represents the updated belief about the reliability parameters, uncertainty intervals or credibility intervals can be calculated. These intervals provide a measure of the uncertainty associated with the estimated reliability metrics.
5. **Integration of Heterogeneous Data:** Bayesian analysis allows for the integration of different types of data from multiple sources. This is particularly relevant in reliability evaluation, where data can come from various sources such as field observations, laboratory testing, and historical records. Bayesian methods provide a unified framework for combining heterogeneous data, resulting in a more robust and comprehensive reliability assessment.

Despite the many benefits of BNs, there are two significant application-related limitations. First off, a lack of sufficient failure data frequently makes it impossible to estimate the CPT. The CPT can be counted using expert judgment as a solution to this issue. The ability of BN to update the conditional probabilities regularly based on new data or evidence helps handle the problem to a considerable extent, even though the effectiveness of this solution is highly dependent on expert capabilities. The second issue of BN that has been extensively studied is its inability to accurately model continuous variables because of the limits of the inference method. All continuous nodes must have some predefined range and intervals defined to achieve uniform discretization. The number of set intervals heavily influences accuracy, but it comes at a high cost due to computational complexity.

Another limitation of BN and DBN is that each node can only represent the state associated with a single observed variable. Therefore, to represent a component state with two or more observed variables, more child nodes need to be added to the network structure, which increases the complexity of the BN. The challenges with the use of BNs are:

- Discretization of continuous Variables.
- Collecting and structuring expert knowledge.
- No support for Feedback loops.
- The quality of the prior beliefs or model is directly correlated to the quality of the outputs produced by the network. If you feel that the system is dependent on a certain variable, then you can only include that variable as a component of a BN.

2.7 Conclusion

In this chapter, the use of BNs in reliability analysis has been demonstrated. Since BNs were first introduced in the early 1980s by Pearl, their use in reliability studies has been the subject of extensive research and has seen successful results. This chapter presents an overview of the reliability evaluation approach using BNs from the standpoint of the objects of evaluation. Because of its user-friendly interaction with domain experts, the modeling framework that BNs form is a practical, useful tool and utilized effectively. In addition, because BNs are found in probability theory, many of the fundamental debates that stymie the development of other modeling frameworks may be avoided. The robust mathematical formulation was applied to build learning approaches in BNs which are effective and efficient.

References

Boyen, X. and Koller, D. (1998). Tractable inference for complex stochastic processes. UAI.

Eliassi, M., Khoshkholgh Dashtaki, A., Seifi, H. et al. (2015). Application of Bayesian networks in composite power system reliability assessment and reliability-based analysis. *IET Generation, Transmission and Distribution* 9 (13): 1755–1764. https://doi.org/10.1049/iet-gtd.2014.0660.

Lauritzen, S.L. and Spiegelhalter, D.J. (1988). Local computations with probabilities on graphical structures and their application to expert systems. *Journal of the Royal Statistical Society: Series B: Methodological* 50 (2): 157–194. https://doi.org/10.1111/j.2517-6161.1988.tb01721.x.

Liu, Y. and Singh, C. (2010). Reliability evaluation of composite power systems using Markov cut-set method. *IEEE Transactions on Power Systems* 25 (2): 777–785. https://doi.org/10.1109/tpwrs.2009.2033802.

Murphy, K. (2002). Dynamic Bayesian networks: representation, inference, and learning. Thesis of the University of California, Berkley.

Weber, P. and Jouffe, L. (2006). Complex system reliability modeling with dynamic object oriented Bayesian networks (DOOBN). *Reliability Engineering and System Safety* 91 (2): 149–162. https://doi.org/10.1016/j.ress.2005.03.006.

3

Evaluation of Basic Reliability Indices Using State Enumeration Method

Rajesh Arya[1], Chandrima Roy[2], Atul Koshti[3], Ramesh C. Bansal[4], and Liladhar Arya[5]

[1] *Department of Information Technology, Acropolis Institute of Technology and Research, Indore, Madhya Pradesh, India*
[2] *Department of Electronics & Communication Engineering, Heritage Institute of Technology, Kolkata, West Bengal, India*
[3] *Department of Electrical Engineering, GES'S R.H. Sapat College of Engineering, Management Studies & Research, Nashik, Maharashtra, India*
[4] *Department of Electrical Engineering, University of Sharjah, Sharjah, United Arab Emirates*
[5] *Department of Electrical Engineering, Shri G.S. Institute of Technology & Science, Indore, Madhya Pradesh, India*

3.1 Introduction

The concept of reliability function is of great significance for nonrepairable or mission-oriented systems. A repairable system once fails to be repaired and then again is ready for service. In such a situation one is interested in knowing the percentage of system in the time is upstate and the failure frequency of the systems. Usually, a constant failure rate is assumed in such analysis. Since the repair time of the system is also a random variable. Repair rate of the system is also treated as a constant i.e. repair density function is also assumed as exponential. Usually, it is expected that a large number of failures in a system occurs, which requires small repair time, whereas few failure will occur, which requires a long repair time. Hence, the plot of the estimate of the repair density function will look as shown in Figure 3.1.

To determine availability function Markov process is utilized. Precisely function of a repair density is defined as

$$f_r(t_r) = \lim_{\Delta t_r \to 0} \frac{P[t_r \leq T_r \leq t_r + \Delta t_r]}{\Delta t_r} \approx$$

$n_r(t_r, \Delta t_r)$: Number of repairs completed between t_r to $t_r + \Delta t_r$
N_r: Total number of repairs carried over in a specified period of study.

3.2 Markov Process

It is a stochastic process that explains a process changing from one state to another. Assume that a process is described by "n" states, which are mutually exclusive with associated probabilities at "t" as $P_1(t), P_2(t), \ldots, P_n(t)$. The Markov process (Billinton & Allan 1983; Ebeling 2000) is defined by following conditional probability equation

Reliability Analysis of Modern Power Systems, First Edition. Edited by R. K. Saket and P. Sanjeevikumar.

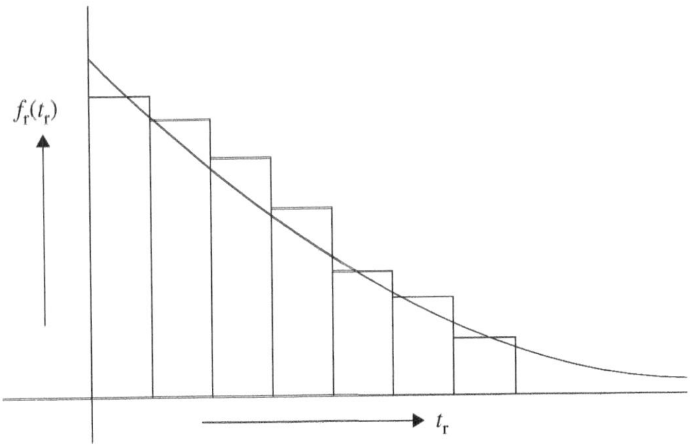

Figure 3.1 Repair density function plot.

$$P\left[X(t+\Delta t) = \frac{j}{X(t)} = i, X(t+\Delta t) = k, X(t-2\Delta t) = 1\right],$$
$$= P\left[X(t+\Delta t) = \frac{j}{X(t)} = i\right] = p_{ij}(t, \Delta t) \tag{3.1}$$

It is obvious that in a Markov process, a state "j" at $(t+\Delta t)$ depend on the occupied state by the process at t i.e. no on earlier states. In this sense, a Markov process is memoryless process. The conditional probability in Eq. (3.1) i.e. $p_{ij}(t, \Delta t)$ are known as transitional probabilities. Further, if transitional probabilities are independent on "t," then the Markov process is called homogeneous/stationary process. In view of the above Eq. (4.1) is written as

$$P\left[X(t+\Delta t) = \frac{j}{X(t)} = i\right] = q_{ij}(t, \Delta t) \tag{3.2}$$

q_{ij} may be defined state as transitions rate i to j. In terms of reliability as shall be made clear, this may be either a failure or repair rate, provided both have exponential probability density functions. Consider the general state transition diagram shown below.

Let us assume $P_1(t), P_2(t), \ldots, P_n(t)$ are state probabilities at time "t." General expression for $p_i(t+\Delta t)$ system of probability in ith state at $[(t+\Delta t)]$ is written below as (Figure 3.2)

$$p_i(t+\Delta t) = \sum_{\substack{k=1 \\ k\neq i}}^{n} q_{ki}\Delta t\, p_k(t) + \left(1 - \sum_{\substack{k=1 \\ k\neq i}}^{n} q_{ki}\Delta t\right) p_i(t)$$

The above equation is rewritten below as

$$p_i(t+\Delta t) = \sum_{k=1}^{n} q_{ki}\Delta t\, p_k(t) + p_i(t) \tag{3.3}$$

Further, the above can be written as

$$\frac{p_i(t+\Delta t) - p_i(t)}{\Delta t} = \sum_{k=1}^{n} q_{ki}\, p_k(t)$$

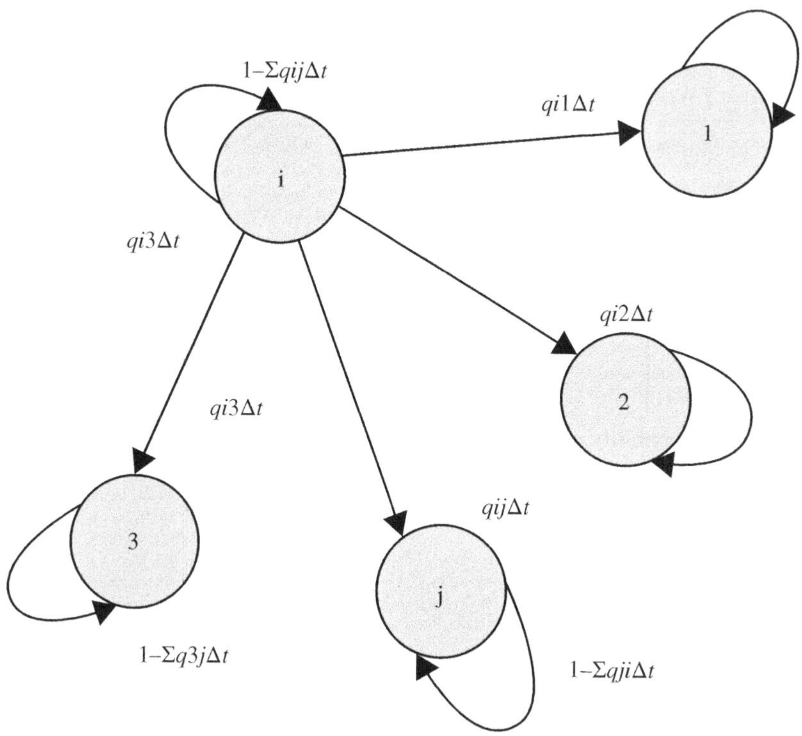

Figure 3.2 General state transition diagram.

Taking the limit as $\Delta t \to 0$

$$\dot{p}_i = \sum_{k=1}^{n} q_{ki}\, p_k(t) \tag{3.4}$$

The following matrix can be used to express Eq. (3.4):

$$\dot{p}^T(t) = p^T A \tag{3.5}$$

Matrix "A" follows from (3.4) as follows:

$$A = \begin{bmatrix} a_{11} & a_{12} & a_{13} & .. & a_{1n} \\ a_{21} & a_{22} & a_{23} & .. & a_{2n} \\ \vdots & \vdots & \vdots & & \vdots \\ a_{k1} & a_{k2} & a_{k3} & .. & a_{kn} \\ \vdots & \vdots & \vdots & & \vdots \\ a_{n1} & a_{n2} & a_{n3} & \dots & a_{nn} \end{bmatrix}$$

Elements of "A" matrix are identified as follows:
aij = rate of departure from ith state to jth state = q_{ij}

aii = −(sum of total transition rate from ith state) = $-\sum_{\substack{j=1 \\ j\neq i}}^{n} q_{ij}$

Matrix "A" is known as the transition intensity matrix.

If one puts $t = r\Delta t$ in Eq. (3.3) to define $p_{ki} = q_{ki}\Delta t$ state k to state i transition probability in an interval, then Eq. (3.3) gets modified as

$$p_i(r+1) = \sum p_{ki}p_k(r) + p_i(r)$$

This represents a discrete Markov process or chain (Billinton & Allan 1996; Lewis 1987) and represented in matrix form as

$$\underline{p}^T(r+1) = \underline{p}^T(r)\,\underline{P} \tag{3.6}$$

$$P = \begin{bmatrix} p_{11} & p_{12} & p_{13} & \cdot\cdot & p_{1n} \\ p_{21} & p_{22} & p_{23} & \cdot\cdot & p_{2n} \\ \vdots & \vdots & \vdots & \vdots & \vdots \\ p_{n1} & p_{n2} & p_{n3} & \cdots & p_{nn} \end{bmatrix}$$

P is known as the transition probability matrix.

$p_{ii} = 1-$(sum of the state i transition probabilities)

$p_{ij} =$ probabilities of transition from state i to j

3.3 Solution of State Equations

When one is interested in long-term solutions to reliability problems, then a steady-state solution of Eqs. (3.5) and (3.6) is required.

3.3.1 Steady State Solution: Continuous Time Case

Long run behavior of Eq. (3.5) is obtained by setting $\dot{p}^T = 0$ i.e. by solving the algebraic matrix eqn.

$$p^T A = 0 \tag{3.7}$$

Since matrix A is singular (3.7), it will have an infinite solution. Hence, $(n-1)$ equations are taken from (3.7), and one more equation as follows

$$\sum p_k = 1 \tag{3.8}$$

Example 3.1

Consider following two two-state continuous Markov process models. Obtain steady-state probabilities.

$$q_{12} = 0.2$$

$$q_{21} = 2$$

The matrix equation is written as follows:

$$[\dot{p}_1\ \dot{p}_2] = [p_1\ p_2]\begin{bmatrix} -0.2 & 0.2 \\ 2 & -2 \end{bmatrix} = 0$$

To get a steady-state solution

$$[p_1\ p_2]\begin{bmatrix} -0.2 & 0.2 \\ 2 & -2 \end{bmatrix} = [0\ 0]$$

$$-0.2\,p_1 + 2\,p_2 = 0, \quad 0.2\,p_1 - 2\,p_2 = 0$$

These two equations are not independent. Hence, one of them is considered.

$$0.2\,p_1 - 2\,p_2 = 0, 2\,p_1 + p_2 = 1$$

Solve the equation

$$p_1 = 0.9090$$
$$p_2 = 0.0909$$

3.3.2 Complete Solution: Continuous Time Case

The complete solution of the matrix equation is required

$$\dot{p}^T(t) = p^T A, \quad p^T(0) = [1 \;\; 0 \ldots 0]$$

3.3.2.1 Laplace Transform Method

$$s\,p^T(s) - p^T(0) = p^T(s)\,A, p^T(s)[s\,I - A] = p^T(0)$$
$$p^T(s) = p^T(0)[s\,I - A]^{-1}, p^T(t) = L^{-1}\{p^T(0)[s\,I - A]^{-1}\}$$

3.3.2.2 Solution by Computing e^{AT}

The solution of state space equation is written as

$$p^T(t) = p^T(0)e^{AT}$$

Matrix exponential can be expressed as

$$e^{AT} = \left[1 + At + \frac{(At)^2}{2!} + \cdots + \frac{(At)^m}{m!} + \cdots\right]$$

For a specific time "t," "m" terms may be considered such that the probabilities do not change significantly by considering $(m + 1)$ terms.

3.3.2.3 Discretization

One of the most powerful methods for solution of state equation is by discretization. The matrix differential equation is discretized as

$$\frac{p^T(t + \Delta t) - p^T(t)}{\Delta t} = p^T(t)A$$

$$p^T\,(t + \Delta t) - p^T(t)\,A\Delta t + p^T\,(t)$$

$$p^T\,(t + \Delta t) - p^T(t)\,[I + A\Delta t]$$

$$p^T(t + \Delta t) - p^T(t)A', A' = I + A\,\Delta t$$

Putting $t = r\Delta t$ and dropping Δt from argument of $p^T(.)$, above is explained as

$$p^T(r + 1) = p^T(r)\,A', \text{Put } r = 0$$
$$p^T(1) = p^T(0)\,A', \text{Put } r = 1$$
$$p^T(2) = p^T(1)\,A' = p^T(0)\,A'^2$$

The general solution is written as

$$p^T(r) = p^T(0)\,A'^r$$

3.3.2.4 Steady State Solution of Discrete Markov Process

$$p^T(r+1) = p^T(r)\,P$$

Long run solution of the above is obtained by letting

$$p^T(r+1) = p^T(r)\,P = \underline{p}^T$$
$$p^T = p^T P, p^T[I-P] = 0\; and \sum p_i = 1$$

3.3.2.5 Complete Solution of Discrete Markov Chain

The following general solution follows from Eq. (3.6)

$$p^T(r) = p^T(0)\,P^r$$

Example 3.2

Consider a two-state continuous Markov process with $\lambda = q_{12} = 0.1$ and $\mu = q_{21} = 10$. Obtain probabilities function for Figure 3.3.

Solution:

The matrix equation is written as follows

$$[\dot{p}_1\ \dot{p}_2] = [p_1\ p_2]\begin{bmatrix} -0.1 & 0.1 \\ 10 & -10 \end{bmatrix}$$

$$[sI - A] = \begin{bmatrix} s+\lambda & -\lambda \\ -\mu & s+\mu \end{bmatrix} = \begin{bmatrix} s+0.1 & -0.1 \\ -10 & s+10 \end{bmatrix}$$

$$[sI - A]^{-1} = \begin{bmatrix} s+\lambda & -\lambda \\ -\mu & s+\mu \end{bmatrix}^{-1}$$

$$[p_1(s)\ p_2(s)] = \frac{1}{s(s+\mu+\lambda)}[1\ 0]\begin{bmatrix} s+\mu & \lambda \\ \mu & s+\lambda \end{bmatrix}$$

$$p_1(s) = \frac{s+\mu}{s(s+\mu+\lambda)}, p_2(s) = \frac{\lambda}{s(s+\mu+\lambda)}$$

Taking Inverse laplace transform given below for correctness

$$p_1(t) = \frac{\mu}{\lambda+\mu} + \frac{\lambda}{\lambda+\mu}e^{-(\lambda+\mu)t}, p_2(t) = \frac{\lambda}{\lambda+\mu} - \frac{\lambda}{\lambda+\mu}e^{-(\lambda+\mu)t}$$

Taking inverse Laplace transform gives

$$p_1(t) = \frac{10}{10.1} + \frac{0.1}{10.1}e^{-10.1t}, p_2(t) = \frac{0.1}{10.1} - \frac{0.1}{10.1}e^{-10.1t}$$

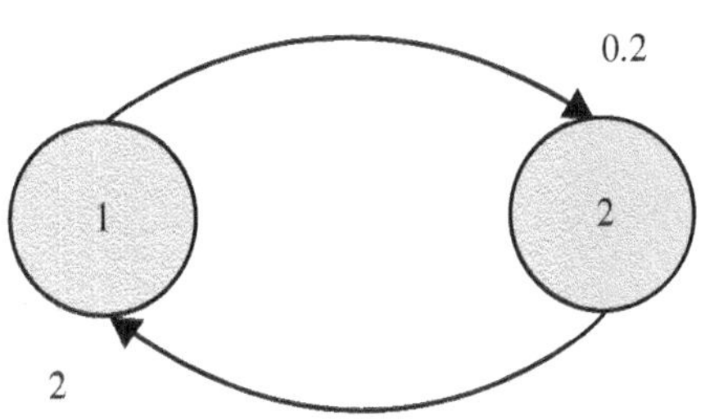

Figure 3.3 Two-state continuous Markov process model.

3.4 Functions of a Single Component's Availability and Unavailability

Consider single components with rates of failure and repair of λ and μ, respectively. Figure 3.4 displays the system's state transition diagram (Arya & Koshti 2008; Arya et al. 2013; Halve et al. 2023).

The states 1 and 2 stand for up and down, respectively. The following matrix form represents the state space equations.

$$[\dot{p}_1 \ \dot{p}_2] = [p_1 \ p_2] \begin{bmatrix} -\lambda & \lambda \\ \mu & -\mu \end{bmatrix}$$

$$p_1(0) = 1, \ p_2(0) = 0$$

$$[sI - A] = \begin{bmatrix} s & 0 \\ 0 & s \end{bmatrix} - \begin{bmatrix} -\lambda & \lambda \\ \mu & -\mu \end{bmatrix}$$

$$= \begin{bmatrix} s+\lambda & -\lambda \\ -\mu & s+\mu \end{bmatrix}$$

$$[sI - A]^{-1} = \begin{bmatrix} s+\lambda & -\lambda \\ -\mu & s+\mu \end{bmatrix}^{-1} = \frac{1}{s(s+\mu+\lambda)} \begin{bmatrix} s+\mu & \lambda \\ \mu & s+\lambda \end{bmatrix}$$

$$[p_1(s) \ p_2(s)] = \frac{1}{s(s+\mu+\lambda)}[1 \ 0] \begin{bmatrix} s+\mu & \lambda \\ \mu & s+\lambda \end{bmatrix}$$

$$p_1(s) = \frac{s+\mu}{s(s+\mu+\lambda)}, p_2(s) = \frac{\lambda}{s(s+\mu+\lambda)}$$

Taking inverse Laplace transform gives

$$P_1(t) = \frac{\mu}{\lambda+\mu} + \frac{\lambda}{\lambda+\mu} e^{-(\lambda+\mu)t}$$

$$P_2(t) = \frac{\lambda}{\lambda+\mu} - \frac{\lambda}{\lambda+\mu} e^{-(\lambda+\mu)t}$$

The state probability $p_1(t)$ is known as availability and $p_2(t)$ is known as unavailability. Figure 3.5 shows the plot of these functions. It is observed that both availability and unavailability approaches to steady state values as given

$$P_{1ss} = \frac{\mu}{\lambda+\mu}, P_{2ss} = \frac{\lambda}{\lambda+\mu},$$

The same figure shows the plot of reliability, which is exponentially decreasing and does not consider repair. No path of return from absorbing states.

Figure 3.4 A system with a single component's state transition diagram.

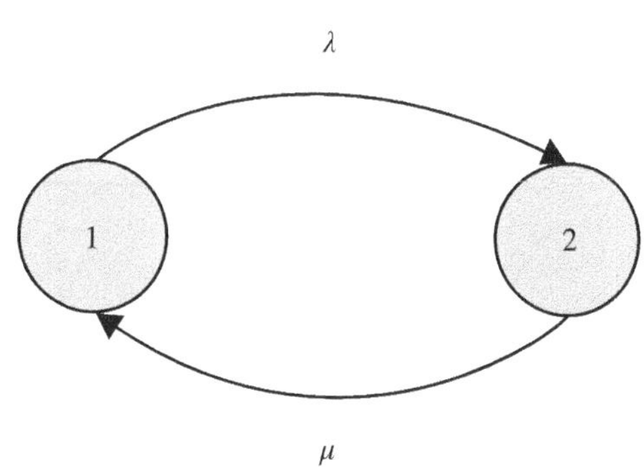

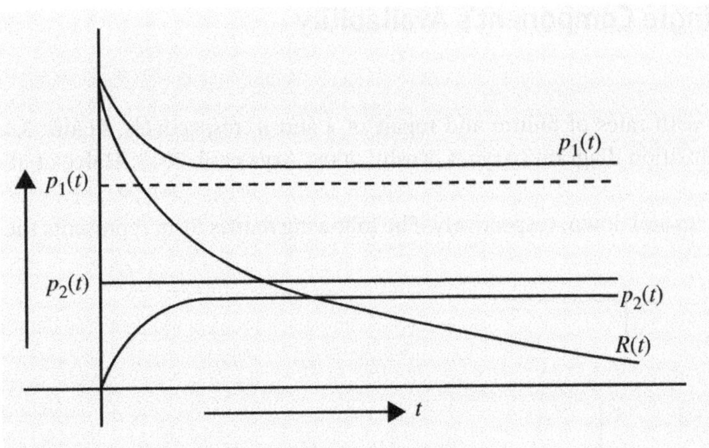

Figure 3.5 Plot of availability and unavailability.

3.5 Two-Component State Model and State Probabilities

Assume that two components "a" and "b" have failure and repair rates as λ_a, λ_b and μ_a, μ_b. The system will have four states, as shown in the diagram (Figure 3.6).

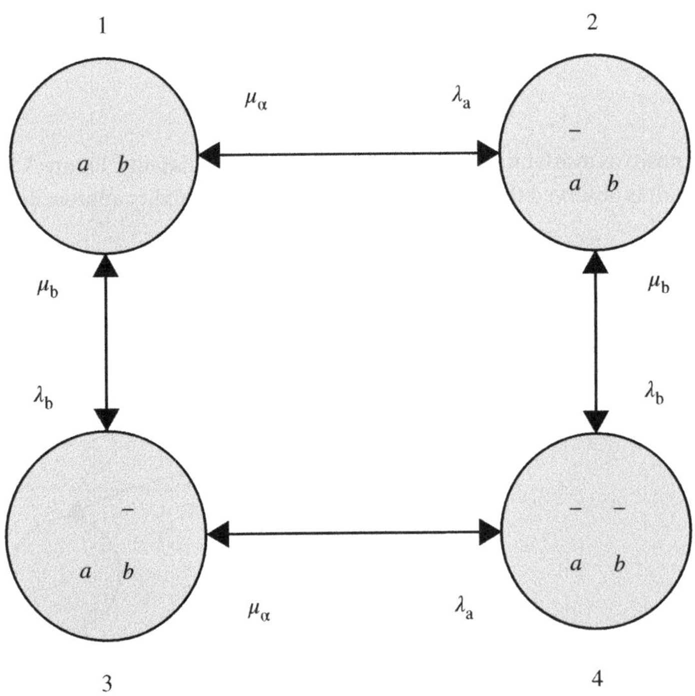

Figure 3.6 Two-component state model.

State space equations are written as follows

$$[\dot{p}_1\ \dot{p}_2\ \dot{p}_3\ \dot{p}_4] = [p_1\ p_2\ p_3\ p_4]\begin{bmatrix} -(\lambda_a+\lambda_b) & \lambda_a & \lambda_b & 0 \\ \mu_a & -(\mu_a+\lambda_b) & 0 & \lambda_b \\ \mu_b & 0 & -(\lambda_a+\mu_b) & \lambda_a \\ 0 & \mu_b & \mu_a & -(\mu_a+\mu_b) \end{bmatrix}$$

State probabilities may be easily evaluated by using the method of discretization. It requires a suitable choice of "Δt." If one does not have experience with the system initially, any value of Δt may be selected, and state probabilities trajectories are obtained. Then, a smaller value of Δt is selected, and then again, state probabilities are obtained. The difference in two calculated state probabilities should be less than the tolerance value specified. Otherwise, further reduce Δt and repeat the calculation of state probabilities till convergence.

State probabilities in the long run are evaluated by solving the following algebraic equations:

$$[p_1\ p_2\ p_3\ p_4]\begin{bmatrix} -(\lambda_a+\lambda_b) & \lambda_a & \lambda_b & 0 \\ \mu_a & -(\mu_a+\lambda_b) & 0 & \lambda_b \\ \mu_b & 0 & -(\lambda_a+\mu_b) & \lambda_a \\ 0 & \mu_b & \mu_a & -(\mu_a+\mu_b) \end{bmatrix} = [0\ 0\ 0\ 0]$$

$$p_1 + p_2 + p_3 + p_4 = 1$$

The state probabilities are obtained as

$$p_1 = \frac{\mu_a\mu_b}{D}, \quad p_2 = \frac{\lambda_a\mu_b}{D}, \quad p_3 = \frac{\mu_a\lambda_b}{D}, \quad p_4 = \frac{\lambda_a\lambda_b}{D}$$
$$D = (\lambda_a + \mu_a)(\lambda_b + \mu_b)$$

If the two components are identical i.e.

$$\lambda_a = \lambda_b = \lambda, \quad \mu_a = \mu_b = \mu$$

The diagram of state transition will reduce Figure 3.7, which is shown below.

State 1 Both units are in up state
State 2 One unit is up and other is down
State 3 Both units are down

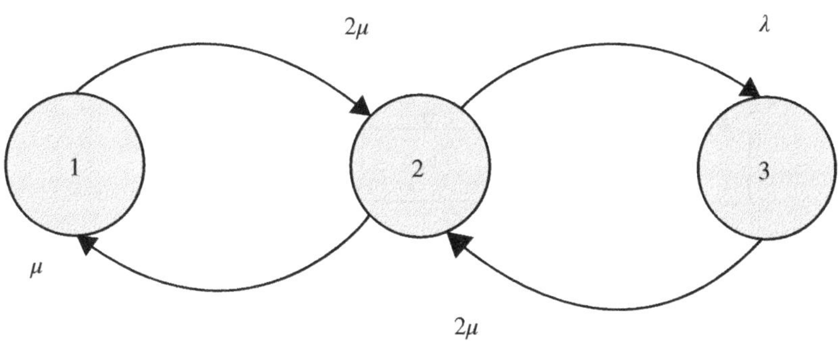

Figure 3.7 Diagram of the reduced form state transition for a two-component system.

State equations are written by inspection

$$[\dot{p}_1\ \dot{p}_2\ \dot{p}_3] = [p_1\ p_2\ p_3]\begin{bmatrix} -2\lambda & 2\lambda & 0 \\ \mu & -(\lambda+\mu) & \lambda \\ 0 & 2\mu & -2\mu \end{bmatrix}$$

Steady-state probabilities are evaluated as follows

$$[p_1\ p_2\ p_3]\begin{bmatrix} -2\lambda & 2\lambda & 0 \\ \mu & -(\lambda+\mu) & \lambda \\ 0 & 2\mu & -2\mu \end{bmatrix} = [0\ 0\ 0], p_1 + p_2 + p_3 = 1$$

Solution gives

$$p_1 = \frac{\mu^2}{(\lambda+\mu)^2}, p_2 = \frac{2\lambda\mu^2}{(\lambda+\mu)^2}, \quad p_3 = \frac{\lambda^2}{(\lambda+\mu)^2}$$

System availability is evaluated by applying system adequacy criterion. For example, at least one of the units is required for the success system, then the availability of a system is given by

$$A_{sys} = p_1 + p_2 = \frac{\mu^2 + 2\lambda\mu}{(\lambda+\mu)^2}$$

Unavailability is given as

$$\overline{A}_{sys} = \frac{\lambda^2}{(\lambda+\mu)^2}$$

3.6 Three-Component State Transition Diagram

Three components of state transition Figure 3.8 are shown below.

If the units are identical, then the above state transition diagram reduces, as shown in Figure 3.9.

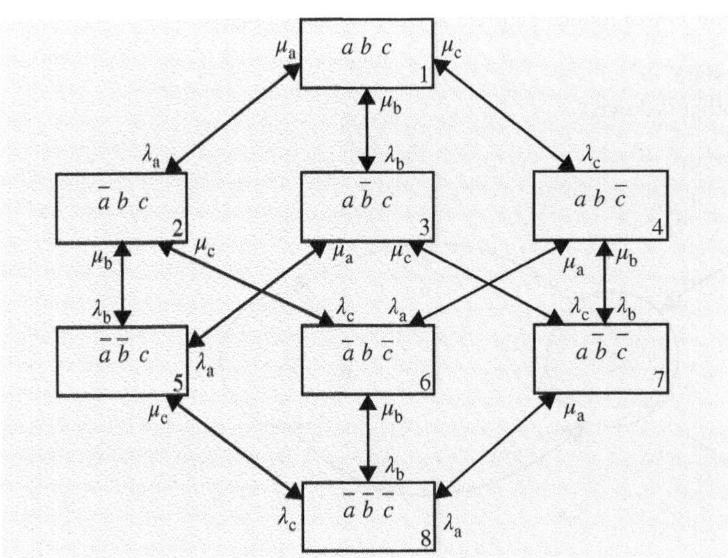

Figure 3.8 Three-component state transition diagram.

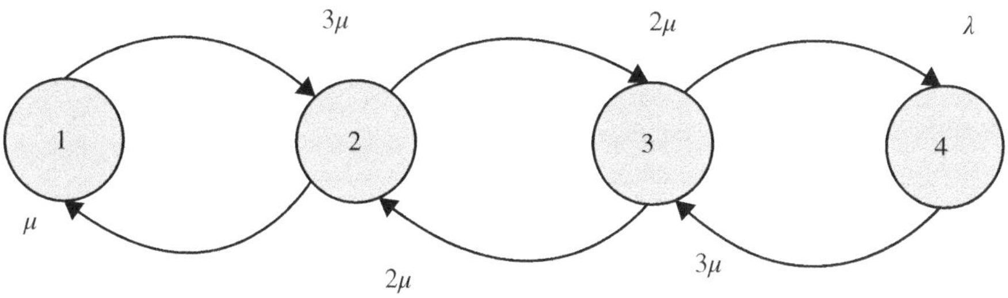

Figure 3.9 Diagram of the reduced state transition for a three-component system.

For steady-state probabilities, state equations are written, and solutions for eight states (different components case) are written (long run):

$$p_1 = \frac{\mu_a \mu_b \mu_c}{D}, \quad p_2 = \frac{\lambda_a \mu_b \mu_c}{D}, \quad p_3 = \frac{\mu_a \lambda_b \mu_c}{D}, \quad p_4 = \frac{\mu_a \mu_b \lambda_c}{D}$$

$$p_5 = \frac{\lambda_a \lambda_b \mu_c}{D}, \quad p_6 = \frac{\lambda_a \mu_b \lambda_c}{D}, \quad p_7 = \frac{\mu_a \lambda_b \lambda_c}{D}, \quad p_8 = \frac{\lambda_a \lambda_b \lambda_c}{D}$$

$$D = (\lambda_a + \mu_a)(\lambda_b + \mu_b)(\lambda_c + \mu_c)$$

If the components are identical, then states are reduced from eight to four, as shown in Figure 3.9:

$$p_1 = \frac{\mu^3}{(\lambda+\mu)^3}, p_2 = \frac{3\mu^2\lambda}{(\lambda+\mu)^3}, \quad p_3 = \frac{3\lambda^2\mu}{(\lambda+\mu)^3}, \quad p_4 = \frac{\lambda^3}{(\lambda+\mu)^3}$$

If at least two units are required from an adequacy viewpoint, the system availability is given as

$$A_{sys} = p_1 + p_2 = \frac{\mu^3 + 3\mu^2\lambda}{(\lambda+\mu)^3}, \overline{A}_{sys} = p_3 + p_4 = \frac{3\lambda^2\mu + \lambda^3}{(\lambda+\mu)^3}$$

3.7 Concept of Frequency and Mean Duration

Assume a component has failure and repair rates are, respectively, λ and μ. MTTF and MTTR are given as $1/\lambda$ and $1/\mu$ respectively. Life length (T_f) and time to repair (T_r) are random variables, and each has an exponential distribution function (Billinton & Allan 1996; Lewis 1987; Koshti et al. 2023). Then cycle time i.e. sums of "T_f" and "T_r" is also a random variable when the process of failure and repair comes in stages. Moreover, there are two random variables that are independent. Hence, as in the case of the standby system, the cycle time ($T_c = T_f + T_r$) has hypo-exponential distribution function (Lewis 1987) and is given as

$$g(t) = \frac{\lambda\mu}{\mu - \lambda}[e^{-\lambda t} - e^{-\mu t}]$$

Average or mean value of cycle time (T_c) is given as

$$T_c = \frac{1}{\lambda} + \frac{1}{\mu}$$

Thus, it can be seen as a statistical cycle, as shown in Figure 3.10.

Cycle time is also known as MTBF i.e. mean time between failures. Average failure frequency or frequency is given as

$$f = \frac{1}{T_c} = \frac{1}{\frac{1}{\lambda} + \frac{1}{\mu}} = \frac{\mu\lambda}{\lambda+\mu} = \left(\frac{\mu}{\lambda+\mu}\right)\lambda = \left(\frac{\lambda}{\lambda+\mu}\right)\mu$$

= (Probability of being in that state) (rate of departure from that state)

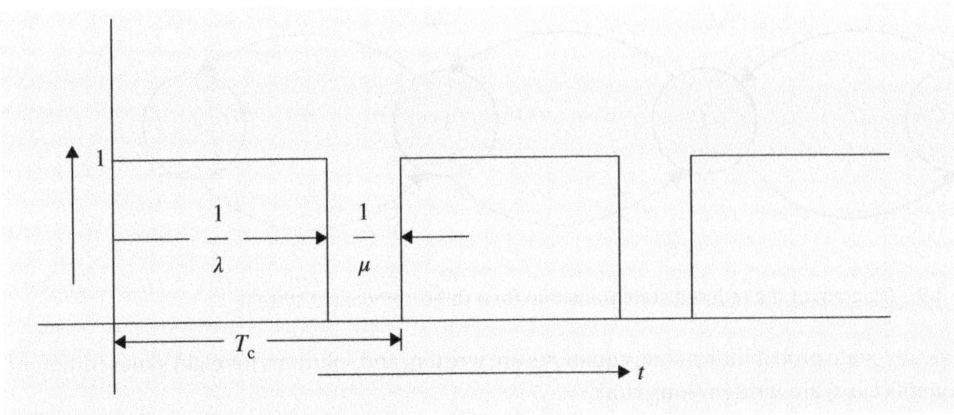

Figure 3.10 Component statistical cycle.

Now consider a general state transition diagram with state probabilities $P_1(t), P_2(t).., P_n(t)$. Further assume $t_1, t_2, \ldots, t_n$ are time spent in these states respectively.

Where,

$$T_0 = t_1 + t_2 + t_3, \ldots + t_n$$

Further assume that $q_1, q_2, q_3, \ldots, q_n$ are total rate of departure from these states, respectively, i.e.

$$q_i = \sum_j q_{ij}$$

Then, the frequency of ith state i.e. state "i" on one side and remaining states on the other side is given as the number of transitions per unit time from ith state:

$$f_i = \frac{q_i t_i}{T_o} = \left(\frac{t_i}{T_o}\right) q_i = p_i q_i$$

= (probability of being in ith state) (total rate of departure from ith state).

3.8 Frequency of Combined Events

Individual frequency of a state is not of importance when large numbers of components are involved in the systems. Transitions per unit of time (frequency) from a collection of events, such as a group of states that correlate to system success, are important under such circumstances. This frequency matches the frequency of failure. Figure 3.11 depicts a generic state transition diagram. The total states are divided into two categories, "W" and "F." Working (up) and failing (down) states, respectively, are represented by the states in "W" and "F." If the system is in a state of "W" set the system is "up" otherwise if the system encounters a state in "F" the system is in "down" state.

In such situations, transitions outside the group are considered. Transitions among states in a group are not considered. Hence, the frequency of transitions from "W" to "F" is written as:

$$f = \sum_{i \in W} p_i \sum_{j \in F} q_{ij} \tag{3.9}$$

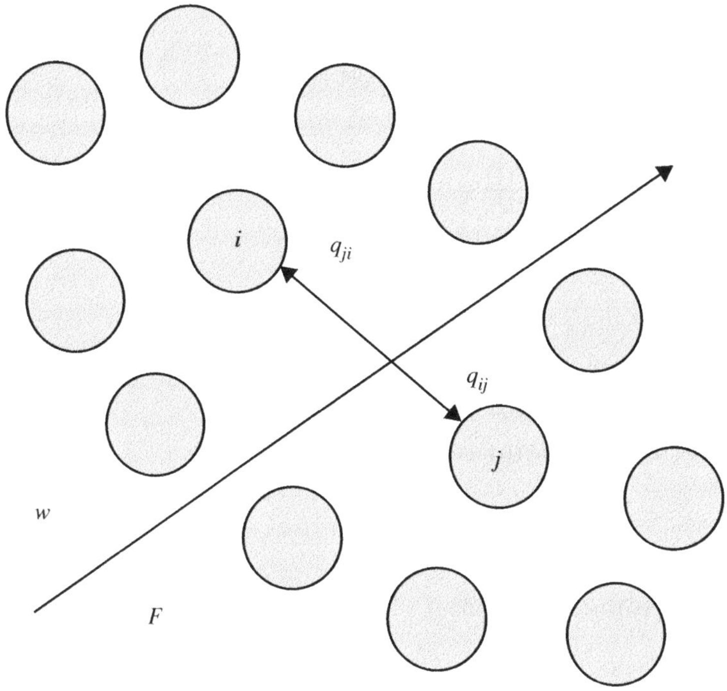

Figure 3.11 General state transition diagram.

Frequency will be the same as calculated as transitions from "F" to "W" and given

$$f = \sum_{j \in F} p_j \sum_{i \in W} q_{ji} \tag{3.10}$$

Example 3.3

A system having two component consists of a $\mu_1 = 9.6$, $\mu_2 = 9.4$, $\lambda_1 = 0.4$ and $\lambda_2 = 0.6$. Obtain (1) f_1, f_2, f_3, f_4 (2) f_{12}, f_{34} (Figure 3.12).

Figure 3.12 General state transition diagram.

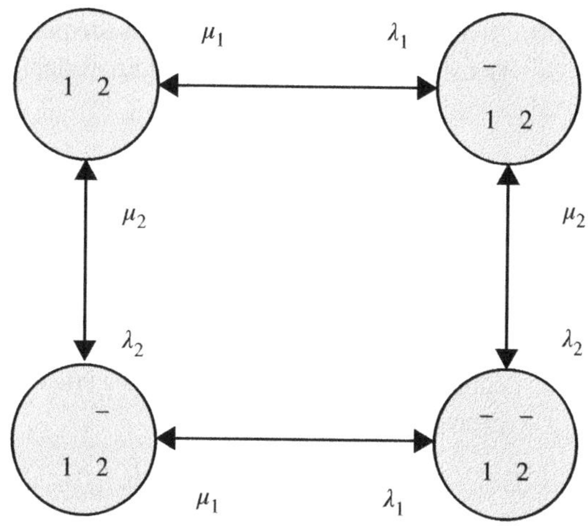

The probabilities in long run are given as

$$p_1 = \frac{\mu_1\mu_2}{D} = 0.94 \times 0.96 = 0.9024, p_2 = \frac{\lambda_1\mu_2}{D} = 0.04 \times 0.94 = 0.0376$$

$$D = (\lambda_1 + \mu_1)(\lambda_2 + \mu_2)$$

$$p_3 = \frac{\mu_1\lambda_2}{D} = 0.96 \times 0.06 = 0.0576, p_4 = \frac{\lambda_1\lambda_2}{D} = 0.04 \times 0.06 = 0.0024$$

$$f_1 = p_1(\lambda_1 + \lambda_2) = (0.94 \times 0.96)(0.4 + 0.6) = 0.94 \times 0.96 = 0.9024$$

$$f_2 = p_2(\mu_1 + \lambda_2) = 0.04 \times 0.94 \times (9.6 + 0.6) = 0.38352$$

$$f_3 = p_3(\lambda_1 + \mu_2) = (0.96 \times 0.06)(0.4 + 9.4) = 0.96 \times 0.06 \times 9.8 = 0.56448$$

$$f_4 = p_4(\mu_1 + \mu_2) = 0.04 \times 0.06 \times (9.4 + 9.6) = 0.04 \times 0.06 \times 19 = 0.0456$$

$$f_{12} = \lambda_2(p_1 + p_2) = 0.6 \times [0.94 \times 0.96 + 0.04 \times 0.94] = 0.564$$

$$f_{34} = \mu_2(p_3 + p_4) = 9.4 \times [\,0.96 \times 0.06 + 0.04 \times 0.06] = 0.564$$

Note:

$$f_{12} = f_{34}$$

This is required since frequency balance must hold.

3.9 State Enumeration Technique for Obtaining Frequency-Duration (FD)

For determining failure frequency, mean up/time of a technique has been explained Figure 3.13 (Arya et al. 2013; Halve et al. 2023; Koshti et al. 2023; Endrenyi 1978).

Step 1: Failure and repair rate is assumed that each component present in the system is provided. Draw a state transition diagram using these states.

Step 2: Solving for steady-state probabilities.
Solve steady-state probabilities

$$p^T A = 0, \sum p_i = 1$$

Step 3: Divide the state using the "system adequacy criterion" space into the "W" and "F" groups. Get steady state availability and unavailability using the following method:

$$p_{up} = \sum_{i \in W} p_i, p_{dn} = \sum_{j \in F} p_j$$

Step 4: The following relations to determine failure frequency

$$f = \sum_{i \in W} p_i \sum_{j \in F} q_{ij}$$

Step 5: The diagram's representation of the general state transition consists of two states and one component where λ_e and μ_e are the rate of failure and repair being equal. The frequency is expressed as follows:

$$f = p_{up}\lambda_e = p_{dn}\mu_e, \lambda_e = \frac{f}{p_{up}}, \mu_e = \frac{f}{p_{dn}}$$

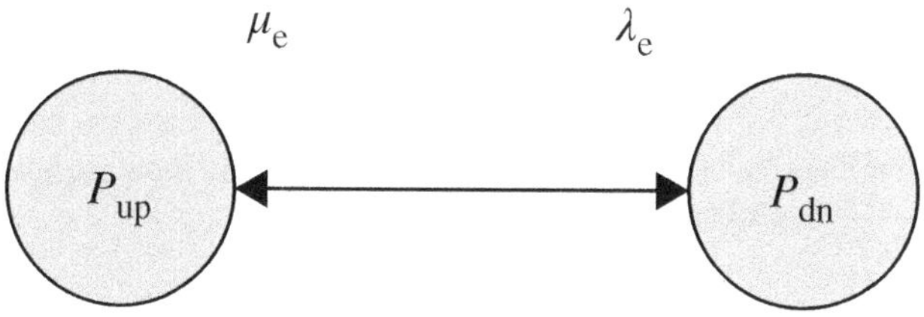

Figure 3.13 Two state one component equivalent state transition diagram.

As for a single component

$$\text{MUT} = \frac{1}{\lambda_e} = \frac{p_{up}}{f} = p_{up}T_c$$

$$\text{MDT} = \frac{1}{\mu_e} = \frac{p_{dn}}{f} = p_{dn}T_c$$

Hence, p_{up}, p_{dn}, f, MUT, and MDT are calculated as reliability index.

Example 3.4

Obtain a system consisting of three identical components having failure and repair rate as λ and μ, respectively. System success requires at least two components in upstate.

(a) Frequency
(b) MUT
(c) MDT

Take $\mu = 9.8$ per year and $\lambda = 0.2$ per year

Solution:

The system will have four states, as shown in Figure 3.9. State probabilities are given as follows

$$p_1 = \frac{\mu^3}{(\lambda+\mu)^3} = 0.98^3 = 0.9411$$

$$p_2 = \frac{3\mu^2\lambda}{(\lambda+\mu)^3} = 3\times 0.98^2 \times 0.02 = 0.057624$$

$$p_3 = \frac{3\lambda^2\mu}{(\lambda+\mu)^3} = 3\times 0.98 \times 0.02^2 = 0.001176$$

$$p_4 = \frac{\lambda^3}{(\lambda+\mu)^3} = 0.02^3 = 0.000008$$

Adequacy criterion is applied and p_{up} and p_{dn} are obtained as

$$p_{up} = p_1 + p_2 = \frac{\mu^3}{(\lambda+\mu)^3} + \frac{3\mu^2\lambda}{(\lambda+\mu)^3} = 0.998724$$

$$p_{up} = \frac{\mu^2(\mu+3\lambda)}{(\lambda+\mu)^3}, p_{dn} = p_3 + p_4 = \frac{3\lambda^2\mu}{(\lambda+\mu)^3} + \frac{\lambda^3}{(\lambda+\mu)^3}$$

$$p_{dn} = \frac{(3\mu+\lambda)\lambda^2}{(\lambda+\mu)^3} = 0.001184$$

Frequency f is calculated as follows

$$f = p_1 \times 0 + p_2 \times 2\lambda$$

$$f = \frac{3\mu^2\lambda}{(\lambda+\mu)^3} \times 2\lambda = \frac{6\mu^2\lambda^2}{(\lambda+\mu)^3} = \frac{6 \times 9.8^2 \times 0.2^2}{1000} = 0.0230496/\text{year}$$

Mean cycle time

$$Tc = 1/f = \frac{(\lambda+\mu)^3}{6\mu^2\lambda^2} = \frac{1000}{6 \times 9.8^2 \times 0.2^2} = 43.38\ \text{year}$$

$$\text{MUT} = \frac{p_{up}}{f} = \frac{\mu^2(\mu+3\lambda)}{6\mu^2\lambda^2}$$

$$= \frac{9.8^2(9.8+0.6)}{6 \times 9.8^2 \times 0.2^2} = 43.33$$

$$\text{MDT} = \frac{p_{dn}}{f} = \frac{\lambda^2(3\mu+\lambda)}{(\lambda+\mu)^3} \times \frac{(\lambda+\mu)^3}{6\mu^2\lambda^2} = \frac{\lambda^2(3\mu+\lambda)}{6\mu^2\lambda^2}$$

$$= \frac{0.2^2(29.4+0.2)}{6 \times 9.8^2 \times 0.2^2} = 0.05$$

Example 3.5

A generator has failure, repair, and re-installation rate as λ, μ, and α, respectively. Obtain expressions for

(a) Availability
(b) Frequency
(c) Mean up time (MUT)
(d) Mean down time (MDT)

Solution:

Figure 3.14 shows the state transition

State 1: Operable state
State 2: Failed state
State 3: Generator repaired but not re-installed for operation

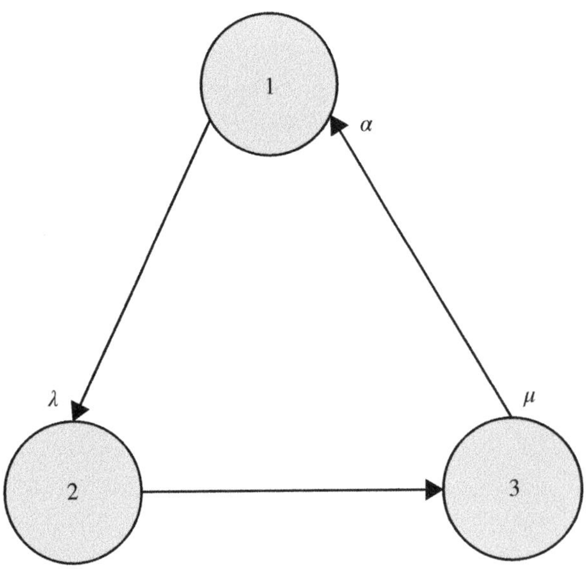

Figure 3.14 State transition diagram for Example 3.5.

Hence states 1 is up state and states 2 and 3 are down state.

$$[\dot{p}_1 \quad \dot{p}_2 \quad \dot{p}_3] = [p_1 \quad p_2 \quad p_3]\begin{bmatrix} -\lambda & \lambda & 0 \\ 0 & -\mu & \mu \\ \alpha & 0 & -\alpha \end{bmatrix}$$

Long run probabilities are calculated by setting derivative terms zero and taking along two equations from the above set of equation

$$-\lambda p_1 + \alpha p_3 = 0, \lambda p_1 - \mu p_2 = 0 \text{ and } p_1 + p_2 + p_3 = 1$$

Solving the above three equations

$$\dot{p}_3 = \frac{\lambda}{\alpha}p_1, \dot{p}_2 = \frac{\lambda}{\mu}p_1, p_1 + \frac{\lambda}{\mu}p_1 + \frac{\lambda}{\alpha}p_1 = 1$$

$$p_1 = \frac{\mu\alpha}{\mu\alpha + \alpha\lambda + \mu\lambda}, p_2 = \frac{\alpha\lambda}{\mu\alpha + \alpha\lambda + \mu\lambda}, p_3 = \frac{\mu\lambda}{\mu\alpha + \alpha\lambda + \mu\lambda}$$

(a) Availability is given as

$$A = p_1 = \frac{\mu\alpha}{\mu\alpha + \alpha\lambda + \mu\lambda}$$

(b) Frequency is given by

$$f = p_1\lambda = \frac{\mu\alpha\lambda}{\mu\alpha + \alpha\lambda + \mu\lambda}$$

(c) Mean up time

$$\text{MUT} = \frac{p_1}{f} = \frac{1}{\lambda}$$

(d) Mean down time

$$\text{MDT} = \frac{p_2 + p_3}{f} = \frac{\mu\lambda + \alpha\lambda}{\mu\alpha\lambda}$$

Take numerical values

$\mu = 9.6/\text{year}$,

$\lambda = 0.4/\text{year}$

$\alpha = 20/\text{year}$

(a) Availability is given as

$$A = p_1 = \frac{9.6 \times 20}{9.6 \times 20 + 9.6 \times 0.4 + 0.4 \times 20} = \frac{192}{203.8} = 0.9421$$

Unavailability

$$\overline{A} = 1 - A = 0.0579$$

(b) Frequency is given by

$$f = \frac{9.6 \times 20 \times 0.4}{203.8} = 0.3768/\text{year}$$

(c) Mean up time

$$\text{MUT} = \frac{p_1}{f} = \frac{0.9421}{0.3768} = 2.50 \text{ year}$$

(d) Mean down time

$$\text{MDT} = \frac{p_2 + p_3}{f} = \frac{0.0579}{0.3768} = 0.15366 \text{ year}$$

Example 3.6

A generator with a 100 MW rating has failure rate and repair rate of 0.2 and 9.8 per year. Obtain-

(a) Availability and unavailability
(b) Frequency
(c) MUT and MDT
(d) Mean capacity
(e) Expected power and energy are not supplied if the load on the generator is constant at 80 MW.

Solution:

(a) Availability is given as

$$A = \frac{9.8}{9.8 + 0.2} = 0.98$$
$$\overline{A} = \frac{0.2}{9.8 + 0.2} = 0.02$$

(b) Frequency is given by

$$f = 0.98 \times 0.2 = 0.196/\text{year}$$

Note: $f < \lambda$ [this will always be the case]

(c) Mean up time

$$\text{MUT} = \frac{A}{f} = \frac{0.98}{0.196} = 5 = \frac{1}{\lambda}$$

Mean down time

$$\text{MDT} = \frac{\overline{A}}{f} = \frac{0.02}{0.196} = 0.102 = \frac{1}{\mu}$$

(d) Mean capacity

$$\overline{c} = \sum c_i p_i = c_1 p_1 + c_2 p_2 = 100 \times 9.8 + 0 \times 0.02 = 98 \text{ MW}$$

$$\text{Variance} = \sigma_c^2$$

$$= \sum (c_i - \overline{c})^2 p_i = (c_1 - \overline{c})^2 p_1 + (c_2 - \overline{c})^2 p_2 = (100 - 98)^2 \times 0.98 + (0 - 98)^2 \times 0.02$$
$$= 196 \text{ MW}^2$$

$$\sigma_c = \sqrt{196} = 14 \text{ MW}$$

(e) Expected power not supplied (EPNS) $= \sum PNS_i \times p_i$

$$= PNS_1 \times p_1 + PNS_2 \times p_2$$

PNS_i power not supplied in ith state

$$= 0 \times 0.98 + 80 \times 0.02 = 160 \text{ MW}$$

Expected energy not supplied (EENS)$=1.6 \times 8760 = 14,016$ MWh/year

3.10 Conclusion

A methodology has been presented in this chapter for evaluating reliability indices of repairable systems employing the Markov process. This methodology gives analysis for the evaluation of frequency failure, MUT, and MDT, along with probabilities of success and probabilities of failure. The Markov process was utilized to evaluate state probabilities based on the state transition diagram. Steady-state solution of state equations has been demonstrated to evaluate availability and unavailability based on state transition diagram of single-component, two-component, and three-component models. Further reduced state transition diagrams are used to evaluate the availability and unavailability of identical components, assuming equal failure rates of components and equal repair rates of components. Illustrative examples are also presented to show the use of state enumeration technique for reliability indices evaluation. Therefore, this chapter presents a methodology to evaluate frequency and duration indices, which will be helpful for power system planning engineers to assess the reliability of the system in the perspective of reliability analysis.

References

Arya, L.D. and Koshti, A. (2008). Distributed generation capacity reliability evaluation using safety index. *Journal of Institution of Engineering (India)* 89: 3–7.

Arya, L.D., Koshti, A., and Choube, S.C. (2013). Frequency-duration analysis of composite distribution system using a non-sequential Monte Carlo simulation. *Electrical Power and Energy Systems* 46: 17–25.

Billinton, R. and Allan, R.N. (1983). *Reliability Evaluation of Engineering Systems*, 2e. Springer.

Billinton, R. and Allan, R.N. (1996). *Reliability Evaluation of Power Systems*, 2e. Plenum Press.

Ebeling, C.E. (2000). *An Introduction to Reliability and Maintainability Engineering*. McGraw Hill Education (India) Edition.

Endrenyi, J. (1978). *Reliability Modeling in Electric Power Systems*. Wiley-Blackwell.

Halve, S.S., Koshti, A., and Arya, R. (2023). A sampling method based on system state transition for distribution system adequacy assessment using distributed generation. *Journal of Operation and Automation in Power Engineering* 11 (4): 249–257.

Koshti, A.S., Verma, A., Arya, R. et al. (2023). Tolerable random interruption duration based reliability estimation of stand alone hybrid renewable energy system (SAHRES) by network reduction and sequential Monte Carlo simulation. *IET Generation Transmission & Distribution* 1–12. (Early view) https://doi.org/10.1049/gtd2.12910.

Lewis, E.E. (1987). *Introduction to Reliability Engineering*. John Wiley & Sons.

4

Methodologies for Reliability Evaluation of Network

Rajesh Arya[1], *Atul Koshti*[2], *Aanchal Verma*[3], *Baseem Khan*[4], *and Liladhar Arya*[5]

[1] *Department of Information Technology, Acropolis Institute of Technology and Research, Indore, Madhya Pradesh, India*
[2] *Department of Electrical Engineering, GES'S R.H. Sapat College of Engineering, Management Studies & Research, Nashik, Maharashtra, India*
[3] *Department of Electrical Engineering, Indian Institute of Technology (BHU), Varanasi, Uttar Pradesh, India*
[4] *Department of Electrical and Computer Engineering, Hawassa University, Hawassa, Ethiopia*
[5] *Department of Electrical Engineering, SGSITS, Indore, Madhya Pradesh, India*

4.1 Introduction

This chapter presents reliability evaluation methodologies which are represented by reliability network or logic diagram. The reliability network may be a series or parallel or series-parallel or a complex configuration. It is to be stressed here that the reliability networks in no way represent the connection of components in a physical system. For example two transmission lines operating in electrically parallel will be considered in series configuration for reliability calculation if successful operation of both is required to meet the demand. The relation used is derived will be applicable for reliability evaluation without time factor or with time dependents function under the assumption that all the component, are put in service simultaneously (this will not be the case in a standby system). Generally, it is assumed that components in system fail independently.

4.2 Series Network

A series network is simplest and block diagram representation of it is shown in Figure 4.1.

Where I = Input, O = Output denote input and output nodes.

In terms of its physical setup, a system configuration can be classified as a series reliability network when it either experiences a failure when a single component fails or remains operational when all its components are functioning properly. This is also known as nonredundant system (Billinton and Allan 1992; Saket et al. 2022). Assuming time-independent reliabilities $r_1, r_2, \dots, r_n$, the reliability of a series system can be expressed as follows:

$$R_s = \prod_{i=1}^{n} r_i \tag{4.1}$$

Sometimes this is also known as static reliability.

Reliability Analysis of Modern Power Systems, First Edition. Edited by R. K. Saket and P. Sanjeevikumar.

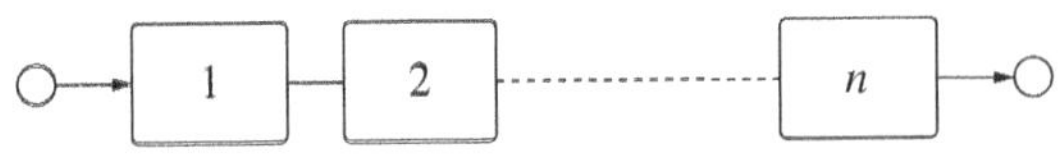

Figure 4.1 Series reliability network.

Now assume $T_1, T_2, \dots, T_n$ are life lengths of these components and $f_1(t), f_2(t), \dots, f_n(t)$ are the failure density function of respective components. Life length "T" of the series system will be minimum of individual life lengths, i.e.

$$T = \min\,[T_1, T_2, \dots, T_n] \tag{4.2}$$

Failure density function of "T" is calculated using following relation

$$(T > t) = (T_1 > t) \cdot (T_2 > t) \cdot \cdots \cdot T_n > t) \tag{4.3}$$

The above relation implies that if $(T > t)$ then all $T_1, T_2, \dots, T_n$ must be greater than "t" (if "T" is minimum). Above relation can be written in terms of probability as follows

$$P(T > t) = P(T_1 > t) \cdot P(T_2 > t) \cdot \cdots \cdot (Tn > t) \tag{4.4}$$

The relationship stated above is expressed using the reliability function as follows:

$$R_s = \prod_{i=1}^{n} r_i(t) \tag{4.5}$$

Failure density function of series system may also obtained using (4.4)

$$1 - F(t) = \prod_{i=1}^{n} (1 - F_i(t))$$

$$F(t) = 1 - \prod_{i=1}^{n} (1 - F_i(t))$$

Failure density function of series system

$$f(t) = \frac{\mathrm{d}}{\mathrm{d}t}\left\{1 - \prod_{i=1}^{n} (1 - F_i(t))\right\} = -\frac{\mathrm{d}}{\mathrm{d}t}\left\{\prod_{i=1}^{n} (1 - F_i(t))\right\}$$

If

$$f_1(t) = f_2(t) = \cdots = f_n(t) = \varphi(t)$$

$$f(t) = -\frac{\mathrm{d}}{\mathrm{d}t}\prod R_i(t) = -\frac{\mathrm{d}}{\mathrm{d}t}R^n(t)$$

$$= nR^{n-1}(t)\,\varphi(t)$$

Let us consider that each component maintains a consistent failure rate denoted by $\lambda_1, \lambda_2, \dots, \lambda_n$

$$R_s(t) = \prod_{i=1}^{n} \mathrm{e}^{-\lambda_i t}$$

$$R_s(t) = \mathrm{e}^{-\lambda_s t}$$

$$\lambda_s = \sum \lambda_i$$

Failure density function of series system

$$f(t) = \lambda_s \mathrm{e}^{-\lambda_s t}$$

Hence, the series system also has exponential failure law. Failure rate is sum of individual failure Rates (Arya et al. 2013; Halve et al. 2023). Further mean time to failure (MTTF) of system

$$\text{MTTF} = \frac{1}{\lambda_s} = \frac{1}{\sum \lambda_i}$$

Reliability importance IR_k of kth component is given as for a specified time

$$\text{IR}_k = \frac{dR_s}{dr_k} = \prod_{i=1 i\neq k}^{n} r_i$$

It is obvious for series system, the reliability importance is highest of "weakest" component. Hence, the component which has the lowest reliability may be improved in design/quality/ strengthening if at all overall system reliability need to be improved.

4.3 Parallel Network

A parallel network consisting of "n" component is shown in Figure 4.2.

The system can fulfill its function as long as a single component is operational. Because all components are interconnected within the system, this configuration is also referred to as an active parallel setup. Probability of failure of the system i.e. unreliability is given by (if there is reliability or independent time factor).

$$Q_p = \prod_{i=1}^{n}(1 - r_i)$$

$$R_p = 1 - \prod_{i=1}^{n}(1 - r_i)$$

Above relation is expanded and written as

$$R_p = \sum_i r_i - \sum r_i r_j + \sum r_i r_j r_k - \cdots + (-1)^{n-1} r_1 r_2 \ldots r_n$$

If component are identical then

$$R_p = 1 - (1 - r)^n$$

Now assume each component has $f_1(t), f_2(t), \ldots, f_n(t)$ as failure density function and corresponding $r_1(t), r_2(t), \ldots r_n(t)$ as reliability functions. Life length of system is given as

$$T = \max[T_1, T_2, \ldots, T_n]$$

Figure 4.2 Parallel reliability network.

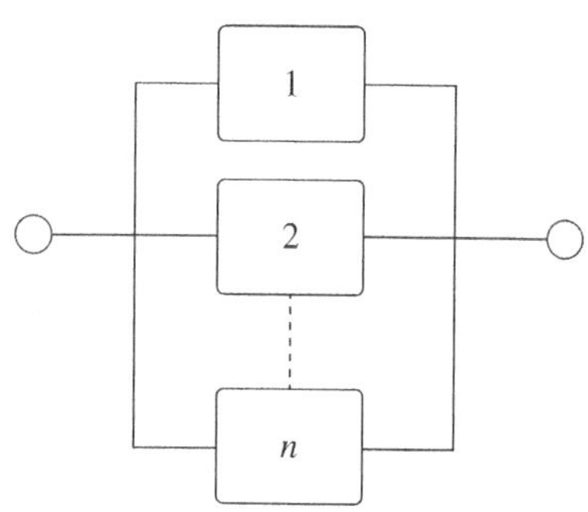

Following is written

$$(T \le t) = (T_1 \le t) \cdot (T_2 \le t) \cdot \dots \cdot (T_n \le t)$$
$$P(T \le t) = P(T_1 \le t) \cdot P(T_2 \le t) \cdot \dots \cdot P(T_n \le t)$$
$$G(t) = F_1(t)F_2(t)...F_n(t)$$

Therefore Cumulative distribution function of parallel network

$$G(t) = \prod F_i(t)$$

Failure law of parallel network

$$g(t) = \frac{d}{dt} \prod F_i(t)$$
$$R(t) = 1 - \prod F_i(t)$$
$$R(t) = 1 - \prod (1 - r_i(t))$$

If each component has identical failure law

$$g(t) = \frac{d}{dt} F^n(t)$$
$$g(t) = nF^{n-1}(t) f(t)$$
$$R(t) = 1 - (1 - r(t))^n$$

Let us assume each component has constant failure rate $\lambda_1, \lambda_2, \dots, \lambda_n$

$$r_i(t) = e^{-\lambda_i t}$$
$$R_p(t) = 1 - \prod_i \left(1 - e^{-\lambda_i t}\right)$$

Failure density function of parallel system

$$g(t) = \frac{d}{dt} \prod \left(1 - e^{-\lambda_i t}\right)$$

Hazard rate of the system (Ebeling 2019) is given as-

$$h(t) = \frac{\frac{d}{dt} \prod \left(1 - e^{-\lambda_i t}\right)}{1 - \prod \left(1 - e^{-\lambda_i t}\right)}$$

It is noted that each component has constant failure rate, but the failure rate is time-dependent. The plot of the hazard rate is given in Figure 4.3 as increasing.

$$\lambda_k = \min \left\{\lambda_1, \lambda_2, \dots, \lambda_n\right\}$$

MTTF of the parallel system is given as

$$\text{MTTF} = \int_0^\infty \left[1 - \prod_i \left(1 - e^{-\lambda_i t}\right)\right] dt$$

$$\text{MTTF} = \int_0^\infty \left[\sum_i e^{-\lambda_i t} - \sum_{j>i} e^{-(\lambda_i + \lambda_j)t} + \sum_{k>j>i} e^{-(\lambda_i+\lambda_j+\lambda_k)t} + \dots + (-1)^n e^{-\sum \lambda_i t}\right] dt$$
$$= \sum_i \frac{1}{\lambda_i} - \sum_{j>i} \frac{1}{(\lambda_i + \lambda_j)} + \sum_{k>j>i} \frac{1}{(\lambda_i + \lambda_j + \lambda_k)} + \dots + (-1)^n \frac{1}{\sum \lambda_i}$$

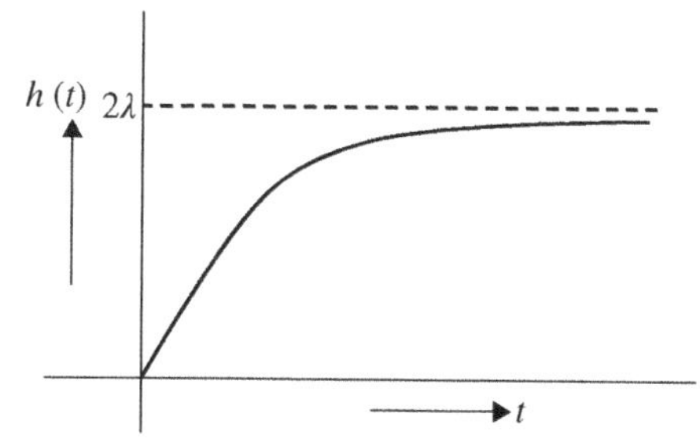

Figure 4.3 Plot of hazard rate versus time.

For example $n = 2$

$$\mathrm{MTTF}_2 = \frac{1}{\lambda_1} + \frac{1}{\lambda_2} - \frac{1}{\lambda_1 + \lambda_2}$$

For $n = 3$

$$\mathrm{MTTF}_3 = \frac{1}{\lambda_1} + \frac{1}{\lambda_2} + \frac{1}{\lambda_3} - \frac{1}{\lambda_1 + \lambda_2} - \frac{1}{\lambda_2 + \lambda_3} - \frac{1}{\lambda_3 + \lambda_1} + \frac{1}{\lambda_1 + \lambda_2 + \lambda_3}$$

If n units are identical having each as constant failure rate λ than expression for MTTF is given as

$$\mathrm{MTTF} = \sum_{k=1}^{n} \frac{{}_nC_k(-1)^{k+1}}{k\lambda}$$

It is true that reliability of parallel system increases as more number of components are connected. But common failure may not permit to achieve ideal results. The cause of common failure includes common operational environment, external catastrophic events, operational errors, common external power source, etc.

The cause of common mode failure (CMF) effects (Allan 2013) is modeled by connecting a hypothetical unit in series with a parallel combination representing the reliability owing common cause failure as shown in Figure 4.4.

λ_1, λ_2 are failure rate of each unit independently and is failure rate due to common cause. The reliability of the system is written as

$$R = \left[e^{-\lambda_1 t} + e^{-\lambda_2 t} + e^{-(\lambda_1+\lambda_2)t}\right] e^{-\lambda_c t}$$

It is necessary that in physical consideration should be small; otherwise, effect of redundancy may be neglected. In other words total failure rate of each component may be written as

$$\lambda_{1T} = \lambda_1 + \lambda_c \quad \lambda_{2T} = \lambda_2 + \lambda_c$$

Putting in previous relation resulting into

$$R = \left[e^{-\lambda_{1T} t} + e^{-\lambda_{2T} t} + e^{-(\lambda_{1T}+\lambda_{2T}-\lambda_c)t}\right]$$

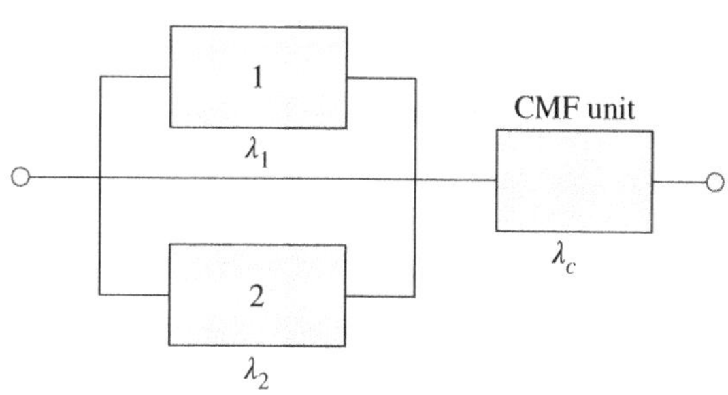

Figure 4.4 Common mode failure.

If units are identical then

$$\lambda_1 = \lambda_2 = \lambda \;\; \lambda_{1T} = \lambda_{2T} = \lambda_T$$

Above relation will be written as

$$R = \left[2e^{-\lambda_T t} - e^{-(2\lambda_T - \lambda_c)t}\right]$$

At times, a factor β is defined as the ratio of common cause failure to total failure rate as

$$\beta = \frac{\lambda_c}{\lambda_T}$$

β-factor may be substituted in above relation and reliability expression is written as follows

$$R = \left[2e^{-\lambda_T t} - e^{-(2\lambda_T - \beta\lambda_T)t}\right]$$
$$R = \left[2e^{-\lambda_T t} - e^{-(2-\beta)\lambda_T t}\right]$$
$$R = \left[2 - e^{-(1-\beta)\lambda_T t}\right] e^{-\lambda_T t}$$

At times, this technique is known β factor technique. General expression for "n" component in parallel accounting common cause failure is written as

$$R(t) = \left[1 - (1 - e^{-\lambda t})^n\right] e^{-\lambda_c t}$$

4.4 Partially Redundant System

This arrangement is also known as k-out of n configuration. In this "n" units are operating in parallel, the success requires at least "k" unit in operation or in upstates, where k is given as

$$1 < k < n.$$

if $k = 1$ it is fully redundant system and $k = n$ it is series system (nonredundant system). k/n system network is represented as shown in Figure 4.5.

Assuming each component is identical and has reliability "r" for a specified duration. Using the binomial distribution probability of "m" success out of "n" unit is written as follows:

$$p\,[X = m] = {}^{n}C_m r^m (1 - r)^{n-m}$$

Since at least "k" units are required for system success. Reliability expression is written as

$$R_{k/n} = \sum_{m=k}^{n} {}^{n}C_m r^m (1 - r)^{n-m}$$

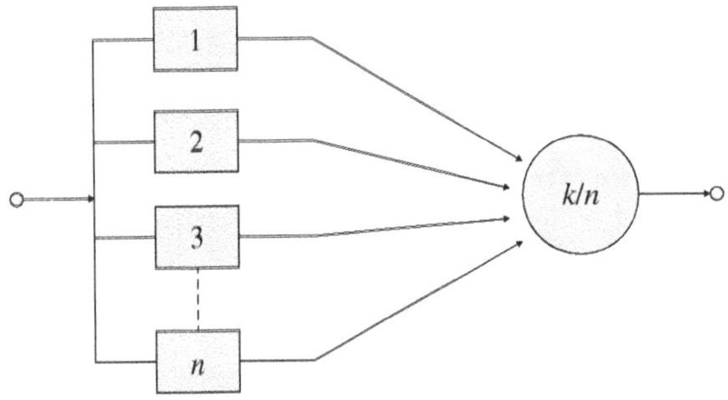

Figure 4.5 k/n partially redundant system.

If each component has failure rate λ, then the reliability is given as $R_{k/n}$

$$R_{k/n} = \sum_{m=k}^{n} {}^{n}C_{m} e^{-\lambda mt}(1 - e^{-\lambda t})^{n-m}$$

Using above failure density function, hazard rate and MTTF of partially redundant system may be calculated. If the units are not identical than Binomial distribution is not applicable. In this situation, a product in the following manner is formed and states where at least "k" units are available are selected. Their probabilities are calculated and added to give the required results.

$$\begin{aligned}&(p_1 + q_1)(p_2 + q_2)...(p_n + q_n)\\&= p_1p_2...p_n + q_1p_2...p_n + p_1q_2...p_n + \cdots + q_1q_2...q_n\end{aligned}$$

In above total events will be $2n$. Success events may be separated to calculate reliability of system.

Exercise 4.1: A system of three components, each having failure rate as $\lambda_1, \lambda_2, \lambda_3$. System success requires at least two components in operation. Obtain (1) reliability function and(2) MTTF.

Solution:

(1) From the product $(r_1 + q_1)(r_2 + q_2)(r_3 + q_3) = r_1r_2r_3 + q_1r_2r_3 + r_1q_2r_3 + r_1r_2q_3 + q_1q_2r_3 + q_1r_2q_3 + r_1q_2q_3 + q_1q_2q_3$

Selecting the events which forms system success

$$\begin{aligned}R_{2/3} &= r_1r_2r_3 + q_1r_2r_3 + r_1q_2r_3 + r_1r_2q_3\\&= e^{-(\lambda_1+\lambda_2)t} + e^{-(\lambda_2+\lambda_3)t} + e^{-(\lambda_1+\lambda_3)t} - 2e^{-(\lambda_1+\lambda_2+\lambda_3)t}\end{aligned}$$

(2)

$$\text{MTTF} = \frac{1}{\lambda_1 + \lambda_2} + \frac{1}{\lambda_2 + \lambda_3} + \frac{1}{\lambda_1 + \lambda_3} - \frac{2}{\lambda_1 + \lambda_2 + \lambda_3}$$

4.5 Reliability Evaluation of Complex Networks

This section deals with the reliability evaluation of a non-series-parallel network (Breneman et al. 2022; Endrenyi 1978). The following four methods are discussed

(1) Event space,
(2) Decomposition,
(3) Tie-set, and
(4) Cut-set.

Simulation-based techniques may also be employed for obtaining solution of the power system reliability (Arya and Koshti 2008; Koshti et al. 2023).

4.5.1 Event Space Method

This is one of the simplest techniques for evaluating the reliability of a network. The method is implemented in the following steps.

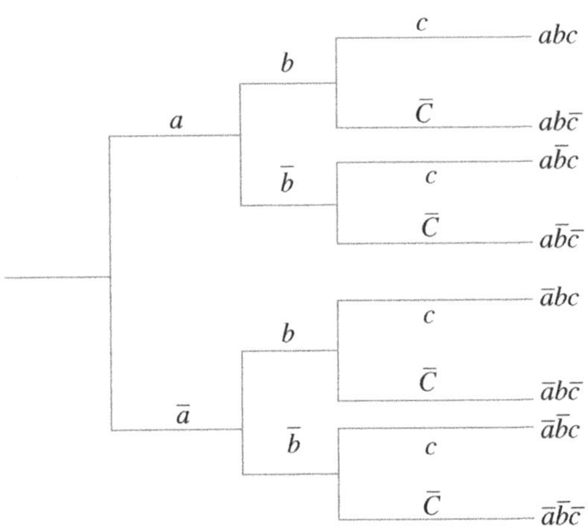

Figure 4.6 Event space tree for three components.

Step 1: Based on the total number of components, enumerate all possible states of the network. Assuming each component has two states (up and down), total no. of states will $2n$. n denotes total no. of components. These events may be identified using a event tree, as shown in Figure 4.6 for three components, namely a, b, c. The event tree will grow in size as the number of component increases.

Step 2: Looking at the network, partition the total number of states in two sets i.e. "W"—success set of event and "F"—fail set of event.

Step 3: Evaluate the probabilities of these events and calculate the system's reliability as outlined below

$$R = \sum_{i \in w} P(E_i)$$

$$\overline{R} = 1 - R = \sum_{j \in f} P(E_j)$$

E_i and E_j are events

$P(E_i)$ the is probability of occurrence of ith event

Note: Simple addition in above relation is due to the event enumerated being mutually exclusive.

Exercise 4.2: Evaluate the reliability of following network using event space method. Assume $r_a, r_b, r_c,$ and r_d are the reliabilities of each component as shown in Figure 4.7.

Solution:

The network will have 16 states written as follows

$$
\begin{aligned}
E_1 &= abcd & E_9 &= a\bar{b}cd \\
E_2 &= ab\bar{c}d & E_{10} &= a\bar{b}\bar{c}d \\
E_3 &= abc\bar{d} & E_{11} &= a\bar{b}c\bar{d} \\
E_4 &= ab\bar{c}\bar{d} & E_{12} &= a\bar{b}\bar{c}\bar{d} \\
E_5 &= \bar{a}bcd & E_{13} &= \bar{a}\bar{b}cd \\
E_6 &= \bar{a}b\bar{c}d & E_{14} &= \bar{a}\bar{b}\bar{c}d \\
E_7 &= \bar{a}bc\bar{d} & E_{15} &= \bar{a}\bar{b}c\bar{d} \\
E_8 &= \bar{a}b\bar{c}\bar{d} & E_{16} &= \bar{a}\bar{b}\bar{c}\bar{d}
\end{aligned}
$$

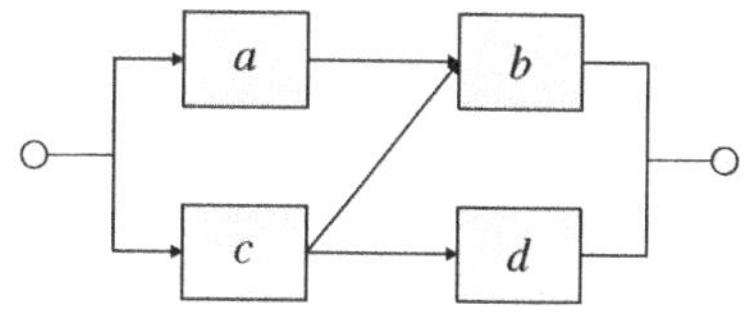

Figure 4.7 Reliability evaluation using event space method.

Matching the event with the network following events are in "w" sets

$$W : [E_1, E_2, E_3, E_4, E_5, E_7, E_9, E_{13}]$$

$$R = P(E_1) + P(E_2) + P(E_3) + P(E_4) + P(E_5) + P(E_7) + P(E_9) + P(E_{13})$$

$$R = P(abcd) + P(ab\bar{c}d) + P(abc\bar{d}) + P(ab\bar{c}\bar{d}) + P(\bar{a}bcd) + P(\bar{a}bc\bar{d}) + P(a\bar{b}cd) + P(\bar{a}\bar{b}cd)$$

$$R = r_a r_b r_c r_d + r_a r_b (1 - r_c) r_d + r_a r_b r_c (1 - r_d) + r_a r_b (1 - r_c)(1 - r_d) + (1 - r_a) r_b r_c r_d$$
$$+ (1 - r_a) r_b r_c (1 - r_d) + r_a (1 - r_b) r_c r_d + (1 - r_a)(1 - r_b) r_c r_d$$

If $r_a = r_b = r_c = r_d = r$
the above relation is simplified as

$$R = r^4 + 4r^3(1 - r) + 3r^2(1 - r)^2$$

$$R = 3r^2 - 2r^3$$

Further assume each component has constant failure rate λ, i.e. $r = e^{-\lambda t}$. The above reliability expression reduces to

$$R(t) = 3e^{-2\lambda t} - 2e^{-3\lambda t}$$

Failure density function of the system is evaluated as follows

$$f(t) = 6\lambda e^{-2\lambda t} - 6\lambda e^{-3\lambda t}$$

And hazard rate is expressed as-

$$h(t) = \frac{6\lambda e^{-2\lambda t} - 6\lambda e^{-3\lambda t}}{3e^{-2\lambda t} - 2e^{-3\lambda t}}$$

MTTF of the system is given as

$$\text{MTTF} = \frac{3}{2\lambda} - \frac{2}{3\lambda} = \frac{5}{6\lambda}$$

The method is simple but as number of component increases states also increases and thus computational burden increases. Total number of states are $2n$, if n denotes total no. of component, but total number of states can be reduced by considering the failure of a small number of components. Since failure of large number of components is unlikely i.e. probability of such events/ states with failure of more than a specific number of component is negligible.

4.5.2 Decomposition Method

In this technique, a key-stone component is selected as "X" and based on this component's state system success is written as follows

$$S = S \cdot X + S \cdot \overline{X}$$

Probability of system success is given as follows

$$P(S) = P(S \cdot X) + P(S \cdot \overline{X})$$

Decomposition is based on that system success will occur either in intersection with "X" working or with intersection "X" has failed. From the above relation, reliability is written as follows

$$P(S) = P(S/X)P(X) + P(S/\overline{X})P(\overline{X})$$

Where, $P(X)$ = Reliability of "X" component
$P(S/X)$ = Probability of system success provided "X" is working.
$P(S/\overline{X})$ = Probability of system success given that "X" has failed ($\overline{X}$)
The above decomposition is shown in Figure 4.8.

As shown in Figure 4.8, the network is decomposed into two networks "A" and "B." In "A" system, success is represented when "X" is working which is shown as "short-circuit," whereas in "B" system success is evaluated while "X" has failed which is shown as "open-circuit" condition. The methodology is explained by evaluating the reliability of the following system shown in Figure 4.9.

Let us assume "X" as key-stone which is component "c"

$$P(C) = P(X) = r_c$$
$$P(\overline{C}) = P(\overline{X}) = (1 - r_c)$$

Decomposed network corresponding to $P(S/X)$ is shown in Figure 4.10.

It is easier to evaluate the $P(S/X)$ from above network. Close observation of above network reveals that reliability of above network is simply "b" and "d" in parallel combination i.e.

$$P(S/X) = r_b + r_d - r_b r_d$$

The original network for $P(S/\overline{X})$ is evaluated as follows

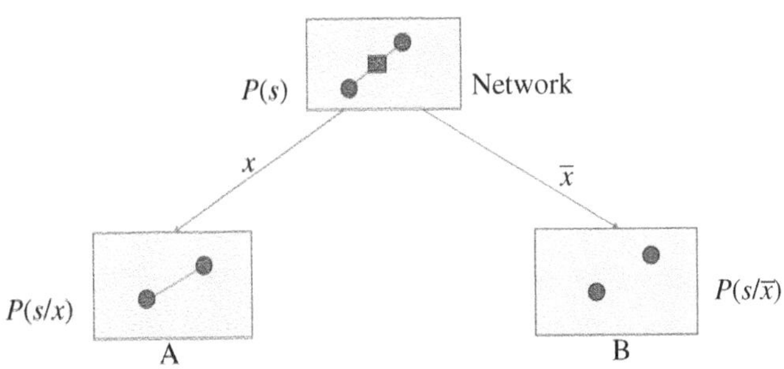

Figure 4.8 Decomposition method illustration.

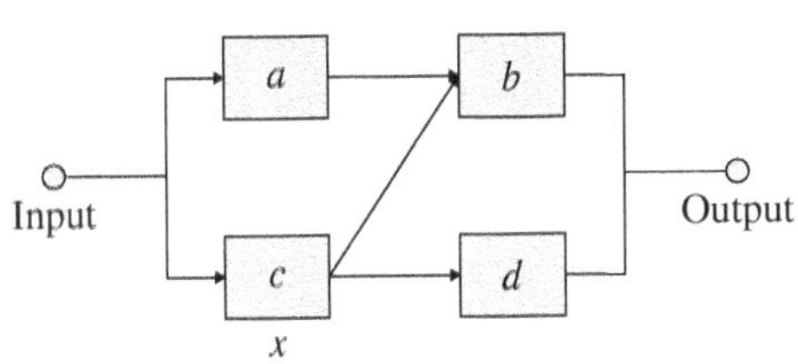

Figure 4.9 Reliability evaluation using decomposition method.

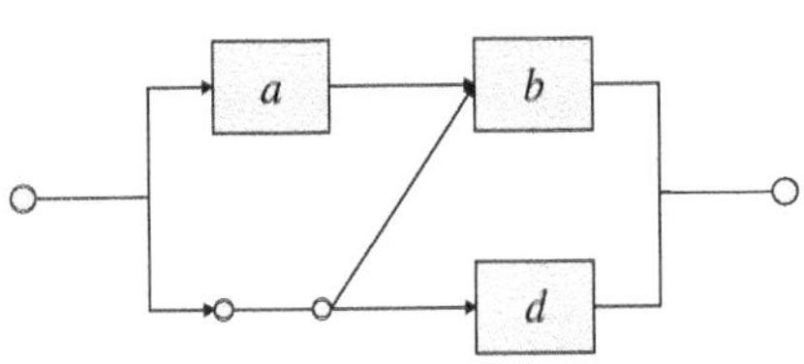

Figure 4.10 Decomposed network for X.

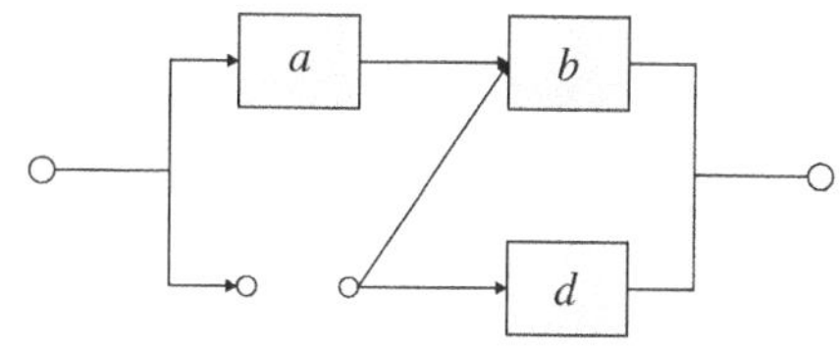

Figure 4.11 Decomposed network for $\overline{X}$.

From Figure 4.11 $P(S/\overline{X})$ is evaluated as follows:

$$P(S/\overline{X}) = r_a r_b$$

Hence over all reliability of system is written as

$$P(S) = (r_b + r_d - r_b r_d)r_c + r_a r_b(1 - r_c)$$

$$P(S) = r_b r_c + r_c r_d - r_b r_c r_d + r_a r_b - r_a r_b r_c$$

Note: If each component has reliability "r" then

$P(S) = 3r^2 - 2r^3$ as obtained using event space method.

Reliability importance of each component is given as follows

$$\frac{\partial P(S)}{\partial r_a} = rb - rbrc = rb(1 - rc)$$

$$\frac{\partial P(S)}{\partial r_b} = r_c(1 - r_d) + r_a(1 - r_c)$$

$$\frac{\partial P(S)}{\partial r_c} = r_b + r_d - r_b r_d - r_a r_b$$

$$\frac{\partial P(S)}{\partial r_d} = r_c - r_b r_c = r_c(1 - r_b)$$

If each component has same reliability then

$$\frac{\partial P(S)}{\partial r_a} = \frac{\partial P(S)}{\partial r_d} = r - r^2$$

$$\frac{\partial P(S)}{\partial r_b} = \frac{\partial P(S)}{\partial r_c} = 2(r - r^2)$$

Hence it is obvious component "b" and "c" has twice reliability importance as compared to "a" and "d." Hence if system reliability is to be improved then it should be improved by investing money in components "b" and "c" both or "b" or "c."

4.5.3 Tie-Set Method

This method is also known as path tracing method. A tie-set is a set of components which provide path from input to output while all of the components in the set are in upstate. Minimal tie-set is a set of minimum number of components which while all of them in upstate provides a path (success). Even if a single component is removed from this "minimal tie-set," the remaining set of component no longer provides a path (success) even if all of them are in upstate. The methodology is explained in the following steps

Step 1: Given a reliable network, obtain all possible minimal tie-sets as

$$\left[T_1, T_2, T_3, \ldots, T_r\right]$$

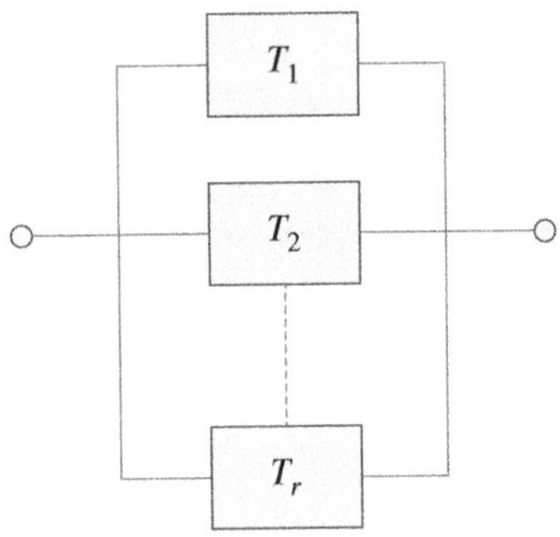

Figure 4.12 Equivalent network using tie sets.

Step 2: For system success, at least one tie-set must be available. Thus equivalent network is drawn in Figure 4.12.

Hence, the reliability expression is written as follows:

$$R = P\left(\sum_{i=1}^{r} T_i\right)$$

$$R = \sum_{i} P(T_i) - \sum_{j>i} P(T_iT_j) + \sum_{k>j>i} P(T_iT_jT_k) + \cdots + (-1)^{r-1}P(T_1T_2......T_r)$$

Consider the complex network used in previous sub-section. Looking at the network Tie-sets are obtained as follows $T_1 = a \cdot b,\ T_2 = c \cdot d,\ T_3 = b \cdot c$

$$R = P(T_1) + P(T_2) + P(T_3) - P(T_1T_2) - P(T_1T_3) - P(T_2T_3) + P(T_1T_2T_3)$$

$$R = r_ar_b + r_cr_d + r_br_c - r_ar_br_c - r_br_cr_d$$

If $r_a = r_b = r_c = r_d$

$$R = 3r^2 - 2r^3$$

4.5.4 Cut-Set Method

A cut-set is defined as a group of components such that the system fails if any of them fails. A minimal cut-set is a set of component which fail system fails; even if a single component is removed from minimal cut-set the remaining components do not form a cut-set. The method is explained in following steps.

Step 1: Identify all possible minimal cut-sets as $C_1, C_2, \ldots, C_T$

Step 2: For system failure at least one cut-set should occur i.e. system will fail even if all the elements in any one of the minimal "cut-set" fails. The equivalent diagram is shown in Figure 4.13.

In view of this unreliability, the system is written as follows:

$$Q = 1 - R = P(C_1 + C_2 + \cdots\cdots + C_r)$$

$$Q = \sum_{i} P(C_i) - \sum_{j>i} P(C_iC_j) + \sum_{k>j>i} P(C_iC_jC_k) + \cdots + (-1)^{r-1}P(C_1C_2......C_r)$$

Let us consider the following example of Figure 4.14

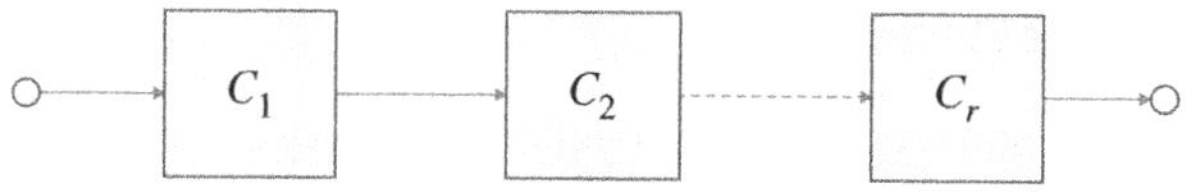

Figure 4.13 Equivalent network using cut-sets.

Figure 4.14 Reliability network.

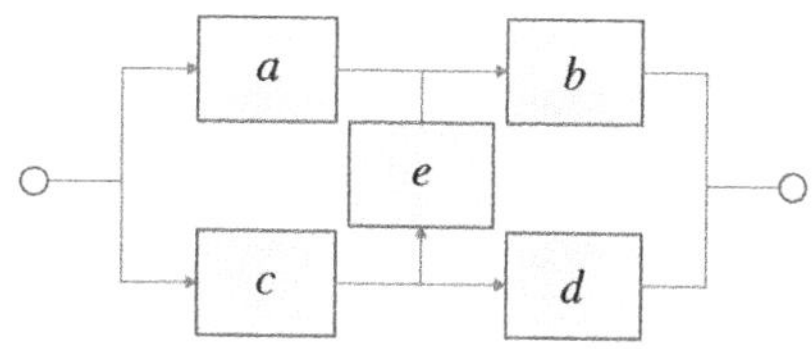

Following cut-sets are identified

$$C_1 = \bar{a}\,\bar{c} \qquad C_2 = \bar{b}\,\bar{d}$$

$$C_3 = \bar{b}\,\bar{c} \qquad C_4 = \bar{c}\,\bar{e}\,\bar{b}$$

$$\begin{aligned} Q = {} & P(C_1) + P(C_2) + P(C_3) + P(C_4) - P(C_1C_2) - P(C_1C_3) - P(C_1C_4) - P(C_2C_3) \\ & - P(C_2C_4) - P(C_3C_4) + P(C_1C_2C_3) + P(C_1C_2C_4) + P(C_2C_3C_4) + P(C_3C_4C_1) \\ & - P(C_1C_2C_3C_4) \end{aligned}$$

If each component has the same reliability r then resulting into

$$Q = 3(1-r)^2 - (1-r)^3 - 2(1-r)^4 + (1-r)^5$$

The cut-set method has the following advantages

(i) The method can be programmed on a digital computer. No doubt tie-set method also can be programmed.

(ii) While one identifies cut-set, various failure modes are directly available. This enables one to strengthen the weak components or devices.

(iii) In cut-set method approximate evaluation of reliability is possible. This may be achieved in one or two following ways

  (a) Limiting the order of cut-sets i.e. neglecting the higher-order cut-sets which contain more than a specified number of elements. Since probabilities of occurrence of such "cuts" may be negligible.

  (b) By truncating the unreliability relation "Q" by retaining only first-order terms as follows.

$$Q \cong \sum P(C_i)$$

$$R \cong 1 - Q$$

Failure probability thus obtained will be on higher side and reliability will be on lower side. Usually in engineering application, each component is highly reliable and above approximations are justified in practice.

No doubt decomposition method is very powerful but it cannot be programmed on computer. Moreover, it may not be used when large numbers of components are present in the system.

4.6 Determination of Tie-Sets

Method for obtaining tie-sets which are required in Tie-Set method of reliability evaluation.

4.6.1 Connection Matrix Method Using Node Elimination

Assuming a reliability network is given, form a connection matrix N defined as follows

$$N_{ij} = 1$$

N_{ij} = Sum of the elements connected between and directed from ith to jth node
= 0 *Otherwise*

Eliminate one by one all intermediate nodes using following formula

$$N'_{ij} = N_{ij} + N_{ik}N_{kj}$$

where "k" is the node being eliminated. Ultimately after eliminating all intermediate nodes a 2×2 matrix will result. All minimal tie-set can be obtained from this matrix from Ninput × output element. In this process at times a nonminimal tie-set may result which should be ignored and identified.

Exercise 4.3: Obtain tie-sets of the following network shown in Figure 4.15 using Node elimination method

Solution:

By inspection connection matrix is written as follows

$$N = \begin{bmatrix} 1 & c & a & 0 \\ 0 & 1 & 1 & d \\ 0 & 0 & 1 & b \\ 0 & 0 & 0 & 1 \end{bmatrix}$$

node $k = 2$ to be eliminated

$$N'_{11} = 1$$
$$N'_{13} = N_{13} + N_{12}N_{23} = a + c \cdot 1 = a + c$$
$$N'_{14} = N_{14} + N_{12}N_{24} = 0 + c \cdot d = cd$$
$$N'_{31} = 0$$
$$N'_{33} = 1$$
$$N'_{34} = N_{34} + N_{32}N_{24} = b + 0 \cdot d = b$$
$$N'_{41} = 0$$
$$N'_{43} = 0$$
$$N'_{44} = 1$$

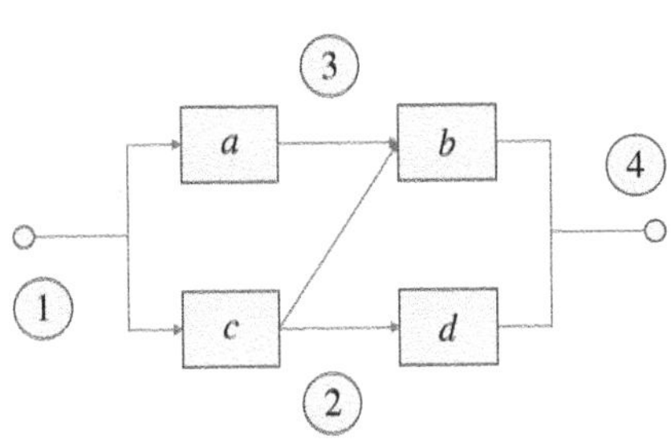

Figure 4.15 Tie-sets using node elimination method.

The resulting matrix is written

$$N = \begin{bmatrix} 1 & a+c & cd \\ 0 & 1 & b \\ 0 & 0 & 1 \end{bmatrix}$$

$$k = 3$$

$$N''_{11} = 1$$

$$N''_{14} = N'_{14} + N'_{13}N'_{34}$$

$$= cd + (a+c) \cdot b$$

$$= cd + ab + bc$$

$$N''_{41} = 0$$

$$N''_{44} = 1$$

$$N'' = \begin{bmatrix} 1 & ab + bc + cd \\ 0 & 1 \end{bmatrix}$$

Tie-sets are obtained as element of N'_{14} i.e.

$$T_1 = ab,\ T_2 = bc,\ T_3 = cd$$

4.7 Method of Obtaining Cut-Set

Once tie-set are obtained, a matrix in the following format is prepared A_{ik} represent an element present in i_{th} path, and kth column is considered for this element. This means $A_{ik} = 1$, if ith tie-set contains kth element and 0 otherwise. Then, if a specific column contains all ones, then the corresponding elements form a first-order cut. Now examine sum of two column's combinations (excluding the first-order cut). Combinations giving one at every place will form second-order cut. Thus, continuing the increased numbers of combinations higher-order cuts are obtained. It is stressed here that while looking at combinations, the combinations of lower order which forms cut should not be considered. The procedure is illustrated with the network shown in Figure 4.16.

The following three tie-sets are obtained

$$T_1 = ab, T_2 = cd, T_3 = bce$$

Following matrix is prepared

Elements / Tie sets	a	b	c	d	e
T_1	1	1	0	0	0
T_2	0	0	1	1	0
T_3	0	1	1	0	1

It is obvious from the above table; there is no column which contains all ones. Hence there is no first-order cut-set.

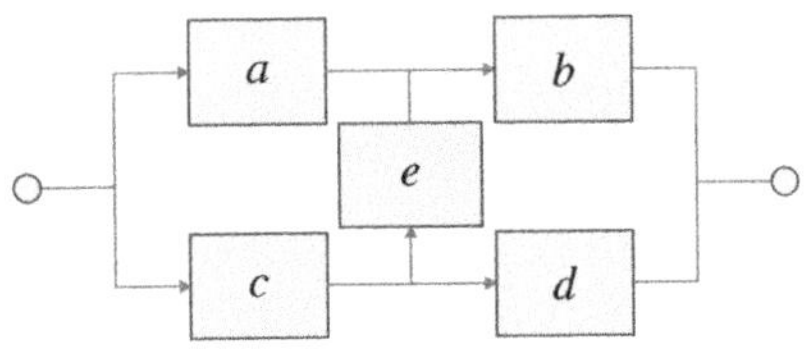

Figure 4.16 Bridge reliability network.

Taking combination of two at a time following second order cut-sets are observed

$$C_1 = \bar{a}\,\bar{c}, \quad C_2 = \bar{b}\,\bar{c}, \quad C_3 = \bar{b}\,\bar{d}$$

Looking at three combinations following one minimal cut-set is observed

$$C_4 = \bar{c}\,\bar{e}\,\bar{b}$$

4.8 Multistate Model

It has been considered that the component has two states i.e. "up" and "down." In numerous scenarios, components may exhibit more than two states. A generator may have "up," "down," and a "derated" states. A diode may function properly; fail by open circuit or short circuit. Thus diode has two failure modes and one state of normal operation. Reliability of such systems are evaluated either by event-space or decomposition method.

4.8.1 Event-Space Method

Consider the example of two diodes operating in parallel for rectification purpose as shown in Figure 4.17.

Reliability of above configuration is required for rectification purpose. Assume P_{n1}, P_{01}, P_{s1} are the probabilities of normal, open circuit, and short circuit for diode one and P_{n2}, P_{02}, P_{s2} are the same probabilities of diode number two. Events may be obtained by product rule or by tree.

$$\begin{aligned}&(P_{01} + P_{s1} + P_{n1})\,(P_{02} + P_{s2} + P_{n2})\\&= P_{01}P_{02} + P_{01}P_{s2} + P_{01}P_{n2} + P_{s1}P_{02} + P_{s1}P_{s2} + P_{s1}P_{n2} + P_{n1}P_{02} + P_{n1}P_{s2} + P_{n1}P_{n2}\end{aligned}$$

In all there are nine states. States corresponding to system success are added and reliability is evaluated as

$$P'_n = P_{n1}P_{n2} + P_{01}P_{n2} + P_{02}P_{n1}$$

Probability of configuration failing by short circuit is given by

$$P'_s = P_{s1}P_{s2} + P_{01}P_{s2} + P_{02}P_{s1} + P_{n1}P_{s2} + P_{n2}P_{s1}$$

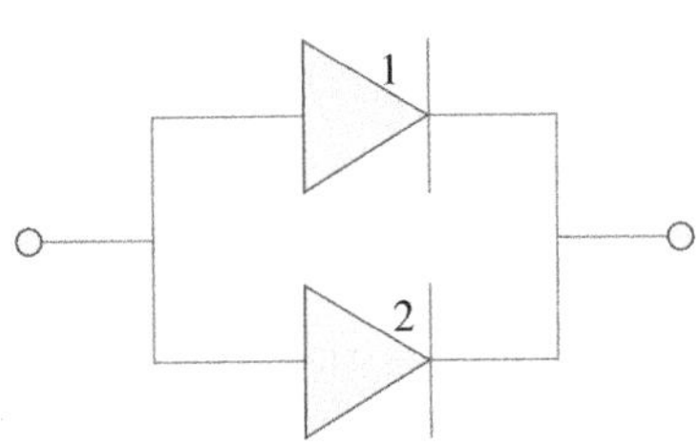

Figure 4.17 Two diodes in parallel.

Probability of configuration failing by open circuit is given as

$$P_o' = P_{o1}P_{o2}$$

In view of this, above configuration is replaced by a single diode shown in Figure 4.18

The equivalent diode has as probabilities of normal, open, and short circuit states. If the diodes are identical i.e.

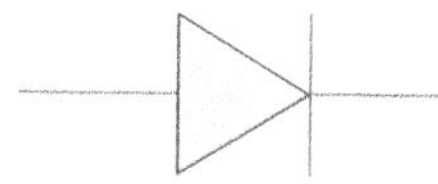

Figure 4.18 Equivalent diode of parallel combination.

$$P_{n1} = P_{n2} = P_n'$$
$$P_{o1} = P_{o2} = P_o'$$
$$P_{s1} = P_{s2} = P_s$$

Then reliability of the configuration is given as

$$P_n' = P_n^2 + 2P_nP_0$$

If configuration is in series as follows (Figure 4.19)

In this case total numbers of states are same i.e. nine as earlier but the equivalent three states are identified as follows

$$P_n'' = P_{n1}P_{n2} + P_{n1}P_{s2} + P_{n2}P_{s1}$$
$$P_o'' = P_{o1}P_{o2} + P_{o1}P_{n2} + P_{o2}P_{n1} + P_{o1}P_{s2} + P_{o2}P_{s1}$$
$$P_s'' = P_{s1}P_{s2}$$

The equivalent diode has above three state probabilities

Again if each diode has same probabilities then

$$P_n'' = P_n^2 + 2P_nP_s$$
$$P_o'' = P_o^2 + 2P_oP_n + 2P_oP_s$$
$$P_s'' = P_s^2$$

It is obvious from expressions of and that for reliability enhancement two diodes should be connected in series if probability of failure of diode by short circuit is more than that of probability of failure by open circuit, otherwise they should be connected in parallel.

4.8.2 Decomposition Approach

Assume the two diodes have been connected in parallel as shown below

Assume diode # 2 as key-stone as X. Then system success "S" and reliability is written as follows

$$S = S \cdot X_0 + S \cdot X_N + S \cdot X_{SC}$$
$$P(S) = P(S/X_0)P(X_0) + P(S/X_N)P(X_N) + P(S/X_{SC})P(X_{SC})$$
$$= p_{n1}p_{02} + (p_{n1} + p_{01})p_{n2} + 0 \cdot p_{s2}$$
$$= P_{n1}p_{02} + p_{n1}p_{n2} + p_{01}p_{n2}$$

Figure 4.19 Two diode in series configuration.

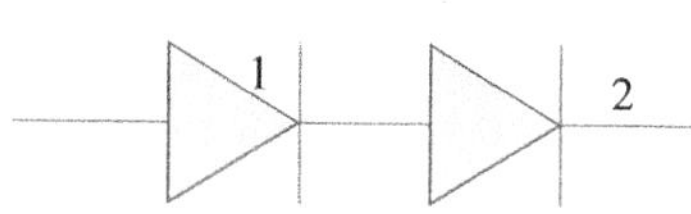

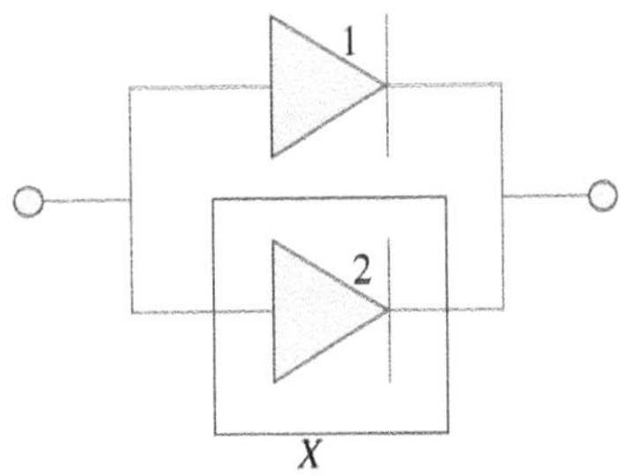

Figure 4.20 Diode in parallel with X as key stone.

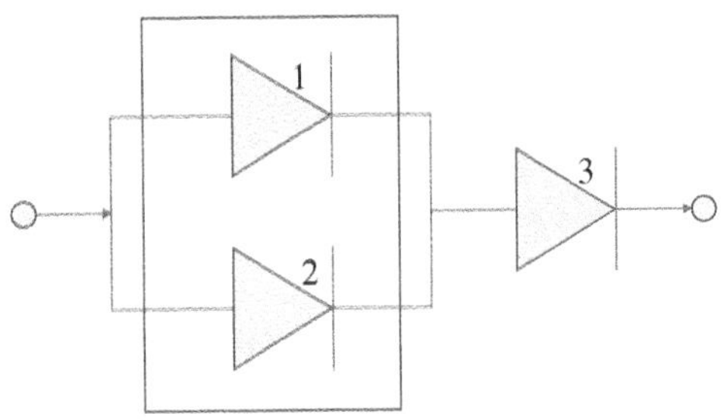

Figure 4.21 Series parallel diode configuration.

Assume again following configuration Parallel combination of diode may be assumed as a single component "X." (Figures 4.20 and 4.21) Then reliability expression is written as

$$P(S) = P(S/X_N) \cdot P(X_N) + P(S/X_{SC}) \cdot P(X_{SC}) + P(S/X_0) \cdot P(X_0)$$
$$= (P_{s3} + P_{n3})(P_{n1} \cdot P_{n2} + P_{01} \cdot P_{n2} + P_{02} \cdot P_{n1}) + P_{n3}(P_{s1} \cdot P_{s2} + P_{01} \cdot P_{s2} + P_{02} \cdot P_{s1} + P_{n1} \cdot P_{s2} + P_{n2} \cdot P_{s1}) + 0 \cdot (P_{01} \cdot P_{02})$$

If all diodes are identical then above reduces to

$$P(S) = (P_s + P_n)(P_n{}^2 + 2P_nP_0) + P_n(P_s{}^2 + 2P_0P_s + 2P_nP_s)$$

4.9 Illustrative Examples

Exercise 4.4: An electrical radial distributor has 10 distributer sections to meet farthest load point. Each distributer section has reliability "r." What should be minimum value of "r" such that farthest load point reliability does not drop below 0.90.

Solution:
This is series system and overall reliability to farthest load point is

$$r^{10} = 0.90$$
$$10 \ln r = \ln 0.90$$
$$\ln r = 0.1 \ln 0.90$$

Gives

$$r = 0.9898$$

Exercise 4.5: An important hospital load to be meet by parallel feeder in which each feeder has reliability 0.9. Determine minimum number of parallel feeder such that reliability to meet these load is 0.99.

Solution:

Reliability expression for "n" component in parallel is

$$R_p = 1 - (1 - r)^n = 0.99$$
$$= 1 - (1 - 0.9)^n = 0.99$$
$$0.1^n = 0.01$$

Taking natural logarithm

$$n \ln 0.1 = \ln 0.01$$
$$n = \frac{0.01}{\ln 0.1}$$
$$\cong 2$$

Exercise 4.6: An administrative block distributed load is supplied by three parallel feeder each having capacity of 50 MW and load is 80 MW. Each feeder has availability of 0.95. Determine the availability of system.

Solution:

To meet a demand of 80 MW of total capacity of three feeders 150 MW at least two feeders must be operable. Hence it is a partially redundant system

$$n = 3,\ k = 2$$

System availability is written as (using binomial distribution)

$$R = r^3 + 3r^2(1 - r) = 3r^2 - 2r^3$$
$$= 3 \times 0.95^2 - 2 \times 0.95^3$$
$$= 0.9025$$

Exercise 4.7: Obtain reliability expression for the following bridge network in Figure 4.22 by all four methods discussed in the chapter. Assume reliability of each component as $r_a = r_b = r_c = r_d = r_e = r = 0.99$

Solution:

(a) Event space method

There are five components and will have 32 states.

$$E_1 = abcde \quad E_9 = a\bar{b}cde$$
$$E_2 = ab\bar{c}de \quad E_{10} = a\bar{b}\bar{c}de$$
$$E_3 = abc\bar{d}e \quad E_{11} = a\bar{b}c\bar{d}e$$
$$E_4 = ab\bar{c}\bar{d}e \quad E_{12} = a\bar{b}\bar{c}\bar{d}e$$

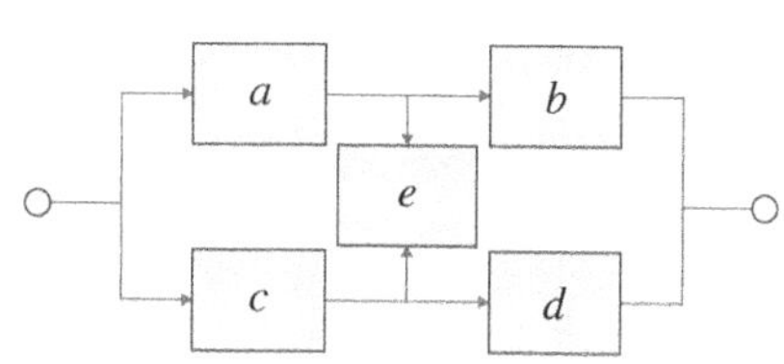

Figure 4.22 Bridge network.

$$
\begin{aligned}
&E_5 = \bar{a}bcde && E_{13} = \bar{a}\bar{b}cde\\
&E_6 = \bar{a}b\bar{c}de && E_{14} = \bar{a}\bar{b}\bar{c}de\\
&E_7 = \bar{a}bc\bar{d}e && E_{15} = \bar{a}\bar{b}c\bar{d}e\\
&E_8 = \bar{a}b\bar{c}\bar{d}e && E_{16} = \bar{a}\bar{b}\bar{c}\bar{d}e\\
&E_{17} = abcd\bar{e} && E_{25} = a\bar{b}cd\bar{e}\\
&E_{18} = ab\bar{c}d\bar{e} && E_{26} = a\bar{b}\bar{c}d\bar{e}\\
&E_{19} = abc\bar{d}\bar{e} && E_{27} = a\bar{b}c\bar{d}\bar{e}\\
&E_{20} = ab\bar{c}\bar{d}\bar{e} && E_{28} = a\bar{b}\bar{c}\bar{d}\bar{e}\\
&E_{21} = \bar{a}bcd\bar{e} && E_{29} = \bar{a}\bar{b}cd\bar{e}\\
&E_{22} = \bar{a}b\bar{c}d\bar{e} && E_{30} = \bar{a}\bar{b}\bar{c}d\bar{e}\\
&E_{23} = \bar{a}bc\bar{d}\bar{e} && E_{31} = \bar{a}\bar{b}c\bar{d}\bar{e}\\
&E_{24} = \bar{a}b\bar{c}\bar{d}\bar{e} && E_{32} = \bar{a}\bar{b}\bar{c}\bar{d}\bar{e}
\end{aligned}
$$

$$
\begin{aligned}
&W\colon \{E_1, E_2, E_3, E_4, E_5, E_7, E_9, E_{10}, E_{13}, E_{17}, E_{18}, E_{19}, E_{20}, E_{21}, E_{25}, E_{29}\}\\
&R = P(E_1) + P(E_2) + P(E_3) + P(E_4) + P(E_5) + P(E_7) + P(E_9) + P(E_{10}) + P(E_{13}) + P(E_{17})\\
&\quad + P(E_{18}) + P(E_{19}) + P(E_{20}) + P(E_{21}) + P(E_{25}) + P(E_{29})\\
&\quad = r_a r_b r_c r_d r_e + r_a r_b(1 - r_c)r_d r_e + r_a r_b r_c(1 - r_d)r_e + r_a r_b(1 - r_c)(1 - r_d)r_e\\
&\quad + (1 - r_a)r_b r_c r_d r_e + (1 - r_a)r_b r_c(1 - r_d)r_e + r_a(1 - r_b)r_c r_d r_e\\
&\quad + r_a(1 - r_b)(1 - r_c)r_d r_e + r_a r_b(1 - r_c)r_d(1 - r_e) + (1 - r_a)(1 - r_b)r_c r_d r_e\\
&\quad + r_a r_b(1 - r_c)r_d(1 - r_e) + r_a r_b r_c(1 - r_d)(1 - r_e) + r_a r_b(1 - r_c)(1 - r_d)(1 - r_e)\\
&\quad + (1 - r_a)r_b r_c r_d(1 - r_e) + (1 - r_a)(1 - r_b)r_c r_d(1 - r_e) + r_a(1 - r_b)r_c r_d(1 - r_e)\\
&if\, r_a = r_b = r_c = r_d = r_e = r = 0.99\\
&R = 2r^5 - 5r^4 + 2r^3 + 2r^2
\end{aligned}
$$

if $r = 0.99$ then

$$R = 0.99979805$$

(b) Decomposition method
The problem can be very easily solved by selecting key stone as component "*e*."

$$
\begin{aligned}
&P(X) = P(e) = r_e\\
&P(\bar{X}) = P(\bar{e}) = (1 - r_e)
\end{aligned}
$$

Then conditional probability P(S/X) is written as follows:

$$
\begin{aligned}
&P(S/X) = (r_a + r_c - r_a r_c) + (r_b + r_d - r_b r_d) - (r_a + r_c - r_a r_c)(r_b + r_d - r_b r_d)\\
&P(S/\bar{X}) = (r_a r_b + r_c r_d - r_a r_b r_c r_d)\\
&R = \left[(r_a + r_c - r_a r_c) + (r_b + r_d - r_b r_d)\right] r_e + \left[(r_a r_b + r_c r_d - r_a r_b r_c r_d)\right] (1 - r_e)\\
&\text{If } r_a = r_b = r_c = r_d = r_e = r
\end{aligned}
$$

Following reliability expression is obtained.
$R = 2r^5 - 5r^4 + 2r^3 + 2r^2$ if $r = 0.99$ then

$$R = 0.99979805$$

(c) Tie-set method

Following minimal tie-sets are obtained.

$$T_1 = ab,\ T_2 = cd,\ T_3 = aed,\ T_4 = ceb$$
$$R = P\left[T_1 + T_2 + T_3 + T_4\right]$$
$$R = P(T_1) + P(T_2) + P(T_3) + P(T_4) - P(T_1T_2) - P(T_1T_3) - P(T_1T_4) - P(T_2T_3) - P(T_2T_4)$$
$$- P(T_3T_4) + P(T_1T_2T_3) + P(T_1T_2T_4) + P(T_2T_3T_4) + P(T_3T_4T_1) + P(T_1T_2T_3T_4)$$

Putting

If $r_a = r_b = r_c = r_d = r_e = r$

$$R = 2r^5 - 5r^4 + 2r^3 + 2r^2$$

if $r = 0.99$ then

$$R = 0.99979805$$

(d) Cut-set method

Following minimal cut-sets are obtained.

$$C_1 = \bar{a}\,\bar{c} \qquad C_2 = \bar{b}\,\bar{d}$$
$$C_3 = \bar{a}\,\bar{e}\,\bar{d} \qquad C_4 = \bar{b}\,\bar{c}\,\bar{e}$$

Unreliability expression is written as

$$Q = 1 - R$$
$$Q = P(C_1) + P(C_2) + P(C_3) + P(C_4) - P(C_1C_2) - P(C_1C_3) - P(C_1C_4) - P(C_2C_3)$$
$$- P(C_2C_4) - P(C_3C_4) + P(C_1C_2C_3) + P(C_1C_2C_4) + P(C_2C_3C_4) + P(C_3C_4C_1)$$
$$- P(C_1C_2C_3C_4)$$
$$Q = 1 - R$$

For equal component reliability

$$R = 2r^5 - 5r^4 + 2r^3 + 2r^2$$

if $r = 0.99$ then

$$R = 0.99979805$$

Exercise 4.8: Consider the following reliability network, each component has same reliability $r = 0.99$. Obtain system reliability (Figure 4.23).

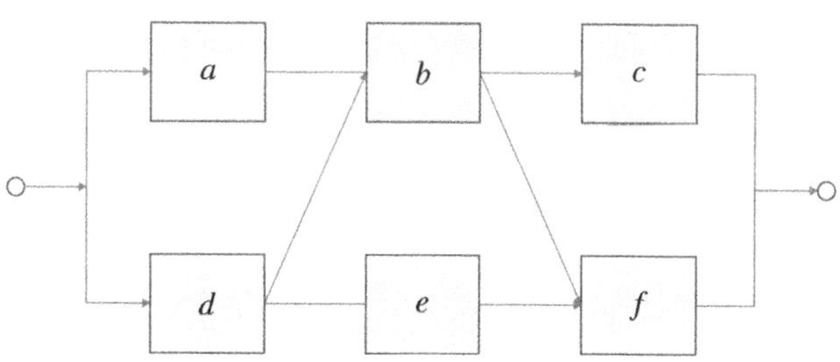

Figure 4.23 Complex reliability network.

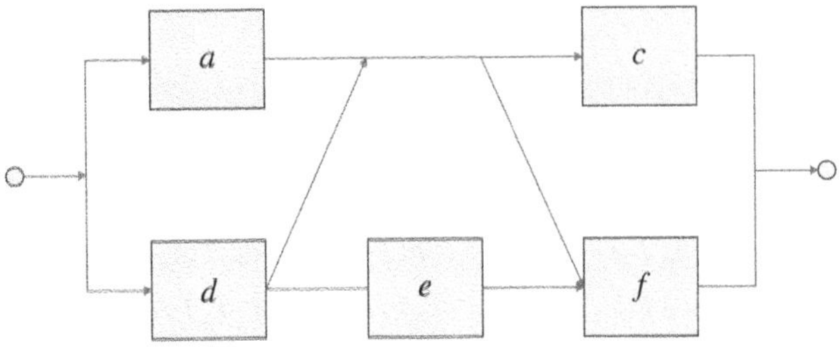

Figure 4.24 Decomposed network with X is working.

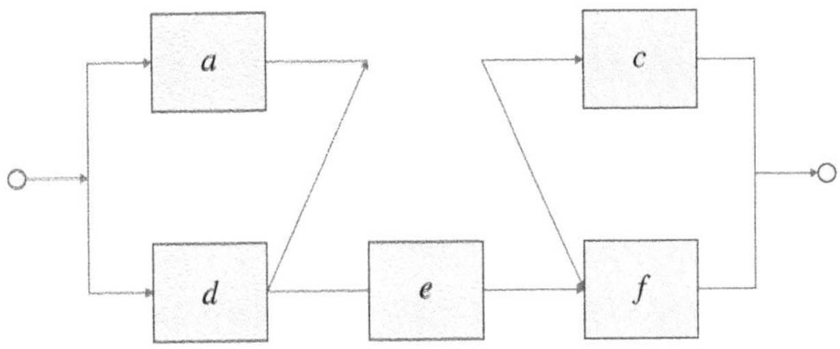

Figure 4.25 Decomposed network with X is failed.

Solution:
Careful examination of the network suggest that the problem can be easily solved by decomposition method selecting component "b" as key stone (Figures 4.24 and 4.25)

$$P(X) = P(b) = r$$
$$P(\overline{X}) = P(\overline{b}) = (1 - r)$$

When X is working then $P(S/X)$

$$P(S/X) = (2r - r^2)^2$$

When X is not working then

$$P(S/\overline{X})$$
$$P(S/\overline{X}) = r^3$$
$$R = (2r - r^2)^2 r + r^3(1 - r)$$
$$R = r^5 - 5r^4 + 5r^3$$

if $r = 0.99$

$$R = 0.999505$$

Note: Reader is advised to solve the same problem by event space, tie-set and cut-set method.

4.10 Conclusions

This chapter has described four basic methodologies for reliability evaluation of systems. Every methodology has been discussed with examples. It is stressed here that each method is applicable in different situations and limitations in some cases. Event space method is not suitable when a large number of components are present in the system. In advance application, state pruning may adopted i.e. truncation of state space. State pruning will require lesser number of states to be enumerated. Decomposition is justified when number of components in the systems are small. Moreover further limitation of the method is that it cannot be programmed on digital computer. Tie-set and cut-set methodologies are effective and both are programmable on digital computer. Usually in practice cut-set method is quite efficient in practice due to the fact approximations are justified. Additional advantage of cut-set method is that, while identifying cut-sets various failure modes are also obtained. Hence cut-set method is a powerful and efficient technique especially for power network problems.

References

Allan, R.N. (2013). *Reliability Evaluation of Power Systems*. Springer Science & Business Media.

Arya, L.D. and Koshti, A. (2008). Distributed generation capacity reliability evaluation using safety index. *Journal of the Institution of Engineers (India)* 89: 3–7.

Arya, L.D., Koshti, A., and Choube, S.C. (2013). Frequency-duration analysis of composite distribution system using a non-sequential Monte Carlo simulation. *International Journal of Electrical Power & Energy Systems* 46: 17–25.

Billinton, R. and Allan, R.N. (1992). *Reliability Evaluation of Engineering Systems*, vol. 792. Springer.

Breneman, J.E., Sahay, C., and Lewis, E.E. (2022). *Introduction to Reliability Engineering*. John Wiley & Sons.

Ebeling, C.E. (2019). *An Introduction to Reliability and Maintainability Engineering*. Waveland Press.

Endrenyi, J. (1978). *Reliability Modeling in Electric Power Systems*. Wiley-Blackwell.

Halve, S.S., Koshti, A., and Arya, R. (2023). A sampling method based on system state transition for distribution system adequacy assessment using distributed generation. *Journal of Operation and Automation in Power Engineering* 11 (4): 249–257.

Koshti, A.S., Verma, A., Arya, R. et al. (2023). Tolerable random interruption duration based reliability estimation of stand alone hybrid renewable energy system by network reduction and sequential Monte Carlo simulation. *IET Generation, Transmission & Distribution* https://doi.org/10.1049/gtd2.12910.

Saket, R.K., Kothari, D.P., and Nagrath, I.J. (2022). *Modern Power System Analysis*. Tata McGraw-Hill Publishing Company.

5

Probabilistic Approach for Standby and Load-Sharing System Reliability Evaluation

Rajesh Arya[1], R. K. Saket[2], Atul Koshti[3], Saad Mekhilef[4], and Pradeep Purey[5]

[1] *Department of Information Technology, Acropolis Institute of Technology and Research, Indore, Madhya Pradesh, India*
[2] *Department of Electrical Engineering, Indian Institute of Technology (BHU), Varanasi, Uttar Pradesh, India*
[3] *Department of Electrical Engineering, GES'S R.H. Sapat College of Engineering, Management Studies & Research, Nashik, Maharashtra, India*
[4] *Department of Electrical Engineering, Swinburne University of Technology, Melbourne, Australia*
[5] *Department of Electronics Engineering, Maharaja Ranjit Singh College of Professional Sciences, Indore, Madhya Pradesh, India*

5.1 Introduction

Reliability evaluation of a standby system needs different considerations and requires a state transition diagram (Hegazy et al. 2003; Arya & Koshti 2008; Costa & Matos 2009; Xu & Singh 2012; Arya & Koshti 2012). Accounting for preventive maintenance, the hazard rate of a component may be treated as constant as an average value during a period. Still nonrepairable systems are being considered. A standby system (Billinton & Li 1993; Wang et al. 2013; Ahmadi-Khatir et al. 2009; Arya et al. 2013; Halve et al. 2023) is one where one of the components is in operation, and upon failure of this component, one of the idle components is switched in for service. It is obvious as the number of idle (standby) components increases, the reliability of the standby system increases. This may not always be true. Under ideal conditions certainly. reliability will increase, but the magnitude of the increase in reliability may reduce owing to imperfect switch and switching (Koshti et al. 2023; Khezri & Mahmoudi 2020; Roy et al. 2022; Ranjbar et al. 2022; Yahaya et al. 2020; Hamzeh & Vahidi 2020; Hassan et al. 2022). These aspects need consideration. Further, the standby component may fail under idle mode. The "switch" which puts the idle component in operation may fail on demand.

5.2 Reliability Evaluation Under Ideal Condition

This section discusses the reliability evaluation of a standby system (Ebeling 2000; Lewis 1987; Endrenyi 1978) under the assumptions

(a) Failure rate of switch is zero
(b) Probability of failing the switch on demand is zero
(c) Failure rate of idle component is zero while in standby mode

A standby system with "r" standby components is shown in Figure 5.1

Reliability Analysis of Modern Power Systems, First Edition. Edited by R. K. Saket and P. Sanjeevikumar.

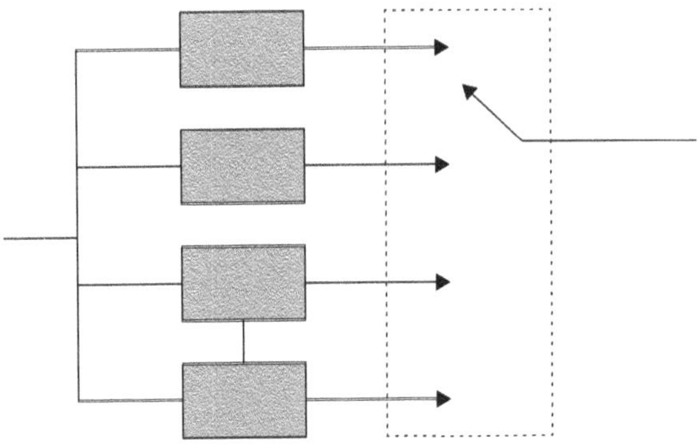

Figure 5.1 Standby system representation.

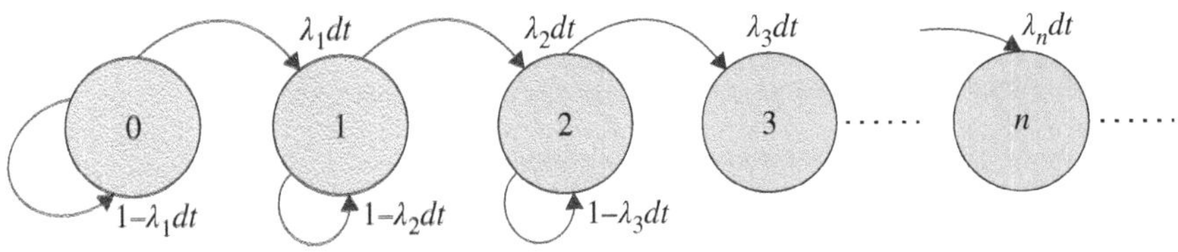

Figure 5.2 Standby system state transition diagram.

Assume that the failure rates of each component as λ_1, λ_2, $\lambda_3 \ldots \lambda_{r+1}$ as seen in the following system's state transition diagram (Figure 5.2).

State-0 Represents no failure i.e. main component is in operation
State-1 Main failed and "first" standby is in operation
State-2 Second standby is in operation [main and first standby has failed]

⋮

⋮

State-n nth standby is in operation [main and (n–1) standby component have failed]

Sate 0 equation is written as follows-

$$P_0(t + dt) = P_0(t)\,(1 - \lambda_1\, dt)$$

(Probability of system being in "0th" state at $t + dt$) = (Probability of system in "0th" state at "t") (probability of the components do not fail in interval t to ($t + dt$))

$$P_0(t) = -\lambda_1\, P_0(t) \tag{5.1}$$

Similarly, equation for state 1 is written as

$$p_1\,(t + \mathrm{d}t) = p_0(t)\,(\lambda_1\, \mathrm{d}t) + p_1\,(t)\,(1 - \lambda_1\, \mathrm{d}t)$$

$$p_1(t + \mathrm{d}t) = p_0(t)\,(\lambda_1 \mathrm{d}t) + p_1(t)\,(1 - \lambda_2 \mathrm{d}t)$$

$$\frac{p_1(t - \mathrm{d}t) - p_1(t)}{\mathrm{d}t} = \lambda_1 p_0 - \lambda_2 p_1$$

$$p_1 = \lambda_1 \dot{p}_0 - \lambda_2 p_1 \tag{5.2}$$

Equation for state 2

$$p_2(t+\mathrm{d}t) = p_1(t)(\lambda_2\,\mathrm{d}t) + p_2(t)(1-\lambda_3\,\mathrm{d}t)$$

$$\frac{p_2(t-dt)-p_2(t)}{dt} = \lambda_1 p_0 - \lambda_2 p_1$$

$$\dot{p}_2 = \lambda_2\,p_1 - \lambda_3\,p_2 \tag{5.3}$$

In general, the equation for nth state is written as follows

$$\dot{p}_n = \lambda_n\,p_{n-1} - \lambda_{n+1}\,p_n \tag{5.4}$$

The above equations are written in matrix form (Endrenyi 1978) as follows

$$\begin{aligned}\dot{p}^T(t) &= p^T A\\ \dot{p}^T &= [\dot{p}_0\ \dot{p}_1\ \dot{p}_2\cdot\ldots]\\ p^T &= [p_0\ p_1\ p_2\cdot\ldots]\end{aligned} \tag{5.5}$$

Matrix "A" follows

$$A = \begin{bmatrix} -\lambda_1 & \lambda_1 & 0 & 0 & 0 & ..\\ 0 & -\lambda_2 & \lambda_2 & 0 & 0 & ..\\ 0 & 0 & -\lambda_3 & \lambda_3 & 0 & ..\\ \vdots & \vdots & \vdots & \vdots & \vdots & \ddots \end{bmatrix}$$

Matrix A may be formed as follows:
$a_{ii} = -$ (Total rate of departure from ith state)
$a_{ij} =$ Total rate of departure from ith state to jth state
In solving the above state equations, the initial condition vector is given as [1 0 0 . . . . 0]
In sequence, the above equation is solved, and if there are "r" standby, then the reliability of the system is given below

$$R(t) = \sum_{i=0}^{r} p_i(t), \overline{R}(t) = 1 - R(t)$$

Example 5.1
Assume the system consists of one main component and one standby component having failure rate as λ_1 and λ_2 respectively.

Solution:
The system's reliability is described as

$$\begin{aligned}R(t) &= p_0(t) + p_i(t)\\ P_0 &= e^{-\lambda_1 t}\\ \dot{P}_1 &= \lambda_1 P_0 - \lambda_2 P_2\\ P_1(t) &= \frac{\lambda_1}{(s+\lambda_1)(s+\lambda_2)}\\ P_1(s) &= \frac{\lambda_1}{(\lambda_2-\lambda_1)(s+\lambda_2)} - \frac{\lambda_1}{(\lambda_2-\lambda_1)(s+\lambda_2)}\end{aligned}$$

Taking inverse Laplace transform gives

$$p_i(t) = \frac{\lambda_1}{(\lambda_2 - \lambda_1)} e^{-\lambda_1 t} - \frac{\lambda_1}{(\lambda_2 - \lambda_1)} e^{-\lambda_2 t}$$

$$R(t) = e^{-\lambda_1 t} \frac{\lambda_1}{(\lambda_2 - \lambda_1)} e^{-\lambda_1 t} - \frac{\lambda_1}{(\lambda_2 - \lambda_1)} e^{-\lambda_2 t}$$

$$R(t) = \frac{1}{(\lambda_2 - \lambda_1)} \left[\lambda_2 e^{-\lambda_1 t} - \lambda_1 e^{-\lambda_2 t}\right]$$

Failure density function

$$f(t) = -\frac{dR}{dt} = \frac{\lambda_1 \lambda_2}{(\lambda_2 - \lambda_1)} \left[e^{-\lambda_1 t} - e^{-\lambda_2 t}\right]$$

Note: It is hypo-exponential failure law.
Further, if two failure rates are equal $\lambda_1 = \lambda_2 = \lambda$, then $R(t), f(t)$ expression reduces to

$$R(t) = (1 + \lambda t)e^{-\lambda t}$$

$$f(t) = \lambda^2 t\, e^{-\lambda t}$$

It is an Erlang distribution function (Billinton & Allan 1983; Billinton & Allan 1996).
In general, for "r" standby system where each component has "λ" failure rate

$$p_n = \frac{(\lambda t)^n}{n!} e^{-\lambda t}$$

Given by a Poisson process, the reliability expression reduces to

$$R(t) = \sum_{n=0}^{r} \frac{(\lambda t)^n}{n!} e^{-\lambda t}$$

5.3 Standby System Reliability Evaluation Under Nonideal Condition

5.3.1 Switching is Imperfect (Endrenyi 1978)

Assume that the system consists of one main and one standby component, each having the same failure rate. Further, assume that P_s is the probability of successful change-over. Figure 5.3 shows the diagram of state transitions.

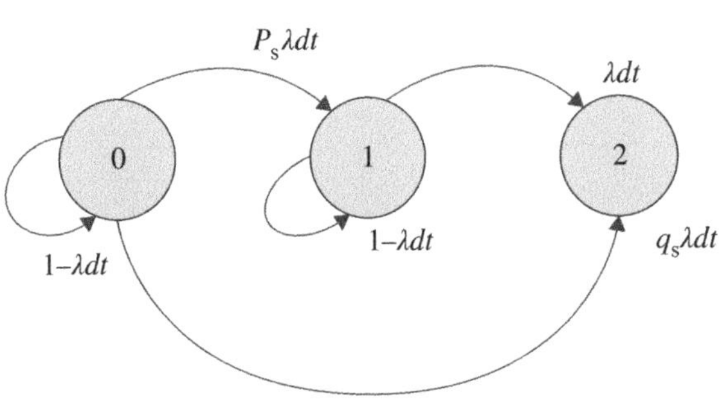

Figure 5.3 Standby system state transition diagram with perfect switching.

State 0 Represents the main component of operation
State 1 Main component has failed change-over has been successful, and standby is operating.
State 2 an absorbing state and is either reached by failure of standby component or unsuccessful change-over when the main component fails.

$$q_s = 1 - p_s$$

State space equation is written as

$$[\dot{p}_0 \ \dot{p}_1 \ \dot{p}_2] = [p_0 \ \ p_1 \ \ p_2] \begin{bmatrix} -\lambda & p_s\lambda & q_s\lambda \\ 0 & -\lambda & \lambda \\ 0 & 0 & 0 \end{bmatrix} \lim_{\delta x \to 0}$$

$$\dot{p}_0 = -\lambda p_0, p_0 = e^{-\lambda t}, \dot{p}_1 = p_s\lambda p_0 - \lambda p_1$$

$$(s+\lambda)\, p_1(s) = \frac{p_s\lambda}{(s+\lambda)}, p_1(s) = \frac{p_s\lambda}{(s+\lambda)^2}$$

Taking inverse Laplace transform

$$p_1(t) = p_s\lambda\, te^{-\lambda t}$$

Reliability is described as

$$R(t) = p_0\,(t) + p_1(t), R(t) = (1 + p_s\,\lambda\, t)\, e^{-\lambda t}$$

$$\text{MTTF} = \frac{1 + p_s}{\lambda}$$

Hence MTTF will reduce depending on the value of p_s and it provides a functional failure density as

$$f(t) = -\frac{dR}{dt}, f(t) = p_s\lambda^2 te^{-\lambda t} + \lambda e^{-\lambda t}(1 - p_s), f(t) = \lambda e^{-\lambda t}[(1 - p_s) + p_s\lambda t]$$

5.3.2 Switching is Imperfect and Switch has a Failure Rate λ_s (Imperfect Switch)

In addition to unsuccessful change-over, the switch while in connection has a failure rate is λ_s. The diagram of the state is shown in Figure 5.4.

State 0 represents the main component is in operation, and standby is in standing mode
State 1 main component has failed, and standby is in operation (with successful) change-over
State 2 an absorbing state which is reached by

(a) by unsuccessful change-over
(b) by failure of switch while main component is in operation

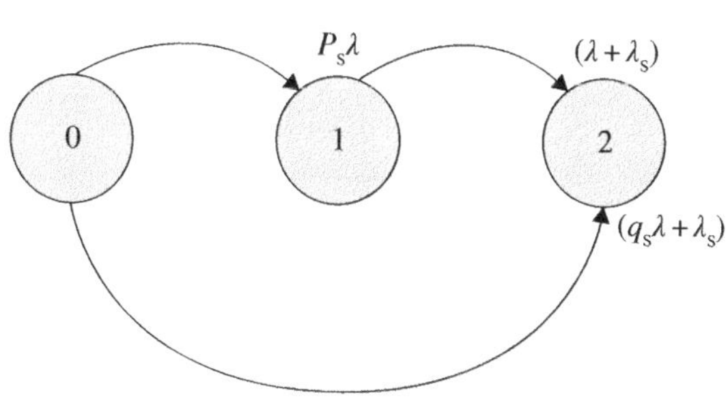

Figure 5.4 Standby system state transition diagram with imperfect switching.

(c) failure of switch while standby is in operation
(d) failure of standby component while in operation

State equation in matrix form is written as

$$[\dot{p}_0\ \dot{p}_1\ \dot{p}_2] = [p_0\, p_1\, p_2]\begin{bmatrix} -(\lambda+\lambda_s) & p_s\lambda & q_s\lambda+\lambda_s \\ 0 & -(\lambda+\lambda_s) & (\lambda+\lambda_s) \\ 0 & 0 & 0 \end{bmatrix}$$

$$\dot{p}_0 = -(\lambda+\lambda_s)p_0, p_0 = e^{-(\lambda+\lambda_s)t}$$

$$\dot{p}_1 = p_s\lambda p_0 - (\lambda+\lambda_s)p_1(s+\lambda+\lambda_s)p_1 = \frac{p_s\lambda}{(s+\lambda+\lambda_s)}$$

$$p_1(s) = \frac{p_s\lambda}{(s+\lambda+\lambda_s)^2}, p_1(t) = p_s\lambda te^{-(\lambda+\lambda_s)t}$$

Reliability is given as

$$R\,(t) = p_0\,(t) + p_1\,(t), R(t) = (1+p_s\lambda t)e^{-\lambda t}e^{-\lambda_s t}$$

The result is expected. The previous result multiplied by the reliability of switch $e^{-\lambda_s t}$ this is equivalent to an imperfect switch in series.

5.3.3 Consider the Standby Components Might Fail While they are in Idle Mode in Addition

λ_i is failure rate of standby components in idle mode. State transition diagram is shown below (Figure 5.5).

State space equations are written as

$$[\dot{p}_0\ \ \dot{p}_1\ \ \dot{p}_2\ \ \dot{p}_3] = [p_0\ \ p_1\ \ p_2\ \ p_3]\begin{bmatrix} -(\lambda+\lambda_s+\lambda_i) & p_s\lambda & \lambda_i & q_s\lambda+\lambda_s \\ 0 & -(\lambda+\lambda_s) & 0 & -(\lambda+\lambda_s) \\ 0 & 0 & -\lambda_s & \lambda_s \\ 0 & 0 & 0 & 0 \end{bmatrix}$$

The following states are described:

State 0 Represents the main component in operation, and standby is in idle mode
State 1 Main component has failed, successful change-over has taken place, and standby component is in operation

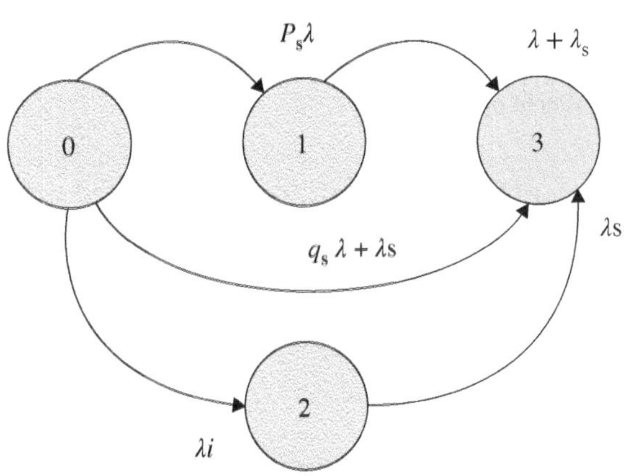

Figure 5.5 Standby system state transition diagram with the assumption of component failure during idle mode.

State 2 Main component is in operation, but the standby component has failed in idle mode
State 3 Absorbing state, which is reached via either unsuccessful change over, switch failure or by standby component failure.

Reliability is given by

$$R(t) = P + P_1 + P_2$$

$$\dot{p}_0 = -(\lambda + \lambda_s + \lambda_i)p_0$$

$$p_0 = e^{-(\lambda+\lambda_s+\lambda_i)t}$$

$$\dot{p}_1 = p_s\,\lambda\, p_0 - (\lambda + \lambda_s)p_1$$

$$(s + \lambda + \lambda_s)p_1 = \frac{p_s\lambda}{(s + \lambda + \lambda_s + \lambda_i)}$$

$$p_1(s) = \frac{p_s\lambda}{(s + \lambda + \lambda_s)(s + \lambda + \lambda_s + \lambda_i)}$$

$$p_1(t) = \frac{p_s\lambda}{\lambda_i}\left[e^{-(\lambda+\lambda_s)t} - e^{-(\lambda+\lambda_s+\lambda_i)t}\right]$$

$$\dot{p}_2 = \lambda_i p_0 - \lambda_s p_2$$

$$p_2(s) = \frac{\lambda_i}{(s + \lambda_s)\,(s + \lambda + \lambda_s + \lambda_i)}$$

Therefore

$$P_2(s) = \frac{\lambda_i}{(\lambda + \lambda_i)\,(s + \lambda_s)} - \frac{\lambda_i}{(\lambda + \lambda_i)\,(s + \lambda + \lambda_s + \lambda_i)}$$

$$p_2(t) = \frac{\lambda_i}{(\lambda + \lambda_i)}\left[e^{-\lambda_s t} - e^{-(\lambda+\lambda_s+\lambda_i)t}\right]$$

Reliability is given as

$$p_0 = e^{-(\lambda+\lambda_s+\lambda_i)t}, p_1(t) = \frac{p_s\lambda}{\lambda_i}\left[e^{-(\lambda+\lambda_s)t} - e^{-(\lambda+\lambda_s+\lambda_i)t}\right]$$

$$p_2(t) = \frac{\lambda_i}{(\lambda + \lambda_i)}\left[e^{-\lambda_s t} - e^{-(\lambda+\lambda_s+\lambda_i)t}\right]$$

Therefore reliability is given as

$$R(t) = e^{-(\lambda+\lambda_s+\lambda_i)\,t} + \frac{p_s\lambda}{\lambda_i}\left[e^{-(\lambda+\lambda_s)\,t} - e^{-(\lambda+\lambda_s+\lambda_i)\,t}\right] + \frac{\lambda_i}{(\lambda + \lambda_i)}\left[e^{-\lambda_s t} - e^{-(\lambda+\lambda_s+\lambda_i)\,t}\right]$$

5.4 Reliability Evaluation of Load-Sharing System (Endrenyi 1978)

Assume two systems are operating in active parallel and sharing the total load/demand. If one of the system fails, the other can meet the demand but operate under stressed conditions, and hence, its failure rate will increase. A load-sharing system's reliability is obtained using state transition diagram. In an earlier analysis of parallel system the two components were operating in parallel and sharing the demand, but when one of them fails, the other meets the demand, but the failure rate remains the same. The diagram of a load-sharing system's state transition is given as follows. If each failure rate is λ_1, λ_2, then the component while both are in operation. λ_1^+ and λ_2^+ are failure rate of these components while only one of them is working. The four-state model of the load-sharing system is given in Figure 5.6.

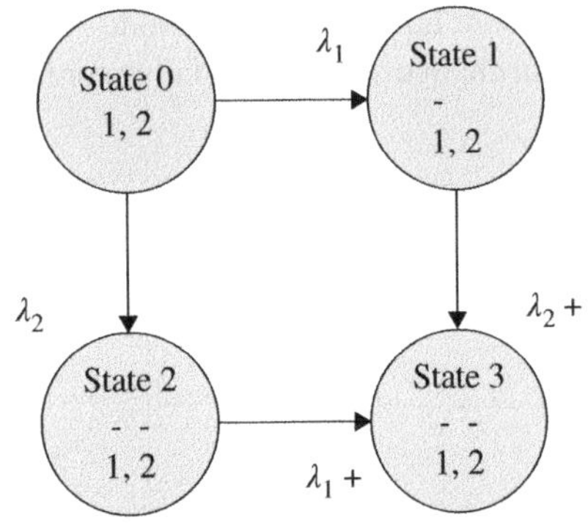

Figure 5.6 Four-state model of the load-sharing system.

State 1 Both elements are operational.
State 2 Component 2 is working; however, component 1 failed.
State 3 Component 1 is functional, whereas component 2 is not.
State 4 both components 1 and 2 have failed.

State space equation of load-sharing system is given as follows:

$$[\dot{p}_0 \quad \dot{p}_1 \quad \dot{p}_2 \quad \dot{p}_3] = [p_0 \quad p_1 \quad p_2 \quad p_3]\begin{bmatrix} -(\lambda_1+\lambda_2) & \lambda_1 & \lambda_2 & 0 \\ 0 & -\lambda_2^+ & 0 & \lambda_2^+ \\ 0 & 0 & -\lambda_1^+ & \lambda_1^+ \\ 0 & 0 & 0 & 0 \end{bmatrix}$$

$$\dot{p}_0 = -(\lambda_1+\lambda_2)p_0, p_0 = e^{-(\lambda_1+\lambda_2)t}, \dot{p}_1 = \lambda_1 p_0 - \lambda_2^+ p_1$$

$$\left(s+\lambda_2^+\right)p_1 = \frac{\lambda_1}{(s+\lambda_1+\lambda_2)}, p_1(s) = \frac{\lambda_1}{(s+\lambda_1+\lambda_2)\left(s+\lambda_2^+\right)}$$

$$p_1(s) = \frac{\lambda_1/\left(\lambda_2^+-\lambda_1-\lambda_2\right)}{(s+\lambda_1+\lambda_2)} - \frac{\lambda_1/\left(\lambda_2^+-\lambda_1-\lambda_2\right)}{\left(s+\lambda_2^+\right)}$$

$$p_1(t) = \frac{\lambda_1}{\left(\lambda_1+\lambda_2-\lambda_2^+\right)}\left[e^{-\lambda_2^+ t} - e^{-(\lambda_1+\lambda_2)t}\right]$$

$$\dot{p}_2 = \lambda_2 p_0 - \lambda_1^+ p_2$$

$$p_2(s) = \frac{\lambda_2}{(s+\lambda_1+\lambda_2)\left(s+\lambda_1^+\right)}$$

$$p_2(t) = \frac{\lambda_1}{\left(\lambda_1+\lambda_2-\lambda_1^+\right)}\left[e^{-\lambda_1^+ t} - e^{-(\lambda_1+\lambda_2)t}\right]$$

Reliability is expressed as follows:

$$R(t) = p_0(t) + p_1(t) + p_2(t)$$

$$R(t) = e^{-(\lambda_1+\lambda_2)t} + \frac{\lambda_1}{\left(\lambda_1+\lambda_2-\lambda_2^+\right)}\left[e^{-\lambda_2^+ t} - e^{-(\lambda_1+\lambda_2)t}\right] + \frac{\lambda_1}{\left(\lambda_1+\lambda_2-\lambda_2^+\right)}\left[e^{-\lambda_1^+ t} - e^{-(\lambda_1+\lambda_2)t}\right]$$

If,

$$\lambda_1^+ = \lambda_1, \lambda_2^+ = \lambda_2$$

Then, the above expression reduces to the expression of parallel system i.e.

$$R\,(t) = e^{-\lambda_1 t} + e^{-\lambda_2 t} - e^{-(\lambda_1+\lambda_2)t}$$

Further, if the component are identically i.e.

$$\lambda_1 = \lambda_2 = \lambda, \lambda_1^+ = \lambda_2^+ = \lambda^+$$

Then, the expression for the reliability of load sharing is given to the system by

$$R(t) = e^{-2\lambda_t} + \frac{2\lambda}{(2\lambda - \lambda^+)}[e^{-\lambda + t} - e^{-2\lambda t}]$$

$$R(t) = \frac{1}{(2\lambda - \lambda^+)}[2\lambda e^{-\lambda^+ t} - \lambda^+ e^{-2\lambda t}]$$

If $\lambda^+ = 2\lambda$, then the following relation is obtained

$$R(t) = (1 + 2\,\lambda\, t)\, e^{-2\lambda t}$$

5.5 Illustrative Examples

This section describes various numerical problems to understand the mathematical interpretation. The following examples are illustrated for the reliability assessment of industrial systems.

Example 5.2

Obtain the reliability of a standby system where main and standby components have failure rates of 0.2 and 0.3 per year, respectively. Estimate the reliability value for 1 year.

Solution:

$$\lambda_M = 0.2, \lambda_S = 0.3$$

$$R(t) = \frac{1}{(\lambda_S - \lambda_M)}\left[\lambda_S e^{-\lambda_M t} - \lambda_M e^{-\lambda_S t}\right]$$

$$R(t) = 10 \times [0.3e^{-0.2t} - 0.2e^{-0.3t}]$$

For one year

$$R(1) = 10 \times [0.3e^{-0.2} - 0.2e^{-0.3}]$$

$$R(1) = 0.97519$$

Example 5.3

A standby system has a failure rate of 0.1/year of main as well as standby component. Obtain reliability for one year if failure rates of the transfer switch are 0.05/year and the probability of successful change-over is 0.95.

Solution:

$$\lambda_M = \lambda_S = \lambda = 0.1$$

$$\lambda_{sw} = 0.05$$

$$p_s = 0.95$$

Reliability expression for the system is written as

$$R(t) = (1 + p_s \lambda t)e^{-(\lambda+\lambda_{sw})t}$$

$$R(t) = (1 + 0.5 \times 0.1 \times t)\, e^{-0.15t}$$

For ($t = 1$ year)

$$R(t) = (1.095)\, e^{-0.15}, R(t) = 0.9424$$

Example 5.4
Obtain a nonredundant system's reliability considering 50 components, and each component has a failure rate of 0.002/year and also has one similar component in spare.

Solution:
Failure rate of the system is given as-

$$\lambda_{SYS} = 50 \times 0.02, \lambda_{SYS} = 1.0/\text{year}$$

If one of the components fails, then it is replaced by the spare component, and again, the failure rate will be 1.0. Hence, the system can be represented as standby system where "both" component has a failure rate of 1.0/year. Then, reliability under ideal conditions is given as

$$R(t) = (1 + \lambda_{sys} t)e^{-\lambda_{sys} t}, R(t) = (1 + t)e^{-t}$$

$$R(1) = 2e^{-1.0}, R(t) = 0.7357$$

Look, if you do not have spare components, then reliability is

$$R(t) = e^{-1.0}, \quad R(t) = 0.3678$$

Example 5.5
Obtain reliability expression for the following system. Each component has a failure rate λ per year. Assume ideal condition (Figure 5.7).

Solution:
This is a series system, where two standby sub-systems are in series. Each sub-system has reliability

$$R = (1 + \lambda t)e^{-\lambda t}$$

Hence, the overall system has reliability

$$R_{SYS} = R^2, R_{SYS} = [(1 + \lambda t)e^{-\lambda t}]^2$$

Example 5.6
Obtain reliability expression for the following system given in Figure 5.8. Each component has failure rate λ per year.

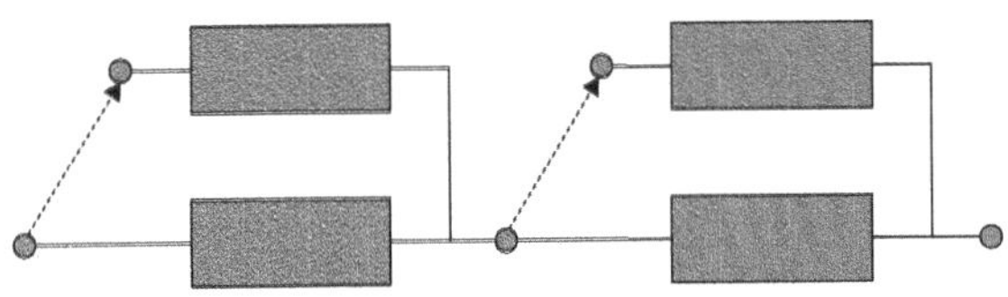

Figure 5.7 Figure for Example 5.5.

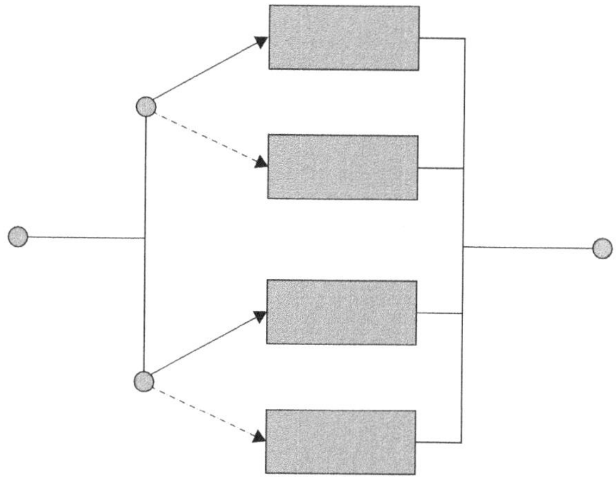

Figure 5.8 Figure for Example 5.6.

Solution:

Hence two standby systems are connected in parallel. Each sub-system has reliability

$$R = (1 + \lambda t)e^{-\lambda t}$$

System reliability overall is described as (Figure 5.9)

$$R_{SYS} = 2\,R - R^2$$
$$R_{SYS} = 2\,(1 + \lambda t)\,e^{-\lambda t} - [(1 + \lambda t)\,e^{-\lambda t}]^2$$
$$R_{SYS} = 2\,(1 + \lambda t)\,e^{-\lambda t} - (1 + \lambda t)^2\,e^{-2\lambda t}$$

Example 5.7

Two generators, each of 100 MW rating, are meeting demand of 150 MW, each having a failure rate of 0.01/year. If one of the generators fails, the other can meet the demand but increase the failure rate $\lambda^+ = 0.04$ per year. Compute the reliability for one year.

Solution:

Reliability expression is written as

$$R(t) = \frac{1}{(\lambda^+ - 2\lambda)}\,[\lambda^+\,e^{-2\lambda t} - 2\lambda\,e^{-\lambda^+ t}]$$
$$\lambda^+ = 0.04,\ \lambda = 0.01$$
$$R(t) = \frac{100}{2}[0.04e^{-0.02t} - 0.02e^{-0.04t}]$$
$$R(t) = 50[0.04e^{-0.02} - 0.02e^{-0.04}]$$
$$R(t) = 50[0.03920 - 0.01921]$$
$$R(t) = 0.9995$$

Example 5.8

A load is supplied for three generators in parallel, having failure rate λ per year. If one of the generators fails, the remaining two generators can meet the demand, but failure rate of each becomes double. If the next generator fails only one generator can meet the demand with failure rate six times. Obtain reliability expression (Figure 5.9).

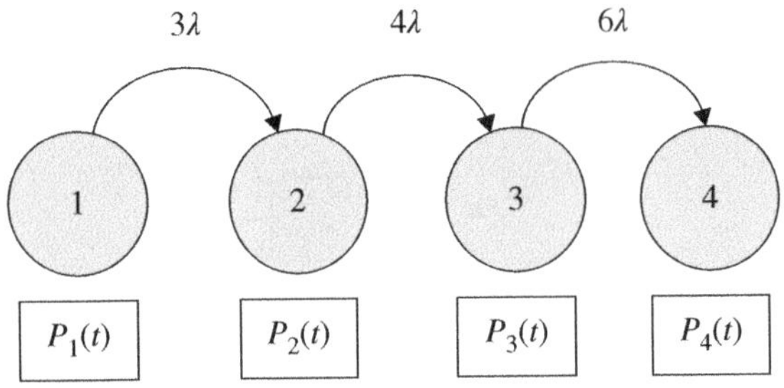

Figure 5.9 Figure for Example 5.8.

Solution:

$$R(t) = p_1 + p_2 + p_3, [\dot{p}_1 \quad \dot{p}_2 \quad \dot{p}_3] = [p_1 \quad p_2 \quad p_3] \begin{bmatrix} -3\lambda & 3\lambda & 0 & 0 \\ 0 & -4\lambda & 4\lambda & 0 \\ 0 & 0 & -6\lambda & 6\lambda \\ 0 & 0 & 0 & 0 \end{bmatrix}$$

$$\dot{p}_1 = 3\lambda p_1, sP_1(s) - 1 = 3\lambda P_1(s)$$

$$P_1(s) = \frac{1}{s + 3\lambda}, p_1(t) = e^{-3\lambda t}, \dot{p}_2 = 3\lambda p_1 - 4\lambda p_2$$

$$sP_2(s) = 3\lambda P_1(s) - 4\lambda P_2(s), (s + 4\lambda)P_2(s) = 3\lambda P_1(s)$$

$$P_2(s) = \frac{3\lambda}{(s + 4\lambda)(s + 3\lambda)}, p_2(t) = 3e^{-3\lambda t} - 3e^{-4\lambda t}$$

$$\dot{p}_3 = 4\lambda p_2 - 6\lambda p_3, sP_3(s) = 4\lambda P_2(s) - 6\lambda P_3(s), (s + 6\lambda)P_3(s) = \frac{4\lambda \times 3\lambda}{(s + 4\lambda)(s + 3\lambda)}$$

$$P_3(s) = \frac{12\lambda^2}{(s + 3\lambda)(s + 4\lambda)(s + 6\lambda)}, P_3(s) = \frac{4}{(s + 3\lambda)} - \frac{6}{(s + 4\lambda)} + \frac{2}{(s + 6\lambda)}$$

$$p_3(t) = 4e^{-3\lambda t} - 6e^{-4\lambda t} + 2e^{-6\lambda t}$$

$$R(t) = p_1(t) + p_2(t) + p_3(t)$$

$$R(t) = e^{-3\lambda t} + 3\,e^{-3\lambda t} - 3\,e^{-4\lambda t} + 4e^{-3\lambda t} - 6\,e^{-4\lambda t} + 2\,e^{-6\lambda t}$$

$$R(t) = 8e^{-3\lambda t} - 9e^{-4\lambda t} + 2e^{-6\lambda t}$$

Example 5.9

A generator meets a load of 200 MW via two parallel transmission lines with 150 MW capacities each. Each transmission line has reliability of 0.99 for 1 year. The generator failure rate is 0.1 per year.

(a) Obtain reliability of the system.
(b) Recalculate the reliability of another generator that is operating as a hot reserve and may take a load in the event of failure of the previous generator. The failure rate of this generator is also 0.1 per year.

Solution:

(a) The logic diagram is shown as (Figure 5.10)

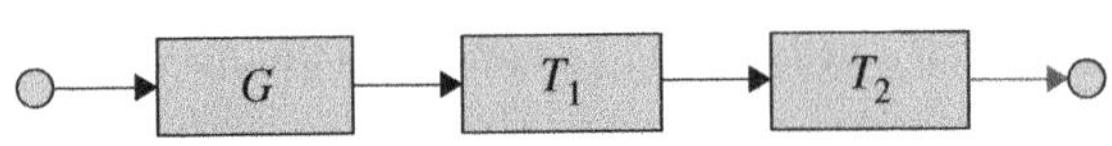

Figure 5.10 Figure for Example 5.9 (a).

Reliability of the system for one year is given as

$$R_G = e^{-0.1t} = e^{-0.1} = 0.9405, R_{T1} = R_{T2} = 0.99$$
$$R = 0.9405 \times 0.99 \times 0.99, R = 0.9217$$

(b) For this case reliability diagram is shown (Figure 5.11).

Figure 5.11 Figure for Example 5.9 (b).

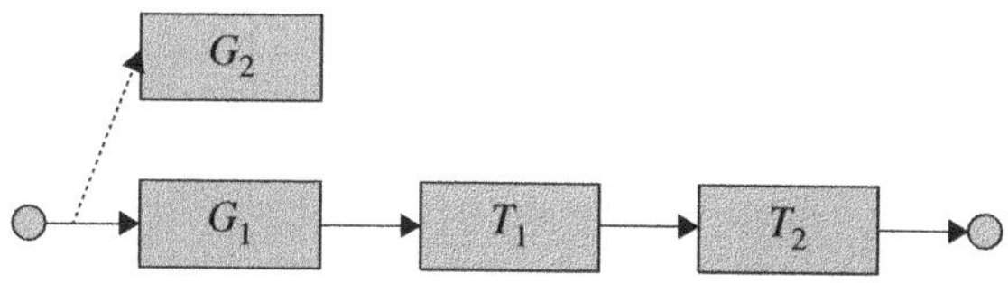

Reliability of generator system as standby

$$R_G = (1 + \lambda t)e^{-\lambda t}, R_G = (1.1)e^{-0.1}, R_G = 0.9953$$

System reliability is

$$R' = R_G R_{T_1} R_{T_2}, R' = 0.9953 \times 0.99 \times 0.99, R' = 0.97549$$

Note: $R' > R$

Example 5.10

A system consists of two components. The main component is in the operation and another component is in standby (spare) mode. Each component has failure, repair, and re-installation rate as λ, μ, and α, respectively. Draw the state transition diagram and write down state probabilities (Figure 5.12).

Figure 5.12 Figure for Example 5.10.

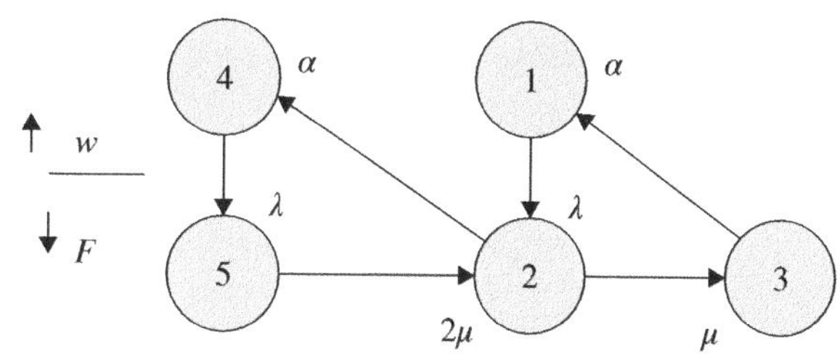

Solution:

Step 1. Main component up standby available
Step 2. Main component down standby available but not installed
Step 3. Main component down main as well as spare component is available but not reinstalled
Step 4. Standby is in operation spare component is down
Step 5. Both main and standby down

States 1 and 4 are up state
States 2, 3, and 5 are down state.
State probabilities in the long run are obtained by solving

$$p^T A = 0, \sum p_i = 1, p_1 = \frac{2\alpha\mu^2(\lambda + \mu) + \alpha^2\mu^2}{\Delta}$$

$$p_2 = \frac{2\lambda\alpha\mu\,(\lambda + \mu)}{\Delta}, p_3 = \frac{2\lambda\mu^2(\lambda + \mu)}{\Delta}, p_4 = \frac{2\lambda\alpha^2\mu}{\Delta}, p_5 = \frac{\lambda^2\alpha^2}{\Delta}$$

$$\Delta = 2\mu[(\alpha + \mu)(\lambda + \alpha)(\lambda + \mu)] + \lambda^2\alpha^2$$

System availability in long run is given as

$$A_{SYS} = p_1 + p_4$$

And system unavailability is given as

$$\overline{A}_{SYS} = p_2 + p_3 + p_5$$

Failure frequency

$$f = p_1\lambda + p_4\lambda, f = (p_1 + p_4)\lambda$$

Putting the p_1 and p_4

$$f = \frac{2\alpha\lambda\mu[(\lambda + \mu)(\alpha + \mu)]}{\Delta}$$

Mean up time is

$$\text{MUT} = \frac{p_1 + p_4}{f} = \frac{1}{\lambda}$$

Mean down time

$$\text{MDT} = \frac{p_2 + p_3 + p_5}{f}$$

$$\text{MDT} = \frac{1}{\alpha} + \frac{\alpha\lambda}{2\mu(\lambda + \mu)(\mu + \alpha)}$$

Take $\lambda = 0.4$, $\mu = 9.6$ and $\alpha = 100$ per year.
System availability

$$A_{\text{SYS}} = p_1 + p_4$$

$$p_1 = 0.88084, p_4 = 0.0378, A_{\text{SYS}} = 0.9188$$

$$\overline{A}_{\text{SYS}} = 1 - A_{\text{SYS}} = 1 - 0.9188$$

$$\overline{A}_{\text{SYS}} = 0.08132$$

$$f = (p_1 + p_4)\lambda = 0.9186 \times 0.4 = 0.36744$$

$$\text{MUT} = \frac{A_{\text{SYS}}}{f} = \frac{0.9186}{0.36744}$$

$$\text{MUT} = 2.5 \text{ years}, \text{MDT} = 0.22 \text{ years}$$

5.6 Conclusion

A methodology has been presented in this chapter for reliability evaluation of standby and load-sharing system. The following cases have been illustrated with expressions for reliability evaluation of a standby system under ideal conditions with (a) assumptions that the failure rate of the switch is zero, (b) probability of failing the switch on demand is zero, and (c) failure rate of idle component is zero in standby mode. Further, standby system reliability under nonideal conditions is evaluated with conditions such as (a) imperfect switch, (b) imperfect switching, and (c) standby components failure during idle mode. Moreover, at the end of the chapter, illustrative exercises are presented to help readers understand the concept. This methodology will give an insight into using the philosophy of reliability evaluation for standby and load-sharing systems, which will be helpful for power system planning engineers to assess the reliability of such systems.

References

Ahmadi-Khatir, A., Fotuhi-Firuzabad, M., and Goel, L. (2009). Customer choice of reliability in spinning reserve procurement and cost allocation using well-being analysis. *Electric Power Systems Research* 79 (10): 1431–1440.

Arya, L.D. and Koshti, A. (2008). Distributed generation capacity reliability evaluation using safety index. *Journal of Institution of Engineers (India)* 89: 3–7.

Arya, L.D. and Koshti, A. (2012). Probabilistic simulation approach for distributed generation (DG) capacity evaluation using artificial neural network representation of load duration curve. *Journal of Institution of Engineers (India)* 93 (1): 57–62.

Arya, L.D., Koshti, A., and Choube, S.C. (2013). Frequency-duration analysis of composite distribution system using a non-sequential Monte Carlo simulation. *International Journal of Electrical Power and Energy Systems* 46: 17–25.

Billinton, R. and Allan, R.N. (1983). *Reliability Evaluation of Engineering Systems*, 2e. Springer.

Billinton, R. and Allan, R.N. (1996). *Reliability Evaluation of Power Systems*, Springer International Edition, 2e.

Billinton, R. and Li, W. (1993). A system state transition sampling method for composite system reliability evaluation. *IEEE Transactions on Power Systems* 8 (3): 761–770.

Costa, P.M. and Matos, M.A. (2009). Assessing the contribution of microgrids to the reliability of distribution networks. *Electric Power Systems Research* 79 (2): 382–389.

C.E. Ebeling, '*An Introduction to Reliability and Maintainability Engineering*', McGraw Hill Education (India) edition, 2000.

Endrenyi, J. (1978). *Reliability Modeling in Electric Power Systems*. Wiley-Blackwell.

Halve, S.S., Koshti, A., and Arya, R. (2023). A sampling method based on system state transition for distribution system adequacy assessment using distributed generation. *Journal of Operation and Automation in Power Engineering* 11 (4): 249–257.

Hamzeh, M. and Vahidi, B. (2020). Reliability evaluation of distribution transformers considering the negative and positive effects of rooftop photovoltaics. *IET Generation, Transmission & Distribution* 14 (15): 3063–3069.

Hassan, R., Das, B.K., and Hasan, M. (2022). Integrated off-grid hybrid renewable energy system optimization based on economic, environmental, and social indicators for sustainable development. *Energy* 250: 123823.

Hegazy, Y.G., Salama, M.M.A., and Chikhani, A.Y. (2003). Adequacy assessment of distributed generation systems using Monte Carlo simulation. *IEEE Transactions on Power Systems* 18 (1): 48–52.

Khezri, R. and Mahmoudi, A. (2020). Review on the state-of-the-art multi-objective optimisation of hybrid standalone/grid-connected energy systems. *IET Generation, Transmission & Distribution* 14 (20): 4285–4300.

Koshti, A.S., Verma, A., Arya, R. et al. (2023). Tolerable random interruption duration based reliability estimation of stand alone hybrid renewable energy system (SAHRES) by network reduction and sequential Monte Carlo simulation. *IET Generation, Transmission & Distribution* 1–12. (Early view) https://doi.org/10.1049/gtd2.12910.

Lewis, E.E. (1987). *Introduction to Reliability Engineering*. John Wiley & Sons.

Ranjbar, H., Saber, H., and Sharifzadeh, M. (2022). Bi-level planning of distributed energy resources into existing transmission grids: pathway to sustainable energy systems. *IET Generation, Transmission & Distribution* 16 (24): 4963–4979.

Roy, P., He, J.B., Zhao, T., and Singh, Y.V. (2022). Recent advances of wind-solar hybrid renewable energy systems for power generation: a review. *IEEE Open Journal of the Industrial Electronics Society* 3: 81–104.

Wang, Y., Guo, C., and Wu, Q.H. (2013). A Cross-entropy-based three-stage sequential importance sampling for composite power system short-term reliability evaluation. *IEEE Transactions on Power Systems* 28 (4): 4254–4263.

Xu, Y. and Singh, C. (2012). Adequacy and economy analysis of distribution systems integrated with electric energy storage and renewable energy resources. *IEEE Transactions on Power Systems* 27 (4): 2332–2341.

Yahaya, A., AlMuhaini, M., and Heydt, G.T. (2020). Optimal design of hybrid DG systems for microgrid reliability enhancement. *IET GTD* 14 (5): 816–823.

Section 2

Reliability-Based Systems Design

6

Physical Reliability Methods and Design for System Reliability

Smriti Singh[1], *Jyoti Maurya*[1], *Eram Taslima*[1], *Bharat B. Sagar*[2], *and R. K. Saket*[1]

[1] *Department of Electrical Engineering, Indian Institute of Technology (BHU), Varanasi, Uttar Pradesh, India*
[2] *Department of Computer Science and Engineering, School of Engineering, Harcourt Butler Technical University, Kanpur, Uttar Pradesh, India*

6.1 Introduction

The impetus behind conducting reliability and maintainability analysis stems from various factors, including the heightened intricacy and advancement of systems, a growing societal emphasis on product quality, the existence of rules and regulations governing product liability, and the necessity to meet performance criteria outlined in government contracts to attain the desired level of reliability (Ebeling 2019). Reliability and maintainability constitute a significant component of the engineering design process (Wang et al. 2004) and assume a crucial role in life cycle assessment, cost–benefit evaluation, studies of operational capability, allocation of resources for repair and facilities, calculation of inventory, need for spare part, replacement decision-making, as well as the formulation of proactive maintenance initiatives.

The prioritized goals of reliability engineering, as per O'Connor and Kleyner's (2012) study are as follows:

- Employing engineering expertise and specialized methodologies to avert or minimize the occurrence of failures.
- Recognizing and rectifying the root causes of failures that transpire despite prevention endeavors.
- Developing strategies to manage failures that arise when their underlying causes remain unaddressed.
- Utilizing techniques to predict the prospective reliability of novel designs and to scrutinize reliability-related data.

In order to achieve these goals, certain reliability methods are used (Saket 2011). Physical reliability methods refer to a set of techniques and approaches used to evaluate and ensure the reliability of physical systems, components, or structures. These methods are employed to assess the capacity of a system to carry out its intended operations without encountering errors and to quantify the probability or likelihood of failure under various operating conditions and stresses (Kothari et al. 2022). Physical reliability methods typically involve the analysis of modes of failure, the identification of potential failure mechanisms, and the establishment of strategies to mitigate or eliminate those failures. These methods can be applied to a wide range of systems, including mechanical, electrical, electronic, and structural systems. In Sections 6.2.1, 6.2.2, etc provides a detailed explanation of the Physical Reliability Methods and the Design for Reliability (DfR) in any system.

Reliability Analysis of Modern Power Systems, First Edition. Edited by R. K. Saket and P. Sanjeevikumar.

The chapter is structured in the following manner: Section 6.1 provides a concise Introduction, Section 6.2 discusses the Reliability methods, Section 6.3 delves into the Design analysis and process, and Section 6.4 serves as the chapter's conclusion.

6.2 Reliability Methods

6.2.1 Block Diagram Analysis

Any reliability design starts with the building up of a model. *Block Diagrams* and *Fault Trees* are used to provide graphical representation which helps in evaluation of the relationship between the various components of the system.

Logical connections between system components and the concept of system failure logic can be depicted using a reliability block diagram (RBD). If the connection is complex, then the construction of the RBD is difficult. After the construction of the block diagram, the process of analysis involves simplifying the overall intricate system into a more basic one, which can then be examined through arrangements of series and parallel components. This method is also known as block diagram decomposition. Some of the methods used to examine a complex system are described in Sections 6.2.1.1, 6.2.1.2, etc.

6.2.1.1 Cut Sets and Tie Sets

Cut set and tie set methods are used to analyze complex reliability block diagrams. A cut set is formed by drawing a line across the system's blocks to emphasize the minimal number of malfunctioning blocks that would lead to system failure. Tie sets (or path sets) are generated by drawing lines through blocks that, if all of them operated correctly, would enable the system to operate.

The employment of cut set and tie set techniques is fit for computer applications. Their utilization is well-suited for examining extensive systems featuring numerous configurations, like control systems in the aviation industry, electricity producing systems, or control and instrumentation systems for large-scale industrial installations.

Figures 6.1 and 6.2 show how cut sets and tie sets are produced. It's noticeable that there are cut sets, which are three in number, and tie sets which are two in number.

The limitations on the reliability of the system, as determined through cut sets and tie sets, are expressed as follows:

$$R_s > 1 - \sum_{j}^{N} \prod_{i}^{n_j} (1 - R_i) \tag{6.1}$$

$$R_s < \sum_{j}^{T} \prod_{i}^{n_j} R_i \tag{6.2}$$

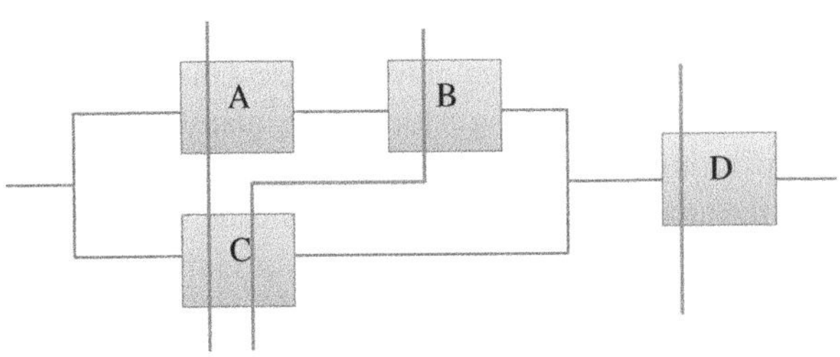

Figure 6.1 Cut sets.

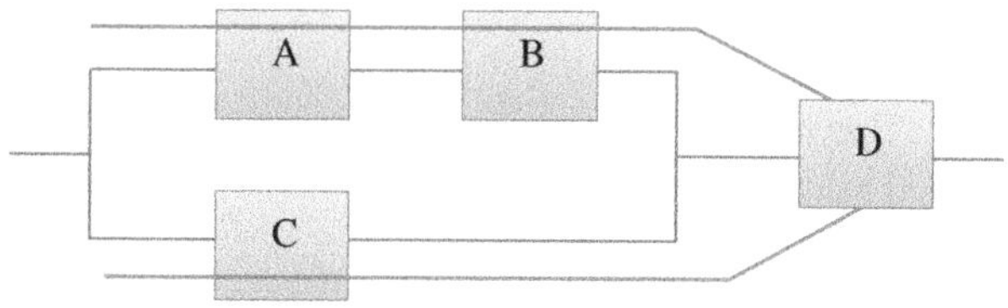

Figure 6.2 Tie sets.

where
N = cut set quantity
T = tie set quantity
n_j = quantity of blocks in the jth cut set or tie set

6.2.1.2 Common Mode Failures

A common mode failure is a shared source of failure that can lead to failed paths in redundant systems. It is important to identify and evaluate a common mode failure since it increases the likelihood of malfunction. There are many cases when more than one unit fails in an industry due to a single cause. The likelihood of failure in redundant systems, when considering only individual failure paths, is notably elevated. Some of the common example of the sources of common mode failure are:

- A sensor system which detects the common mode failure.
- An indicator system which notifies the path failure.
- A common power supply.
- An act of maintenance or operation that is frequently encountered on many different paths.
- Software that assists different paths.
- When one element fails, the remaining elements or redundant units in series are more likely to fail.

Hence, the occurrence of common cause failures holds significant importance in reliability analysis and necessitates inclusion for precise reliability forecasting. Various methods have been developed that consider the common mode failure for reliability evaluation. An approach of this kind operates under the premise that all common mode failures linked to the redundant system are consolidated into a single hypothetical unit, which is then integrated in series with the redundant system. Instances of such redundant systems encompass a parallel network, a k-out-of-n unit system, and a bridge configuration.

6.2.1.3 Enabling Events

Enabling events are one which will cause a high level failed event. Quite similar to common failure modes, it is difficult to predict an enabling event. Some of the common examples of enabling event are:

- Incorrect control settings.
- Disabled warning systems, due to false warnings.
- Following wrong procedures by operating and maintenance personnel.
- Temporarily unavailable standby element.

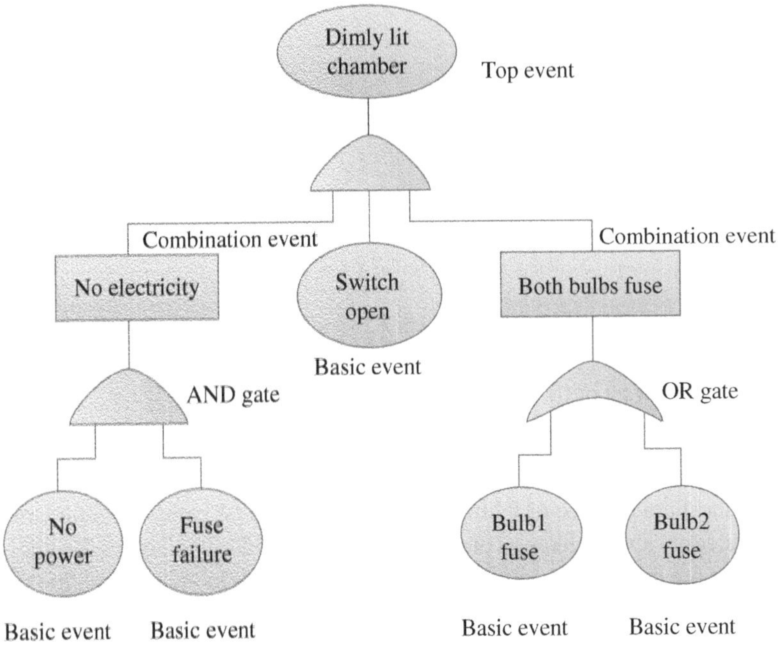

Figure 6.3 Fault tree.

6.2.2 Fault Tree Analysis

Fault tree analysis (FTA) is a systematic and graphical method used in reliability engineering, risk assessment, and safety analysis to evaluate the possible system failure modes and spot out the underlying causes that could lead to those failures. It is particularly useful for understanding and mitigating complex systems' reliability issues and safety concerns.

The primary purpose of FTA is to identify the chain of events or conditions that could lead to an undesired event, often referred to as the "top event" or "undesired outcome." It furnishes a systematic method for examining the logical connections among diverse events and conditions that contribute to the unfolding of the primary event. Figure 6.3 shows a FTA diagram. It roots out the possible causes of failure through tree branching and focuses on prevention.

In 1962, H.A. Watson created FTA while working at Bell Laboratory, as part of a contract with the U.S. Air Force Ballistic System Division. This contract aimed to assess the Minuteman I International Ballistic Missile (ICBM) Launch Control System (Ericson and Ll 1999), (Vesely 2002).

Standard symbols are used to define the basic events corresponding to such logical connections and for construction of a fault tree. Figure 6.4, shows the symbols used in the construction of a fault tree. The AND and OR function gates are used to depict the symptoms of failure, down to the real root cause. Finally, from all levels of failure, the top event probability is quantified using FTA.

As shown in Figure 6.3, failure of a component is depicted using a circle, signifying a basic event. The failure can happen due to one or more failure events (failure of sub-components). Components can thus be arranged in series (OR Gate) or parallel (AND Gate) fashion. Components are linked in series if the breakdown of either component results in the collapse of the whole system. Conversely, if the failure of one component doesn't lead to the failure of the entire system, then the blocks are linked in a parallel arrangement. In cases where system blocks are arranged in a sequential

Fault tree symbol	Meaning	Fault tree symbol	Meaning
BASIC EVENT	A basic fault tree event which is independent of other event and does not require further development.	SWITCH	Is used to include or exclude a part of Fault Tree, which may or may not be required.
BASIC EVENT	A basic event which depends on lower events developed as separate fault tree.	AND GATE	Allows failure only when all inputs fail.
BASIC EVENT	A basic event which depends on lower event, but not developed downwards.	OR GATE	Allows failure in case of any one of the input fails.
COMBINATION EVENT	An event which is the result of combination of basic events through the input logic gates.	INHIBIT GATE	Defines a causal relationship between one fault and another. The output event is produced directly from the input event if the indicated condition is satisfied.
OUT IN TRANSFERRED EVENT	A line from the apex of the triangle indicates a transfer in; A line from the side denotes a transfer out.		

Figure 6.4 Fault tree symbol.

configuration, and R_i be the reliability of component C_i, $i = 1, 2, \ldots n$ then the system reliability is given by:

$$R_{\text{system}} = R_1 * R_2 * \cdots * R_n = \prod_{i=1}^{n} R_i \tag{6.3}$$

Similarly, for system components connected in parallel, system reliability is as follows:

$$R_{\text{system}} = 1 - \prod_{i=1}^{n} (1 - R_i) \tag{6.4}$$

Different FTA is constructed for each defined top event. It is difficult to construct an FTA for a large complex system. Drawing and evaluating them manually is an impractical task. In such cases, computer programs are used to generate and evaluate FTAs. In this way, an effective, economic and easy iterative analysis is provided by the assistance of computer programming. FTA is an essential part of reliability analysis as it considers single as well as multiple failure event.

6.2.2.1 Steps of FTA

Following are the basic steps to be followed for a FTA:

1. Level 1 is defined by the top failure event.
2. Level 2 defines the input symptoms which are the contributing cause for the top event.
3. Levels 3, 4 or more are defined by subsequently breaking down the failures with additional gate levels.
4. Finalize the diagram.

6.2.2.2 Applications of FTA

Given are the main applications of FTA:

- To breakdown failure into more branches.
- FTA is generally applied when a backward approach identifies the potential cause of failure.
- To define inter-relationship between the cause and failure.

Some of the recent applications of FTA for reliability evaluations are: Kang et al. (2019) where FTA finds application in the reliability assessment of floating offshore wind turbines. In Badida et al. (2019) risk assessment of oil and natural gas pipelines is done.

Fuzzy fault tree analysis (FFTA) (Kuzu et al. 2019; Singh et al. 2022) is an extension of traditional FTA that incorporates the concepts of fuzzy logic and uncertainty into the analysis process. Fuzzy logic enables the portrayal of uncertain or ambiguous data, a common occurrence in real-world scenarios where exact probabilities or values may be difficult to determine. FFTA is particularly useful when dealing with systems that involve subjective judgments, incomplete data, or qualitative information.

6.2.3 State Space Analysis (Markov Analysis)

Every framework or constituent has the potential to exist in one of two conditions: active or inactive. The active state denotes operational status, while the inactive state signifies a malfunction. Employing state space analysis, it becomes possible to determine the probabilities linked to these conditions. Among state space analysis techniques, Markov analysis is the most widely recognized method, applicable under the following presumption:

- The transition process from one state to another should exhibit homogeneity, with a consistent probability that remains unchanged.
- The future state of the system does not depend on the previous state except the immediate one that precedes the future state.

This technique bears the name of a Russian mathematician, Andrei Andreyevich Markov (1856–1922). Markov analysis is implemented for the analysis of complicated system such as power generation and communication system. It is generally used for reliability analysis of systems where block diagram analysis cannot be applied.

In recent times, this method has been used to model the volatile nature of renewable energy systems (Theristis and Papazoglou 2013; Singh et al. 2023), so that it is possible to execute the reliability analysis. In Peiravi et al. (2020), the authors have solved an optimization problem based on Markov model using redundancy strategy. Markov model is used for the reliability analysis of wind turbine in Li et al. (2019).

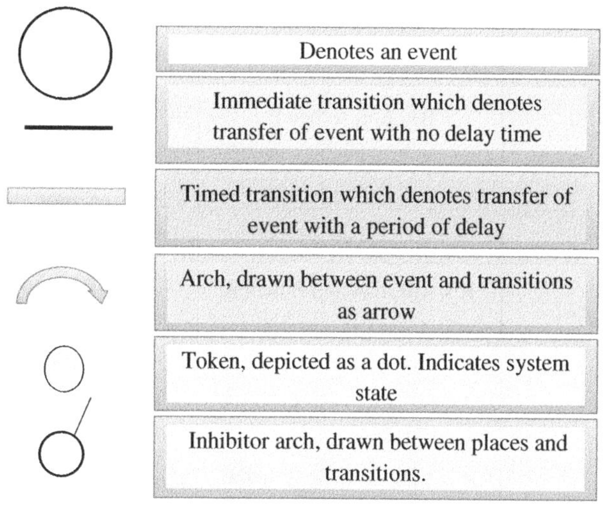

Figure 6.5 Petri nets basic symbols.

6.2.3.1 Advantages, Limitations, and Application of Markov Analysis

- The process of transitioning from one state to another must be homogeneous, and its probability must remain constant.
- The future state of the system does not depend on the previous state except the immediate one that precedes the future state.

6.2.4 Petri Nets

Petri nets is an extension of state space analysis technique. Carl Adam Petri introduced this technique in 1962. Petri nets describes the existing relation between condition and event related to failure event using graphical and mathematical analysis. In late 1970s, stochastic Petri nets (SPN) or time Petri nets came into the picture. SPN deals with the flaws of Markov chain analysis. It models the state of the component such that its state can be inferred from previous states, unlikely, in the Markov chain process. In this way, Petri nets are often used as a pre-processor in modeling, which is converted internally to Markov state space. The dependence of the Markov method on constant rate of occurrence is a major drawback of the Markov method. To overcome this disadvantage, the transition process is simulated using Monte Carlo method. In software design, original Petri nets are employed, whereas in reliability engineering, timed Petri nets are utilized.

Following are the fundamental symbols in Petri nets:

Figure 6.6 lists the basic structures depicting the logical relations for Petri nets. Petri nets is an efficient tool for system modeling, monitoring of failure, simulation and reliability analysis. Dynamic behavior can also be observed form Petri nets.

6.2.4.1 Fault Tree and Petri Net Transformation

A fault tree can be correlated with Petri nets, as shown in Figure 6.7. This is known as conversion from fault trees to Petri nets.

Figure 6.8 shows the Petri net conversion of a fault tree.

In recent literature, Petri nets find its application in reliability evaluation. The authors in Kabir and Papadopoulos (2019) present a review of Bayesian Network and Petri nets in reliability assessment, safety and risk analysis. Stochastic PetriNets are used to model the critical systems for safety

Petri nets	Logic relation	Boolean function
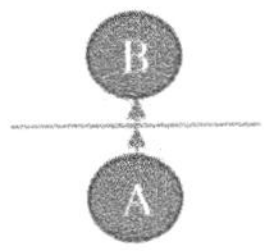	TRANSFER	B = A, If A then B
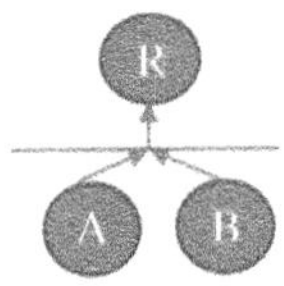	AND	R = A.B, If A and B then R
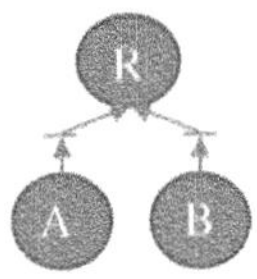	OR	R = A+B, If A OR B then R
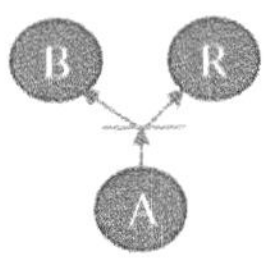	TRANSFER AND	B = R = A, If A then B and R
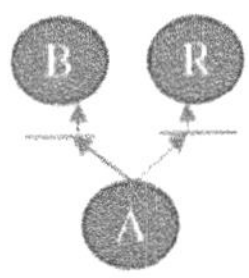	TRANSFER OR	B+R = A, If A then B OR R
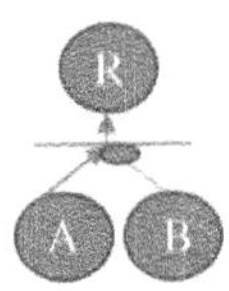	INHIBITION	R = A.B', If A AND B' then R

Figure 6.6 Logical relations with Petri Nets.

of Nuclear Power Plants to check for reliability qualification in Jyotish et al. (2023). In Kumar et al. (2022), the authors have employed a Petri Nets-based method to forecast the reliability of software in industrial contexts. In Kumar et al. (2020), performance analysis of milk processing systems is done using PetriNets.

6.2.5 Failure Mode Effects and Criticality Analysis (FMECA)

Failure modes effects and criticality analysis (FMECA) also known as Failure modes and effects analysis (FMEA) is a widely used reliability method in system design. FMECA examines the

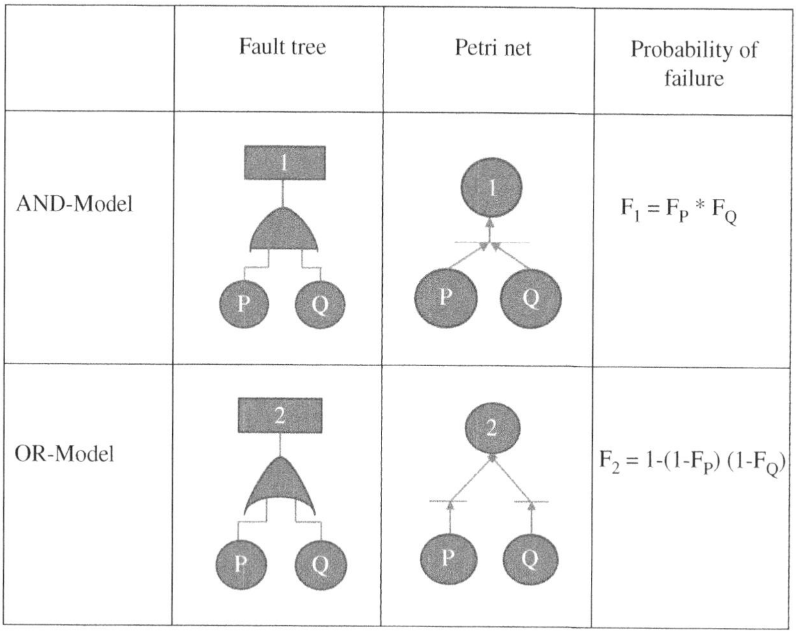

Figure 6.7 Correlation between fault tree and Petri Net.

Figure 6.8 Petri net transformation of Fault Tree of Figure 6.3.

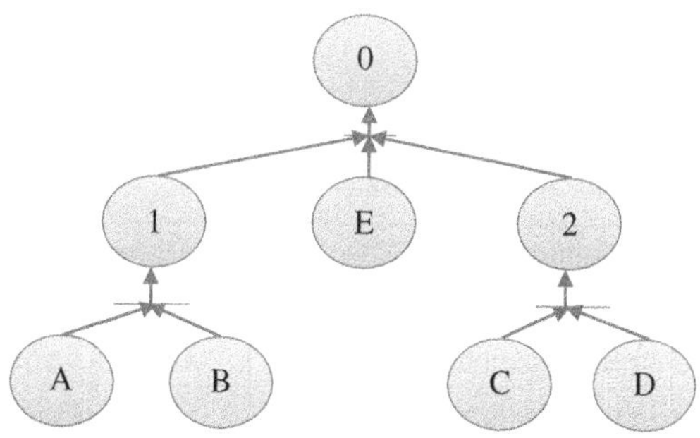

potential modes of failure within the system, classifies them by severity and also determines the effect of these failures on the operation of the system as a whole.

FMEA was initially introduced for military purposes in the late 1940s by the US Armed Forces, as noted by Dhillon (2004). Subsequently, its application extended to aerospace and rocket development. Table 6.1 lists some of the prospective failure modes experienced by the industries where the product is manufactured. The motive of FMEA is to reduce failures, beginning with the highest priority failure. Few of the common causes that require special attention are depicted in Table 6.2.

Failure Modes are ways through which an error or defect can occur in the design or item that affects the consumer. *Effect Analysis* refers to the assessment of the consequences of these failures. Table 6.3 shows the possible effects of failure.

Prior to commencing the FMEA process, a worksheet is prepared that encompasses crucial system specifications. This worksheet systematically outlines all components and their respective functions. Figure 6.9 shows a FMEA Worksheet.

Variety of industries which find application of FMEA include semiconductor processing, software, healthcare, and food services. FMEA combined with Advanced Product Quality Planning

Table 6.1 Failure modes.

Corrosion	Broken	Burrs	Omitted	Wrapped
Bent	Crack	Melted	Sharp Edges	Not fulfilled
Discolor	Porous	Shortness	Discoloring	Leaking
Defective	Delay	Bound	Eccentricity	Open Circuited
Out of Tolerance	Damage	Tightness	Misassemble	Brittle
Misalignment	Blistered	Roughness	Grounding	Deformity

Table 6.2 Causes of failure.

Inadequate venting	Damaged part	Missed operation
Handling damage	Incorrect tooling	Misalignment
Improper surface preparation	Improper torque	Improper tool setup
Out of tolerance	Inadequate gating	Clamping
Inadequate control system	Inadequate control system	Tool damage/blurs
Inadequate holding	Material failure	Overload capacity
Assembly error	Incorrect speed/feeds	Heat treatment shrinkage

Table 6.3 Effects of failure.

Brake failure	Air/water leakage	Excess interest
Oil carryover	Wrong carryover	Wrong material
Poor coordination	Delay in delivery/payment	Unable to assemble
Poor valve performance	Customer dissatisfaction	Oil carry over
Inadequate service life	Abrupt use of material	Contamination-valve system

S. No.	Process Step/ Input	Potential Modes of Failure	Potential Effects of Failure	SEVERITY (1-10)	Potential Causes	OCCURRENCE (1-10)	Current Controls	Detection (1-10)	RPN	Action Recommended
1										
2										
3										
4										
5										
6										

Figure 6.9 A sample of FMEA worksheet.

(APQP) serves as a risk mitigation tool. To assess the risk, risk priority number (RPN) is used. Following step are followed in determining RPN:

1. Rating of the effect of failure in the order of severity.
2. Determining the likelihood of failure occurrence.
3. Likelihood of pre-detection of failure.

$$\text{RPN} = \text{Severity} \times \text{Occurrence} \times \text{Detection} \tag{6.5}$$

1. **Severity:** Severity determines all the failure modes. A malfunctioning mode within a single component initiates a malfunctioning mode within another component. Failure effect results from the failure mode. Every effect is allotted a severity number (S) which varies from 1 to 10. 1 represents no danger and 10 represents critical. With the help of these numbers the engineers rank the failure modes and their impacts. If severity number is 9 or 10, the failure modes need to be eliminated and the design has to be changed. Such a high severity number generally results in an effect that is injurious to the consumer and may cause litigation.
2. **Occurrence:** This step takes into consideration the frequency with which a particular cause is resulting in failure. It is essential to recognize and record all major factors contributing to failure modes using technical terminology. The occurrence rank is given to the failure mode from 1 to 10. If the occurrence is high (greater than 4 or greater than 1 and the severity number is 9 or 10), action must be taken.
3. **Detection:** Detection refers to the quantification of the probability that a malfunction can be detected. This is done using testing and monitoring. Once the detection number (D) is known, the ability to remove or detect the failure on time can be prioritized. D assesses the possibility that the failure might go unnoticed. The higher the detection number, the higher the chance that the detection of failure is escaped or the chance of detection is low.

RPN rating ranges from 1 to 10. Higher the risk or severity, higher will be the RPN rating. RPN can thus be used to sort out the problems in the priority order; so that corrective actions can be made.

The Automotive Industry Action Group (AIAG) mandates the incorporation of FMEA within the automotive APQP process. Additionally, AIAG provides a comprehensive manual outlining the application of this method. Furthermore, the American Society for Quality (ASQ) offers support for FMEA practices.

Criticality rating of component can also be given by the failure mode criticality number:

$$C_m = \beta \alpha \lambda_p t \tag{6.6}$$

Where

β = the likelihood of function or mission loss given specific conditions.

α = failure mode ratio (for an item, $\alpha = 1$).

λ_p = item failure or hazard rate.

t = operating or at-risk time of item.

6.2.5.1 Steps in Performing an FMECA

The first and foremost step is to collect all the information regarding the design. These include Computer-Aided Engineering (CAE) data, drawings, full specifications, stress analysis, test results, etc. Reliability prediction information must be available for criticality analysis, if not available it needs to be generated. For complete analysis functional block diagram and RBD needs to be prepared. The FMECA process should commence once initial design details become accessible.

The authors in Mzougui and El Felsoufi (2019) propose a modified FMEA to improve the reliability of any item. FMEA is applied to the semiconductor industry in Cabanes et al. (2021). Healthcare risk analysis is done using FMEA in Liu and Liu (2019). The authors in Balaraju et al. (2019) use a Fuzzy-FMEA-based risk assessment method of a Load-Haul-Dumper machine.

6.2.5.2 Advantages of FMEA

The benefits of employing FMEA encompass:

- Enhancements are made to the product's quality, safety, and reliability.
- The image of the organization manufacturing the product is improved.
- User is more satisfied.
- Time and cost for system development is minimized.
- Future failures can be reduced.
- Identifies and eliminates the failure modes early.
- Minimizes the possibility of same type of failure again and again.

6.2.6 Accelerated Life Testing (ALT)

Accelerated Life Testing (ALT) is a testing methodology used to assess the reliability, durability, and lifespan of a product or component in a shorter time frame than its actual expected life under normal operating conditions. This testing approach involves subjecting the product or component to elevated levels of stress, such as increased temperature, humidity, voltage, mechanical load, or other environmental factors, to accelerate the aging and wear-out processes. The purpose of ALT is to predict and identify potential failures and weaknesses in the product more quickly, allowing manufacturers to make design improvements and ensure that the product meets reliability and safety standards before it reaches the market.

In Woo et al. (2020), the authors outline the reliability-focused design of a household compressor that undergoes recurrent internal stress. The book Balakrishnan (2023) brings forward the recent advances in reliability and life testing. In Diao et al. (2019), the authors have described the utilization of ALT and modeling capacity degradation for a $LiCoO_2$-graphite cell.

6.2.7 Reliability Apportionment

Individual reliability of subsystems can be evaluated against overall system reliability using Reliability Apportionment which is sometimes necessary. It holds a significant role during the designing process of a system, especially when a desired system reliability is expected or when the system is large enough. The subsystem specifications have to be decided based on the requirement of individual reliability of the sub-subsystem. This activity is known as a reliability "budgeting."

Reliability Apportionment starts with the drawing of the RBD for the system. Each block is assigned the requisite reliability, taking into account, the complex nature, different operating conditions, and risk associated with each block.

Forcina et al. (2020) present a comprehensive overview of the reliability allocation techniques employed in recent times through a systematic literature review. Maheve and Peter (2021) put light on its various applicability and limitations. In Choudhary and Monisha (2020), reliability estimation of power supply modules of aerospace is done using ARINC Reliability Apportionment Technique. The authors in Shen et al. (2022) provide a cost minimization model for reliability allocation of mechanical system.

6.3 Design Analysis and Process

Reliability is an essential part of today's marketplace. As a result, there is a need to establish a team that supports the entire process of design and development. The three basic activities that the reliability team will undertake are:

1. Design for Reliability
2. Reliability Verification
3. Analytical Physics

With these activities, a robust reliability program can be built up.

DfR takes precedence as the initial and most crucial step in product development. It makes use of knowledge about the physics of failure (PoF) to design the prospective problems. The next activity is Reliability Verification. In this, the objectives for customer reliability are ensured through verification studies and demonstration tests. Analytical Physics, which is the third activity, collects information about the PoF of the product. In this way, comprehending how a product will fail and the reason behind it is essential to the design of a product that meets the customer's expectations.

6.3.1 Design for Reliability

DfR is a technique that begins at an early stage of the design process and is incorporated into every subsequent stage. It sets reliability goals for the design team. It makes use of several tools and practices to ensure reliability of product. The primary stages of the DfR Program comprise the subsequent measures, as shown in Figure 6.10 (sequential phases represented in block format):

- Identify,
- Design,
- Analyze,
- Verify,
- Validate, and
- Monitor and Control.

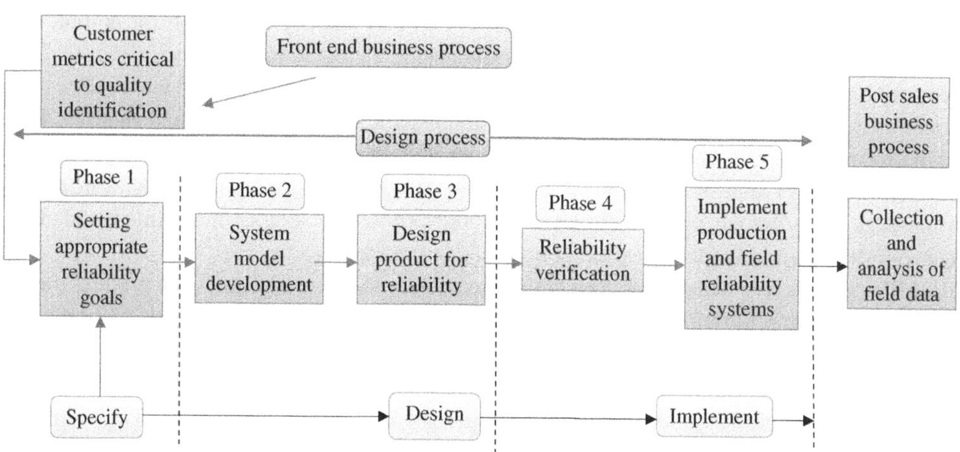

Figure 6.10 Design for reliability process.

Phase 1, known as the identification phase, involves establishing reliability goals. Phases 2 and 3 encompass the design and analysis stages, focusing on modeling the system with reliability considerations. Phase 4 involves verifying and quantifying the design. Phase 5, validation, entails conducting functional and environmental tests on the system.

The reliability goals are defined quantitatively in the identify phase. Tools that help in the definition are: *Goal Setting, Bench-marking, Gap Analysis, Reliability Program Plan Writing, and Quality Function Deployment (QFD).*

In the design phase stage, specific design activities are started. These activities include circuit layout, mechanical drawing, selection of components/suppliers etc. Some of the important tools used in this phase are: *FMEA, Tolerance evaluation, Reliability prediction, Evaluating Tolerance and trade off cost.*

The third phase in the DfR is the analysis phase. In this phase, a rough estimate of the reliability is done. Tools used in this phase are *RBD, Finite Element Analysis (FEA), Physics of failure (PoF), Reliability Prediction, FMEA, Stress-Strength Analysis, Design Review by Failure Mode (DRBFM), and Warranty Analysis of the existing products.*

The next phase is the Verify phase. In this, the hardware prototype is built. All the test results are quantified. The prototype is ready for testing and thorough analysis. Tools used in the Verify phase include: *Accelerated Life Testing (ALT), Highly Accelerated Life Testing (HALT), Test to Failure (Life Data Analysis), Degradation Analysis, Reliability Growth Process, and Design Review Based on Test Results (DRBTR).*

The functional and environmental testing of the system is done in the Validation phase. It may happen that there is modification in the design in order to make the system more robust. Tools used in the process include: *Design Validation (Including ALT and Reliability Demonstration) and Process Validation.*

To keep the variation within limits, the Control and Monitor phase is used. Tools implemented for this phase are: Control Chart and process capability Studies, *Human Reliability, Continuous Compliance, Ongoing Reliability Testing (ORT), Audits, and Field Return Analysis.*

6.3.1.1 Design for Reliability Tools

DfR tools are used in the development of the product, such that the design is robust and matches the customer's requirement.

1. **Failure Mode and Effect Analysis**: Failure mode and effect analysis (FMEA) is a major tool to integrate reliability in the design of products. FMEA can identify the reason for failure, the gravity of the failure, and the likelihood of malfunction.
2. **Competitive Benchmarking**: Competitive benchmarking measures how and where the product stands against its competitors. This is done by using a set of predetermined metrics. The design process with such benchmarking makes sure that all the significant design aspects have been included. It is always important to assess the competition. This guarantees a product's cutting-edge performance, optimal choice of materials, robust packaging, mechanical and electrical soundness, and the identification of potential cost concerns.
3. **Reliability Predictive Modeling**: Reliability predictive modeling estimates the product mean time between failure (MTBF) and mean time to failure (MTTF). These indices hold a significant role in comprehending the viability of the design's capability to achieve the reliability objectives required to satisfy the needs of consumers. Also, the evaluation of these metrics directs and assists in the trade-off design of the product so that the best design approach can be adopted. Although, reliability predictive modeling provides an excellent comparative result; it is known to have limited accuracy.
4. **Physics of Failure**: It is via the concept of PoF that we gain insight into the material's physical attributes, processes, and technologies employed in the design, and the interplay of these

attributes with the potentially detrimental conditions encountered by the design throughout the product's lifecycle.

5. **Design Studies**: Design studies find out the capabilities of platforms and products to meet life's requirements. The data provided by design studies is used to calculate the precise MTBF or MTTF. The design maturity testing platform (with no failure) is designed to meet the customer's need.

6.3.1.2 Design for Reliability Metrics

1. **Risk Assessment**: Risk assessment is performed to manage and keep the level of DfR activity at an appropriate level. Some inquiries that should be posed during the initial stages of program development include:
   - Will this product use any new technology with an unproven track record of reliability?
   - Will the new design be different from the old one (approximately more than 30% new content)?
   - Will there be a different environmental exposure to the product?
   - Are there any new requirements for this product?
   - Are there new applications for this product?
   - Is there any new material used in the design process?
   - Will there be any new suppliers for the products?
   - Will the product be manufactured at a different location?
   - Are there any other changes in the product design compared to the old one that affect reliability? If the answer is yes, then the risk is high and more DfR tools will be required to design the product development.
2. **Assessing Program using IEEE 1624 Scoring System**: Reliability assessment is a detailed analysis of the organization techniques and practices used in the development of a reliable product. It results in a reliability program plan (RPP) and captures the present situation of the organization. The outcomes of the reliability assessment are mapped onto a reliability maturity matrix. IEEE 1624 scoring system defines the key reliability practices through assessment methodology. Some of these practices are:
   - Reliability requirements and strategic planning
   - Training and development
   - Reliability assessment
   - Testing for reliability
   - Supply chain management
   - Monitoring and examining field information
   - Confirmation and authentication
   - Enhancements in reliability
3. **Assessing Program using AIAG Scoring System**: AIAG stands for AIAG which also contains assessment methodologies enumerated below:
   - Planning for reliability
   - Reliability-centric design
   - Reliability modeling and prediction
   - Reliability of mechanical components and systems
   - Concepts of statistics
   - Analyzing and reporting the failures
   - Analysis of reliability Data
   - Reliability testing
   - Manufacturing reliability

6.3.2 Reliability Verification

The third building block activity in the design process is reliability verification. It involves ensuring that a product or system functions as intended and remains operational over its expected lifespan, even in the presence of various stresses, environmental conditions, and potential failures (Wasserman 2002). A structured method for conducting reliability testing begins by establishing a reliability objective, followed by executing tests that are directly tied to performance in order to assess the product's reliability. In contemporary industries, it is imperative that a reliability verification test explicitly establishes their connection to the overall reliability performance of the product, as well as their influence on warranty expenses and customer contentment.

Reliability verification studies, also known as reliability testing, can be done effectively only when a proper environmental tool is used to simulate the design weakness. The equipment used for testing facilities includes thermal shock chambers, accelerated life test chambers, vibration systems, and humidity chambers. Data collection methods play a significant role in grasping the intricacies of the design during the testing phase of reliability verification. The authors in Wen et al. (2023) introduce an accelerated reliability verification approach for testing the reliability of the motor drive mechanism within the precision corn seed-metering device.

6.3.3 Analytical Physics

Physical analysis, or failure analysis, also called analytical diagnostics, plays a major role in reliability engineering. It helps in understanding and determining the reason for failure and finding corrective action for it. Repeated failures can be minimized by identifying the underlying reasons for failure. Similarly, finding out the PoF gives a better understanding of the product and its limitations. Customer satisfaction increases as problem resolution improves and abnormalities are eliminated. Some of the major factors for finding the reason behind failure are: *corrosion, contamination, effects of stress strain, diffusion, etc.*

Failure in a system may also be due to human errors. This cannot be ignored as an important contributing factor. In Saket et al. (2011), the authors introduce a novel methodology that employs a probabilistic approach to conduct biorhythmic analysis, aiming to mitigate aviation accidents. The methodology is crafted using Gaussian distribution techniques to assess the reliability of the aviation system while accounting for the influence of biorhythmic factors on the pilot. The theory of biorhythms predominantly employs scientific techniques to map the patterns or cycles that impact the internal processes of the body and human conduct. This theory is especially relevant to physical, emotional, and intellectual capacities. According to the biorhythm theory, three statistical cycles commence at birth and persist consistently throughout one's lifetime.

6.3.4 Reliability Life Cycle

Figure 6.11 shows the phases of reliability life cycle. It comprises of several phases, each of which plays a critical role in ensuring the reliability and longevity of a system or product.

These phases can be summarized as follows:

- **Specification for Reliability:** Specification for reliability: reliability specifications are crucial in ensuring that products and systems meet the desired levels of performance and reliability.

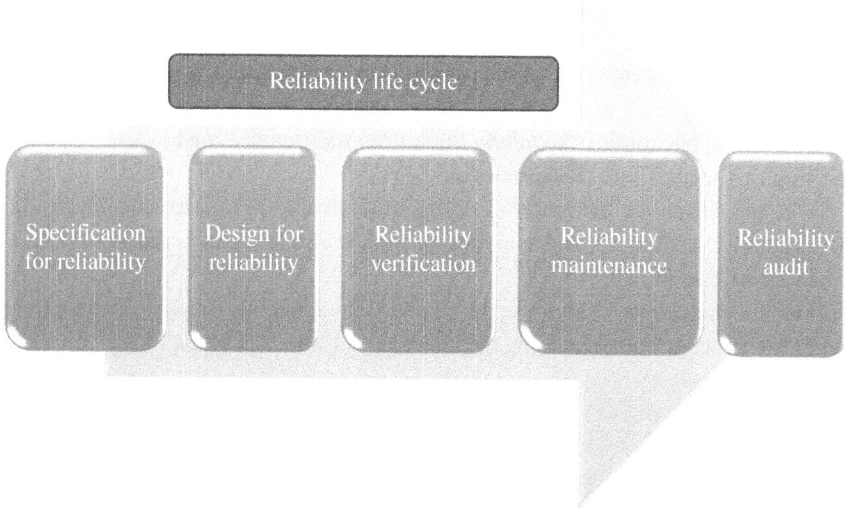

Figure 6.11 Phases of reliability life cycle.

They provide a quantitative and qualitative framework for designing, testing, and maintaining reliable solutions.

- **Design for Reliability:** In this the product is designed incorporating reliability specifications into the design process. The risk of costly failures is reduced, customer satisfaction is improved.
- **Reliability Verification:** This confirms that a product or system meets its specified reliability requirements and standards.
- **Reliability Maintenance:** Reliability maintenance is a proactive approach to manage and preserve the reliability of system. Its primary goal is to prevent failures, reduce downtime, extend the life of asset, and optimize the performance.
- **Reliability Audit:** Reliability audit evaluates whether the organization is effectively managing reliability to meet its objective and compliance requirements.

The reliability life cycle consists of a series of interconnected process steps aimed at ensuring that a product or system maintains its reliability and performance throughout its operational lifespan. These steps help organizations proactively manage reliability considerations.

Figure 6.12 shows the various process steps followed by a product in its entire life cycle along with the tools and methods corresponding to these steps.

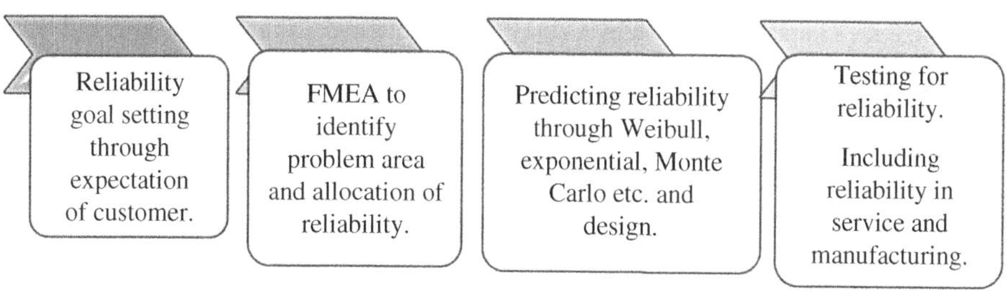

Figure 6.12 Process Steps with related tools and methods in Reliability Life Cycle.

6.4 Conclusions

The chapter offers an extensive overview of physical reliability methods, which are used to evaluate system reliability. It delves into each method's description, outlines their respective procedures, and highlights their recent applications in scholarly literature. Additionally, it addresses both the benefits and constraints associated with these methods.

Furthermore, the chapter explores the realm of design analysis and processes, encompassing topics such as DfR, reliability verification, and analytical physics. It also provides a concise overview of the tools and metrics utilized in DfR.

Ultimately, the chapter concludes by examining the phases of the reliability life cycle, delineating the steps involved in this process, and presenting the relevant tools for each phase.

Bibliography

Badida, P., Balasubramaniam, Y., and Jayaprakash, J. (2019). Risk evaluation of oil and natural gas pipelines due to natural hazards using fuzzy fault tree analysis. *Journal of Natural Gas Science and Engineering* 66: 284–292.

Balakrishnan, N. (2023). *Recent Advances in Life-Testing and Reliability.* CRC Press.

Balaraju, J., Raj, M.G., and Murthy, C.S. (2019). Fuzzy-FMEA risk evaluation approach for LHD machine–a case study. *Journal of Sustainable Mining* 18 (4): 257–268.

Cabanes, B., Hubac, S., Le Masson, P., and Weil, B. (2021). Improving reliability engineering in product development based on design theory: the case of FMEA in the semiconductor industry. *Research in Engineering Design* 32: 309–329.

Choudhary, K. and Monisha, S. (2020). Aerospace power supply modules reliability estimation using ARINC reliability apportionment technique. In: *Reliability, Safety and Hazard Assessment for Risk-Based Technologies: Proceedings of ICRESH 2019*, 825–832. Springer.

Dhillon, B.S. (2004). *Reliability, Quality, and Safety for Engineers.* CRC Press.

Diao, W., Saxena, S., and Pecht, M. (2019). Accelerated cycle life testing and capacity degradation modeling of LiCoO2-graphite cells. *Journal of Power Sources* 435: 226830.

Ebeling, C.E. (2019). *An Introduction to Reliability and Maintainability Engineering.* Waveland Press.

Ericson, C.A. and Ll, C. (1999). Fault tree analysis. *System Safety Conference*, Volume 1, Orlando, Florida, USA, 1–9.

Forcina, A., Silvestri, L., Di Bona, G., and Silvestri, A. (2020). Reliability allocation methods: a systematic literature review. *Quality and Reliability Engineering International* 36 (6): 2085–2107.

Jyotish, N.K., Singh, L.K., Kumar, C., and Singh, P. (2023). Batch deterministic and stochastic Petri nets modeling for reliability quantification for safety critical systems of nuclear power plants. *Nuclear Engineering and Design* 404: 112191.

Kabir, S. and Papadopoulos, Y. (2019). Applications of Bayesian networks and Petri nets in safety, reliability, and risk assessments: a review. *Safety Science* 115: 154–175.

Kang, J., Sun, L., and Soares, C.G. (2019). Fault tree analysis of floating offshore wind turbines. *Renewable Energy* 133: 1455–1467.

Kothari, D.P., Nagrath, I.J., and Saket, R.K. (2022). *Modern Power System Analysis.* New Delhi: Tata McGraw-Hill Publishing Company.

Kumar, N., Tewari, P.C., and Sachdeva, A. (2020). Performance assessment of milk processing system using Petri nets. *SN Applied Sciences* 2: 1–12.

Kumar, K., Sumit, Kumar, S. et al. (2022). Predicting reliability of software in industrial systems using a Petri net based approach: a case study on a safety system used in nuclear power plant. *Information and Software Technology* 146: 106895.

Kuzu, A.C., Akyuz, E., and Arslan, O. (2019). Application of fuzzy fault tree analysis (FFTA) to maritime industry: a risk analysing of ship mooring operation. *Ocean Engineering* 179: 128–134.

Li, J., Zhang, X., Zhou, X., and Lu, L. (2019). Reliability assessment of wind turbine bearing based on the degradation-Hidden-Markov model. *Renewable Energy* 132: 1076–1087.

Liu, H.-C. and Liu, H.-C. (2019). FMEA for proactive healthcare risk analysis: a systematic literature review. In: *Improved FMEA Methods for Proactive Healthcare Risk Analysis*, 15–45. Singapore: Springer.

Maheve, M. and Peter, P. (2021). Applicability and limitations of reliability allocation methods. In: *2021 Annual Reliability and Maintainability Symposium (RAMS)*, 1–4. IEEE.

Mzougui, I. and El Felsoufi, Z. (2019). Proposition of a modified FMEA to improve reliability of product. *Procedia CIRP* 84: 1003–1009.

O'Connor, P. and Kleyner, A. (2012). *Practical Reliability Engineering*. John Wiley & Sons.

Peiravi, A., Ardakan, M.A., and Zio, E. (2020). A new Markov-based model for reliability optimization problems with mixed redundancy strategy. *Reliability Engineering & System Safety* 201: 106987.

Saket, R.K. (2011). Reliability evaluation of defence support systems. In: *Innovations in Defence Support Systems-2: Socio-Technical Systems*, 241–286. Springer.

Saket, R.K., Kaushik, W.C.S.P., and Singh, C.G. (2011). Biorhythmic analysis to prevent aviation accidents. In: *Innovations in Defence Support Systems-2: Socio-Technical Systems*, vol. 338 (ed. L.C. Jain, E.V. Aidman, and C. Abeynayake), 207–240. Berlin, Heidelberg: Springer-Verlag.

Shen, L., Zhang, Y., Zhao, Q. et al. (2022). A reliability allocation methodology for mechanical systems with motion mechanisms. *IEEE Systems Journal* 16 (4): 5596–5607.

Singh, K., Kaushik, M., and Kumar, M. (2022). Integrating α-cut interval based fuzzy fault tree analysis with bayesian network for criticality analysis of submarine pipeline leakage: a novel approach. *Process Safety and Environmental Protection* 166: 189–201.

Singh, S., Saket, R.K., and Khan, B. (2023). A comprehensive review of reliability assessment methodologies for grid-connected photovoltaic systems. *IET Renewable Power Generation* 17 (7): 1859–1880.

Theristis, M. and Papazoglou, I.A. (2013). Markovian reliability analysis of standalone photovoltaic systems incorporating repairs. *IEEE Journal of Photovoltaics* 4 (1): 414–422.

Vesely, B. (2002). Fault Tree Analysis (FTA): Concepts and Applications. *NASA HQ*.

Wang, W., Loman, J., and Vassiliou, P. (2004). Reliability importance of components in a complex system. In: *Annual Symposium Reliability and Maintainability, 2004-RAMS*, 6–11. IEEE.

Wasserman, G. (2002). *Reliability Verification, Testing, and Analysis in Engineering Design*, vol. 153. CRC Press.

Wen, C., Zhang, J., Zheng, K. et al. (2023). Accelerated verification method for the reliability of the motor drive mechanism of the corn precision seed-metering device. *Computers and Electronics in Agriculture* 212: 108163.

Woo, S.-w., Pecht, M., and O'Neal, D.L. (2020). Reliability design and case study of the domestic compressor subjected to repetitive internal stresses. *Reliability Engineering & System Safety* 193: 106604.

7

Design for Maintainability and Availability Analysis for System Design

Jyoti Maurya[1], Om P. Bharti[2], K. S. Anand Kumar[3], and R. K. Saket[1]

[1] *Department of Electrical Engineering, Indian Institute of Technology (BHU), Varanasi, Uttar Pradesh, India*
[2] *Department of Electrical Engineering, Government Polytechnic, Ghazipur, Uttar Pradesh, India*
[3] *Electrical Division, CSIR National Aerospace Laboratory, Bangalore, Karnataka, India*

7.1 Introduction

It is clearly essential that systems under maintenance are structured to facilitate the easy completion of tasks, ensuring that the expertise required for diagnosing issues, conducting repairs, and performing routine maintenance is not excessively intricate. This consideration should encompass the skill set and training of anticipated maintenance staff and users (Kothari et al. 2022).

The overarching objective for a completed system (be it a car, aircraft, avionics device, or computer) should encompass its capability to function and fulfill its intended function for the entire projected duration of its planned design lifespan (Diatte et al. 2022). Consequently, while developing a system, we must avoid the mindset of just delivering the system and disengaging. Instead, a supplier's system should offer continuous support throughout the entire operational lifespan of the product. This involves several concepts illustrated in Figure 7.1. Achieving supportability necessitates a skillful integration of reliability, maintainability, logistics, operations engineering, and safety engineering. This integration ensures that the system remains accessible and functional for its intended purpose throughout its designated operational lifespan.

Maintainability plays a crucial role in ensuring the system's successful maintenance and continuous support, as well as its adaptability for modifications and improvements throughout its operational lifespan (Sarita et al. 2023). Through system analysis and considerations of maintainability, the alignment of project development with a broader systems perspective becomes evident. The focus of maintainability procedures and documentation lies in guiding the process of upkeep and highlighting the typical documentation generated for extensive software programs. Through maintainability analysis, the process of making decisions among various alternative components while considering trade-offs becomes apparent. When equipment or machinery malfunctions, an eventuality given that no design can be entirely infallible, it becomes crucial to promptly address the issue for its speedy restoration, allowing for its continued utilization. Indeed, users are often primarily interested in its availability, denoting the portion of time the equipment remains functional as required, akin to household television sets.

Reliability Analysis of Modern Power Systems, First Edition. Edited by R. K. Saket and P. Sanjeevikumar.

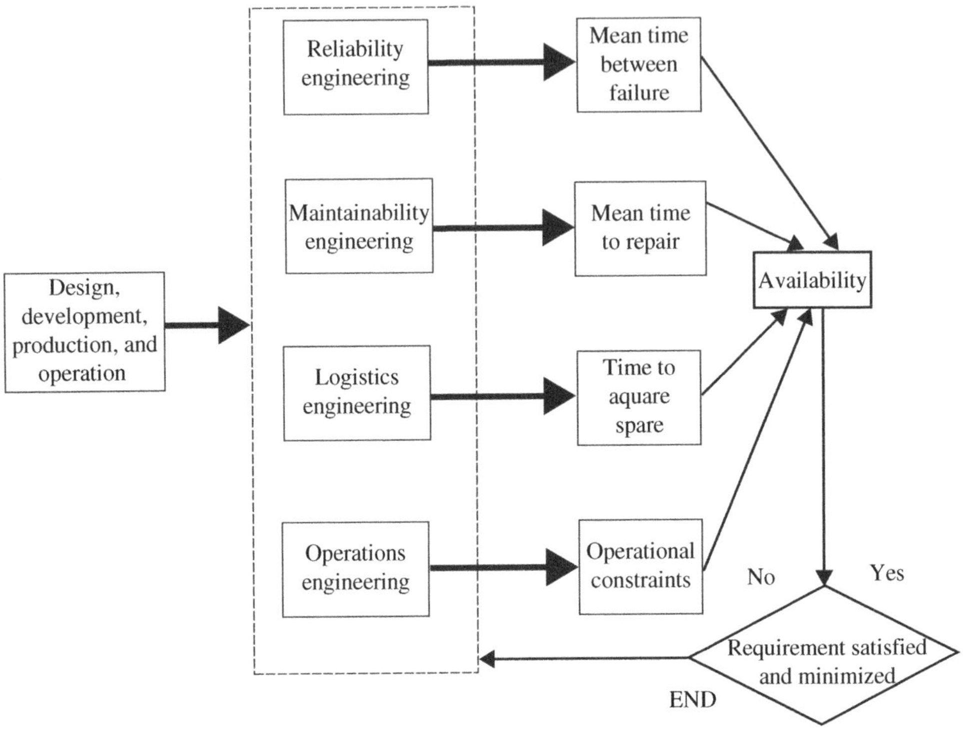

Figure 7.1 Requirement of system supportability.

Paying customers, such as those who cover their fares, constitute the primary income stream for an airline. Consequently, the majority of commercial aircraft need to operate for extensive hours daily to remain financially viable (Sayed et al. 2020). Irrespective of the intricacy involved, all equipment must possess favorable maintainability, which gauges the speed at which any dip in performance is recognized, the issue is pinpointed, repairs are executed, and a confirmation is conducted to reinstate the equipment's regular functionality. Integrating maintainability into the initial design is crucial, as with other reliability factors. Trying to enhance it later by making adjustments to already manufactured equipment won't yield a satisfactory outcome.

Enhancements that boost maintainability can often be seamlessly integrated into the initial design without incurring notable extra expenses. Nevertheless, the designer needs to exercise caution in enhancing maintainability, as doing so might inadvertently compromise reliability. For instance, replacing soldered connections with plugs and sockets to expedite the removal of assemblies could potentially result in decreased dependability at times (O'Connor 2010).

After an introduction, this chapter is structured in the following manner: Section 7.2 includes elements of maintainability, Section 7.3 delves availability of the systems, and Section 7.4 serves as the conclusion of the chapter.

7.2 Elements of Maintainability

We need to factor in the necessary tasks for sustaining the system right from the initial stages of our design. The system should either remain operational throughout the mission without any failures, or the failed components must be substituted. If our system lacks impeccable dependability and is not immune to wear and tear, we need to address the subsequent inquiries:

- Which components are susceptible to malfunction, and how will their faults be identified? For instance, in the case of a cathode ray tube (CRT) screen that doesn't display an image, is the issue due to screen failure, power supply failure, or the computer ceasing to transmit screen data?
- Is it possible to easily diagnose a range of issues? How fast can the problem be pinpointed? Is data from sporadic malfunctions retrievable for later analysis? In cases where a problem can't be isolated or the system lacks adequate diagnostic tools, the process of restoring the system might become a lengthy endeavor.
- When can we expect the system to be restored? Is the system designed with modular components for easy replacement? Are the components stacked with numerous attachment points? Can software also assist in identifying issues?
- Will a defective unit be discarded or undergo repairs? If repairs are needed, where should the process take place? Additionally, what specific tools and personnel are required to carry out the repair task?
- Would specialized parts be accessible for the purpose of repairing the device? Will specialized components like traveling wave tubes or low-noise amplifiers be produced continuously if they are necessary for unit repairs? Will the original manufacturer handle the unit repairs? Will the supplier sustain their operations if repair agreements are reached, considering potential logistical challenges?

All these inquiries need to be tackled during the product development phase. While a few of these matters intersect with logistics (which deals with providing and sustaining a system over its complete lifespan), they all require attention. The system's maintenance approach and the plans for ensuring easy maintenance should both be contemplated in the initial stages of the product's design. To achieve this, we should start by examining certain definitions. Which components are susceptible to malfunction, and how will their breakdowns be identified? As an illustration, in the case of a CRT monitor failing to display an image, potential culprits could include a defective screen, a malfunctioning power supply, or a cessation of screen data transmission from the computer.

7.2.1 Maintainability and System Engineering

Figure 7.2 illustrates a global representation of a continuous research endeavor with a long-term scope, like the space program, and its integration of maintainability. The original purpose of the horizon mission methodology (HMM) was to examine pioneering space technologies. HMMs are imaginary space missions characterized by criteria for performance that exceed the capabilities of current space technologies. These missions are designed to promote abstract thinking by steering clear of uncomplicated forecasts and variations of existing expertise (Lalli and Packar 2018).

This framework has been created to offer a structured and analytical method for evaluating and identifying the technological needs of breakthrough technologies (BTOs). Additionally, it aims to evaluate the extent to which these technologies could bring about groundbreaking capabilities for advanced space missions. Consequently, we can conceptualize space programs, or similarly impactful scientific endeavors, as unified and comprehensive initiatives directed toward overarching objectives, such as achieving milestones like lunar landings, manned missions to Mars, or the establishment of a sustained human presence on the moon.

The fundamental idea of the program assumes a singular unchanging objective. This involves gathering well-tested tools to move closer to achieving the objective. Another element of the work

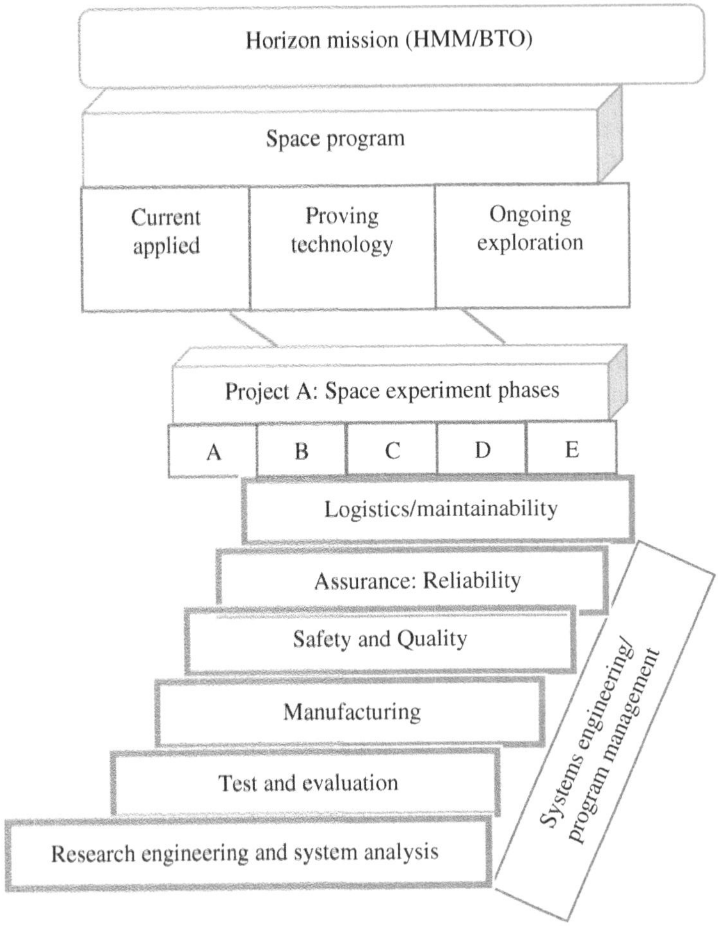

Figure 7.2 System engineering and maintainability.

involves advancing technology and components, as well as continuously exploring the distant possibilities of what is known to lie ahead. At the level of a specific project, diverse fields of expertise come together to plan, construct, roll out, and manage the designated project. As exemplified, one of these fields is maintainability (Zhang et al. 2016). Extending the diverse efforts for maintainability across the different phases of a project leads to the visual representation displayed in Figure 7.3. The National Aeronautics and Space Administration (NASA) employs five stages to define a mission. Our aim is to integrate our maintainability program throughout all five stages. The figure illustrates the descriptions of the tasks involved.

In Sections 7.2.2 and 7.2.3 delineate the various actions. The pivotal aspect is the early introduction of the project's maintainability concept within the program. If this step is neglected, long-term missions will entail higher costs and more periods of inactivity. Granted, the initial expenses during early development may increase, but the overall expenditure will be reduced. Several projects, in their pursuit of securing budgetary approval for a new system, have failed to consider maintainability and integrated diagnostics (Muralidharan and Syamsundar 2012). Nevertheless, this consistently leads to higher overall costs.

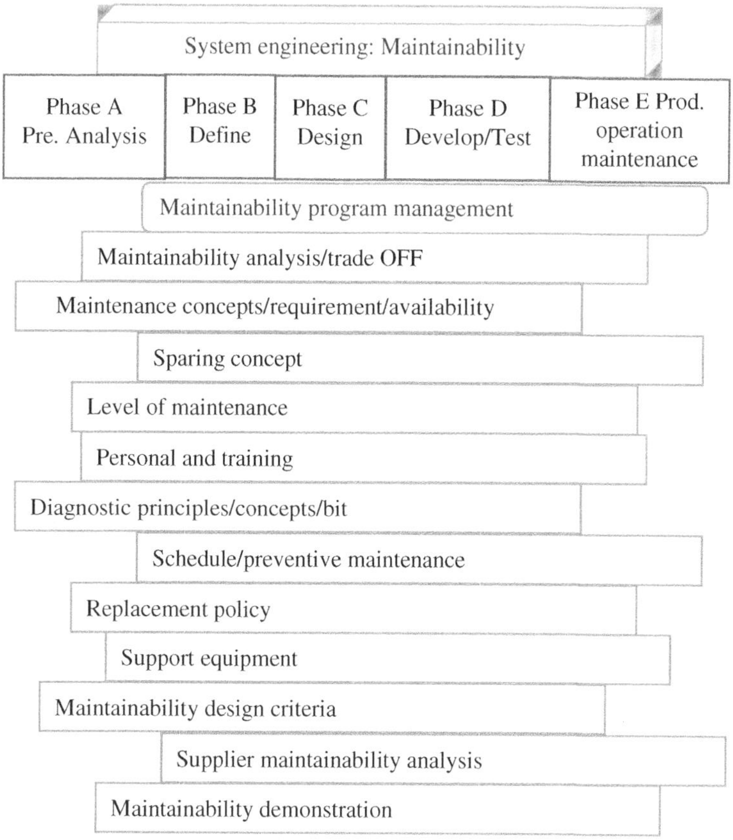

Figure 7.3 System engineering and maintainability.

The interconnected nature of the project's diverse activities, as well as the flow of work and information between the operations, reliability, and logistics units, can be understood by the maintainability of the systems engineering process as illustrated in Figure 7.4. The concept of maintainability evolves through an assessment of system operation and mission prerequisites. This concept is also shaped by the reliability of components and the numerous evaluations of reliability conducted. Subsequently, this maintenance analysis collaborates with design engineering to formulate a design that is both reparable and maintainable (Leslie 1986).

Ultimately, data and requirements related to maintainability are transmitted to the logistics sector, facilitating the creation of effective support resource initiatives. The outcome of the maintenance analysis also holds significant importance for logistical support analysis. The Logistics Support Analysis Record (LSAR) and the development of Support Resources contribute to the formulation of a comprehensive plan encompassing several elements:

- Designing suitable facilities for equipment housing and ground operations.
- Procuring ground support equipment.
- Devising a logistics plan and other associated activities.
- Generating technical publications containing equipment operation and maintenance instructions.
- Identifying necessary personnel and training prerequisites for equipment maintenance, repair, and support.

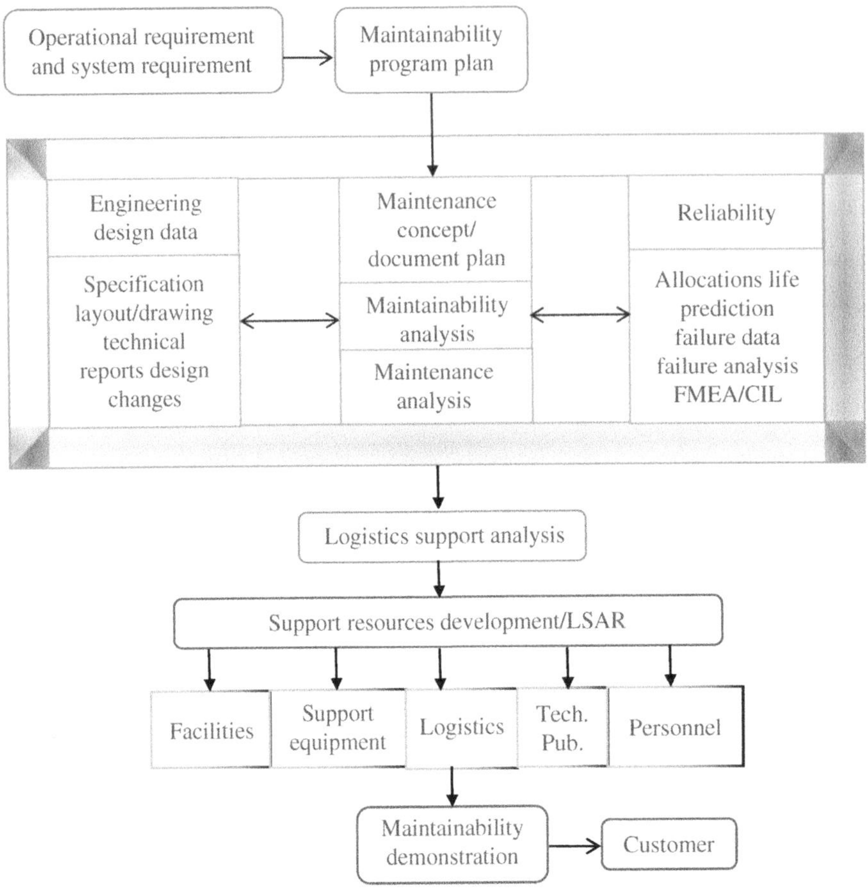

Figure 7.4 Maintainability in the system engineering process.

Ultimately, a maintainability demonstration is conducted to assess the actual time required for diagnosing and physically replacing a line replaceable unit (LRU) or an orbital replaceable unit (ORU).

7.2.2 Maintainability Analysis Process

The assessment of maintainability (Figure 7.5) also considers three factors:

- Estimated time for corrective maintenance multiplied by the number of failures.
- Estimated time for preventive maintenance (PM) multiplied by the number of scheduled PMs.
- Estimated time for replacing limited-life items multiplied by the number of scheduled change-outs. By incorporating these values, an estimation of the total maintenance time per period is derived.

Moreover, the potential downtime due to maintenance allows us to predict the system's availability. The overall availability, along with other indicators of maintainability, is adjusted as the design progresses and as the Failure Mode and Effects Analysis/Critical Items List (FMEA/CIL) and supplier maintainability program data become more refined. The insights garnered from the

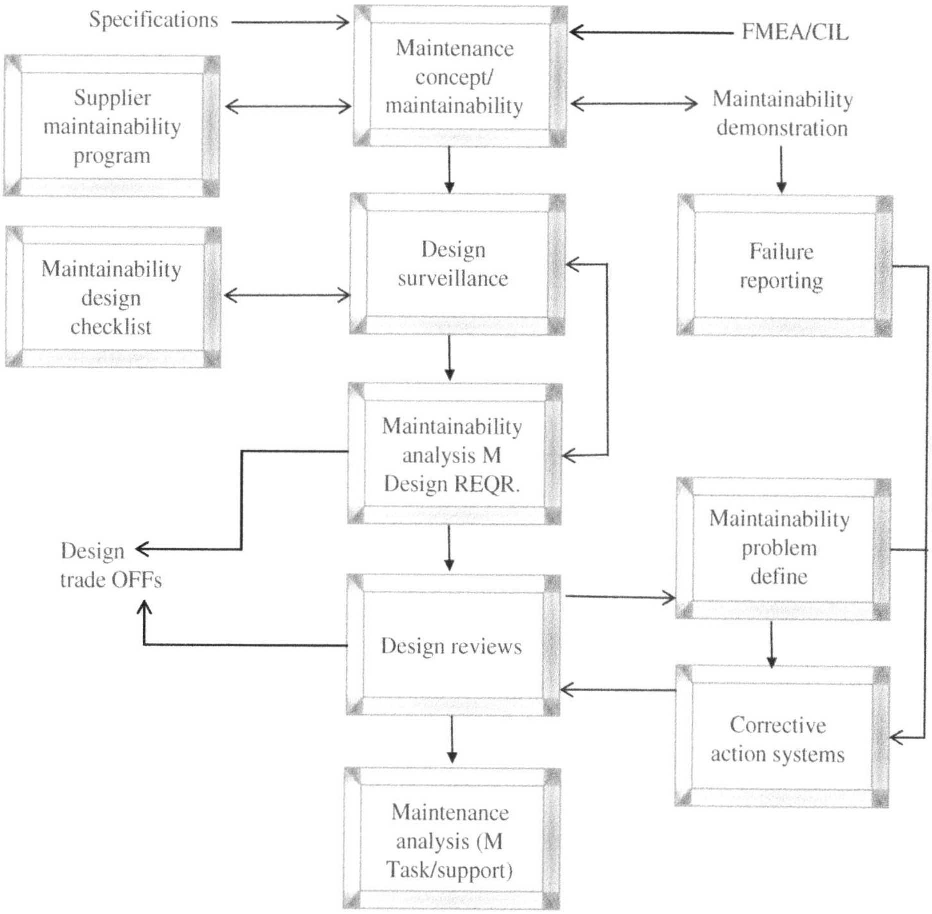

Figure 7.5 Maintainability analysis process.

maintainability analysis are employed by project management to evaluate the overall maturity of the design and its capacity to fulfill program objectives.

7.2.3 Maintainability Analysis Mathematics

The availability needs can be achieved through the utilization of an extremely reliable system, a system that is easily repairable and possesses an ample reserve of spare parts, or a fusion of both approaches. The degree of system employment and the specific mission profile will similarly influence the stipulations for system availability. Several instances from NASA and other sources that showcase continuous and intermittent mission requirements have been given as follows:

- Continuous operations
  1. Spacecraft (LEO)
  2. Space station
  3. Air traffic control system
- Intermittent operation—on demand
  1. Emergency vehicle

2. Research fighter
3. Shipboard Gatling gun

- Intermittent operation—schedule
  1. Space experiment
  2. CAT scan I MRI equipment in hospital
  3. Pace Shuffle main engines

7.3 Availability of the Systems

Availability is the likelihood of an item being accessible when required, represented as the proportion of time the item is usable. This aspect is determined for a repairable item through its failure rate, denoted as A, in conjunction with its rate of repair or replacement. The steady-state availability is the proportion of the total period that the item is available. The steady-state availability of a simple unit with a constant failure rate X and a constant mean repair rate, where MTTR is mean time to repair

$$A = \frac{\mu}{\lambda + \mu} = \frac{\text{MTBF}}{\text{MTBF} + \text{MTTR}} \tag{7.1}$$

The likelihood of the item being available at time t is equivalent to its immediate availability:

$$A = \frac{\mu}{\lambda + \mu} + \frac{\lambda}{\lambda + \mu} * \exp[-(\lambda + \mu)t] \tag{7.2}$$

As the value of t increases, availability tends to approach a steady state. When assessing design choices, it is usually more insightful to focus on system nonavailability. The steady-state unavailability of system is illustrated by Eq. (7.3):

$$= 1 - A = \frac{\mu}{\lambda + \mu} \tag{7.3}$$

and instantaneous unavailability of system is illustrated by Eq. (7.4):

$$= \frac{\mu}{\lambda + \mu} - \frac{\lambda}{\lambda + \mu} * \exp[-(\lambda + \mu)t] \tag{7.4}$$

If there is a need for planned maintenance that involves taking the system offline, this downtime should be taken into account when calculating the system's availability. The availability of extra units for replacement repairs is often a relevant aspect to think about, which depends on past usage of spare units and the rate at which replacement units can be repaired. In intricate systems like power plants, satellites, chemical plants, and radar stations, the emphasis is on availability. While strong reliability is necessary, it's not sufficient to guarantee the system's availability when required. Equally important is the ability to swiftly perform repairs and essential routine maintenance tasks, ideally without causing a system shutdown. Consequently, ensuring maintainability is a vital aspect of designing for optimal availability, often necessitating a balance between reliability and maintainability attributes. Several electronic systems, such as those found in built-in test equipment (BITE), exemplify this approach.

The heightened intricacy may diminish dependability and result in erroneous failure alerts. Nevertheless, BITE can substantially curtail maintenance durations and consequently enhance accessibility by promptly pinpointing defect locations (This is not the sole rationale for BITE adoption. It can also obviate the necessity for external testing apparatus and training for diagnostic procedures, among other benefits.), redundancy also influences availability. The overall availability could see a notable rise if backup systems can be repaired and refurbished while the primary system carries out the vital function. Certain system configurations exhibit characteristics of reliability and availability in steady-state. These configurations showcase notable enhancements in reliability and steady-state availability through the implementation of redundancy. Nevertheless, these scenarios are relatively uncomplicated due to the underlying assumption of a consistent failure rate. Furthermore, for the standby redundancy situation, it is assumed:

1. The changeover system's dependability is one.
2. There are no common-cause failures.
3. Failures are immediately discovered and remedied.

Certainly, these conditions might not consistently hold, particularly when dealing with standby equipment. Such equipment needs to undergo periodic inspections to confirm its operational status. The determination of its availability is based on the frequency of these tests. While methods like BITE are at times employed for electronic devices, they do not guarantee a perfect detection rate for all issues. In practical situations, these variables need to be taken into account, leading to intricate analyses.

7.3.1 Availability of Repairable Systems

Maintaining a system's ease of repair and its ability to undergo regular PM directly affects how accessible the system is. The duration required for troubleshooting issues and conducting routine upkeep can temporarily render the system unavailable. Consequently, a strong connection exists between reliability and maintainability, each impacting the other, while also jointly influencing the system's availability and associated expenses. When the system has settled into a stable state, once any temporary variations have diminished and maintenance procedures are expected to proceed uniformly:

$$\text{Availability } (A) = \frac{\text{MTBF}}{\text{MTBF} + \text{MTTR}} * \frac{\text{PMcycle}}{\text{PMcycle} + \text{PMtime(total)}} \tag{7.5}$$

The PM cycle refers to the period during which all PM chores are finished. The visual representation of the aforementioned availability, maintainability, and reliability criteria can also be observed in Figure 7.5, depicting availability as a function of mean time between failures (MTBF) and MTTR. The boundary of the "solution space" for the system and mission requirements is determined by the 0.990 availability line, a minimum MTBF of 300 hours, and a maximum MTTR of five hours. Additionally, the constant availability lines within this graph are generated using average values of MTBFs and MTTRs. MTTR and MTBF are commonly considered variables distributed with either an exponential or normal distribution.

7.3.2 Availability of Nonrepairable Systems

Mean time to failure (MTTF) for nonrepairable systems is the average time it takes for a system or component to fail, and it cannot be repaired or restored once it fails. MTTF is a measure of reliability and is typically expressed in units of time, such as hours, days, or years. It represents the expected lifespan of the system or component before it experiences a failure that renders it permanently non-operational. Availability for nonrepairable systems can be calculated in terms of MTTF and MTBF using the following formulas:

$$A = \frac{\text{MTTF}}{\text{MTBF} + \text{MTTR}} \tag{7.6}$$

A is the availability of the system. MTTF is the mean time to failure, which represents the average time until a failure occurs. MTTR is the mean time to repair, which is not applicable to nonrepairable systems because they cannot be repaired once they fail. For nonrepairable systems, MTTR is essentially zero because there are no repair actions taken.

Problems

7.1 A power component has an MTTF = 34 hours and an MTTR = 2.5 hours. What is the steady-state availability? If the MTTR is reduced to 1.5 hours, what MTTF can be tolerated without decreasing the steady-state availability of the power component?

Solution:

Steady-state availability = $\frac{\text{MTTF}}{\text{MTTF+MTTR}} = \frac{34}{34+2.5}$ = 0.931 MTTR is reduced to 1.5 hour

Now,

$$\frac{\text{MTTF}}{\text{MTTF} + 1.5} = 0.931$$

$$\text{MTTF} = 0.931(\text{MTTF} + 1.5)$$

$$(1 - 0.931)\text{MTTF} = 0.931 \times 1.5$$

$$\text{MTTF} = \frac{0.931 \times 1.5}{1 - 0.931} = 20.239\ \text{hours}$$

7.2 A power system protection relay has a constant failure rate of 0.2 per day. Once it has failed, the mean time to repair is 2.5 days with constant repair rate. What are the point availability at the end of two days the interval availability over a two-day mission, starting from zero and the steady-state availability? (i) If two protective relays operate in series, compute the availability at the end of two days. (ii) If they operate in parallel, compute the steady-state availability of the system. (iii) If one relay operates in a standby made with no failure in standby, what is the steady-state availability?

Solution:

$$\lambda = 0.2\ \text{per day};\ \frac{1}{\mu} = 2.5\ \therefore\ \mu = 0.4\ \text{per day}$$

(i) The point availability

$$A_e(t) = \frac{\mu}{\lambda + \mu} \mp \frac{\lambda}{\lambda + \mu} e^{-(\lambda+\mu)t}$$

$\therefore$

$$A_p(2) = \frac{0.4}{0.2 + 0.4} + \frac{0.2}{0.2 + 0.4} e^{-(12)+0.4)\times 2} = 0.65 \pm (0.33 \times 0.30) = 0.7590$$

$$A_j(T) = \frac{\mu}{\lambda + \mu} + \frac{\lambda}{(\lambda + \mu)^2 \times T}\{1 - e^{-(\lambda+\mu)T}\}$$

$$\therefore A_l(T) = \frac{0.4}{0.2 + 0.4} + \frac{0.2}{(0.2 + 0.4)^2 \times 2}\{1 - e^{-(0.2+0.4)\times 2}\}$$

$$0.66 + 0.28 \times (1 - 0.30) = 0.8560$$

$$A(\infty) = \frac{\mu}{\lambda + \mu} + \frac{0.4}{0.2 + 0.4} = 0.66$$

(ii) $A_s(2) = \{A_p(2)\}^2 = (0.7590)^2 = 0.5760$
(iii) $A_s(\infty) = 1 - \{1 - A(\infty)\}^2 = 1 - \{1 - 0.66\}^2 = 0.84$
(iv) For the standby redundant system

$$A_s(\infty) = \frac{\lambda_1\mu + \mu^2}{\mu^2 + \lambda_1\mu + \lambda_1\lambda_2}; \text{ Here } \lambda_1 = \lambda_2 = 0.2 \text{ and } \lambda = 0.4$$

$$A_s(\infty) = \frac{(0.2)(0.4) + (0.4)^2}{(0.4)^2 + (0.2)(0.4) + (0.2)(0.2)} = 0.8571$$

7.4 Conclusion

The primary objective of a system maintenance program aligns with the mission success, a paramount goal of every NASA System Reliability and Quality Assurance (SR&QA) office. An effectively devised maintainability program offers increased availability at reduced expenses, leading to the creation of designs featuring easily serviceable (and assembled) modules. When maintenance is taken into account, it curtails the tendency to opt for lower-cost components at the expense of reliability unless there are valid maintainability trade-offs. Moreover, maintainability analysis compels the consideration of potential obsolescence and the necessity for upgrades, all while diminishing overall maintenance hours and the total cost of ownership. The following are the maintainability program advantages:

- Ensures system preparedness and accessibility.
- Defines the parameters that influence the trade-offs affecting the genuine reliability value.
- Ensures a precise evaluation of maintenance and logistics program costs.
- Facilitates the expansion of the system.
- Minimizes the need for maintenance personnel.
- Lowers operational expenses.
- Decreases the overall cost of ownership.

Bibliography

Diatte, K., O'Halloran, B., and Van Bossuyt, D.L. (2022). The integration of reliability, availability, and maintainability into model-based systems engineering. *Systems* 10 (4).

Kothari, D.P., Nagrath, I.J., and Saket, R.K. (2022). *Modern Power System Analysis*. New Delhi: Tata McGraw-Hill Publishing Company.

Lalli, V.R. and Packar, M.H. (2018). Designing for Maintainability and System Availability.

Leslic, R.E. (1986). *System Analysis and Design, Methods and Inventions*. Englewood Cliffs, NJ: Prentice-Hall.

Muralidharan, K. and Syamsundar, A. (2012). *Stastical Methods for Quality, Reliability and Maintainability*, Eastern Economy Edition. New Delhi: PHI Learning Private Limited.

O'Connor, P.P. (2010). *Practical Reliability Engineering*. Wiley India Edition.

Sarita, K., Saket, R.K., and Khan, B. (2023). Reliability, availability, and condition monitoring of inverters of grid-connected solar photovoltaic systems. *IET Renewable Power Generation* 17 (7): 1635–1653.

Sayed, A., EL-Shimy, M., El-Metwally, M., and Elshahed, M. (2020). Impact of subsystems on the overall system availability for the large scale grid-connected photovoltaic systems. *Reliability Engineering & System Safety* 196: 106742.

Zhang, X., Sun, L., Sun, H. et al. (2016). Floating offshore wind turbine reliability analysis based on system grading and dynamic FTA. *Journal of Wind Engineering and Industrial Aerodynamics* 154: 21–33.

8

Genetic Algorithm and Artificial Neural Networks in Reliability-Based Design Optimization

Heeralal Gargama, Sanjay Kumar Chaturvedi, and Rajiv Nandan Rai

Subir Chowdhury School of Quality and Reliability, Indian Institute of Technology Kharagpur, Kharagpur, West Bengal, India

8.1 Introduction

In the realm of reliability-based design optimization (RBDO), the traditional constraints undergo a conversion into probabilistic constraints (as outlined by Rao 1992), and probability distributions are frequently employed to depict the randomness and uncertainty in design variables. To assess the likelihood of structural design failure, it is essential to construct a joint probability density function (*pdf*) encompassing all problem parameters and design variables. However, creating a joint *pdf* can prove to be a daunting task, and in some cases, it may seem nearly impossible, particularly when dealing with intricate computational requirements and limited data accessibility in complex engineering systems (Kaymaz and McMahon 2005). To address this challenge and to estimate failure probability more efficiently, the researchers have developed *approximate reliability techniques* and *advanced simulation methods* (Haldar and Mahadevan 2000; Valdebenito and Schuëller 2010). These techniques have been integrated in various ways within optimization procedures, resulting in the development of several RBDO approaches in recent years, as outlined by Deb et al. (2009).

In order to attain accurate approximations of failure probabilities, methods such as the second-order reliability method (SORM) or first-order reliability method (FORM) are utilized to ensure an elevated level of precision. In this process of conducting reliability analysis through these techniques, the initial step entails converting the performance or limit state function from its original space of random variables into a space of standardized normal variables. In the context of the FORM, the limit state function is approximated using a first-order Taylor series expansion centered at the design point, which is typically known as the Most Probable Point (MPP) of failure. The approximation results in a linear function within the U-space. Conversely, SORM utilizes a second-order surface to approximate the limit state surface (Youn and Choi 2004a). The process of locating the MPP involves addressing a constrained optimization challenge and various methods that can be utilized to accomplish this objective, including:

(a) Performance measure approach (Youn and Choi 2004b);
(b) Reliability index approach (Yu et al. 1998);
(c) Fast performance measure approach (Du and Chen 2001);
(d) Fast reliability index approach (Deb et al. 2009).

Researchers have integrated the aforementioned methods into optimization algorithms in diverse ways, leading to the development of various methodologies that include:

Reliability Analysis of Modern Power Systems, First Edition. Edited by R. K. Saket and P. Sanjeevikumar.

(a) RBDO using double-loop techniques;
(b) RBDO using single-loop techniques;
(c) RBDO using decoupled techniques.

The above approaches allow designers to efficiently address the challenges posed by uncertainty and randomness in complex engineering systems. Nevertheless, the FORM and SORM come with certain limitations (Youn and Choi 2004a; Rahman and Wei 2008):

(a) **Nonlinearity:** These methods may not provide adequate accuracy when dealing with highly nonlinear performance functions;
(b) **Multiple MPPs**: If multiple MPP exist within the system, FORM and SORM methods can produce erroneous failure probability estimates;
(c) **Derivative Requirements**: FORM and SORM necessitate the availability of higher-order derivatives, yet many engineering problems lack analytical derivatives, making their application challenging.

To address these challenges in RBDO, simulation techniques have been employed, including direct integration using optimization algorithms and meta-model approximations.

Direct integration involves computing gradient information using Monte Carlo simulation (MCS). This gradient information can then be efficiently utilized within gradient-based optimization algorithms to solve RBDO problems, as demonstrated by Jensen et al. 2009. MCS-based RBDO techniques are known for their ease of implementation, robustness, and accuracy, which persist unaffected by the problem's dimension or intricacy. Furthermore, the demand for a substantial number of MCS cycles to estimate low failure probabilities can be mitigated through various methods of sampling, which include importance sampling, stratified sampling, latin hypercube sampling (LHS), etc.

The MCS technique for evaluating the probability of failure (*pof*) is recognized for its ease of implementation and robustness. However, it does come with certain limitations (Gargama et al. 2014a,b):

(a) **Numerical Noise:** MCS can introduce numerical noise due to its reliance on a limited sampling size, which may affect the precision of results;
(b) **Low Target Probability:** In RBDO, when aiming for a low target *pof*, accuracy of MCS can deteriorate further with a limited sample size;
(c) **Zero Probability:** In cases where the *pof* is extremely low, MCS may erroneously provide a zero *pof*.

Meta-models address the above challenges in conjunction with RBDO methods thereby offering an attractive approach to reduce computational burden. Meta-models are typically combined with the design of experiments (DOE) to create approximate functions that depict performance responses within the given boundaries of design space. Among various meta-models, the response surface method (RSM) and its other variants have proven effective, as demonstrated by their successful utilization with different inverse reliability measures (Foschi et al. 2002; Qu and Haftka 2003; Yoo et al. 2011). For smaller problems with few variables, polynomial response surfaces are particularly appealing due to their simplicity and efficacy (Wang and Shan 2007). Nonetheless, this approach may not be well-suited for problems featuring highly nonlinear responses or a large number of random parameters/variables. Accurately representing all peaks and valleys can pose a challenge. Fortunately, there are alternative meta-models suitable for addressing highly nonlinear and large-scale optimization problems, including kriging, ANNs, radial basis functions, support vector regression, and more (Zhu et al. 2011).

The ANNs represent a promising paradigm in this context. They can function as surrogate models, providing simplified approximations of the limit state or objective function, which proves computationally efficient when assessing failure probabilities. They can also be particularly useful in combination with methods like first/second-order reliability methods or sampling methods. Following this, double-loop structures in RBDO have been utilized for optimization purposes, as exemplified by the research of Chau 2007, Cardoso et al. 2008, and Gondal and Lee 2012.

In this chapter, the author aims to approximate an accurate design response capable of estimating relatively very low probabilities of failure. To achieve this, we employ the concept of an equivalent safety factor known as the PSF. It is important to note that obtaining *pof* using approximated functions is only meaningful if their accuracy is high. To obtain more accurate reliability estimates, given the extremely nonlinear behavior of the electromagnetic interference (EMI) shielding design issue, we suggest PSF as a surrogate measure of the *pof* and use of ANN to approximate PSF[1].

In the proposed approach, the author begins by generating various design points throughout the entire design space. Subsequently, calculate the PSF at each of these design points using MCS. These design vectors, in conjunction with their respective PSF values, are subsequently employed for the training of an ANN. Following the ANN training, it is applied to the design optimization problem to achieve the desired PSF value while adhering to the specified design variable constraints. To address the optimization task, we employed a real-coded GA as outlined in the work by Deb et al. (2007) This specific real-coded GA approach is illustrated through the design of electromagnetic shields for military applications operating within the frequency range of 7.99–12.5 GHz in this chapter.

The chapter is illustrated as follows. The theoretical backgrounds required for the reliability-based optimization have been introduced in Section 8.2. Wherein the definition of performance-based reliability is presented in Section 8.2.1. RBDO formulation, PSF calculation, MCS approach, and ANN approach are presented in Sections 8.2.2–8.2.5, respectively. The GA deployed for the optimization is explained in Section 8.2.6. Section 8.3 presents the RBDO methodology stepwise to solve the case study undertaken. A case study of shielding structure design has been illustrated in Section 8.4. Equations used for the evaluation of shielding effectiveness are summarized in Appendix 8.A.

8.2 Reliability-based Design

8.2.1 Performance-based Reliability

In the system or product development lifecycle, the primary goal is to ensure that the intended functions are achieved with satisfactory performance levels. The design approach utilized has a direct impact on both the functional attributes and the performance outcomes. The system's design capacity (also referred to as "capacity" or "resistance" in the literature) is analogized to the system's strength, while the customer's requirements (referred to as the load or demand) are analogized to the stress. This analogy allows us to calculate performance-based reliabilities using the stress-strength theory (Tsai and Chang 2008). The performance requirements can alternatively be articulated as an "*acceptable level of damage defined by acceptable probability of failure.*" To be more

---

1 The concept of PSF was initially introduced by Birger (1970) and has been evaluated in comparison to conventional measures of *pof* and safety indices for application of reliability-based design by Qu and Haftka 2003. Their research showed that PSF maintains accuracy, especially in the low *pof* region, and exhibits lesser variation compared to traditional failure probability or safety indices. Consequently, approximating the limit state function using PSF leads to better accurate results than using traditional failure probabilities.

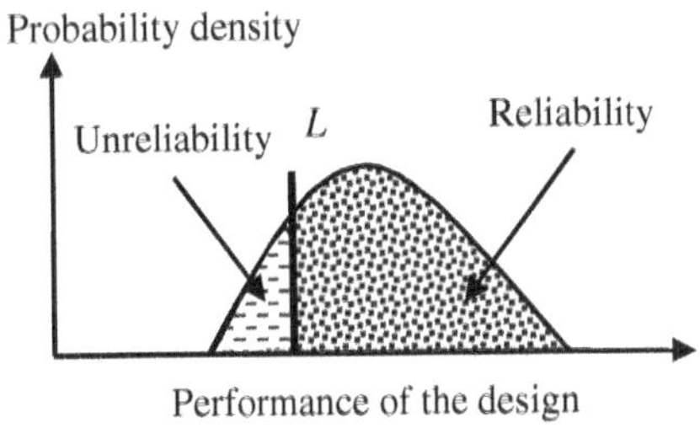

Figure 8.1 Reliability assessments using "bigger the better" PI.

precise, performance-based design (PBD) can be defined as follows; "*Design specifically intended to limit the consequences of one or more perils to a predefined acceptable level*" (Frydrýšek 2008). The classification of PBD into four distinct categories is determined by the characteristics of performance indices (PI), as outlined by Tsai and Chang (2008):

(a) **Nominal the Best PI:** In this category, the notion is that the provided performance is considered better when it aligns more closely with the target value.
(b) **Smaller the Better PI:** The category signifies that provided performance is deemed better when it is smaller, with the optimal target value being zero.
(c) **Bigger the Better PI:** This classification implies that superior performance is associated with larger values, with an emphasis on achieving the highest possible level of strength or a related attribute. Consequently, the optimal target value in this scenario can be a positive maximum.
(d) **Zero Point Proportional PI:** In this category, the objective is to have the output and input exhibit a closer, constant proportional relationship for better performance. The ideal outcome in this category may be represented as a straight line passing through the origin.

In this chapter, the author focuses on the (c) "bigger the better" PI for reliability evaluation, as depicted in Figure 8.1. In this context, when evaluating the design performance against a predefined threshold value labeled as "*L*," the reliability of the structures can be described as the likelihood that the performance of the design surpasses the specified value. For the "bigger the better" PI, the reliability evaluation interval would be from L to ∞.

8.2.2 RBDO Problem

Let $\boldsymbol{B}$ represent a vector comprising independently distributed random design variables $\{B_1, B_2, \ldots, B_N\}$, and let $\boldsymbol{C}$ denote a vector containing independently distributed random parameters $\{C_1, C_2, \ldots, C_M\}$. Additionally, $\boldsymbol{D}$ represents a vector of design variables with precise specifications (i.e., deterministic variables). Each parameter C_l, $l = 1, 2, 3, \ldots, M$ and random variable, $B_j, j = 1, 2, 3, 4, \ldots, N$ is characterized by its *pdf*, along with respective vectors indicating standard deviation and mean, denoted as μ_{B_j}, μ_{C_l} and, SD_{B_j}, SD_{C_l}, respectively. As a result, the optimization problem exhibits stochasticity arising from two distinct sets, namely $\boldsymbol{B}$ and $\boldsymbol{C}$, while both $\boldsymbol{B}$ and $\boldsymbol{D}$ serve as decision variable vectors. During the optimization progression, the standard deviation of a particular design variable or parameter may be held constant or varied based on a fixed coefficient of variation, based on the inherent randomness it possesses.

The general RBDO problem formulation, seeking the optimal solution to achieve a desired or target reliability (R) for any design of the system, can be articulated as (Gargama et al. 2014a):

$$
\begin{aligned}
&\text{Maximize} f(\mu_B, \mu_B, \mathbf{D}) \\
&\text{subject to } Pr\,(E_i(\mathbf{B}, \mathbf{C}, \mathbf{D}) \leq 0) > (1-\alpha_i) \quad \text{or} \quad R_i, \quad i = 1, 2, \ldots, I \\
&\boldsymbol{B}_L \leq \mu_B \leq \boldsymbol{B}_U \\
&\boldsymbol{D}_L \leq \boldsymbol{D} \leq \boldsymbol{D}_U \\
&\alpha_i/R \in [0, 1]
\end{aligned} \tag{8.1}
$$

In the above formulation (8.1), design variables bounds are represented by the subscripts U and L, while $\alpha_i = 1 - R_i$, denotes the probability of meeting the ith constraint. To be noted that the $E_i(\boldsymbol{B}, \boldsymbol{C}, \boldsymbol{D})$ represents the ith performance function/performance criteria. The limit state function for the ith performance criteria are defined as $E_i(\boldsymbol{B}, \boldsymbol{C}, \boldsymbol{D}) = 0$, which serves as the boundary distinguishing between success and failure regions within the given design parameter constraints. A design is considered success if $E_i(\boldsymbol{B}, \boldsymbol{C}, \boldsymbol{D}) > 0$ and failure otherwise. The joint probability of achieving feasible solutions in the presence of randomness or uncertainty, for all I constraints is denoted term $Pr(\cdot)$ denotes. These constraints, $Pr(E_i(\boldsymbol{B}, \boldsymbol{C}, \boldsymbol{D}) \leq 0)$, are commonly referred to as chance constraints and pose challenges in solving RBDO problems as they are not straightforward to address.[2]

The *pof*, denoted as p_{f_i}, for the solution with respect to the ith constraint is determined through a multi-dimensional integration across the failure domain, as follows:

$$
p_{f_i} = \int_{G_i(B,C,D)\leq 0} f_{\mu_X,\mu_P}(B, C)\, dBdC \tag{8.2}
$$

Here f_{μ_B,μ_C} represents the joint *pdf* of all random design variables and parameters. In many engineering applications, the joint *pdf* might not have a readily available explicit expression, even when the *pdf* of individual variables and parameters are known. In such cases, it becomes necessary to employ simulation and approximation methods, such as SORM, FORM, MCS, ANN, RSM, etc. Simulation and approximation approach used in this chapter are presented in Sections 8.2.4 and 8.2.6, respectively. The next section will go through how to compute the PSF, an alternative index related to the factor of safety, which improves the precision of probability calculations, especially in scenarios with low failure probabilities.

8.2.3 Probability Sufficiency Factor

Traditional design methods commonly utilize cautious scaling factors like safety factors (SF) or safety margins to guarantee the security of design. The conventional SF is defined as follows:

$$
SF_i = g_{R_i}/g_{S_i} \tag{8.3}
$$

These SF are calculated using the nominal resistance, i.e., capacity or strength, denoted as g_{R_i}, and the nominal load, i.e., demand or stress, denoted as g_{S_i}, corresponding to the ith performance criterion. However, relying solely on these SF may not be adequate, as they do not provide insight into the actual safety margin in design reliability.

Hence, in the context of RBDO, reliability constraints are articulated using the *pof* and are structured as a stipulation on the SF, as follows:

$$
p(SF_i \leq 1) = p_{tf_i} \tag{8.4}
$$

2 Most of the research in the field of RBDO is dedicated to efficiently and accurately estimating these constraints.

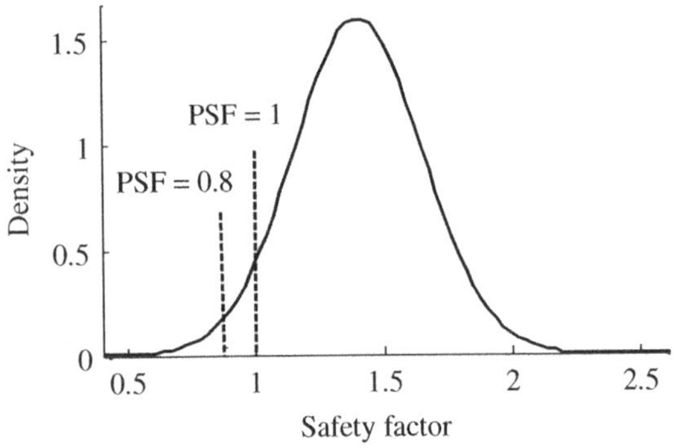

Figure 8.2 Pdf of the SF.

where p_{tf_i} is the target *pof*. PSF for an RBDO application with the target *pof* denoted as p_{tf_i} is solution to the following equation:

$$p(SF_i \leq PSF_i) = p_{tf_i} \tag{8.5}$$

Note: *Figure 8.2 illustrates the schematic pdf of the SF for a single performance criterion let's call it* $E_1(\boldsymbol{B}, \boldsymbol{C}, \boldsymbol{D}) = 0$. *In this figure, the region beneath the curve on the left side of the vertical line* $SF \leq 1$ *represents the pof. It is worth noting that by employing the inverse transformation described in Eq. (8.5), we can determine a specific value for the SF corresponding to a given value of* p_{tf}. *For example, consider the region to the left of the vertical line where PSF = 0.8 (equivalent to SF = 0.8), indicating an acceptable pof, the value 0.8 indicates that achieving this probability is possible through either a 25% increase in capacity or a 20% reduction in load (calculated as 1/0.8-1).*

8.2.4 MCS Approach

In MCS, the goal is to capture the probabilistic behavior of a system by repeatedly sampling each random variable and parameter from their respective representative/estimated *pdf*. The MCS process involves the following steps:

(a) Express the problem by incorporating all the parameters and random variables.
(b) Construct *pdfs* for each input random variable and parameter. These *pdfs* are commonly derived from engineering estimations or the insights of experts.
(c) Generate random samples from respective *pdfs* of each random variable and parameter.
(d) Evaluate the system's response for each set of values generated in Step (c).
(e) Extract probabilistic data by tallying the occurrences of simulation cycles where the performance function fails. In other words, calculate the fraction of simulations where $E_i(\boldsymbol{B}, \boldsymbol{C}, \boldsymbol{D}) \leq 0$ occurs.

This fraction provides an unbiased estimator of the *pof* from the MCS. It is determined as follows:

$$p_{f_i} = \frac{1}{N} \sum_{b=1}^{N} G_b(B_b, C_b) = \frac{Q_{f_b}}{N} \tag{8.6}$$

Here, p_{f_i} represents the approximated *pof* for the *i*th performance criteria, and the value of indicator $G_b(B_b, C_b) = 1$ if $E_i(\boldsymbol{B}, \boldsymbol{C}, \boldsymbol{D}) \leq 0$ and 0, otherwise. The symbol N and Q_{f_b} denotes combined count of simulation trials and the cumulative number of simulations resulting in failure, respectively. The accuracy of MCS typically relies on the total number of simulation cycles. Each individual experiment or simulation within MCS obeys to a binomial distribution, characterized by a certain probability of failure or success, and is considered independent. By employing a normal distribution approximation to the binomial distribution, one can compute the anticipated error linked to the number of cycles in the simulation. As a result, when considering a 94.99% confidence interval, one can approximate the percentage deviation in the estimated *pof* as:

$$\text{error}\% = \sqrt{\left(1 - p_{\mathrm{f}}^{\mathrm{T}}\right) / \left(N \times p_{\mathrm{f}}^{\mathrm{T}}\right)} \times 200 \tag{8.7}$$

where p_f^{T} is the true probability of the failure. A significant limitation of MCS is the necessity for a substantial sample size, denoted as "N," when calculating low failure probabilities. To mitigate the computational expenses associated with estimating these low failure probabilities, various systematic sampling techniques have been devised. These methods include importance sampling, Subset Simulation, LHS, Line Sampling, and the Horseracing Simulation Method, etc. (Zuev 2009). Here, the LHS is applied as a DOE tool.

8.2.5 Latin Hyper Cube Design

The design of Experiments generates a collection of input design vectors within the design space, often referred to as the experimental region. Subsequently, each set of values of the design vectors is used to assess the limit state function. This procedure can be divided into two types: (i) physical experiments and (ii) computer-generated experiments, depending on the nature of the simulation model employed, which can be either deterministic and/or stochastic. The deterministic model generates response values by applying mathematical relationships that align with the input design or decision variables. Consequently, these responses are free from random errors and exclusively encompass systematic errors, often arising from approximated models or model uncertainty.

To minimize systematic errors, various space-filling designs are suggested, including LHS designs, factorial design, and central composite design. Among these, LHS design stands out as a favored space-filling design and is particularly well-suited for DOE in computer-based experiments. It excels at ensuring dense coverage of each variable's entire range within the design space (Myers et al. 2009). LHS design is capable of handling a substantial number of variables and has been found to provide accurate estimates of output (i.e., means, variances, and distribution function) compared to other sampling approaches such as random and stratified sampling (McKay et al. 1979).

Using the LHS procedure, the generation of a set of sampling points k from the design variables $k_1, k_2, \dots, k_n$ is carried out by the following:

(a) Divide the range of all the variable into k distinct, non-overlapping intervals, with the boundaries of these intervals determined based on equal probability of occurrences.
(b) Randomly select one value from individual interval, taking into account the probability density within that interval.
(c) Pair up the k values obtained for k_1 with k values from k_2 in a random manner. Continue this pairing process by combining each pair with k values from k_3, and so on, until the last design variable is included.

As stated earlier, the nature of the EMI shielding problem is highly nonlinear, we suggest PSF as a surrogate measure of the *pof* and use of ANN to approximate PSF.

8.2.6 Artificial Neural Networks

ANNs have found extensive utility in numerous practical applications within engineering and science due to their numerous benefits. These advantages include the capacity to accumulate expertise through learning from examples and their inherent ability to generalize (Gondal and Lee 2012). Single-layer ANNs have limitations when it comes to learning nonlinear functions. Therefore, additional layers of neurons are introduced between the output and input layers, and these layers are trained effectively to model any well-defined function with remarkable precision.

Multilayer feedforward ANNs, often referred to as multilayer perceptrons, have gained significant recognition and have proven successful in tackling a wide array of challenging problems, including function approximation, pattern recognition, data classification, etc. A multilayer perceptron consists of multiple layers, each comprising its set of neurons, connections with associated weights (represented as a weight matrix), a bias vector, and an output vector. The layers responsible for producing the final output are referred to as output layers, while the intermediary layers are known as hidden layers. As highlighted by Hornik (1991), multilayer perceptron networks equipped with sigmoid activation functions and linear activation functions in their hidden layers output layers, respectively, possess remarkable power, enabling them to approximate any continuous function with a high degree of precision.

In this chapter, the multilayer perceptron networks have an output layer comprising a single neuron which represents the PSF with respect to the network's inputs. In the hidden layer, the determination of the number of neurons is achieved through a trial-and-error process, aiming to obtain a multilayer perceptron model with the lowest mean squared error (MSE) and a linear correlation coefficient (R-value) that approximates ≈1. For the training of the multilayer perceptron networks, the back error propagation approach is employed. This technique minimizes the MSE by adjusting the values of the interconnected weights. The equation employed for computing the MSE is as depicted below:

$$\mathrm{MSE} = \frac{1}{C}\sum_{l=1}^{C}(t_l - a_l)^2 \tag{8.8}$$

In this formula, t_i represents the target output, a_i signifies the network's output for the ith training dataset, and C denotes the total number of training data samples.

8.2.7 Real-coded GA

After approximating the design response surface in terms of PSF using ANN, an appropriate algorithm is employed to solve the optimization problem presented in Eq. (8.1). GA optimizers typically demand users to make several upfront choices prior to initiating the optimization process. These choices encompass parameters like population size (N), crossover probability (p_c), mutation probability (p_m), selection strategy, specific characteristics of the fitness function, as well as the types of crossover and mutation to be employed. According to previous considerations rooted in schema processing (Goldberg et al. 1992), it was proposed that the population size should grow in tandem with the complexity of the problem. Nevertheless, it is essential to recognize that the appropriate population size should also factor in the signal-to-noise ratio inherent in a given problem. To address this issue, Deb (2000) advocated for a straightforward calculation method: $N = 10n$, where n denotes the variable number within the problem. Furthermore, comparatively larger populations offer increased genetic diversity, leading to expedited convergence.

Regarding the probability of crossover, it has been observed that a value around 0.7 tends to yield optimal results across a diverse range of problems. However, a higher crossover probability p_c can enhance the speed of exploration and serve as a primary means for a GA to uncover novel and improved solutions. In the realm of real-coded GAs, it is worth noting that the mutation operator is the sole mechanism responsible for generating new values for variables. Consequently, it is necessary to set a considerably higher mutation probability p_m compared to binary-coded GAs (typically ranging from 0.01 to 0.1) (Vukovć and Sopta 1999).

The real-coded GA retains the identical evaluation procedure and reproduction operator as the binary-coded genetic algorithm. Nevertheless, it diverges when it comes to the crossover and mutation processes. Without delving into intricate specifics, here is an overview of the steps involved in performing mutation and crossover on the population during Steps 5 and 6 of the executed real-coded GA.

Step 1: Choose N, p_c, p_m, tournament size (T_n) and maximum generation number ($G_{\max}$).

Step 2: Create initial population of solutions by employing a uniform distribution method within the predefined range of design parameters.

Step 3: Calculate the fitness function (as described by Deb 2000) by first computing the objective function and checking for constraint violations in each solution vector, denoted as $\vec{x}$,

$$F(\vec{x}) = \begin{cases} f(\vec{x}), \text{if } g_j(\vec{x}) \geq 0 & \forall j = 1, 2, \ldots, m, \\ f_{\max} + \sum_{j=1}^{m} \langle g_j(\vec{x}) \rangle, & \text{otherwise.} \end{cases} \tag{8.9}$$

Here, $F(\vec{x})$, $f(\vec{x})$ and $g_j(\vec{x})$ are the fitness function, objective function and number of constraints ($j = 1, 2, \ldots, m$), respectively. The value of bracket penalty operator $\langle g_j(\vec{x}) \rangle = g_j(\vec{x})$, when $g_j(\vec{x})$ is zero and negative, otherwise. The objective function's maximum value, denoted as $f_{\max}$, represents the most unfavorable feasible solution within the entire population. In the case of a minimization problem, it is common to transform and utilize the fitness function in the following manner:

$$F_T(\vec{x}) = \frac{1}{1 + F(\vec{x})} \tag{8.10}$$

Step 4: Utilize the tournament selection operator with replacement to pick favorable strings from the population for the purpose of creating the mating pool.

(a) **Pairs formation:** Generate a random set of competitors and subsequently create couple of competitors, ensuring that there are total of T_n pairs.

(b) **Winner selection:** The winner among the pairs of competitors is determined by identifying the string associated with the competitor possessing the best fitness, denoted as $F_{\mathrm{T}}(\vec{x})$.

(c) Repeat this method T_n rounds to obtain the population size equal to N.

Step 5: Perform simulated binary crossover (SBX) operations on randomly chosen pairs from the population. SBX manipulates offspring solutions around the parental solutions by leveraging a distribution index. To initiate this process, first, pick parents for the crossover operation. Generate a series of random numbers ranging from 0 to 1, with the count of random values matching the number of strings in the population, represented as N. Examine each random number to see if it is less than or equal to a predefined threshold, p_{c}. If it meets this condition, select the corresponding string for crossover; otherwise, place the strings directly into an intermediate population. To compute the two child solutions, c_1 and c_2, from the selected parents, p_1 and p_2, adhere to the following steps:

(i) First, within the range of 0 to 1 a random number u is generated. Next, determine the $\bar{\beta}$ ordinate using the polynomial probability distribution function according to the instructions provided below:

$$\begin{cases} (2u)^{1/(\eta_c+1)}, & \text{if } u \le 0.5 \\ \left(\frac{1}{2(1-u)}\right)^{1/(\eta_c+1)}, & \text{otherwise.} \end{cases}$$

The distribution index, η_c, is a parameter that can take on any positive value. A smaller value of the distribution index allows to create the child solutions farther away from the parent solutions, while a larger value brings them closer. For solving the problem, undertaken in this chapter, we have considered $\eta_c = 1$.

(ii) The child solutions are *calculated* as:

$$c_1 = 0.5[(p_1 + p_2) + \bar{\beta}|p_2 - p_1|], c_2 = 0.5[(p_1 + p_2) - \bar{\beta}|p_2 - p_1|]$$

For calculating the c_1 and c_2 within the given constraints of p^L and p^U, $\bar{\beta}$ needs to be changed as:

$$\bar{\beta} = \begin{cases} (\alpha u)^{1/(\eta_c+1)}, \text{if } u \le \frac{1}{\alpha} \\ \left(\frac{1}{2-\alpha u}\right)^{1/(\eta_c+1)}, \text{otherwise,} \end{cases}$$

where $\alpha = 2 - \beta^{-(\eta_c+1)}$ and $\beta = 1 + \frac{2}{c_2-c_1} \min[(p_1 - p^L), (p^U - p_2)]$. Here, it is assumed that $p_1 < p_2$, moreover the above equation may be modified for $p_1 > p_2$.

Step 6: With mutation probability of p_m, perform mutation operation. To determine the mutation value for the constraint variables, we establish a perturbation factor denoted as $\bar{\delta}$ in the following manner:

$$\begin{cases} \left[2u + (1-2u)(1-\delta)^{\eta_m+1}\right]^{1/\eta_m+1} - 1 & \textit{if } u \le 0.5; \\ 1 - \left[2(1-u) + 2(u-0.5)(1-\delta)^{\eta_m+1}\right]^{1/(\eta_m+1)} & \text{otherwise,} \end{cases}$$

where u represents a random value ranging from 0 to 1, η_m is the distribution index, $\delta = \min[(p - p^L), (p^U - p)]/(p^U - p^L)$ and the maximum permissible perturbance in the parent value p is $\Delta U^L{}_{\max}$. Subsequently, the mutated child solution is calculated as follows:

$$c = p + \bar{\delta}\Delta_{\max}$$

Step 7: Substitute the previous population with the newly generated population.

Step 8: Check for the termination criteria, which may be defined as reaching a maximum number of generations $G_{\max}$, achieving a desired fitness level, or reaching a specified time limit for calculation. If any of these termination criteria are satisfied, stop the optimization process. Otherwise, repeat the optimization process starting from Step 3.

8.3 RBDO Methodology Using PSF and ANNs

After presenting the several aspects of RBDO problems in earlier sections and for the sake of brevity, the entire optimization methodology described in this chapter follows these subsequent steps:

Step 1: Generate a collection of design points that encompass the entire design space.
Generate a collection of design points using the LHS technique within a specified range of design variables, given in Section 8.2.5 as a DOE method from the design space.

Step 2: Perform probabilistic assessment.
  (a) Define *pdf* of each parameter and design variable.
  (b) Specify the desired output performance and the acceptable *pof* for that performance.
  (c) Assess the PSF by utilizing MCS at every design point generated through DOE, as outlined in Section 8.2.3 and Section 8.2.4, respectively.

Step 3: Approximate the PSF using ANN back error propagation architecture (Section 8.2.6) corresponding to design points generated via DOE.
In this chapter, ANN is developed and trained to offer a cost-effective means of verifying the violation of probabilistic constraints necessary for conducting reliability analysis. The ANN's training process involves using a database consisting of various input design variable combinations (input) and their corresponding PSF values, which were evaluated during *Step 2*.

Step 4: Perform RBDO using GA (Section 8.2.7):
  (a) Specify the objective function that needs to be minimized;
  (b) Specify the ranges for design variables and deterministic constraints;
  (c) Utilize an approximate ANN for checking probabilistic constraints.

Step 5: Verify the convergence conditions, ensuring they meet the reliability requirements. If they are met, halt the process; otherwise, narrow down the design space and recommence from Step 1.

8.3.1 Case Study: EMI Shielding Design for the Required Load of 80 dB

In this chapter, the transmission line modeling described in Appendix 8.A is used to calculate the SE of the planar multilayer shielding structure. An alternative formulation to the RBDO problem presented in Eq. (8.1) could be to attain the target *pof* by solving the constraint optimization problem with the decision variable vectors. Consequently, the design problem, which centers around a single performance criterion, can be articulated as the task of minimizing the discrepancy between the outputs of the trained ANN. This discrepancy pertains to the error between the approximated PSF ($\mathrm{PSF_a}$) and the target PSF ($\mathrm{PSF_t}$). Therefore, the problem statement for the shielding design can be articulated as follows:

$$\begin{aligned}
&f(\varepsilon_1, \mu_1, d_1, \ldots \varepsilon_n, \mu_n, d_n) = Min.\ |\ (\mathrm{PSF_t} - \mathrm{PSF_a})\ | \\
&\text{Subject to } d_1 + \cdots + d_n = d \text{ or } \le d \\
&\varepsilon_{1L}, \ldots \varepsilon_{nL} \le \varepsilon_1, \ldots \varepsilon_n \le \varepsilon_{1U}, \ldots \varepsilon_{nU} \\
&\mu_{1L}, \ldots, \mu_{nL} \le \mu_1, \ldots \mu_n \le \mu_{1U}, \ldots \mu_{nU} \\
&d_{1L}, \ldots, d_{nL} \le d_1, \ldots d_n \le d_{1U}, \ldots d_{nU}
\end{aligned} \tag{8.11}$$

where the desired total thickness of the shielding structure is denoted as d (in μm). Constrained optimization techniques can be applied in both the formulations, i.e., Eqs. (8.1) and (8.11). To implement RBDO formulation (8.11), sets of input design vectors are generated from the design space. These vectors are used to compute the PSF using (8.5) equation for each set of values. Typically, identifying local behaviors in the design space necessitates a larger number of analyses compared to detecting global responses. To ensure robust coverage and prevent potential inaccuracies near the corners of the design space, we generate 500 sampling points. This strategy helps avoid issues that might arise from using LHS near certain corners, ultimately ensuring better accuracy in the vicinity of these design points.

Table 8.1 Distribution properties of parameters and variables.

Parameters and random variables	Distribution type
σ_1, σ_3	Normal distribution with coefficients of variation of 0.02
d_1, d_3	Normal distribution with standard deviation of 3
Frequency	Uniform distribution with limits from 8 GHz to 12.5 GHZ
θ	Uniform distribution with limits from 0 to 89°
Polarization	Uniform distribution for occurrence of TE or TM

The objective is to determine the design variable values that yield an acceptable/target *pof* represented as p_{tf}, ensuring that the shielding responses remain below a specified limit. In the shielding design problem, the failure mode occurs when the SE in decibels (dB) falls short of the required SE (dB) due to variability in the shielding manufacturing and the randomness in the material properties. Therefore, it is assumed that the thickness and conductivity of the third and first layers within the three-layered shielding structure adhere to independent normal distributions. Other external random parameters, such as polarization, frequency, and incidence angle, are presumed to conform to uniform distributions. A summary of the probabilistic attributes representing the variability of these parameters and variables is given in Table 8.1.

When setting p_{tf} equal to 0.001, we conducted one million MCS cycles to collect the PSF dataset for 500 designs. According to Eq. (8.7), the calculated percentage error, based on one million simulation cycles, is 6.29%. Consequently, there is a 95% likelihood that the estimated p_{tf}, obtained through these one million MCS cycles, will lie within the range of $1 \times 10^{-3} \pm 6.3214 \times 10^{-5}$.

The dataset, which comprises 500 design points along with their corresponding PSF, is approximated using a two-layered multilayer perceptron neural network. To divide the dataset for training and evaluation, we employed a random split, allocating 70% of the data for training, 15% for validation, and another 15% for testing purposes. The training of the neural network was carried out in batch mode, where all training set inputs were applied before updating the weights. This method accelerates the training procedure while also resulting in a reduced error compared to incremental training. For the ANN training, we opted for the Levenberg-Marquardt algorithm, which is known for its speed and stability. Subsequently, we identified a network configuration with 13 neurons in the hidden layer that demonstrated the best validation performance and correlation coefficient. The MSE calculated for this network reached 1.9081×10^{-6} at epoch 125, as illustrated in Figure 8.3a. Figure 8.3b depicts the *R*-value. Notably, the training process appeared optimal, as there was no significant increase in the test curve before the validation curve started to rise, indicating the absence of overfitting.

The design problem described in Eq. (8.11) with its associated constraints can now be formulated as follows:

$$
\begin{aligned}
&f(\sigma_1, d_1, \sigma_3, d_3) = \text{Minimize}|(1 - PSF_a)| \\
&\text{subject to } d_1 + d_3 = 500, 30 \leq \sigma_1 \leq 10{,}000, \\
&30 \leq \sigma_3 \leq 10{,}000, 10 \leq d_1 \leq 300, \\
&10 \leq d_3 \leq 300, d_2 = 125, \\
&\varepsilon_1 = 0, \varepsilon_2 = 3.1, \\
&\varepsilon_3 = 0, \sigma_2 = 0, \mu_1, \mu_2, \mu_3 = \mu_0.
\end{aligned}
$$

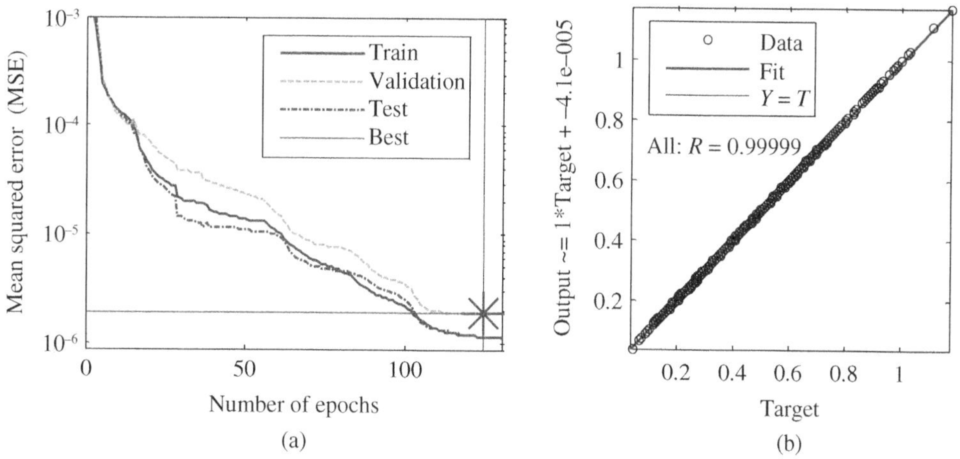

Figure 8.3 (a) Learning curves and (b) regression plot of all design points of the data set.

Table 8.2 Reliability-based design parameters for EMI shielding structure.

Material parameters	Acceptable/ target *pof*	Achieved *pof*	SE (dB)
$\sigma_1 = 7815.8408$, $d_1 = 252.3906$	1×10^{-3}	1.0406×10^{-3}	107.1588
$\sigma_3 = 8419.1538$, $d_3 = 247.6144$			

In order to streamline optimization process, we utilized a real-coded GA with specific parameter settings: with a population size of $N = 100$, a tournament size of $T_n = 4$, crossover probability of $p_c = 0.899$, mutation probability of $p_m = 0.01$, and maximum number of generations $G_{\max} = 100$. The outcome of this optimization is displayed in the first column of Table 8.2, which includes both an acceptable probability of failure and the actual achieved probability. The fourth column presents the relevant SE when considering electromagnetic (EM) waves at an 8 GHz frequency, assuming a perfectly optimized design point and excluding any uncertainty. Notably, considering the uncertainty in the parameters and design variables, this analysis indicates that there is a requirement for an extra margin of around 27.16 dB in the practical shielding design based on the specified scenario and distribution parameters of the random variables.

8.4 Conclusion

In this chapter, an RBDO approach has been presented and demonstrated using cases related to the design of shielding structures utilized in military applications. Utilizing RBDO in the design of these structures allows for the deliberate inclusion of uncertainties originating from both the process of manufacturing and external random variable's impact. The methodology commenced through the generation of the design points within design space utilizing LHS and computed the PSF through the MCS approach. The PSF served as an additional safety measure, enhancing the accuracy of probability calculations in regions of low failure likelihood. To approximate the PSF values associated with LHS-generated design points, Artificial Neural Networks are employed. Although constructing the deterministic response database and training the ANN can be time-intensive, one of the significant advantages is that the predicted PSF values align closely with acceptable tolerances and can be generated efficiently with minimal computational resources during the design optimization phase. The design parameters are used to estimate the *pof*, and this estimated probability is then compared to the desired or target *pof*, showing that the design successfully aligns with the specified requirements.

Appendix 8.A

Evaluation of Electromagnetic Shielding Effectiveness

Now, let's contemplate a scenario where a plane wave is propagating in the z-direction and impinges upon a multilayer planar dielectric medium with varying thicknesses represented by $d_1, d_2, \dots, d_n$. This occurs at an arbitrary incident angle θ_0, which implies the wave is incident at an angle other than normal (i.e., oblique incidence). The boundaries between these layers are denoted as $Z(0)$, $Z(1), \dots, Z(n)$, and this arrangement is depicted in Figure 8.A.1 for reference.

Each individual medium is assumed to exhibit homogeneity and isotropy, characterized by specific constitutive parameters for each layer, denoted as permittivity (ε_k), permeability (μ_k), and thickness (d_k). Notably, the 0th and $(n+1)$th regions are considered semi-infinite and may possess distinct properties. The plane wave fields within each layer are decomposed into two polarization types: transverse electric (TE) and transverse magnetic (TM). It is important to emphasize that the expressions for transmission coefficients vary depending on the polarization type of the electromagnetic waves under consideration (Schulz et al. 1988). Assessing the shielding effectiveness under normal incidence conditions (i.e., when $\theta_0 = 0$) is simply a specific instance within the broader framework of oblique incidence (Oraizi and Afsahi 2007).

The characteristics of the kth layer, including permittivity, wave number (k_k), propagation constant (γ_k), and characteristic impedances (η_{kTE} and η_{kTM}), are provided as follows:

$$\varepsilon_k = \varepsilon'_k - j\varepsilon''_k \tag{8.A.1}$$

$$k_k = \omega\sqrt{\mu_k \varepsilon_k} = \beta_k - j\alpha_k \tag{8.A.2}$$

where $\omega = 2\pi f$ and f is the frequency of incident wave in hertz (Hz). The real and imaginary part of the kth layer permittivity is denoted by ε'_k and ε''_k, respectively, where $\varepsilon''_k = \frac{\sigma_k}{\omega\varepsilon_0}$, σ_k is the electrical conductivity and ε_0 is the permittivity of free space (8.854×10^{-12} farads per meter (Fm^{-1})). Permeability is denoted by $\mu_k = \mu'_k - j\mu''_k$, where μ'_k and μ''_k is the real and imaginary part of the kth layer permeability, respectively. In the medium left to the shield, the wavenumber is real

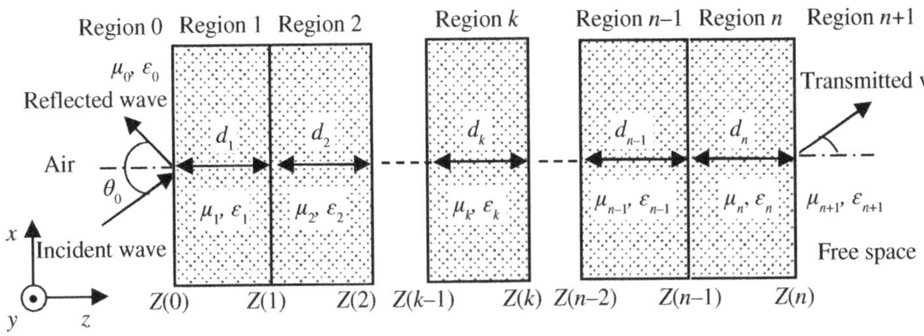

Figure 8.A.1 Oblique incidence of plane wave on a planar multilayer shield.

with the components $k_x = k_0 \sin\theta_0$, $k_z = k_0 \cos\theta_0$ with $k_0 = \omega\sqrt{\mu_0\varepsilon_0}$ where μ_0 is the permeability of the free space ($4\pi \times 10^{-7}$ Henries per meter (Hm^{-1})) and in the lossy medium, $k_{kx} = k_x$ and $k_{kz} = \beta_{kz} – j\alpha_{kz} = \sqrt{{k_k}^2 - {k_{kx}}^2}$:

$$\gamma_k = jk_{kz} = \alpha_{kz} - j\beta_{kz} \tag{8.A.3}$$

where α_{kz} is the attenuation constant and is measured in nepers per meter (Np/m), whereas β_{kz} is the phase constant and is measured in radians per meter (rad/m):

$$\eta_{kTE} = (\omega\mu_k/k_{kz}), \eta_{kTM} = (k_{kz}/\omega\mu_k) \tag{8.A.4}$$

For the multilayer shield input impedance seen at the left boundary of the kth layer toward the right side is given as:

$$Z_{in}(d_k) = \eta_{k(\mathrm{TE/TM})} \frac{Z(d_k)\cos h(\gamma_k d_k) + \eta_{k\left(\frac{\mathrm{TE}}{\mathrm{TM}}\right)} \sin h(\gamma_k d_k)}{Z(d_k)\sin h(\gamma_k d_k) + \eta_{k(\mathrm{TE/TM})}\cos h(\gamma_k d_k)}, k = 1, 2, \ldots, n \tag{8.A.5}$$

The transmission coefficient, which is defined as the ratio of the transmitted field to the incident field, is provided as follows:

$$T_m = P_m \prod_{k=1}^{n} \frac{\exp[-d_k(\gamma_k - \gamma_0)]}{[1 - q_k \exp(-2\gamma_k d_k)]} \tag{8.A.6}$$

where m represents either E or H field and

$$P_E = \frac{2^{n+1}\eta_{1(\mathrm{TE/TM})}\eta_{2(\mathrm{TE/TM})} \cdots \eta_{n(\mathrm{TE/TM})} Z_L}{(\eta_0 + \eta_{1(\mathrm{TE/TM})})(\eta_{1(\mathrm{TE/TM})} + \eta_{2(\mathrm{TE/TM})}) \ldots (\eta_{n(\mathrm{TE/TM})} + Z_L)},$$

$$P_H = \frac{2^{n+1}\eta_0\eta_{1(\mathrm{TE/TM})} \cdots \eta_{n(\mathrm{TE/TM})}}{(\eta_0 + \eta_{1(\mathrm{TE/TM})})(\eta_{1(\mathrm{TE/TM})} + \eta_{2(\mathrm{TE/TM})}) \ldots (\eta_{n(\mathrm{TE/TM})} + Z_L)},$$

$$q_k = \frac{(\eta_{k(\mathrm{TE/TM})} - \eta_{(k-1)(\mathrm{TE/TM})})(\eta_{k(\mathrm{TE/TM})} - Z(d_k))}{(\eta_{k(\mathrm{TE/TM})} + \eta_{(k-1)(\mathrm{TE/TM})})(\eta_{k(\mathrm{TE/TM})} + Z(d_k))}, \text{for } k = 1, 2, \ldots, n.$$

Using Eq. (9.17), the total SE is determined as:

$$S_\mathrm{m} = -20 \log_{10} |T_\mathrm{m}| . \tag{8.A.7}$$

References

Birger, I.A. (1970). *Safety Factors and Diagnostics, Problems of Mechanics of Solid Bodies*, 71–82. Leningrad: Sudostroenve (in Russian).

Cardoso, J.B., de Almeida, J.R., Dias, J.M., and Coelho, P.G. (2008). Structural reliability analysis using Monte Carlo simulation and neural network. *Advances in Engineering Software* 39 (6): 505–513.

Chau, K.W. (2007). Reliability and performance-based design by artificial neural network. *Advances in Engineering Software* 38 (3): 145–149.

Deb, K. (2000). An efficient constraint handling method for the genetic algorithm. *Computer Methods in Applied Mechanics and Engineering* 186 (2–4): 311–338.

Deb, K., Karthik, S., and Tatsuya, O. (2007). Self-adaptive simulated binary crossover for real-parameter optimization. *Proceedings of the Genetic and Evolutionary Computation Conference (GECCO-2007), UCL London*, 1187–1194.

Deb, K., Gupta, S., Daum, D. et al. (2009). Reliability based optimization using evolutionary algorithms. *IEEE Transaction on Evolutionary Computation* 13 (5): 1054–1074.

Du, X. and Chen, W. (2001). A most probable point-based method for efficient uncertainty analysis. *Journal of Design and Manufacturing Automation* 4 (1): 47–66.

Foschi, R.O., Li, H., and Zhang, J. (2002). Reliability and performance-based design: a computational approach and applications. *Structural Safety* 24 (2–4): 205–218.

Frydrýšek, K. (2008). Performance-based design applied for a beam subjected to combined stress. *Annals of the Faculty of Engineering Hunedoara-International Journal of Engineering* Fascicule 2 (Tome VI): 129–134.

Gargama, H., Chaturvedi, S.K., and Thakur, A.K. (2014a). Reliability-based design optimization of electromagnetic shielding structure using neural networks and real-coded genetic algorithm. *Proceedings of the Institution of Mechanical Engineers, Part C: Journal of Mechanical Engineering Science* 228 (18): 3471–3481.

Gargama, H., Chaturvedi, S.K., and Thakur, A.K. (2014b). Reliability-based design optimization scheme for designing electromagnetic shielding structures. *Journal of Electromagnetic Waves and Applications* 28 (6): 765–776.

Gondal, Z.R. and Lee, J. (2012). Reliability assessment using feed-forward neural network-based approximate meta-models. *Proceedings of the Institution of Mechanical Engineers, Part O: Journal of Risk and Reliability* 226 (5): 448–454.

Goldberg, D.E., Deb, K., and Clark, J.H. (1992). Genetic algorithms, noise, and the sizing of populations. *Complex Systems* 6: 333–362.

Haldar, A. and Mahadevan, S. (2000). *Reliability Assessment Using Stochastic Finite Element Analysis*. New York: John Wiley & Sons, Inc.

Hornik, K. (1991). Approximation capabilities of multilayer feedforward networks. *Neural Networks* 4 (2): 251–257.

Jensen, H., Valdebenito, M., Schuëller, G., and Kusanovic, D. (2009). Reliability-based optimization of stochastic systems using line search. *Computer Methods in Applied Mechanics and Engineering* 198 (49–52): 3915–3924.

Kaymaz, I. and McMohan, C.A. (2005). A response surface method based on weighted regression for structural reliability analysis. *Probabilistic Engineering Mechanics* 20 (1): 11–17.

McKay, M.D., Bechman, R.J., and Conover, W.J. (1979). A comparison of three methods for selecting values of input variables in the analysis of output from a computer code. *Technometrics* 21 (2): 239–245.

Myers, R.H., Montgomery, D.C., and Anderson-Cook, C.M. (2009). *Response Surface Methodology: Process and Product Optimization Using Designed Experiments*, 3e. Hoboken, New Jersey: John Wiley & Sons, Inc.

Oraizi, H. and Afsahi, M. (2007). Analysis of planar dielectric multilayers as FSS by transmission line transfer matrix method (TLTMM). *Progress in Electromagnetics Research* 74: 217–240.

Qu, X. and Haftka, R.T. (2003). Deterministic and reliability-based optimization of composite laminates for cryogenic environments. *AIAA Journal* 41 (10): 2029–2036.

Rao, S.S. (1992). *Reliability-Based Design*. New York: McGraw-Hill.

Rahman, S. and Wei, D. (2008). Design sensitivity and reliability-based structural optimization by univariate decomposition. *Structural and Multidisciplinary Optimization* 35 (3): 245–261.

Schulz, R.B., Plantz, V.C., and Brush, D.R. (1988). Shielding theory in practice. *IEEE Transactions on Electromagnetic Compatibility* 30 (3): 187–201.

Tsai, Y.-T. and Chang, H.-C. (2008). Reliability-based optimum design for mechanical problems using genetic algorithms. *Proceedings of the Institution of Mechanical Engineers, Part C: Journal of Mechanical Engineering Science* 222: 1791–1799.

Valdebenito, M.A. and Schuëller, G.I. (2010). A survey on approaches for reliability-based optimization. *Structural and Multidisciplinary Optimization* 42 (5): 645–663.

Vukovć, S. and Sopta, L. (1999). Binary-coded and real-coded genetic algorithm in pipeline flow optimization. *Mathematical Communications* 4: 35–42.

Wang, G.G. and Shan, S. (2007). Review of metamodeling techniques in support of engineering design optimization. *Journal of Mechanical Design* 129 (4): 370–380.

Yoo, K.-S., Eom, Y.-S., Park, J.-Y. et al. (2011). Reliability-based topology optimization using successive standard response surface method. *Finite Elements in Analysis and Design* 47 (7): 843–849.

Youn, B.D. and Choi, K.K. (2004a). Selecting probabilistic approaches for reliability-based design optimization. *AIAA Journal* 42 (1): 124–131.

Youn, B.D. and Choi, K.K. (2004b). An investigation of nonlinearity of reliability-based design optimization approaches. *Journal of Mechanical Design* 126 (3): 403–411.

Yu, X., Chang, K.H., and Choi, K.K. (1998). Probabilistic structural durability prediction. *AIAA Journal* 36 (4): 628–637.

Zhu, P., Zhang, Y., and Chen, G. (2011). Metamodeling development for reliability-based design optimization of automotive body structure. *Computers in Industry* 62 (7): 729–741.

Zuev, K. (2009). Advanced stochastic simulation methods for solving high-dimensional reliability problems. Ph.D. thesis, The Hong Kong University of Science and Technology, http://hdl.handle.net/1783.1/5947.

9

Parametric Estimation Models for Minimal and Imperfect Maintenance

Rajiv Nandan Rai, Sanjay Kumar Chaturvedi, and Heeralal Gargama

Subir Chowdhury School of Quality and Reliability, Indian Institute of Technology, Kharagpur, Kharagpur, West Bengal, India

9.1 Introduction

Maintenance encompasses a series of activities aimed at ensuring the functionality and operational condition of equipment, systems, or assets. These tasks involve inspections, repairs, replacements, and upgrades, all with the goal of preventing failures, reducing downtime, and prolonging the lifespan of the assets. To conduct reliability analysis on maintained systems, many industries have established different facilities to maintain, repair, and overhaul (MRO) these systems. Before delving into the rest of the content of this chapter, the authors find it necessary to provide a brief introduction to MRO and its basics, providing a fundamental understanding (Keith Mobley 2014). This primer on MRO will facilitate comprehending all the reliability assessment and analysis techniques discussed later in this chapter, specifically tailored for maintained systems. MRO entails life cycle maintenance, encompassing preventive maintenance (PM) carried out on a routine basis, plan maintenance when the system is out-of-service, and repairs in correction, overhaul or rebuilds for equipment damage. Industries primarily utilize MRO for costly, long-lasting products or equipment, such as massive machines used in manufacturing, generation of electrical power, infrastructure and marine boats, vehicles used in mass transit, and vehicles and systems used in military.

Over time, systems used in industries naturally deteriorated due to long-term use, continuous use, and exposure to environment, causing harm due to different factors. System deterioration eventually leads to its failure, which can cause various issues related to their safety, damage, quality, and machine downtime unplanned. In the past, maintenance was viewed as a challenging task managed reactively after failures occurred. However, it is now widely recognized as a critical element of asset management and a vital commercial function. Industries are increasingly realizing the impact of well-planned maintenance interventions on productivity and reliability, resulting in a rise in PM activities that better align with other industrial processes, like scheduling of production and management of spare parts. For example, companies in the process and chemical sectors can significantly enhance profitability by inhibiting unscheduled breakdowns.

The continuous automation of production processes and growing market competition have heightened awareness of the importance of effective maintenance planning. The primary objective of maintenance professionals is to keep infrastructure, facility provider equipments, and tools in optimal and efficient operating condition, thus preventing unexpected downtime or failure of equipment. This is where the system's reparability and maintainability play an essential role. The terms "maintenance" and "repair" are used interchangeably, but these differ in meaning in the

Reliability Analysis of Modern Power Systems, First Edition. Edited by R. K. Saket and P. Sanjeevikumar.

business of asset management. Repairs need to do restorative work when any break or damage in asset occurs, or it ceases to function properly. On the other hand, maintenance includes the task of routine basis repair correction and prevention, which is performed on assets to extend their life expectancy by avoiding damage. Examples of maintenance activities include routine cleaning, grease trap maintenance, air conditioning unit upkeep, painting, and regular inspections.

An overhaul is a comprehensive procedure of maintenance conducted on industrial equipment or machines with the aim of preserving its functionality. Typically, maintenance service providers carry out machinery overhauls, and the frequency of overhauls can be agreed upon, often planned annually for regular maintenance. Overhauls usually commence with a thorough inspection by skilled maintenance personnel, monitoring the machine's operation while in use.[1] Subsequently, disassembly of the equipment takes place for further inspection and subsequent stages of overhauling such as repairs/replacement. A trained technician proficiently examines the machine to identify parts requiring replacement or repair. Based on issues either identified, damaged parts of the machine are replaced or machine is kept fixed, making overhauling a more efficient alternative to replace entire equipment. However, replacement of parts may entail excessive delay if they need to be acquired from the manufacturer. After successfully substituting any necessary spare parts, the equipment is reassembled. Reassembling followed by testing is a crucial step in proper functioning of the equipment. Testing ensures that the reassembled equipment operates correctly, and if any issues arise, the procedure may need to restart from the inspection stage.

Most intricate systems, such as automobiles, communication, civil and military aviation, medical, locomotives, and more, undergo repairs once they experience failures. When these systems are put into operation, they are influenced by three main factors: operations, maintenance, and logistics. These factors must be purposefully managed based on the owners' commercial plans. Evaluation of performance parameters, characteristics, and reliability under these conditions becomes a significant concern. Key areas of interest include assessment of anticipated failures during system operation and period of warranty, maintenance of reliability threshold at minimum and within a specified interval, address of rate of wear and tear, identification of optimal replacement, determination of overhaul timings, and minimization of cost of complete life cycle.

Traditional process whether parametric or nonparametric for analysis of accelerated test data and reliability life, relies on a sample randomly drawn from a single population, by assuming that the data collected for reliability analysis is identically distributed (i.i.d) and independent. This assumption may hold true for the first failure occurs in units, that are identical, stemming from the same design process, same manufacturing process, and operating in similar environment. The life data for such items include time taken in a single failure (or in a very first failure or suspension) when items are reparable. Recent literature on reliability covers such analysis of such data in plenty, where times of failure of equipment are modeled via appropriate life distributions. However, in case of systems that are repairable, successive failures occur in a single system, which often violates the assumption of independence and identical distribution. Consequently, methods of reliability assessment and analysis based on statistical approaches for repairable and nonrepairable systems differ.

To address the characteristics, of complex systems, which are repairable and drawn for their reliability, a process is employed rather than a distribution. For systems, which are repairable, the time spent until the next failure occurs depends on the distribution of life and probability of the first failure and after the first failure, the impact of actions done to maintain a system is taken into consideration. The Power Law Process (PLP) (Rigdon & Basu 2000) is a widely used model for this

1 Regular inspections and frequent checks play a crucial role in preventing significant damage in older and more complex machines.

purpose. Moreover, if the time taken in the first failure follows the Weibull type of distribution, then the power law model adequately captures the phenomenon of minimal repair and effectively governs each failure that occurs successively. In other words, the very first failure is addressed by the Weibull type of distribution, and subsequent failures are addressed by PLP for a repairable system. Consequently, the PLP can be seen as a generalization of the Poisson process and an extension of the Weibull type of distribution. The PLP offers the advantage of computational simplicity by providing practical solutions. Practical solutions so obtained are well accepted and adopted in real-world applications.

9.2 Maintenance Actions on Maintained Systems

Maintenance actions are of utmost importance in the total technical life (TTL) of a repairable system, significantly impacting its overall reliability, availability, downtime, and operational costs. When we deal with the systems, which are repairable, it becomes crucial to comprehend different actions of maintenance usually they perform. These maintenance actions can be broadly categorized into two main groups, as shown in Figure 9.1.

Reactive or Corrective Maintenance (CM): Corrective (or unplanned or unscheduled) type of maintenance usually perform tasks on repairable systems to restore it to operating condition in smallest possible time from failure state. It involves repairing of those components, which are responsible for failure of a system. CM (Jezdimir Knezevic 1997) is done generally in the following three steps:

Detection of a Problem: first step is to locate the component failed and then to detect the problem.
Repairing of Faulty Component(s): next step is to perform repairing or replacing actions on faulty part of a system.
Verification: Once the component(s) is (are) repaired, it is verified by checking whether its operation is operating successfully or not.

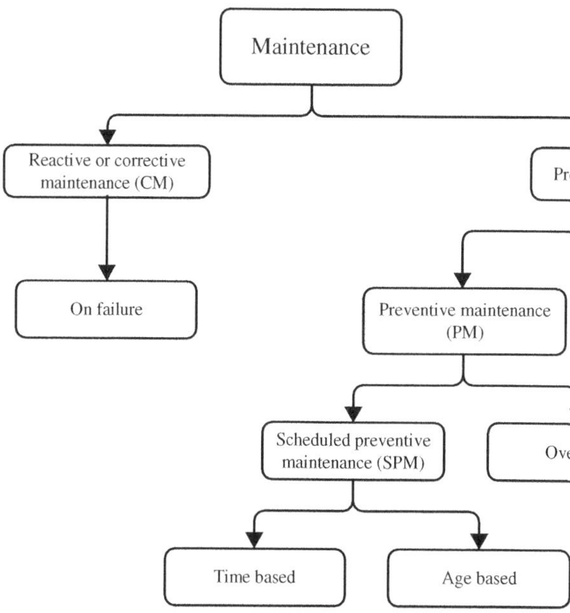

Figure 9.1 Maintenance actions on maintained systems.

Proactive Maintenance (PM): Proactive maintenance is a maintenance approach focused on detecting and addressing potential causes of equipment failure in advance, thus preventing such failures from occurring. The primary objective of proactive maintenance is to enhance asset reliability and minimize the chances of downtime. It is essential to recognize that wear and tear are natural components of equipment lifecycles and categorized into two: preventive type of maintenance (PM) and predictive type of maintenance.

Preventive Maintenance: Preventive maintenance (PM) (or scheduled maintenance) involves carrying out regularly scheduled maintenance tasks to proactively avoid future unforeseen failures during system operation. In simple terms, it entails addressing issues before they lead to a breakdown or malfunction. The tasks are carried out either on fixed time of operation or age of an item. Maintenance actions are scheduled based on wear out profile and characteristics of failure. Overhaul PM activities are important and necessary one to perform a task on systems, which are possible to repair TTL. In overhauling, we disassemble the whole system, inspect each component and again reassemble the system. Overhauling rejuvenates the system and brings the system close to as good as new condition. It is called "capital maintenance" in industrial terminology.

Predictive Maintenance: Predictive maintenance is done check operational conditions of equipment during servicing, enabling the estimation of optimal maintenance timing. This approach offers cost-saving advantages over preventive maintenance based on time and performed on routine basis, since the task of maintenance is performed at the time of need only. Thus, it is called maintenance based on condition (CBM), executed based on estimations of an item's degradation state. However, CBM is recommended, mainly for critical systems, because it may not be cost-justifiable or feasible to apply on all the systems. The primary goal of predictive maintenance is to facilitate well-timed scheduling of corrective actions and prevent unexpected equipment failures. The key advantage lies in obtaining the right information about the equipment's current condition, leading to prolonged equipment lifespan, increased safety of plant, reductions in accidents due to environmental impacts, and optimized handling of spare parts. What sets both the strategies apart is reliance of predictive type of maintenance on real-time condition data of equipment rather than relying on mean or expected life statistics solely to predict the future maintenance time. Machine learning (ML) approaches are commonly adopted for defining the actual condition of the system and forecasting its future states.

Several essential components are necessary for the successful implementation of predictive maintenance, including health-indicating parameter(s), sensors, their types, collection of data, its preprocessing, fault detection, time-to-failure prediction, early fault detection, resource optimization, and maintenance scheduling. Predictive maintenance has also emerged as a driving force for enhancing productivity and achieving "just-in-time" manufacturing practices. In this chapter, our focus is solely on the investigation of corrective actions of maintenance for further analysis.

9.3 Classifications of Imperfect Maintenance Categories

9.3.1 Perfect, Imperfect, and Minimal Repairs

A repairable system, after losing its functionality, is restored to a functional state, without any need to replace complete system. The repairing extent depends on factors like how critical was the component failed, what is the status of operation a system, risk indices etc. Consequently, the level of repair of the system is decided by the management. Two extremes of the repair are perfectly repaired and minimally repaired. If it is perfectly repaired, it gets restored to "*as-good-as-new*"

(AGAN) equipment, like it is being replaced by brand new system. This type of repair is typically is performed for highly critical components that could otherwise compromise the system's operation, safety, and the well-being of personnel working with it. On the other hand, a system is deemed minimally repaired, if it is restored to "*as-bad-as-old*" (ABAO) state. This type of repair is undertaken when there is a pressing need for the system to function for a limited time or when the system is scheduled for PM activities shortly or when it will be discarded soon.

Imperfect repair encompasses any repair that falls outside the categories of perfect and minimal repairs. In everyday systems, the majority of repairs are imperfect, meaning that the system is neither fully restored nor left in a bad state from "as-good-as-new" to "ABAO."

Figure 9.2 (Rajiv et al. 2020) illustrates the distinctions between perfect repair, minimal repair, and general repair. In the case of perfect repair, the system is brought back to an "AGAN" state, indicated by a zero starting point on the scale of time, demonstrating that the degradation of performance has been entirely reversed. On the other hand, minimal repair leaves the system's age unchanged after the repair action, meaning there is no restoration of life below its previous age. Regarding general repair, it renews a portion of the system's life, allowing it to function again but not fully restoring it to an "ABAO" or "AGAN" state. Instead, the system is brought back to an intermediate state between these two extremes.

Figure 9.3 (Rajiv et al. 2020) summarizes various techniques for repairable and nonrepairable systems, respectively. Counting is said to be *renewal process* (RP) in which the intervals between occurrences are independent stochastically and have the same distribution, which can be any arbitrary life distribution. In an RP, the *time between failures* (TBF) can be characterized by a single distribution, and the repair frequency remains constant. However, when the frequency of repairs increases (indicating deteriorating systems) or decreases (indicating system improvement), it leads to nonrenewal behavior, which in turn influences the corresponding maintenance costs. In *homogeneous Poisson process* (HPP), the random variables are exponential and are independent and identically distributed (i.i.d) On the other hand, in *nonhomogeneous Poisson process* (NHPP)

Figure 9.2 Types of repairs.

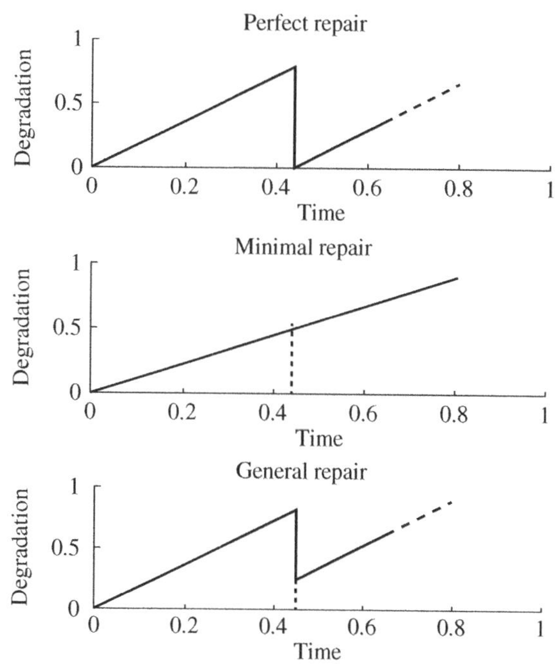

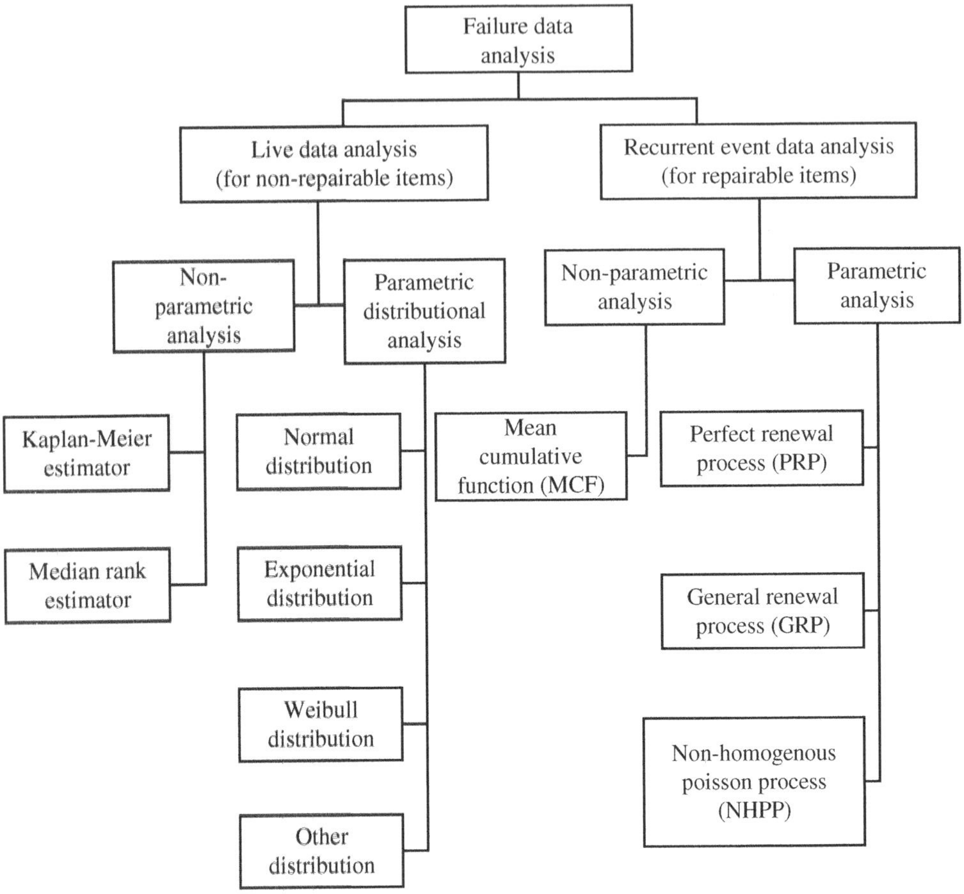

Figure 9.3 Various techniques for reliability analysis.

(Rigdon & Basu 2000; Rajiv et al. 2020) the random variables are neither independent statistically nor distributed identically.

The NHPP finds frequent application in modeling repairable systems that undergo minimal repair. The generalized renewal process (GRP) provides a framework that encompasses both extremes of repair, ranging from RP to NHPP. The GRP proves especially valuable in studying the patterns of failure of a specific unit and comprehending the impact of repair actions on the system's age. An ideal example of a system that suits the GRP model is one that receives repairs after a failure but is only partially rejuvenated, achieving neither an RP nor an NHPP condition. In such cases, it is essential to exercise caution when analyzing the data, as overlooking the actual behavior may underestimate or overestimate engineering metrics.

Analyzing failure-repair scenarios using parametric methods necessitates a certain level of statistical expertise, proficiency in solving equations of complex nature, and validate the assumptions of distributions. Additionally, equations require an iterative approach or specialized software to provide solutions rejecting analytical approach. Moreover, parametric type of approaches demand computations and may not be easily understandable for beginners or individuals with average statistical knowledge.

Whether system is able to repair or not able to repair, analysis of events follow a path from nonparametric to versatile models, with complete analysis on graphical ground. The selection of appropriate method relies heavily on data available and specific questions we aim to address.

9.4 Parametric Reliability Estimation Models for Maintained Systems

Having described the classification of maintenance categories in preceding section, we are going to discuss now parametric methods for estimation of reliability estimation with different performance parameters for the cases of perfectly, minimally and imperfectly done maintenances.

Modeling of reliability for the systems that are repairable involves employing various parametric and nonparametric methods. Here, we are focusing on parametric methods primarily considering CM. Classification of these methods of maintenance is done, based on the restoring capability of the system to its operational condition, namely: perfect, minimal, and imperfect maintenance (Rigdon & Basu 2000; Rajiv et al. 2020; Rai & Bolia 2014). Now, let's discuss parametric approaches for all categories of maintenance. In the literature, one can find several parametric approaches available for each maintenance category. However, our focus lie on HPP, NHPP and GRP in subsequent sections. In the upcoming section, essential fundamental definitions related to repairable systems are introduced.

9.4.1 Definitions

Consider few terms and definitions related to repairable systems that are reproduced from (Rigdon & Basu 2000; Rajiv et al. 2020).

(1) **Point Process**: A stochastic model used to possess randomness inherent that describes time of occurrence of events. These occurrences are thought of as point on the time continuum. In general, neither the time period lapse between occurrences are distributed identically nor it is independent. Therefore, "occurrences in time" is actually a time over which system stayed failed.

(2) **Counting Random Variable**: Let $N(t)$ denotes failure numbers occurred in the interval $[0, t]$ when N has its argument an interval $N\,(a, b]$, it will result number of failures within that interval. Here, N is the counting random variable. Failure number within $(a, b]$ is given by:

$$N\,(a, b] = N\,(b) - N(a) \tag{9.1}$$

$$\text{Suppose, } N(t_1) = k_1, N\,(t_2) = k_2, \ldots .N\,(t_n) = k_n \tag{9.2}$$

Note:

- For all value of k, If $P(N(t, t+s] = k)$ is independent of t a point process shows stationary increments.
- A point process shows independent increment if for all n and for all $a_1 < b_1 \leq a_2 < b_2 \leq \ldots a_n \leq b_n$, the random variables $N\,(a_1, b_1]$, $N\,(a_2, b_2]$, ……. , $N\,(a_n, b_n]$, are independent. In other words, $P\left(N\,\left(a_1, b_1\right] = k_1, \ldots\ldots, N\,\left(a_n, b_n\right] = k_n\right) = \prod_{i=1}^{n} P\left(N\,\left(a_i, b_i\right] = k_i\right)$.

(3) **Mean Function of Point Process**: It is the expected value of counting random variable $N(t)$ through time, i.e., the expected number of failures through time,

$$\Lambda(t) = E(N(t)) \tag{9.3}$$

(4) **Rate of Occurrence of Failures**: When $\Lambda(t)$ is differentiable then ROCOF is defined as:

$$u(t) = \frac{d}{dt}\Lambda(t) \tag{9.4}$$

(5) **Intensity Function**: Probability of failure over a small interval divided by interval length is called intensity function of a point process, i.e.,

$$u(t) = \lim_{\Delta t \to 0} \frac{P(N(t, t+\Delta t] \geq 1)}{\Delta t} \tag{9.5}$$

- *The intensity and ROCOF functions are measures of reliability a repairable system is having. Both the functions are equal provided that failures simultaneously do not occur. (The simultaneous failures occurs at discontinuous mean function $\Lambda(t)$.)*
- *Intensity and hazard rate functions are of similar kind. Notwithstanding, The intensity function is unconditional probability of a failure over a small interval of time divided by interval length. Similarly, hazard function is conditional probability of occurrence of one failure over a small interval, divided interval length. Further, for a wearing out system, probability of failure within interval $(t_0, t_0 + \Delta t_0)$ conditioned on survival past time t_0 will be < than the probability of failure in a later interval $(t_1, t_1 + \Delta t_1)$ conditioned on survival past time t_1.*
- *Comparing hazard and intensity function, the hazard function shows the limit of conditional probability; while the intensity function is not.*
- *The ROCOF (Rate of Occurrence of Failures) or intensity function of many repairable systems exhibits a bathtub-shaped curve, resembling the hazard rate curve often associated with non-repairable systems. Despite the similarity in shape, they are different in interpretations. In Figure 9.4, we can observe intensity function of bathtub-shaped. In the initial phase of the ROCOF curve or intensity function, the system is prone to experience infant mortality failures, which occur due to deficiencies in design and poor quality of manufacturing. With rectification of infant mortality failures, intensity function decreases gradually. Subsequently, the ROCOF remains constant, indicating that only random failures would occur during a phase. Finally, as the system ages, deterioration sets in, leading to a reduction in the TBF due to wear and tear in the system during this phase.*

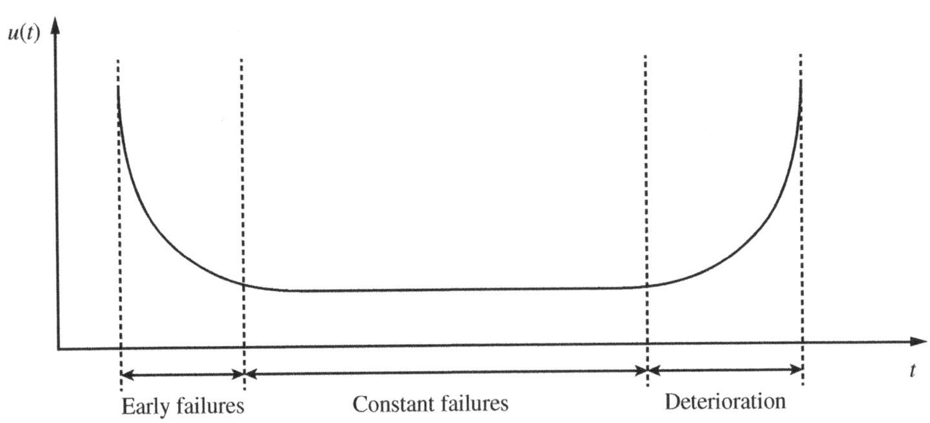

Figure 9.4 Bathtub-shaped intensity function.

(6) **Complete Intensity Function**: Consider conditional probability for some models given the failure history of process. Let $\mathcal{H}_t$ history of failure process through time t. which is represented by a set of failure times $\{t_i : i = 1, 2, \ldots, N(t)\}$. therefore the intensity function is:

$$u(t) = \lim_{\Delta t \to 0} \frac{P(N(t, t + \Delta t] \geq 1 \,|\mathcal{H}_t)}{\Delta t} \tag{9.6}$$

9.4.2 Parametric Analysis Approaches

9.4.2.1 Renewal Process

The random events are modeled stochastically and represents RP. Assuming RP is the time between the successive arrivals of events that are independent and distributed identically. If a maintainable system in operation can be restored to AGAN state after each breakdown and if the probability density function governing the TBF remains consistent across consecutive failures, then the sequence of failures can be classified as an RP. A special case is HPP with function u. In other words, if TBF, $X_1, X_2 \ldots\ldots$ are independent and distributed identically in an exponential way and randomly varying, then $N(t)$ corresponds to an HPP with u. The number of failures that are expected in $[0, t]$, would be:

$$E[N(t)] = \int_0^t u \, \mathrm{d}t = ut \tag{9.7}$$

Given that the intensity function remains unchanged, the HPP is unsuitable for representing systems that degrade or enhance over time, necessitating careful consideration before its application. In such cases, opting for a Poisson process characterized by a varying intensity function could serve as a more suitable and effective alternative.

9.4.2.2 Nonhomogeneous Poisson Process

A Poisson process (NHPP) of nonhomogeneous nature is characterized by an intensity function varying nature. Let $N(t)$ be the cumulative number of failures observed in cumulative test time t, and $u(t)$, the failure intensity. Under the NHPP model, $u(t)\Delta t$ is the probability of a failure occurring over the interval $[t, t + \Delta t]$ for small Δt. Thus the expected number of failures within interval $[0, t]$ is given by:

$$E[N(t)] = \int_0^t u(t)\mathrm{d}t \tag{9.8}$$

$u(t)$ is approximated by Power Law Model, i.e.,

$$u(t) = a \times b \times t^{b-1}; a > 0, b > 0 \tag{9.9}$$

where a is scale and b is shape parameter. For $b = 1$, $u(t) =$ a, a stable system. For $b > 1$, $u(t)$ is increasing and indicating that system is deteriorating whereas when $b < 1$, $u(t)$, is decreasing indicating that system is improving.

The power law model finds practical foundation in minimal repair and popular one. It is used to model the situation when a failed system repairing is just enough to operate the system again. If the time taken in first failure is distributed Weibull type, then each failure occur in succeeding stages is modeled by PLP. So, PLP model represents an extension of Weibull distribution. The expected number of failures then becomes:

$$E[N(t)] = \int_0^{\mathrm{t}} u(t)\mathrm{d}t = at^{\mathrm{b}} \tag{9.10}$$

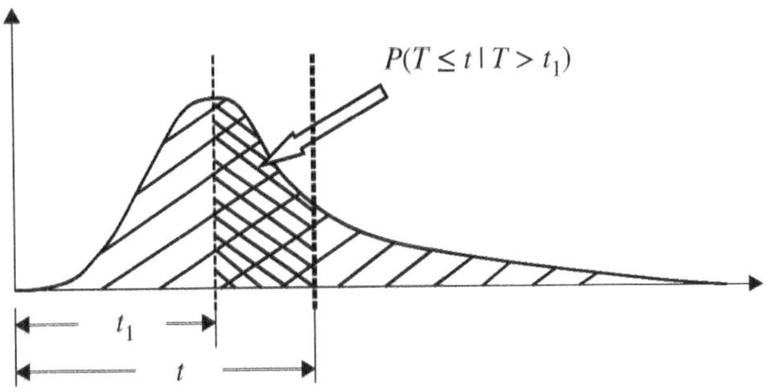

Figure 9.5 Conditional probability of occurrence of failures.

The inter-arrival times between failures occur successively and follow a conditional distribution of probability Weibull type. The arrival of ith failure up to $(i-1)^{th}$ failure is conditional on the cumulative operating time. The system retains the condition of ABAO after the $(i-1)^{th}$ repair. Thus, the repair process does not restore any added life to the component or system.

Model parameters, of conditional probability are as given by (refer Figure 9.5):

$$P(T \leq t \mid T > t_1) = \frac{F(t) - F(t_1)}{R(t_1)} = 1 - \frac{R(t)}{R(t_1)} \tag{9.11}$$

where $F(\cdot)$ is probability of failure and $R(\cdot)$ is reliability of a component at the respective time. Thus,

$$F(t_i) = 1 - \exp\left[a(t_{i-1})^{\mathrm{b}} - a(t_i)^{\mathrm{b}}\right] \tag{9.12}$$

By using above two equation the probability density function of the ith failure given that $(i-1)^{th}$ failure occurred at time t_{i-1} can be obtained as:

$$f(t_i \mid t_{i-1}) = \mathrm{abt}_i^{\mathrm{b}-1} \exp\left[-a\left\{(t_i)^{\mathrm{b}} - (t_{i-1})^{\mathrm{b}}\right\}\right] \tag{9.13}$$

NHPP model parameters can be estimated by method called maximum likelihood estimation (MLE) given below. The likelihood function is defined as:

$$L = \prod_{i=1}^{n} f(t_i \mid t_{i-1});$$

where n is number of failures.

The likelihood function is given by:

$$L = a^n b^n e^{-\mathrm{a}t^{*b}} \prod_{i=1}^{n} t_i^{b-1} \tag{9.14}$$

where,

$$t^* = \begin{cases} t_n & \text{if the test is failure terminated} \\ t > t_n & \text{if the test is time terminated} \end{cases}$$

Natural logarithm both sides give of Eq. (9.14)

$$\ln L = n \ln a + n \ln b - \mathrm{a}t^{*b} + (b-1)\sum_{i=1}^{n} \ln t_i \tag{9.15}$$

Differentiate (9.15) with respect to a and equated to zero,

$$\frac{\partial(\ln L)}{\partial a} = \frac{n}{a} - t^{*b} = 0$$

$$\hat{a} = \frac{n}{t^{*b}} \tag{9.16}$$

Similarly, taking first derivative of (9.15) with respect to b and equated to zero gives,

$$\frac{\partial(\ln L)}{\partial b} = \frac{n}{b} - \mathrm{a}t^{*b} * \ln t^* + \sum_{i=1}^{n} \ln t_i = 0$$

$$\hat{b} = \frac{n}{n \ln t^* - \sum_{i=1}^{n} \ln t_i} \tag{9.17}$$

The above notion can be extended to a fleet of K identical systems then,

$$L = \prod_{l=1}^{K} \left[a^{n_l} b^{n_l} e^{-\mathrm{a}t^{*b}} \prod_{i=1}^{n_l} t_i^{b-1} \right]$$

$$\ln L = \ln a \sum_{l=1}^{K} n_l + \ln b \sum_{l=1}^{K} n_l - \mathrm{a}t^{*b} + (b-1) \sum_{l=1}^{K} \sum_{i=1}^{n_l} \ln t_i \tag{9.18}$$

where, $l = 1, 2, \ldots K$.

Differentiating Eq. (9.18) with respect to a and b and equated to zero, we get

$$\hat{a} = \frac{\sum_{l=1}^{K} n_l}{K * t^{*b}} \tag{9.19}$$

$$\hat{b} = \frac{\sum_{l=1}^{K} n_l}{\sum_{l=1}^{K} \sum_{i=1}^{n_l} \ln\left(\frac{t^*}{t_i}\right)} \tag{9.20}$$

9.5 NHPP: Illustrative Example

Example 9.1

Consider the simulated Data for K=3 Systems operated for T=2000 hours (Crow 1974) in Table 9.1. *Estimate the Intensity Function, Instantaneous MTBF, Cumulative MTBF using NHPP. Comment on the results.*

Solution:

Here, $n = 34$ and $K = 3$. Using Eqs. (9.19) and (9.20), the values obtained are:

$$b = 0.453, a = 0.36224, u(t) = 0.36224 \times 1.56 \times t^{-0.547},$$

Instantaneous MTBF $= \frac{1}{u(t)}$, MTBF ($t = 2000$ hours) $= 389.56$ hours,

$$\text{CMTBF} = b \times \text{MTBF} = 176.47 \text{ hours.}$$

From Figure 9.6, it can be observed that the intensity function is decreasing with time (also evident from the value of $b = 0.453$ which is less than 1). Hence the systems are in infant mortality period of their lifecycle and improving.

Table 9.1 Failure times in hours for three systems.

System #1	1.2, 55.6, 72.7, 111.9, 121.9, 303.6, 326.9, 1568.4, 1913.5
System #2	1.4, 35.0, 46.8, 65.9, 181.1, 712.6, 1005.7, 1029.9, 1675.7, 1787.5, 1867.0
System #3	0.3, 32.6, 33.4, 241.7, 396.2, 480.8, 588.9, 1043.9, 1136.1, 1288.1, 1408.1, 1439.4, 1604.8

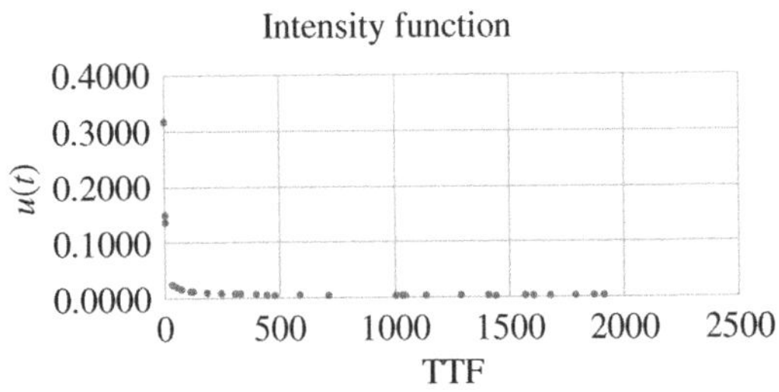

Figure 9.6 Intensity function curve for Example 9.1.

Example 9.2
The aircraft cabin's temperature is controlled by the air-conditioning system, which guarantees a pleasant environment for both passengers and crew. This system can adjust the air temperature, providing cooling or heating as necessary, depending on external factors and the desired comfort level. The air-conditioning system works in tandem with the aircraft's pressurization system, ensuring proper cabin pressure at different altitudes to avoid discomfort and potential health concerns for those on board. Table 9.2 provides the TBF data of 13 Aircraft Air-Conditioning Systems with the tail number of aircraft mentioned in the first column of the Table 9.2 (Blischke & Prabhakar Murthy 2000).

Estimate the Intensity Function, Instantaneous MTBF, Cumulative MTBF using NHPP. Comment on the results for the following cases: (1) All failures, all systems (2) Selected failures, all systems, and (3) Selected failures, main systems

Solution:
Case 1. All failures all Systems: Here $n = 212$, $K = 13$. Using Eqs. (9.19) and (10.20), the computed values obtained are: $b = 0.6633$, a=0.0928, $u(t) = 0.0928 \times 0.6633 \times t^{-0.3367}$, instantaneous MTBF = $\frac{1}{u(t)}$, MTBF ($t = 2422$ hours) = 230.458 hours, CMTBF = 152.863 hours.

Figure 9.7, shows that the intensity function decreases with time as also evident from the value of $b = 0.6633$ which is less than 1. Hence the Aircraft Air-Conditioning Systems are in infant mortality period of their lifecycle.

Case 2. Selected failures (Excluding system failures occurrences at the same time), all Systems

Using Eqs. (9.19) and (9.20), the computed values obtained are: $b = 0.8315$, a=0.0209, $u(t) = 0.0209 \times 0.8315 \times t^{-0.1685}$, instantaneous MTBF = $\frac{1}{u(t)}$, MTBF ($t = 2422$ hours) = 215.78 hours and CMTBF = 179.42 hours and

Figure 9.8, shows that the intensity function decreases with time as also evident from the value of $b = 0.8315$ which is less than 1. Hence the Aircraft Air-Conditioning Systems are still in infant mortality period of their lifecycle.

Case 3. Selected failures, Main systems (Tail numbers 7908, 7909 and 7910).

Using Eqs. (9.19) and (9.20), the computed values obtained are: b=1.1, a=0.00458, $u(t) = 0.00458 \times 1.1 \times t^{0.1}$, Instantaneous MTBF = $\frac{1}{u(t)}$, MTBF (t = 2422 hours) = 91.90 hours and CMTBF = 92.81 hours.

Table 9.2 Time between failures of aircraft air-conditioning system.

7907	194, 15, 41, 29, 33, 181
7908	413, 427, 485, 522, 622, 687, 696, 865, 1312, 1496, 1532, 1733, 1851, 1885, 1916, 1934, 1952, 2019, 2076, 2138, 2145, 2167, 2201
7909	90, 100, 160, 346, 407, 456, 470, 494, 550, 570, 649, 733, 777, 836, 865, 983, 1008, 1164, 1474, 1550, 1576, 1620, 1643, 1705, 1835, 2043, 2113, 2214, 2422
7910	74, 131, 179, 208, 710, 722, 792, 813, 842, 1228, 1287, 1314, 1467, 1493, 1819
7911	55, 320, 56, 104, 220, 239, 47, 246, 176, 182, 33, 15, 104, 35
7912	23, 261, 87, 7, 120, 14, 62, 47, 225, 71, 246, 21, 42, 20, 5, 12, 120, 11, 3, 14, 71, 11, 14, 11, 16, 9, 1, 16, 52
7913	97, 51, 11, 4, 141, 18, 142, 68, 77, 80, 1, 16, 106, 206, 82, 54, 31, 216, 46, 111, 39, 63, 18, 191, 18, 163, 24
7914	50, 44, 102, 72, 22, 39, 3, 15, 197, 188, 79, 88, 46, 5, 5, 36, 22, 139, 210, 97, 30, 23, 13, 14
7915	359, 9, 12, 270, 603, 3, 104, 2, 438
7916	50, 254, 5, 283, 35, 12
7917	130, 493
8044	487, 18, 100, 7, 98, 5, 85, 91, 43, 230, 3, 130
8045	102, 209, 14, 57, 54, 32, 67, 59, 134, 152, 27, 14, 230, 66, 61, 34

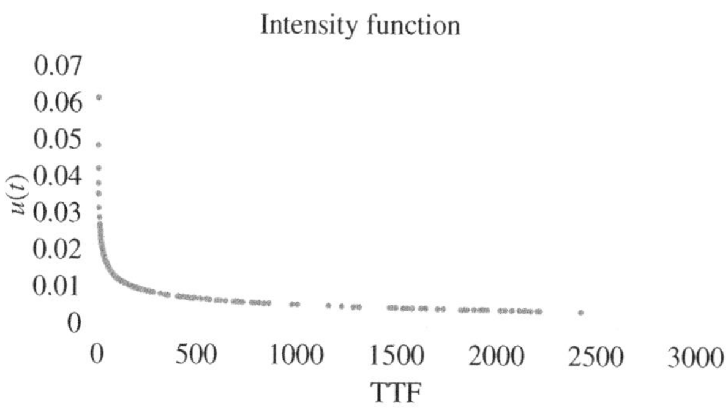

Figure 9.7 Intensity function curve for Example 9.2, Case 1.

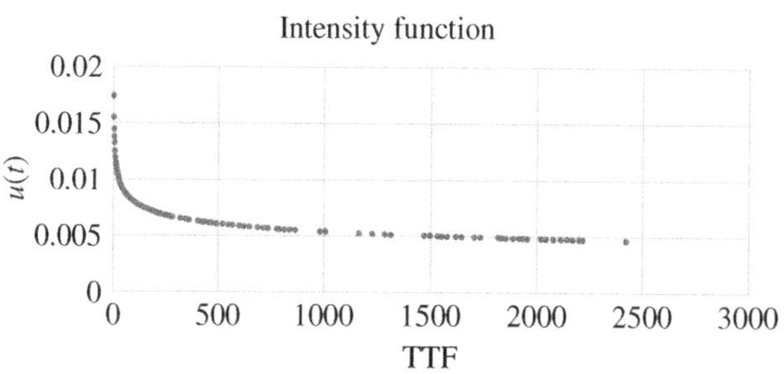

Figure 9.8 Intensity function curve for Example 9.2, Case 2.

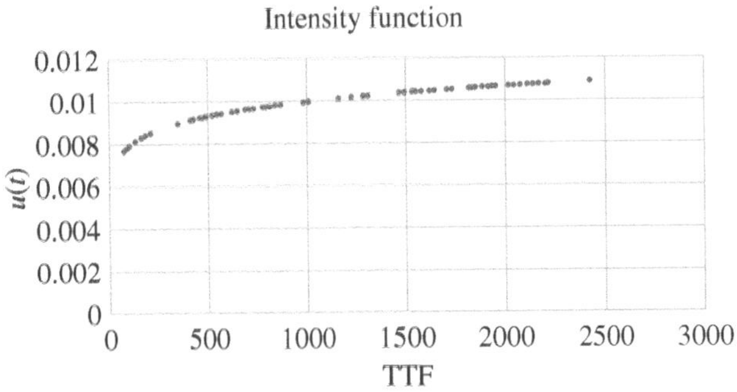

Figure 9.9 Intensity function curve for Example 9.2, Case 3.

From Figure 9.9 it can be observed that when main systems are selected then the intensity function shows an increase with time as also evident from the value of $b = 1.10$ which is greater than 1. Hence the Aircraft Air-Conditioning Systems are facing slight wear out.

9.6 Generalized Renewal Process

As per the outlined framework GRP is rooted in two overarching strategies for addressing repairs that are imperfect : Arithmetic Reduction of Age (ARA) and Arithmetic Reduction of Intensity (ARI). Within ARA models, the reparative impact is elucidated by presuming a decrease in the actual system age, resulting in the attainment of a designated "virtual age." Among these, Kijima's virtual age models, which rest on the ARA concept, hold prominence as widely referenced and efficacious constructs in existing literature. Conversely, in the ARI approach, the reparative influence is evaluated based on alterations in the failure intensity pre- and post-failure. A concise introduction to ARA models is presented in the subsequent section.

9.6.1 Arithmetic Reduction of Age Models

At the core of this category of models lies the concept that repairs have a rejuvenating effect on the system. The actual age of the system is denoted by its operational time, t, while the virtual age is established as a constructive function of the real age, potentially influenced by previous failures. The foundational premise underpinning Kijima's virtual age models (Kijima 1989; Kijima & Sumita 1986) stems from the notion that repair interventions contribute to a reduction in the system's age.

To understand these models, consider observation time of repairable system from time $t_0 = 0$ and $t_1, t_2 \ldots$ are failure times in success. The TBF is $X_n = t_n - t_{n-1}$. Effectiveness of repair is accounted by the virtual age concept in the following way. If repair effectiveness (RE) is q and V_n represents virtual age after nth repair of the system with $V_0 = 0$. Kijima - I model (Rajiv et al. 2020) assume that the nth repair can remove the damage incurred only during the time between $(n-1)$th and nth failure, yielding virtual age as:

$$V_n = V_{n-1} + qX_n \tag{9.21}$$

$$V_i = V_{i-1} + qX_i \tag{9.22}$$

$$V_i = q\sum_{j=1}^{i} X_j \tag{9.23}$$

where the distribution of $X_n$is given by:

$$Pr\{X_n < x \mid V_{n-1} = y\} = \frac{F(x+y) - F(y)}{1 - F(y)} \tag{9.24}$$

However, in practice the nth repair action may also decrease all damage accumulated up to nth failure, yielding the Kijima - II model for virtual age:

$$V_n = q(V_{n-1} + X_n) \tag{9.25}$$

$$V_i = q(V_{i-1} + X_i) \tag{9.26}$$

$$V_i = \sum_{j=1}^{i} q^{i-j+1} X_j \tag{9.27}$$

where the distribution of X_n is given by Eq. (9.24).

In this model, after repair $q = 0$ converges to AGAN condition and modeled by RP. If we assume $q = 1$ means restored component to the same condition what it was before the repair, i.e., ABAO condition and modeled as NHPP. The RE (i.e., q) is index represents effectiveness and repair quality. If q within interval $0 < q < 1$ represent a state of the system better than old condition of the system but worse than new. For $q > 1$, condition of a system is worse than old. The assumptions taken in Kijima models are given below:

- TTFF distribution is known and is estimated using Weibull distribution MLEs form data available.
- The failures is viewed as a point processes if repair time is negligible.

GRP model parameters are estimated using MLE. Approach given in Yanez et al. (2002) is widely used for GRP parameter estimation currently and most widely. The inter-arrival of failures are assumed to follow the Weibull distribution, and $f(.)$ and $F(.)$ of the time to ith failure is given by:

$$f((t_i \mid t_{i-1})) = \left[ab(V_{i-1} + X_i)^{b-1}\right] \times \exp\left[a\left\{(V_{i-1})^b - (V_{i-1} + X_i)^b\right\}\right]$$

$$F(t_i) = 1 - \exp\left[a\left\{(V_{i-1})^{\mathrm{b}} - (V_{i-1} + X_i)^b\right\}\right]$$

In above equations V_i could be either from Kijima - I or Kijima - II model.

The likelihood, log-likelihood and MLEs for the failure terminated (single and multiple repairable systems) and time terminated (single and multiple repairable systems) cases for system data are given in the following sub-sections for both Kijima - I and Kijima - II models.

9.6.1.1 Kijima - I Model

The likelihood, log-likelihood functions and MLEs for failure terminated single repairable system data set are given as follows (Rajiv et al. 2020):

$$L = \prod_{i=1}^{n} \left[ab\left(q\sum_{j=1}^{i-1} X_j + X_i \right)^{b-1} \right] \times \exp\left[a\left\{ \left(q\sum_{j=1}^{i-1} X_j \right)^{b} - (V_{i-1} + X_i)^b \right\} \right] \tag{9.28}$$

And, by taking log on both sides of Eq. (9.28).

$$\ln L = n\log(b) + n\log a + (b-1) \times \sum_{i=1}^{n} \log\left(q\sum_{j=1}^{i-1} X_j + X_i\right) + \sum_{i=1}^{n}\left[a\left\{\left(q\sum_{j=1}^{i-1} X_j\right)^b - \left(q\sum_{j=1}^{i-1} X_j + X_i\right)^b\right\}\right] \tag{9.29}$$

To obtain failure terminated MLEs, differentiate (9.29) with respect to a, b and q and equate to zero.

The likelihood, log-likelihood functions and MLEs for time terminated single repairable system data set are given as follows:

$$L = \prod_{i=1}^{n}\left[ab\left(q\sum_{j=1}^{i-1} X_j + X_i\right)^{b-1}\right] \times \exp\left[a\left\{\left(q\sum_{j=1}^{i-1} X_j\right)^b - \left(q\sum_{j=1}^{i-1} X_j + X_i\right)^b\right\}\right] \times \exp\left[a\left\{\left(q\textstyle\sum_{j=1}^{n} X_j\right)^b - \left(T - t_n + q\textstyle\sum_{j=1}^{n} X_j\right)^b\right\}\right] \tag{9.30}$$

Taking log on both sides of Eq. (9.30), we obtain

$$\ln L = n\log(b) + n\log a + (b-1)\sum_{i=1}^{n}\log\left(q\sum_{j=1}^{i-1} X_j + X_i\right) + \sum_{i=1}^{n} \times \left[a\left\{\left(q\sum_{j=1}^{i-1} X_j\right)^b - \left(q\sum_{j=1}^{i-1} X_j + X_i\right)^b\right\}\right] + \left[a\left\{\left(q\sum_{j=1}^{n} X_j\right)^b - \left(T - t_n + q\sum_{j=1}^{n} X_j\right)^b\right\}\right] \tag{9.31}$$

To obtain time terminated MLEs, differentiate the above logarithm Eq. (9.31) of the likelihood function with respect to a, b, and q and equate to zero.

The likelihood, log-likelihood functions and MLEs for failure terminated multiple repairable system data set are given as follows:

$$\mathrm{L} = \prod_{l=1}^{\mathrm{K}}\left[\prod_{i=1}^{n_l}\left[ab\left(q\sum_{j=1}^{i-1} X_{l,j} + X_{l,i}\right)^{b-1}\right] \times \exp\left[a\left\{\left(q\sum_{j=1}^{i-1} X_{l,j}\right)^b - \left(q\sum_{j=1}^{i-1} X_{l,j} + X_{l,i}\right)^b\right\}\right]\right] \tag{9.32}$$

where K is number of systems ($l = 1, 2, \dots, K$). Taking log on both sides of Eq. (9.32)

$$\ln L = \sum_{l=1}^{\mathrm{K}} n_l \log(b) + \sum_{l=1}^{\mathrm{K}} n_l \log a + (b-1)\sum_{l=1}^{\mathrm{K}}\sum_{i=1}^{n_l}\log\left(q\sum_{j=1}^{i-1} X_{l,j} + X_{l,i}\right) + \sum_{l=1}^{\mathrm{K}}\sum_{i=1}^{n_l}\left[a\left\{\left(q\sum_{j=1}^{i-1} X_{l,j}\right)^b - \left(q\sum_{j=1}^{i-1} X_{l,j} + X_{l,i}\right)^b\right\}\right] \tag{9.33}$$

To get failure terminated MLEs, differentiate the above logarithm of the likelihood function with respect to a, b and q and equate to zero. The likelihood, log-likelihood functions and MLEs for time terminated multiple repairable system data set are given as follows:

$$L = \prod_{l=1}^{K}\left[\prod_{i=1}^{n_l}\left[ab\left(q\sum_{j=1}^{i-1}X_{l,j} + X_{l,i}\right)^{b-1}\right] \times \exp\left[a\left\{\left(q\sum_{j=1}^{i-1}X_{l,j}\right)^{b} - \left(q\sum_{j=1}^{i-1}X_{l,j} + X_{l,i}\right)^{b}\right\}\right]\right.$$
$$\left.\times \exp\left[a\left\{\begin{matrix}\left(q\sum_{j=1}^{n_l}X_{l,j}\right)^{b} - \\ \left(T - t_{1,n} + q\sum_{j=1}^{n_l}X_{l,j}\right)^{b}\end{matrix}\right\}\right]\right] \tag{9.34}$$

Taking log on both sides of Eq. (9.34), we obtain:

$$\ln L = \sum_{l=1}^{K} n_l \log(b) + \sum_{l=1}^{K} n_l \log a + (b-1)\sum_{l=1}^{K}\sum_{i=1}^{n_l}\log\left(q\sum_{j=1}^{i-1}X_{l,j} + X_{l,i}\right)$$
$$+ \sum_{l=1}^{K}\sum_{i=1}^{n_l}\left[a\left\{\left(q\sum_{j=1}^{i-1}X_{l,j}\right)^{b} - \left(q\sum_{j=1}^{i-1}X_{l,j} + X_{l,i}\right)^{b}\right\}\right]$$
$$+ \sum_{l=1}^{K}\left[a\left\{\left(q\sum_{j=1}^{n_l}X_{l,j}\right)^{b} - \left(T - t_{l,n} + q\sum_{j=1}^{n_l}X_{l,j}\right)^{b}\right\}\right] \tag{9.35}$$

To obtain time terminated MLEs, differentiate the above logarithm of the likelihood function with respect to a, b and q, and equate to zero.

9.6.1.2 Kijima - II Model

The likelihood, log-likelihood functions and MLEs for failure terminated single repairable system data set are given as follows (Rajiv et al. 2020):

$$L = \prod_{i=1}^{n}\left[ab\left(\sum_{j=1}^{i-1}q^{i-j}X_j + X_i\right)^{b-1}\right] \times \exp\left[a\left\{\left(\sum_{j=1}^{i-1}q^{i-j}X_j\right)^{b} - \left(\sum_{j=1}^{i-1}q^{i-j}X_j + X_i\right)^{b}\right\}\right] \tag{9.36}$$

Taking log on both sides of Eq. (9.36)

$$\ln L = n\log(b) + n\log a + (b-1)\sum_{i=1}^{n}\log\left(\sum_{j=1}^{i-1}q^{i-j}X_j + X_i\right)$$
$$+ \sum_{i=1}^{n}\left[a\left\{\left(\sum_{j=1}^{i-1}q^{i-j}X_j\right)^{b} - \left(\sum_{j=1}^{i-1}q^{i-j}X_j + X_i\right)^{b}\right\}\right] \tag{9.37}$$

To obtain failure terminated MLEs, differentiate the above logarithm of the likelihood function with respect to a, b and q, and equate to zero.

The likelihood, log-likelihood functions and MLEs for time terminated single repairable system data set are given as follows:

$$L = \prod_{i=1}^{n}\left[ab\left(\sum_{j=1}^{i-1} q^{i-j}X_j + X_i\right)^{b-1}\right] \times \exp\left[a\left\{\left(\sum_{j=1}^{i-1} q^{i-j}X_j\right)^b - \left(\sum_{j=1}^{i-1} q^{i-j}X_j + X_i\right)^b\right\}\right]$$

$$\times \exp\left[a\left\{\begin{array}{l}\left(\sum_{j=1}^{n} q^{n-j+1}X_j\right)^b - \\ \left(T - t_n + \sum_{j=1}^{n} q^{n-j+1}X_j\right)^b\end{array}\right\}\right] \tag{9.38}$$

Taking log on both sides of Eq. (9.38), we obtain:

$$\ln L = n\,\log(b) + n\log a + (b-1)\sum_{i=1}^{n}\log\left(\sum_{j=1}^{i-1} q^{i-j}X_j + X_i\right)$$

$$+ \sum_{i=1}^{n}\left[a\left\{\left(\sum_{j=1}^{i-1} q^{i-j}X_j\right)^b - \left(\sum_{j=1}^{i-1} q^{i-j}X_j + X_i\right)^b\right\}\right]$$

$$+ \left[a\left\{\left(\sum_{j=1}^{n} q^{n-j+1}X_j\right)^b - \left(T - t_n + \sum_{j=1}^{n} q^{n-j+1}X_j\right)^b\right\}\right] \tag{9.39}$$

To obtain time terminated MLEs, differentiate the above logarithm of the likelihood function with respect to a, b and q, and equate to zero.

The likelihood, log-likelihood functions and MLEs for failure terminated multiple repairable system data set are given as follows:

$$L = \prod_{l=1}^{K}\left[\prod_{i=1}^{n_l}\left[ab\left(\sum_{j=1}^{i-1} q^{i-j}X_{l,j} + X_{l,i}\right)^{b-1}\right] \times \exp\left[a\left\{\left(\sum_{j=1}^{i-1} q^{i-j}X_{l,j}\right)^b - \left(\sum_{j=1}^{i-1} q^{i-j}X_{l,j} + X_{l,i}\right)^b\right\}\right]\right] \tag{9.40}$$

where, K is number of systems ($l = 1, 2, \ldots\ldots, K$). Taking log on both sides of Eqs. (9.40) and (9.40), we obtain:

$$\ln L = \sum_{l=1}^{K} n_l \log(b) + \sum_{l=1}^{K} n_l \log a + (b-1)\sum_{l=1}^{K}\sum_{i=1}^{n_l}\log\left(\sum_{j=1}^{i-1} q^{i-j}\; X_{l,j} + X_{l,i}\right)$$

$$+ \sum_{l=1}^{K}\sum_{i=1}^{n_l}\left[a\left\{\left(\sum_{j=1}^{i-1} q^{i-j}X_{l,j}\right)^b - \left(\sum_{j=1}^{i-1} q^{i-j}X_{l,j} + X_{l,i}\right)^b\right\}\right] \tag{9.41}$$

To obtain failure terminated MLEs, differentiate the above logarithm of the likelihood function with respect to a, b and q, and equate to zero.

The likelihood, log-likelihood functions and MLEs for time terminated multiple repairable system data set are given as follows:

$$L = \prod_{l=1}^{K}\left[\left[\prod_{i=1}^{n_l}\left[ab\left(\sum_{j=1}^{i-1} q^{i-j}X_{l,j} + X_{l,i}\right)^{b-1}\right]\right]\right.$$

$$\times \exp\left[a\left\{\left(\sum_{j=1}^{i-1} q^{i-j}X_{l,j}\right)^b - \left(\sum_{j=1}^{i-1} q^{i-j}X_{l,j} + X_{l,i}\right)^b\right\}\right]\right]$$

$$\times \exp\left[a\left\{\begin{matrix}\left(\sum_{j=1}^{n_i} q^{n-j+1}X_{l,j}\right)^b - \\ \left(T - t_{l,n} + \sum_{j=1}^{n_i} q^{n-j+1}X_{l,j}\right)^b\end{matrix}\right\}\right]\right]\right] \quad (9.42)$$

Taking log on both sides of Eq. (9.42), we obtain

$$\ln L = \sum_{l=1}^{K} n_l \log(b) + \sum_{l=1}^{K} n_l \log a + (b-1)\sum_{l=1}^{K}\sum_{i=1}^{n_l} \log\left(\sum_{j=1}^{i-1} q^{i-j}X_{l,j} + X_{l,i}\right)$$

$$+ \sum_{l=1}^{K}\sum_{i=1}^{n_l}\left[a\left\{\left(\sum_{j=1}^{i-1} q^{i-j}X_{l,j}\right)^b - \left(\sum_{j=1}^{i-1} q^{i-j}X_{l,j} + X_{l,i}\right)^b\right\}\right]\right]$$

$$+ \sum_{l=1}^{K}\left[a\left\{\left(\sum_{j=1}^{n_i} q^{n-j+1}X_{l,j}\right)^b - \left(T - t_{l,n} + \sum_{j=1}^{n_i} q^{n-j+1}X_{l,j}\right)^b\right\}\right]\right] \quad (9.43)$$

To obtain time terminated MLEs, differentiate the above logarithm of the likelihood function with respect to a, b and q, and equate to zero.

Note: The MLE equations so obtained from Kijima - I and Kijima - II models cannot be easily solved as these are complex and exhibit nonlinearity. These can be solved in MATLAB or alike. An easier way to obtain estimators is by maximizing the log-likelihood functions. There are various methods to maximize the objective function and any off the shelf software can be used for this.

9.6.1.3 Virtual Age based Reliability Metrics

Transforming reliability metrics from the actual time domain to a virtual time domain greatly simplifies mathematical computations when estimating parameters of reliability. The virtual time scale is converted back to the original real-time scale. Using virtual time scale all the equations for MTBF, reliability, availability, intensity function, and expected number of failures (Rajiv et al. 2020; Rai & Bolia 2014) are formulated and presented in Table 9.3 for ready reference.

9.7 GRP: Illustrative Examples

Example 9.3

Military aircraft utilize aero engines with formidable thrust capabilities to facilitate rapid ascent and withstand substantial "G" forces during maneuvers. These engines are meticulously engineered to avert surge and stall phenomena induced by the back pressure generated by rocket and missile launches, which create considerable turbulence ahead of the engine. These power plants endure extensive aerodynamic and thermal strains, thus rendering them susceptible to recurrent failures. The recorded failure times in hours for one such aero engine, with a time interval between overhauls set at 550 hours, are detailed below (Rajiv et al. 2020):

203, 477, 318, 536, 494, 213, 303, 525, 345, 299, 154, 230, 132, 321, 123, 351, 188, 49. 02, 267, 548, 380, 61, 160, 375, 550, 174, 176, 257, 102, 81, 541, 518, 533, 547, 299, 208, 326, 451, 349, 152, 509,

Table 9.3 Virtual age-based reliability metrics.

S. no.	Parameter	Equation	
1	Intensity function	$u(v_i) = a \times b \times (v_i)^{b-1}$	(9.44)
2	Mean time between failures (MTBF)	$\text{MTBF}(v_i) = \dfrac{1}{u(V_i)}$	(9.45)
3	Reliability	For first failure, $R(v_1) = \exp\{-a(v_1)^{b}\}$ (Note: $v_1 = qt_1$)	(9.46)
		For subsequent failures, $R(v_i \mid v_{i-1}) = \dfrac{R(v_i + v_{i-1})}{R(v_{i-1})} = \exp a\left[(v_{i-1})^{b} - (v_i + v_{i-1})^{b}\right]$	(9.47)
4	Availability	$A(v_i) = \dfrac{\text{MTBF}(v_i)}{\text{MTBF}(v_i) + \text{MTTR}}$	(9.48)
		where MTTR stands for mean time to repair	
5	Expected number of failures	$E[N(t)] = \int_0^{V_i} u(V_i)\text{d}V$ $i = 1, 2, \ldots n$	(9.49)

249, 325, 261, 328, 48, 19, 142, 200, 426, 90, 522, 446, 338, 55, 549, 84, 342, 162, 250, 368, 96, 431, 14, 207, 324, 546.

The maintenance team's task of determining if the aero engine experiences degradation within a 550-hour span is of utmost importance. This assessment will also aid in evaluating the effectiveness of the current PM strategy. Additionally, the team is keen on determining the characteristic lifespan of the aero engine. Estimate repair effectiveness index, shape and scale parameters using GRP Kijima - I model. Plot the intensity function curve. If MTTR = 528 hours of an aero engines. Plot availability curve. What will be MTBF and availability of an aero engines at $t = 550$ hours?

Solution:

Maximizing Eq. (9.33), we obtain the following results:

$$a = 0.00022, b = 1.35, q = 0.75, \text{MTTR} = 528 \text{ hours (given)}$$

Using Eqs. (9.44), (9.45) and (9.48), Intensity function, MTBF and availability can be estimated. The intensity function equation from the values obtained works out to be

$$u(v_i) = 0.00022 \times 1.35 \times (v_i)^{0.35}$$

The intensity function curve is plotted at Figure 9.10.

$$\text{MTBF } (t = 550 \text{ hours}) = 394.5602 \text{ hours}$$

The availability curve is plotted at Figure 9.11. Availability ($t = 550$ hours) = 0.4277.

Example 9.4

For GRP application, let a system tested for $T = 395.2$ hours with 56 times failure as given in Table 9.4 below. The first failure is recorded at 0.7 hours, the second failure is recorded 3 hours

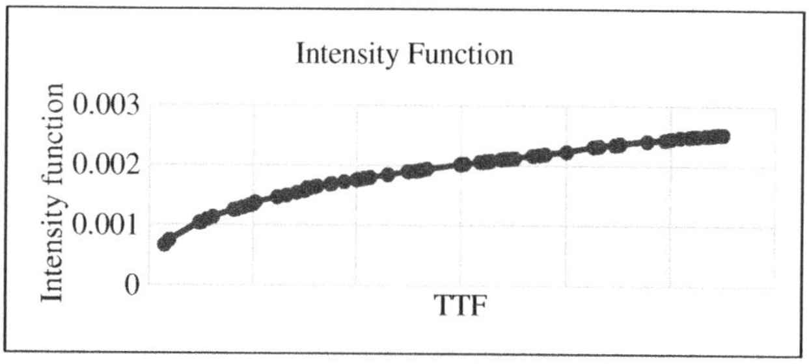

Figure 9.10 Intensity function plot for Example 9.3.

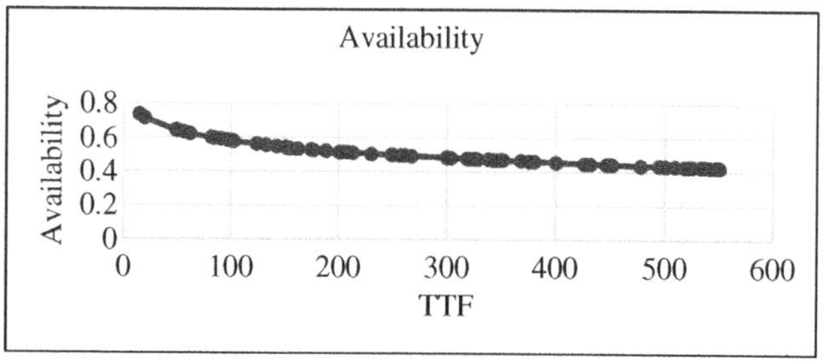

Figure 9.11 Availability plot for Example 9.3.

Table 9.4 Time to failure Data (hrs) of Example 9.4.

0.7	0.63	125	244	315	366
3.7	72	133	249	317	373
1	99	151	250	320	379
1	99	163	260	324	389
1	100	164	263	324	394
2	102	174	273	342	395.2
4	112	177	274	350	
5	112	191	282	355	
5	120	192	285	364	
5	121	213	304	364	

later at 3.7. The last failure occurred at 395.2 hours into the test and thereafter system was removed from the test. This failure data is truncated. Find out values of a, b, and q Kijima MLEs (Rajiv et al. 2020).

Solution:

Here $n = 56$, $K = 1$, and $T = 395.2$ hours

Maximizing Eq. (9.33), we obtain the following results for Kijima - I model: $b = 0.9372$, $a = 0.2061$, $q = 1.0$

Maximizing Eq. (9.37), we obtain the following results for Kijima - II model: $b = 0.24725$, $a = 0.89442$, $q = 0.93$

Table 9.5 Time to failure data (hrs) of Example 9.5.

System 1	System 2	System 3	System 4	System 5	System 6
2227.08	772.9542	900.9855	411.407	688.897	105.824
2733.229	1034.458	1289.95	1122.74	915.101	
3524.214	3011.114	2689.878			
5568.634	3121.458	3928.824			
5886.165	3624.158	4328.317			
5946.301	3758.296	4704.24			
6018.219		5052.586			
7202.724		5473.171			

Example 9.5
Suppose $K = 6$ systems are observed during $[0, T_i]$ hours $= 1, \dots, k$. That is, the data are time truncated with $T_1 = 8760$, $T_2 = 5000$, $T_3 = 6200$, $T_4 = 1300$, $T_5 = 2650$, $T_6 = 500$. Failure data are given in Table 9.5 below. This failure data set is from multiple repairable systems. Find out a, b and q values for the given data set using Kijima MLEs (Rajiv et al. 2020).

Solution:
Here, $K = 6$, $T_1 = 8760$, $T_2 = 5000$, $T_3 = 6200$, $T_4 = 1300$, $T_5 = 2650$, $T_6 = 500$.

Maximizing Eq. (9.35), we obtain the following results for Kijima - I model: $b = 1.238$, $a = 0.00018$, $q = 0.10$

Maximizing Eq. (9.43), we obtain the following results for Kijima - II model: $b = 1.358$, $a = 0.000068$, $q = 0.55$

9.8 Conclusion

A repairable system has the potential to land in conditions beyond the extremes of "As Good As New" (AGAN) and "As Bad As Old" (ABAO) after repair. These conditions include being better than old yet not AGAN, being superior to new, and conversely, being worse than old. To enhance accuracy in analyses and predictions, the GRP emerges as a valuable tool for diminishing the modeling uncertainty stemming from repair assumptions.

This chapter took a concise look at fundamental terminology linked to repairable systems, followed by an exploration of modeling and analysis within the repairable systems context, utilizing the GRP-based ARA concept through Kijima - I and Kijima - II virtual age models. It furnished likelihood and log-likelihood functions, along with maximum likelihood estimators (MLEs) for NHPP, Kijima - I, and Kijima - II virtual models, encompassing both failure and time-terminated data for single and multiple repairable systems. Moreover, reliability metrics based on virtual age were presented to streamline mathematical computations. To foster a better grasp of the content, all models have been illustrated through solved examples. Additional resources on ARI models are suggested for readers with a keen interest in the subject. Though this chapter is oriented mainly on Kijima based ARA models and its literature that provide a broad spectrum of the available models on imperfect repair (Rajiv et al. 2020).

Practice Problems

9.1 A Regional transit bus of six-year-old experiences minimal repair upon failure. Which had intensity function $u(t) = 0.0464\, t^{2.1}$ with t measured in years. Estimate following: Ans: (a) 21.56 years. (b) 2.37

(a) MTBF (instantaneous)

(b) Expected number of failures over the coming year.

9.2 Consider the TBF of a compressor presented in Table Problem 9.2 below: (a) Estimate the shape and scale parameters. (b) Also estimate the REI. Use Kijima - I model for this failure terminated compressor failure time's data (Yanez et al. 2002). Ans: (a) $\beta = 1.62, \alpha = 3072$ hours (b) $q = 0.70$.

Table Problem 9.2 Time between failures for a compressor.

No. of failures	Time between failures (h)	No. of failures	Time between failures (h)
1	3456	13	360
2	1584	14	998
3	236	15	656
4	516	16	180
5	1820	17	244
6	452	18	1528
7	432	19	44
8	1264	20	3064
9	3072	21	324
10	384	22	1528
11	2448	23	348
12	32	24	336

References

Blischke, W.R. and Prabhakar Murthy, D.N. (2000). *Reliability Modeling, Prediction and Optimization.* Canada: John Wiley & Sons.

Crow, L.H. (1974). Reliability analysis for complex, repairable systems. *SIAM Reliability & Biometry* 379–410.

Jezdimir Knezevic, D.R. (1997). *Systems Maintainability.* London: Chapman & Hall.

Keith Mobley, R. (2014). *Maintenance Engineering Handbook.* New York: Mc Graw Hill.

Kijima, M. (1989). Results for Repairable Systems with General Repair. *Journal of Applied Probability* 26 (1): 89–102.

Kijima, M. and Sumita, U. (1986). A useful generalization of renewal theory: counting processes governed by non-negative markovian increments. *Journal of Applied Probability* 23 (1): 71–88.

Rai, R.N. and Bolia, N. (2014). Availability based optimal maintenance policies for repairable systems in military aviation by identification of dominant failure modes. *I Mech E Part O: Journal of Risk and Reliability* 228 (1): 52–61.

Rai, R.N. and Bolia, N. (2014). Availability based methodologies for optimal maintenance policies in military aviation. *International Journal of Performability Engineering* 10 (6): 641–648.

Rajiv, N.R., Chaturvedi, S.K., and Bolia, N. (2020). *Repairable Systems Reliability Analysis: A Comprehensive Framework*. USA: John Wiley & Sons.

Rigdon, S.E. and Basu, A.P. (2000). *Statistical Methods for the Reliability of Repairable Systems*. New York: Wiley.

Yanez, M., Joglar, F., and Modarres, M. (2002). Generalized renewal process for analysis of repairable systems with limited failure experience. *Reliability Engineering and System Safety* 77: 167–180.

Section 3

Reliability Analysis of Transmission Systems

10

Transmission System Reliability Evaluation Including Security

Pushpendra Singh[1], *Rajesh Arya*[2], *Lakhan Singh Titare*[3], *Mohd. Tauseef Khan*[1], *and Sharat Chandra Choube*[4]

[1]*Department of Electrical Engineering, Rajkiya Engineering College, Banda, Uttar Pradesh, India*
[2]*Department of Information Technology, Acropolis Institute of Technology and Research, Indore, Madhya Pradesh, India*
[3]*Department of Electrical Engineering, Government Engineering College, Ujjain, Madhya Pradesh, India*
[4]*Department of Electrical & Electronics Engineering, University Institute of Technology, RGPV, Bhopal, Madhya Pradesh, India*

10.1 Introduction

The power system security analysis and monitoring plays a crucial role in modern energy management systems (EMS). However, implementing this analysis in real time poses challenges for power system engineers. It is essential for operational staff of a power system to be aware of any disruptions or contingencies that may lead to limit breaches and emergency conditions in order to ensure safe operation. Security assessments provide alerts to system operator about the both secure and unsecured characteristics of the operating states, particularly in the event of unexpected incidents. This allows operators to take appropriate control or corrective actions within a timely manner. Conducting a comprehensive examination of every potential contingency in real time is impractical due to time constraints. Contingency selection is employed to identify the most critical scenarios that have the potential to disrupt the system and their likelihood of occurrence. Two common techniques used for ranking techniques and screening approaches include contingency selection. Ranking methodologies involve assigning a rough severity ranking to contingencies based on performance index values. Performance indicators are quickly evaluated and expressed in terms of network characteristics, providing a relative measure of contingency severity. Screening techniques utilize approximations of network solutions, such as AC/DC load flow, distribution parameters, local solution techniques, linearized load flow, etc., to identify critical situations. These techniques aim to reduce the number of unforeseen circumstances by screening and identifying contingencies that may cause significant system disruptions. By employing both ranking methodologies and screening techniques, power system operators can prioritize their attention and resources on the most critical contingencies, allowing for efficient and effective real-time security analysis and monitoring.

Contingency evaluation is a critical task for power system planning and operation engineers. During planning, contingency analysis helps assess the requirement for transmission expansion based on the power system's performance to meet increased demand or generation. Operational analysis of contingencies enables system operators to maintain the power system at safe operating level, ensuring that device loads remain within secure limits and consumers receive energy

Reliability Analysis of Modern Power Systems, First Edition. Edited by R. K. Saket and P. Sanjeevikumar.

that meets acceptable quality standards. The screening and rating of contingencies aim to identify potential contingency that could leads to system instability or violations of voltage stability criteria. Various techniques and methods have been proposed for voltage security analysis using performance indicators (PI) (Vaahedi et al. 1999; Naik et al. 2015; Singh et al. 2021; Singh et al. 2023). Wang et al. (2020) introduced an incremental-risk based technique for filtering significant contingency events in transmission network operation and design. Choube et al. (2001) proposed a technique based on singular value decomposition for contingency selection. Arya et al. (2002) utilized the concept of vanishing eigenvalues to develop a strategy for contingency selection. Zhu and Liu (2022) proposed a heuristic method based on the triangle count approach for selection of the branch contingency during the screening phase. Heidari et al. (2022) proposed a utilizing clustering and similarity, a complete method for online contingency screening, providing a real-time list of potential outcomes that may lead to system instability. Coelho et al. (2022) suggested a topological technique to test critical transmission lines with buses for various transmission line contingencies. Amjady and Esmaili (2003) presented a method to effectively assess the status of voltage stability in both the current situation and the post-contingency phase. Zhao and Singh (2018) developed a contingency ranking technique as a pre-selection technique for reliability assessment in line switching operations. Nan et al. (2021) offered an N-1 contingency selection approach based on graph theory to efficiently detect the most important contingencies for security-constrained part commitment decisions. Hagh et al. (2018) introduced a sequential technique considering N-1 contingency criteria to assess system security under post-outage transient stability conditions. Armaghani et al. (2020) presented an approach for online vulnerability assessments with N-1 line outage in the large transmissions system, particularly relevant for the estimation of post-contingency phases involving wind power resources. These techniques and methods contribute to the assessment and selection of contingencies, enabling power system engineers to make informed decisions for secure and reliable power system operation.

Operators of control rooms can make use of the technique Ni et al. (2003) provided for evaluating online security using probabilistic risk-based indices. Based on the Bayes classifier, Kim and Singh (2005) developed a method for evaluating the security of power systems. Verma and Shrivastava (2005) presented an effective method for formulating the voltage control area considering the impact of contingencies, including network topology changes, P-V, Q-coupling, and their effects. Zarate et al. (2006) developed a fast calculation approach using nonlinear programming methods to assess security margins against voltage collapse. Arya et al. (2006) introduced techniques for assessing voltage collapse risk using ANN and managing voltage security margins in advance. Srivani and Swarup (2008) proposed a method for static security evaluation using the ward network. Berizzi et al. (2009) utilized fuzzy logic for online evaluation of voltage security at the present operating point. Bahmanyar and Karami (2014) developed an voltage stability margins may be monitored online using an ANN-based technique. Lee et al. (2010) introduced a new indicator called the "stability region area for the voltage" that provides valuable information for load shedding, contingency screening, and ranking. ANN-based approaches have received significant attention in the field of security assessment (Javan et al. 2013) due to their ability to handle data in parallel while maintaining high accuracy and speed. These techniques and methods contribute to the analysis and assessment of voltage stability, enabling operators and engineers to monitor and ensure the security of the power system in real-time.

Indeed, deterministic criteria used in power system planned construction often overlook the inherent randomness and uncertainty present in various components such as site resources, system behavior, and customer expectations. To address this limitation, probabilistic approaches can be employed to account for the inherent uncertainty in these factors. However, the interpretation and utilization of probabilistic reliability indices can pose challenges for power system planners and designers. One way to tackle this challenge is by incorporating deterministic factors into probabilistic assessments using the concept of well-being. By considering deterministic elements alongside probabilistic analysis, a more comprehensive understanding of system reliability and performance can be achieved. This approach allows for a better evaluation of the impact of uncertainties on system behavior and the overall well-being of the power system. By combining deterministic and probabilistic considerations, power system planners and designers can make more informed decisions, considering both the average system performance and the associated uncertainties. This integrated approach helps address the limitations of purely deterministic criteria and provides a more realistic and robust assessment of power system reliability and performance.

This book chapter introduces a novel technique for evaluating the security of a power system, considering uncertainty in control variables and line parameters. The proposed techniques consist of two main part. In the first part, a Monte Carlo simulation is utilized to determine the system's security index, considering both the availability and unavailability of transmission lines. This simulation helps in assessing the system's security under different scenarios and provides a ranking of lines based on their security index values. The second part focuses on evaluating the using the cut-set method for the probabilistic insecurity index (PISI). This approach considers single and double-line outages and examines the impact of these outages on voltage stability limits. Continuous power flow analysis is utilized to determine the voltage stability limits when specific lines are out of service. To further enhance the accuracy and applicability of the technique, a multi-layer recurrent neural network (RNN) is trained using the obtained PISI values for various single and double-line outages at different loading scenarios. This trained RNN can then be used to predict the PISI for any given operating conditions, enabling online applications of the proposed technique. By combining Monte Carlo simulation, cut-set analysis, and RNN modeling, this technique offers a comprehensive approach to evaluate the security of power systems considering uncertainties. It provides valuable insights for system operators in real-time decision-making and enhances the overall security assessment process.

10.2 Problem Formulation

Real-time system operation in power systems involves dealing with uncertainties in control variables and system parameters, which can be considered as arbitrary variables. Transmission line parameters, in particular, are prone to errors for various reasons. One of the reasons for inaccuracies in computed transmission line parameters is the truncation of the Taylor series expansion, which is a mathematical approximation commonly used in calculations. Simplified modeling assumptions, such as assuming a flat earth, fully transposed lines, or rounded conductors, can also contribute to inaccuracies. Human errors during the data input phase can lead to significant mistakes in parameter values. This can occur due to manual data management issues, where incorrect values are entered into the system. Weather factors, such as temperature variations, can affect the sagging of conductors and introduce additional uncertainties. Furthermore, frequency fluctuations can also impact the accuracy of parameter values. Studies have shown that these mistakes in parameter values can range from 5 to 10%. In cases where human data entry errors occur, the magnitude of the mistakes can be even larger. However, in the field of system planning and operation, it is generally acceptable for a few percentage errors to be present, considering the overall complexity and uncertainties in the system. To account for parameter uncertainty, power systems are typically designed with a high stability margin and a sufficiently low likelihood of voltage collapse. This ensures that even with the worst-case consequences of parameter uncertainty, the system remains stable and secure. Overall, uncertainties and errors in control variables and system parameters are inevitable in real-time power system operation. However, through proper system design, modeling techniques, and margin considerations, the impacts of these uncertainties can be mitigated to maintain the power system stable and reliable.

The following are the failure probabilities taking into consideration the control variable and line parameter uncertainties:

$$Pf = \left[\sum_{i=1}^{n} P(Cs_i) - \sum_{\substack{i=1 \\ j=2}}^{n,m} (P(Cs_i) * P(Cs_j)) \right] \tag{10.1}$$

maintain the power system stable and reliable. The following limitations under the base case scenario and the following anticipated loading scenario, which takes into account the ambiguities in the control variables and line parameters, are applicable:

(i) Power flow restrictions:

$$F(V, \delta) = 0 \tag{10.2}$$

(ii) Constraints on the generated reactive power:

$$Q^k_{g,\min} \le Q^k_g \le Q^k_{g,\max} \tag{10.3}$$

$$k = 1, 2, \ldots, NG$$

(iii) Constraints on the generated active power:

$$P^k_{g,\min} \le P^k_g \le P^k_{g,\max} \tag{10.4}$$

$$k = 1, 2, \ldots, NG$$

(iv) Constraints on load bus voltages:

$$V^i_{\min} \le V^i \le V^i_{\max} \tag{10.5}$$

$$i = (NG + 1), \ldots \ldots, NB$$

10.3 Monte Carlo Simulation for Evaluation of the Security Index: With and Without Considering the Absence of Transmission Lines

In power systems, various variables and parameters, such as PV-bus voltages, line charging susceptances, line resistances, and line reactances, can be considered as random variables due to their inherent uncertainties. These uncertainties arise from factors such as measurement errors, model inaccuracies, environmental conditions, and other sources of variability. Determining the static voltage stability limits and a continuous power flow approach can be employed. This approach involves evaluating the power flow equations for different sampled values of parameters and variables, considering their probabilistic nature. By considering a range of possible values for these random variables, the static voltage stability limit is established as a random variable as well. The following steps involve the implementation of Monte Carlo simulation:

Step 1: Using the corresponding probability density functions, obtain the sample vector $\underline{z} \in Z$. For constant power factor of loads, the "Z" vector comprises resistances to the line, reactances to the line, susceptance to line charging, and generated bus voltages.

Step 2: Perform step 1 repeatedly, saying "N" each time.

Step 3: Outage on a certain line, $i = 1$

where,

$$i = 1, 2, 3, \ldots \ldots, NL.$$

Step 4: With the use of the continuous power flow approach (Ajjarapu and Christy 1992), for each sampled condition "Z" determine the static voltage stability limit.

Step 5: At step 4, all "N" static voltage stability requirements are met. Get the value of $X(Z_j)$ for $j = 1, 2, 3, 4, \ldots\ldots, N$, which indicates whether voltage collapse has occurred or not. If the nominal load level (S_d) for this sampled condition exceeds the static voltage stability limit, then $X(Z_j) = 1$.

$= 0$, Otherwise

Step 6: Calculate the security index as follows without taking the absence of a transmission line into account:

$$\hat{p}_j = \left[\sum_{j=1}^{N} X(Z_j)\right] / N \tag{10.6}$$

Step 7: Calculate the security index as follows, accounting for the absence of a transmission line SI

$$SI_j = \hat{p}_j^* \overline{A}_j \tag{10.7}$$

Step 8: For various line outage scenarios that reoccur starting with step 5, get estimates for $\hat{p}_j$ and SI_j.

Step 9: For "N" times, Steps 1 through 8 are repeated.

It is obvious that a lot of samples may be produced using steps 1 through 9 and utilized to rank of the transmission lines.

The two-state models for transmission lines functioning are commonly use to represent the operational status of a transmission line. In this model, each transmission line is considered to have two states: the down state and the up state. The up state represent the normal operating condition of the transmission line, where it is available and capable of transmitting power. The transmission line is considered to be functioning and operational in this state. On the other hand, the "down" state represents a failure or planned outage of the transmission line. In this state, the transmission line is unavailable and cannot be used for power transmission. The down state can be due to various reasons such as equipment failure, maintenance, or other planned events. The transmission line failure and repair rates are depicted in Figure 10.1 as λ_j and μ_j, respectively. By analyzing the failure and repair rates of the transmission line, system operators can assess the reliability and availability of the transmission infrastructure. This information is crucial for planning and managing the transmission network, ensuring its resilience and minimizing disruptions in power supply.

This model is employed to determine the long-term availability and unavailability of transmission lines. They have the following expressions (Rubinstein 1981):

$$A_j = \mu_j/(\lambda_j + \mu_j), \overline{A_j} = \lambda_j/(\lambda_j + \mu_j) \tag{10.8}$$

The letters A_j, and A_j stand for the transmission line's availability and unavailability, respectively.

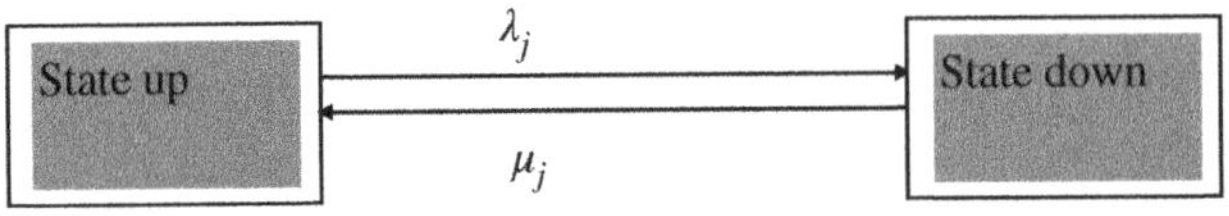

Figure 10.1 For every transmission line, there is a two-state model.

10.4 Evaluation of the Load Flow's Minimal Eigenvalue Jacobian

Load flow's minimal eigenvalue a critical proximity indication is the Jacobian for evaluating the voltage stability of a certain system. The system load has an inverse relationship with this indicator's value. Here is how Jacobian is rated:

$$\begin{bmatrix} H & N \\ M & L \end{bmatrix} \begin{bmatrix} \Delta\delta \\ \Delta V \end{bmatrix} = \begin{bmatrix} \Delta P_g \\ \Delta Q_g \end{bmatrix} \tag{10.9}$$

Equation (10.9) may be used to solve for the incremental power flow at each solution point. $\begin{bmatrix} H & M \\ N & L \end{bmatrix} \begin{bmatrix} \Delta\ \delta \\ \Delta\ V \end{bmatrix} = \begin{bmatrix} \Delta\ P \\ \Delta\ Q \end{bmatrix}$ when the load flow is used Jacobian J' is

$$J' = \begin{bmatrix} H & N \\ M & L \end{bmatrix} \tag{10.10}$$

The following describes the voltage collapse condition (Kundur 1994; Singh et al. 2017; Kothari et al. 2022).

$\lambda_{\min}[H] \neq 0$

$\lambda_{\min}[J'] = 0$

$J' = \begin{bmatrix} H & M \\ N & L \end{bmatrix}$ $\lambda_{\min[H]} \neq 0$ λ_{min} - is minimum eigen value of $[H]$ or $[J]$.

For the i^{th}eigen value, the modal solution to the incremental power flow Eq. (10.9) may be stated as follows.

$$\Delta V_{m,i} = \xi_i / \lambda_i \tag{10.11}$$

λ_i - is the ith eigenvalue of the Jacobian for load flow.

$\Delta V_{m,i}$ is the change in the load bus voltage vector as a result of modal variation, and ξ_i is the right eigen vector corresponding to λ_i.

$$\Delta V_{m,i} = \sum_i (\alpha_i \xi_i)/\lambda_i$$

$$\Delta V_i = \sum_i \Delta V_{m,i} = \sum_i \frac{\alpha_i \xi_i}{\lambda_i} \tag{10.12}$$

According to Eq. (10.12), all of the load flow Jacobian's eigenvalues should be positive for the system to retain a stable voltage. When the minimal eigenvalue is extremely near to zero, a very little change in load can cause a very significant change in voltage. From the perspective of load flow, the operating point is thought of as voltage unstable. Thus, the minimum eigen value's size must be big enough to provide a sufficient stability buffer. The steps below are used to determine the load flow Jacobian's minimum eigen value:

Step 1: With the continuous power flow method, perform AC power flow and get load flow Jacobian $[J']$

Step 2: To determine the inverse load flow Jacobian.

$$[J'I] = [J']^{-1}$$

Step 3: Select $E_0 = [1, 0, \ldots\ldots, 0]^T$
Step 4: The iteration count is set to $k = 1$.

10.5 Evaluation of Schur's Inequality

As stated in (Arya et al. 2012), Schur's inequality is as follows:

$$\lambda_{\max} \leq \sqrt{\sum_{ij} a_{ij}^2} \tag{10.13}$$

where, a_{ij} - is the ij^{th} load flow element $[J']$ for Jacobian.

$\lambda_{\max}$-maximum load flow eigen value Jacobian.

The magnitude of the most significant eigenvalue is less than or equal to the square root of the squares of each matrix component. The lower constraint on the lowest eigenvalue of the load flow Jacobian is derived using Eq. (10.13). The following is the sensitivity matrix $[S]$

$$[S] = [J']^{-1} \tag{10.14}$$

Now, the higher limit on the highest eigenvalue of $[S]$ is provided using inequality Eq. (10.13) as follows:

$$S\lambda_{\max} \leq \sqrt{\sum_{ij} s_{ij}^2} \tag{10.15}$$

where $\boldsymbol{s_{ij}}$ is the element of $[\boldsymbol{S}]$ and $\boldsymbol{S\lambda_{\max}}$ is the greatest eigenvalue of $[\boldsymbol{S}]$. The matrix theory reveals that:

$$\boldsymbol{J'\lambda_{\min} = 1/(S\lambda_{\max})} \tag{10.16}$$

The load flow Jacobian's minimum eigenvalue is $J'\lambda_{\min}$. Inequality Eq. (10.15) may be stated as follows using Eq. (10.16):

$$J'\lambda_{\min} \geq 1/\left(\sqrt{\sum_{ij} s_{ij}^2}\right) = \tau \tag{10.17}$$

A proximity indicator (τ) and a lower limit on the lowest eigenvalue of the load flow Jacobian are what the right-hand side of Eq. (10.17) actually represents. The size of this proximity indicator represents the separation between voltage collapse and the measured operating point, which has been used to verify voltage stability.

10.6 Evaluation of the PSI and the Cut-set Approach

An effective method for evaluating the reliability of power system is a cut-set method. The technique is easy to program on a digital computer and provides a quick and effective check of any general grid. Cut-sets also help identify the unique and identifiable types of failures in a system, as they are closely related to the types of system failures. In determining a cut set, one selects the line or group of lines that is most important. A minimal cut set is a group of transmission lines where if one line fails, the entire system fails; however, as long as all components in the set are functioning properly, the system is not affected. For a system to fail, every element of a minimum cut-set must be in the failure state. A power network's failure likelihood can be expressed as follows:

$$Pf = [P(Cs_1 + Cs_2 + \ldots + Cs_n) - P(Cs_1Cs_2 + Cs_1Cs_3 + \ldots + Cs_nCs_m)] \tag{10.18}$$

where the minimal cut-sets Cs_1, Cs_2, …. , Cs_n, and Cs_m are. Although this exact evaluation is theoretically always feasible, it is a laborious and time-consuming procedure that is prohibitively expensive with power networks.

In order to get around this issue, failure probability evaluations typically make use of approximations, which slow down the assessment process but increase speed. Keeping the first-order in terms within the expansion of Eq. (10.18), the failure probability may be roughly calculated as follows. This is because the error added is typically small and is within the range of the statistics on the availability of transmission lines for power systems, which have high repair rates and low failure rates for each transmission line.

$$Pf = \left[\sum_{l=1}^{n} P(Cs_l) - \sum_{\substack{i=1 \\ j=2}}^{n,m} (P(Cs_i) * P(Cs_j)) \right] \tag{10.19}$$

$i \neq j$

This chapter takes into account cut sets up to the second order. This is because there is very little chance that more than two transmission lines will go down. Under typical loading conditions, a cut-set of a higher order may not be a minimum cut-set, which is necessary in Eq. (10.18). In order to calculate the probability of failure (Pf) under the aforementioned approximation, Eq. (10.19) is used.

When a given load is compared to the static voltage stability limit in the base case and extra line loss situations, the lowest cut-set (first & second order) is determined. The continuing flow of power approach for used to compute the static voltage stability limit during single or double-line outages (Arya and Verma 1996). An acceptable stability margin must be maintained for proper system operation. The results of ongoing power flow may be used to build a capacity table is available, which offers permissive loads & matching network condition. The capacity availability for various load levels may thus be used to calculate the cut-set. Equation (10.19) can be used to determine the probability of failure.

The PISI of the transmission system is the name given to this failure likelihood. It is further emphasized that system factors and control variables affect load ability limit. In order to account for uncertainties, failure probability, or PISI, is determined. The following is the suggested algorithm for computing PISI:

Step 1: Calculate the static voltage stability limit using the base case load flow method.

Step 2: To establish the stability limit for each dependent circumstance, for single and double-line outages, execute continuous power flow.

Step 3: Plan for the possibility of capacity failure table displaying the stability limit and potential scenarios.

Step 4: For the ith loading condition, determine the cut-sets for the different peak load considerations: $Cs_1(i)$, $Cs_2(i)$, $C_{s3}(i)$, $Cs_4(i)$, ……, $Cs_r(i)$.

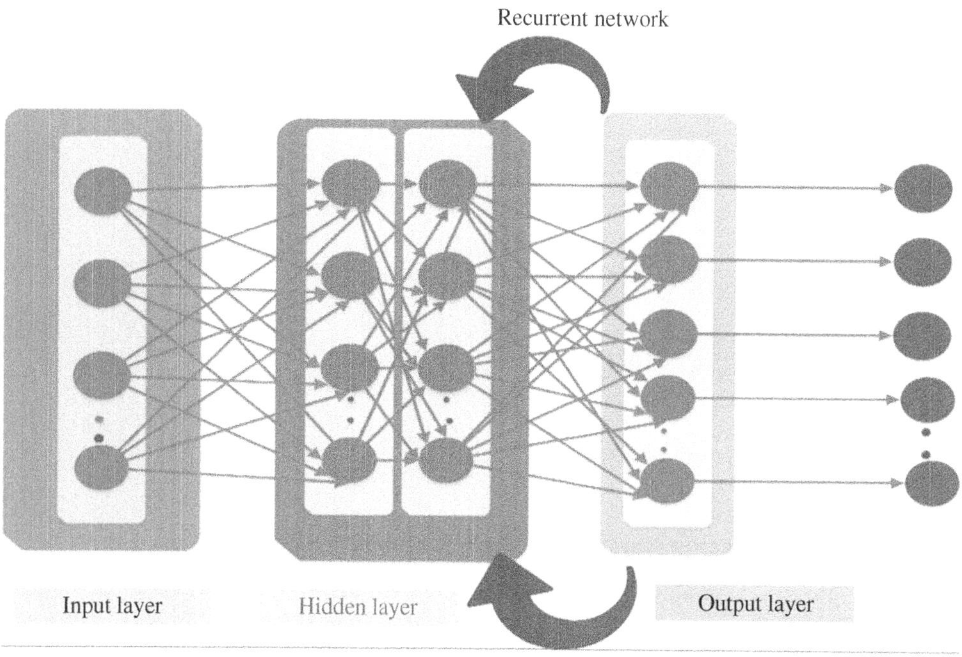

Figure 10.2 Recurrent neural network.

where

$$(i = 1, 2, 3, ..NL)$$

Step 5: Use the equation to get the probability of failure (Pf), using relation (10.10).
Step 6: Stop.

10.7 Recurrent Neural Network (RNN) Assessment of Probabilistic Insecurity

In a recurrent neural network (RNN), as depicted in Figure 10.2, the architecture includes a looped design that allows the network to retain the information from the previous input. This recurrent structure enables the RNN to process sequential data and utilize prior knowledge when making predictions or decisions. The inputs of the RNN are the total system load, generator bus voltage, and the line parameters (resistances, reactances, and susceptance). Line parameters are equal to the number of units in the input layer. The output layer consists of a single unit that generates the probabilistic insecurity index (PISI) as the result of the RNN's computation. Figure 10.3 (Sherstinsky 2020) illustrates the structure of a conventional RNN, where each repeating module in the network contains a single activation layer utilizing the hyperbolic tangent function. This activation layer performs calculations based on the input data & previously learned information, enabling the RNN to capture temporal dependencies and patterns in the sequential data. RNNs are commonly employed in applications involving sequential data, such as voice recognition and time series forecasting, where the order and context of the data play a crucial role. The equations (Haykin 2003;

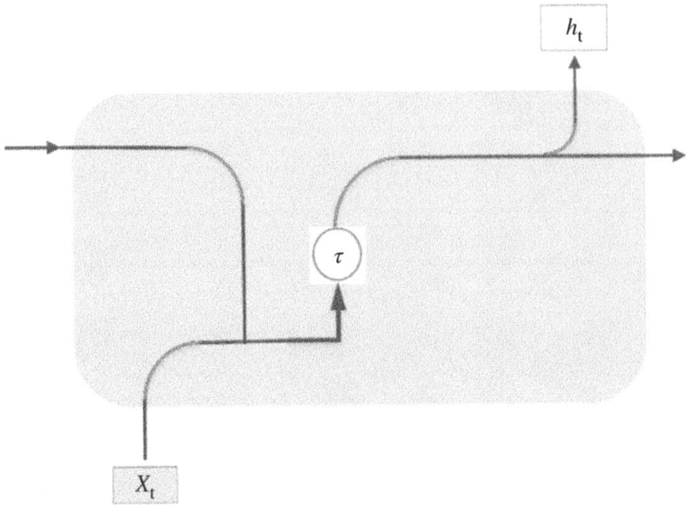

Figure 10.3 Repeated module in a standard RNN. Source: Adapted from Haykin (2003) and Ajitha et al. (2022).

Ajitha et al. 2022) below represent a basic formulation of a simple RNN:

$$h^{(t)} = \sigma(W_{hx} * X + W_{hh} * h^{(t-1)} + b_h) \tag{10.20}$$

where,

- W_{hx}—is the matrix of weights between the i/p and the hidden layers.
- W_{hh}—is the weight matrix that recurs between neighboring time steps' hidden layers.
- X—is an element of the input vector.
- b_h and b_y—bias vectors that let every node to learn and offset.
- $h^{(t)}$—is the hidden layer's neuronal output.

$$[X]^T = [R_1, R_2, \ldots., R_{NL}, X_{L1}, X_{L2}, \ldots., X_{LNL}, Bc_1, Bc_2, \ldots., Bc_{NL}, V_1, V_2, \ldots., V_{NG} \& S_d]^T \tag{10.21}$$

$$Y^{(t)} = sof\max(W_{yh} * h^{(t)} + b_y) \tag{10.22}$$

10.8 Results and Discussions

In the 6-bus 7-line, 14-bus 20-line, and 25-bus 35-line IEEE standard test systems, the proposed method has been implemented [Appendix 10.A]. Resistances to the line, reactances to the line, susceptance to line charges, voltage of the PV bus and loads bus power factor make up the vector of to be sampled random variables "Z." Both the load bus and PV-bus voltage should be between the ranges of 0.80 pu – 1.05 pu and 0.95 pu – 1.10 pu, respectively.

10.8.1 IEEE Six-Bus System

There are 2-PV buses and 4-PQ buses in this system. Lines 1 through 7 are presumed to be available at 0.99636033, 0.99636033, 0.99658703, 0.99545481, 0.99545481, 0.99658703, and 0.99658703

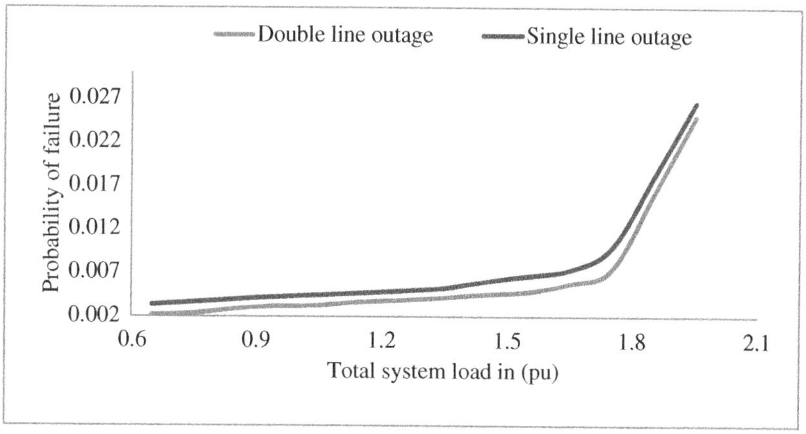

Figure 10.4 Probability of failure shown against different levels of load for $\tilde{V}_1 = 0.9625$ pu, $\tilde{V}_2 = 0.9517$ pu, $R_T = 1.4307$ pu, and $X_T = 3.6482$ pu for six-bus system.

(Billinton and Allan 1984). Using Eqs. (10.6) and (10.7), determine the security index with and without taking transmission line availability into account, and then arrange the transmission lines in severity order. The value of the security index has determined the line's severity. Table 10.1 compares the contingency ranking of lines based on the security index. The ranking was performed considering various factors such as the availability of the line, voltage performance index, minimum eigenvalue, Schur's inequality value, and line flow performance index (Verma and Niazi 2012; Singh et al. 2018). Equation (10.19) was used to calculate the PISI, which takes line uncertainties into consideration, is the probability of failure characteristics and control factors under different loading conditions. Static voltage stability constraints derived using the predictor-corrector approach (Arya and Verma 1996) were utilized. The PISI values were determined, which provide an assessment of the system's voltage stability. Table 10.2 lists the inputs, values for the input layer in an RNN, PISI values generated from trained networks and cut-set calculations that take into account single and double-line outages, as well as the % error. The buried layer's neuron count was increased to 17. The input layer's variables include system load ($R_1 - R_7, X_1 - X_7, V_1 - V_2$, and S_d) voltage of generator buses, and line characteristics (line resistances, reactances, and charging susceptances). In addition to being utilized in training, trained RNN has undergone limited testing. Percentage errors along with the test instances in Table 10.3, the PISI computed the uses of trained network and cut-set approach taking into account the single and double-line outage is displayed. The ratio of training to testing instances is not governed by any rules.

1000 training examples were chosen for the current study in order to represent a wide variety of power system operating situations. In addition, outside of training cases, testing examples have been chosen from this range. 1000 test instances have been used to test the trained network. Table 10.3 only lists a select few with the greatest % error magnitudes. The estimated probability of failure is plotted against different load levels in Figure 10.4, with $\tilde{V}_1 = 0.9625$ pu, $\tilde{V}_2 = 0.9517$ pu, $R_T = 1.4307$ pu, and $X_T = 3.6482$ pu taking into account single and double-line outages. The percentage error in the failure probability as determined by trained RNN is less than $\pm 1\%$.

10.8.2 IEEE 14-Bus System

Eleven load buses and three generator buses make up this system. It is assumed that lines 1 through 20 have the following availability values: 0.99545486, 0.99627723, 0.996336033,

Table 10.1 Ranking of contingencies for the six-bus, seven-line system accounting unpredictability based on the security index with and without taking into account the availability of transmission lines, the voltage performance index, the lowest eigen value, the value of Schur's inequality, and the performance index for line flows ($S_d = 0.7401$ pu).

	Security index without taking into account the absence of a transmission line		Transmission line availability consideration while evaluating security index		Index of voltage performance (PI_V)		Minimum eigen value (λ_{min})		Schur's inequality value (τ)		Index of line flow performance (PI_{MW})	
S. no.	$\hat{p}$	**Line number**	**SI**	**Line number**	PI_V	**Line number**	λ_{min}	**Line number**	τ	**Line number**	PI_{MW}	**Line number**
1.	0.8936	7	0.003252	7	0.0089	7	0.2138	7	0.1849	7	0.8405	7
2.	0.4139	5	0.001506	5	0.3187	2	0.2140	5	0.2002	2	0.7162	5
3.	0.3611	2	0.001232	2	0.3264	5	0.2218	2	0.2037	5	0.6317	2
4.	0.1217	1	0.000553	1	0.4295	6	0.2577	1	0.2381	6	0.5641	1
5.	0.0997	6	0.000453	6	0.6056	1	0.4553	6	0.3432	1	0.5299	6
6.	0.0086	4	0.000029	4	0.6571	4	0.4765	3	0.3748	4	0.5073	3
7.	0.0032	3	0.000011	3	0.6708	3	0.4849	4	0.3788	3	0.4364	4

Table 10.2 For the six-bus test system, training instances were used to train RNN, PISI produced using the cut-set approach, and RNN.

Sr. no.	Total resistance of the line (pu)	Total reactance of the line (pu)	Voltage on the PV bus (pu)		Overall system load (pu)	Outage on a single line			Double-line outage		
						PISI calculated using the cut-set technique	PISI calculated using the RNN	% Error	PISI calculated using the cut-set technique	PISI calculated using the RNN	% Error
	R_T	X_T	V_1	V_2	S_d						
1	1.4667176	3.5405501	0.9647450	1.0855388	1.35	0.022203691	0.022227383	−0.106703691	0.021884957	0.021908688	−0.108436011
2	1.4723193	3.5616766	1.0843165	0.9579196	1.00	0.019951132	0.020002480	−0.257369584	0.019632398	0.019683788	−0.261759334
3	1.4032147	3.6191689	0.9847865	1.0155034	1.10	0.026780936	0.026753588	0.102116022	0.026462202	0.026434895	−0.241678302
4	1.4622588	3.6280009	1.0067415	0.9657292	1.10	0.023364876	0.023338060	0.114769662	0.023046142	0.023019367	0.116179858
5	1.4225665	3.5652915	1.0536958	0.9839854	1.00	0.01632311	0.016345186	−0.135242228	0.016004376	0.016026493	−0.138195707
6	1.4115554	3.5919799	0.9718635	1.0230430	1.40	0.026765424	0.026815515	−0.187146623	0.026446690	0.026496820	−0.189550538
7	1.3876319	3.6286525	0.9524766	1.0177007	1.10	0.018628132	0.018684821	−0.304321099	0.018309398	0.018366128	−0.309841752
8	1.4308110	3.5267878	1.0277902	0.9727332	1.00	0.023367966	0.023343840	0.103244472	0.023049232	0.023025148	0.104490897
9	1.4169602	3.6333706	0.9992843	0.9581896	1.85	0.026777846	0.026809088	−0.116671358	0.026459112	0.026490392	−0.118218883
10	1.4157303	3.6547260	1.0057756	1.0033403	1.80	0.026781967	0.026822293	−0.150570174	0.026463233	0.026503596	−0.152525520
11	1.4736682	3.6101888	1.0844996	1.0781777	1.70	0.019929694	0.019983579	−0.270374176	0.019610960	0.019664882	−0.274959809
12	1.4536465	3.6148612	0.9764595	0.9620212	1.00	0.015228614	0.015268276	−0.260446692	0.014909880	0.014949584	−0.266291191
13	1.4895729	3.6302745	1.0661153	0.9846217	1.05	0.019951132	0.019991827	−0.203972790	0.019632398	0.019673133	−0.207491116
14	1.4062879	3.6742212	1.0634236	1.0439772	1.40	0.019975202	0.019991816	−0.083173261	0.019656468	0.019673121	−0.084719108
15	1.4764972	3.6514651	0.9601764	0.9566103	1.70	0.026809901	0.026786442	0.087502874	0.026491167	0.026467745	0.088412517

Table 10.3 Test cases created for the 6-bus system utilizing the cut-set technique to validate the trained network.

Sr. no.	Total resistance (pu)	Total reactance (pu)	Voltage on the PV bus (pu)		Total system load (pu)	Single line outage			Double-line outage		
						PISI calculated using the cut-set technique	PISI calculated using the RNN	% Error	PISI calculated using the cut-set technique	PISI calculated using the RNN	% Error
	R_T	X_T	V_1	V_2	S_d						
1	1.4764972	3.6514651	0.9601764	0.9566103	1.65	0.021691506	0.02171474	−0.107112442	0.021559552	0.021582802	−0.107841196
2	1.4367804	3.6242044	0.9750771	1.0908399	1.45	0.019683454	0.019731262	−0.242884410	0.0195515	0.019599324	−0.244605042
3	1.4157303	3.6547260	1.0057756	1.0033403	1.25	0.023182217	0.023237565	−0.238752311	0.023050263	0.023105628	−0.240190538
4	1.4658493	3.5779551	1.0584796	0.9726279	1.15	0.026469674	0.026442507	0.102633762	0.026337720	0.026310570	0.103084486
5	1.4944772	3.5577229	1.0553987	1.0555223	1.80	0.026473749	0.026467346	0.024184745	0.026341795	0.026335408	0.024247355
6	1.4600478	3.5391203	1.0607959	1.0188818	1.85	0.021132127	0.021107740	0.115400543	0.021000173	0.020975802	0.116052008
7	1.4047314	3.5561451	0.9564821	0.9963408	1.75	0.016304293	0.016324544	−0.124207942	0.016172339	0.016192606	−0.125318601
8	1.3982844	3.6256708	1.0947264	1.0508896	1.65	0.023144236	0.023198757	−0.235568442	0.023012282	0.023066818	−0.236987690
9	1.4152725	3.5614248	1.0983200	1.0694296	1.70	0.016248296	0.016304573	−0.346353580	0.016116342	0.016172634	−0.349286516
10	1.4529598	3.5946015	0.9890439	0.9810556	1.05	0.018620484	0.018677469	−0.306034907	0.018488530	0.018545532	−0.308311016
11	1.4622588	3.6280009	1.0067415	0.9657292	1.05	0.023178096	0.02315436	0.102405757	0.023046142	0.023022423	0.102918230
12	1.4047314	3.5561451	0.9564821	0.9963408	1.80	0.026494156	0.02653479	−0.153368359	0.026362202	0.026402851	−0.154195393
13	1.4046704	3.5794307	1.0941954	1.0988784	1.70	0.019759457	0.019813717	−0.274603914	0.019627503	0.019681779	−0.276529728
14	1.4528591	3.6040027	0.9994537	0.9628956	1.00	0.019725126	0.019766744	−0.210991175	0.019593172	0.019634807	−0.212499922
15	1.4169602	3.6333706	0.9992843	0.9581896	1.80	0.026591066	0.026567732	0.087752429	0.026459112	0.026435793	0.088130980

Table 10.4 Ranking of contingencies for the 14-bus, uncertainty in the 20-line method of accounting based on the security index with and without taking into account the availability of transmission lines, the voltage performance index, the lowest eigen value, the value of Schur's inequality, and the line flow performance index ($S_d = 3.2057$ pu).

S. no.	$\hat{p}$	Line number	SI	Line number	PI_V	Line number	λ_{min}	Line number	τ	Line number	PI_{MW}	Line number
1.	1.0000000	10	0.0037617	10	0.0532	15	0.1300	10	0.1247	15	0.9753	8
2.	0.9835935	8	0.0030776	8	0.2075	10	0.2117	15	0.1627	10	0.7549	15
3.	0.9825183	15	0.0032821	13	0.2468	8	0.2861	8	0.2201	8	0.6938	10
4.	0.8957811	3	0.0027724	1	0.3283	13	0.3774	3	0.279	1	0.6265	3
5.	0.8693352	13	0.0016593	15	0.3751	3	0.3848	13	0.2912	13	6024	13
6.	0.8276328	1	0.0025521	3	0.4087	1	0.3899	2	0.2915	3	0.4537	16
7.	0.8267000	2	0.0014862	2	0.4164	2	0.4150	1	0.2934	16	0.4162	1
8.	0.8176957	16	0.0041268	12	0.4237	16	0.4246	12	0.308	12	0.3697	2
9.	0.6674546	12	0.0023984	16	0.4204	4	0.4400	6	0.3116	2	0.3678	12
10.	0.6607942	4	0.0042791	4	0.4365	12	0.4484	16	0.3126	6	0.2548	4
11.	0.6546000	9	0.0020839	6	0.4429	6	0.4529	9	0.3218	4	0.2457	9
12.	0.5964156	6	0.0029421	9	0.4538	9	0.4764	4	0.3265	9	0.2194	6
13.	0.5279563	17	0.0039513	17	0.4576	17	0.4809	5	0.3311	11	0.2081	5
14.	0.4967308	11	0.0035999	11	0.4832	5	0.4828	17	0.3358	17	0.1949	11
15.	0.4861905	5	0.0027908	5	0.4884	11	0.4911	20	0.3383	5	0.1508	17
16.	0.4556678	20	0.0022592	20	0.4925	20	0.4913	11	0.3393	18	0.1407	20
17.	0.3656452	18	0.0015340	18	0.5001	18	0.4918	19	0.3425	20	0.1383	19
18.	0.3269835	7	0.0011212	7	0.5017	19	0.4960	18	0.3486	7	0.1241	7
19.	0.3060000	19	0.0015552	19	0.5037	7	0.5075	7	0.3512	19	0.1005	18

Table 10.5 For the 14-bus test system, training instances were used to train RNN, PISI produced using the cut-set approach, and RNN.

Sr. no.	Total resistance of the line e (pu)	Total reactance of the line (pu)	Total susceptance of the line (pu)	Voltage on the PV bus (pu)			Total system load (pu)	Single line outage			Double-line outage		
								PISI calculated using the cut-set technique	PISI calculated using the RNN	% Error	PISI calculated using the cut-set technique	PISI calculated using the RNN	% Error
	R_T	X_T	Bc_T	V_1	V_2	V_3	S_d						
1	1.6042	4.53288	0.2849	1.0985	1.0867	1.0357	3.80	0.051198207	0.050073527	2.19671664	0.050187265	0.051221869	−2.06148649
2	1.6090	4.48502	0.2911	1.0846	1.0080	0.9644	3.00	0.038371241	0.038895992	−1.36756387	0.038360351	0.038344849	0.04041105
3	1.5875	4.37574	0.2929	1.0713	1.0543	0.9603	3.70	0.068012148	0.066988146	1.50561669	0.066324552	0.066232799	0.13833890
4	1.6804	4.52455	0.3550	1.0559	1.0102	0.9667	3.50	0.073982852	0.074841303	−1.16033752	0.073602433	0.073970831	−0.50052405
5	1.6727	4.53445	0.3537	1.0930	1.0749	1.0366	3.20	0.020096817	0.020690922	−2.95621350	0.018478687	0.018667433	−1.02142648
6	1.5556	4.63175	0.3287	1.0768	1.0768	1.0079	3.90	0.027705047	0.027833410	−0.46332120	0.027366816	0.027618466	−0.91954552
7	1.6457	4.55146	0.2835	0.9716	0.9614	0.9523	2.70	0.070073542	0.070135728	−0.08874461	0.070043845	0.070098245	−0.07766603
8	1.5498	4.45702	0.2756	0.9954	0.9747	1.0190	2.90	0.032044691	0.032063200	−0.05776055	0.032021164	0.032054873	−0.10527244
9	1.5897	4.50539	0.3194	1.0153	0.9995	0.9817	2.30	0.011411055	0.011261375	1.31170690	0.011097978	0.011403143	−2.74973381
10	1.6804	4.52455	0.3550	1.0559	1.0102	0.9667	3.00	0.057929880	0.058511169	−1.00343578	0.057575880	0.058019316	−0.77017713
11	1.6496	4.49989	0.2837	1.0334	1.0206	1.0107	2.50	0.023621750	0.023611395	0.04383530	0.022975380	0.023607145	−2.74974790
12	1.5937	4.50971	0.3045	1.0927	1.0355	1.0911	3.60	0.050502396	0.048997483	2.97988455	0.047567311	0.048950519	−2.90789566
13	1.5630	4.52398	0.3604	1.0927	1.0513	1.0231	3.70	0.043050652	0.042890121	0.37288964	0.042608465	0.042949935	−0.80141350
14	1.6611	4.59325	0.3765	1.0868	1.0778	1.0360	3.00	0.030837685	0.030732716	0.34039355	0.030711506	0.030209043	1.63607398

Table 10.6 Test cases created for the 14-bus system utilizing the cut-set technique to validate the trained network.

Sr. no.	Total resistance of the line (pu)	Total reactance of the line (pu)	Total susceptance of the line (pu)	Voltage on the PV bus (pu)			Total system load (pu)	Single line outage			Double-line outage		
								PISI calculated using the cut-set technique	PISI calculated using the RNN	% Error	PISI calculated using the cut-set technique	PISI calculated using the RNN	% Error
	R_T	X_T	Bc_T	V_1	V_2	V_3	S_d						
1	1.6804	4.5245	0.3550	1.0559	1.0102	0.9667	3.30	0.0739828520	0.074982862	−1.35167868	0.0727800820	0.074170974	−1.91108902
2	1.5889	4.5819	0.3185	1.0894	1.0698	0.9637	2.80	0.0149570180	0.014919278	0.25232268	0.0136388773	0.013428926	1.53935804
3	1.6811	4.5088	0.3066	1.0582	1.0110	1.0552	3.50	0.0670110200	0.065672101	1.99805785	0.0656672440	0.065562060	0.16017651
4	1.6611	4.5932	0.3765	1.0868	1.0778	1.0360	3.70	0.0598030410	0.058998621	1.34511562	0.0563377080	0.056274471	0.11224566
5	1.5608	4.4150	0.3487	1.0275	0.9885	0.9937	3.20	0.0367757500	0.037776034	−2.71995640	0.0362458115	0.036422585	−0.48770794
6	1.5937	4.6211	0.3576	1.0889	0.9650	0.9885	3.50	0.0287576780	0.029326858	−1.97922870	0.0285766392	0.028542475	0.11955184
7	1.6811	4.5088	0.3066	1.0582	1.0110	1.0552	3.30	0.0429028730	0.043094332	−0.44626157	0.0426598144	0.042976752	−0.74294112
8	1.6042	4.5329	0.2849	1.0985	1.0867	1.0357	3.60	0.0331167070	0.033616714	−1.50983411	0.0328075000	0.033200258	−1.19715961
9	1.5937	4.5097	0.3045	1.0927	1.0355	1.0911	3.40	0.0325348010	0.032934479	−1.22846344	0.0325348010	0.032969857	−1.33720130
10	1.5327	4.6164	0.3303	1.0639	0.9759	0.9527	2.80	0.0596568890	0.058609378	1.75589338	0.0596568890	0.058557856	1.84225586
11	1.6351	4.4319	0.3364	1.0550	1.0788	0.9504	4.00	0.0203750950	0.020775105	−1.96322980	0.0202975266	0.020100544	0.97047483
12	1.5937	4.5097	0.3045	1.0927	1.0355	1.0911	3.30	0.0238195870	0.023920014	−0.42161585	0.0237105532	0.023915106	−0.86270971
13	1.6384	4.5724	0.3241	1.0323	0.9942	1.0311	3.80	0.0318673690	0.032549425	−2.14029715	0.0315572695	0.032162061	−1.91648942
14	1.5630	4.5240	0.3604	1.0927	1.0513	1.0231	3.40	0.0430506520	0.042050654	2.32284001	0.0428745883	0.042314125	1.30721658

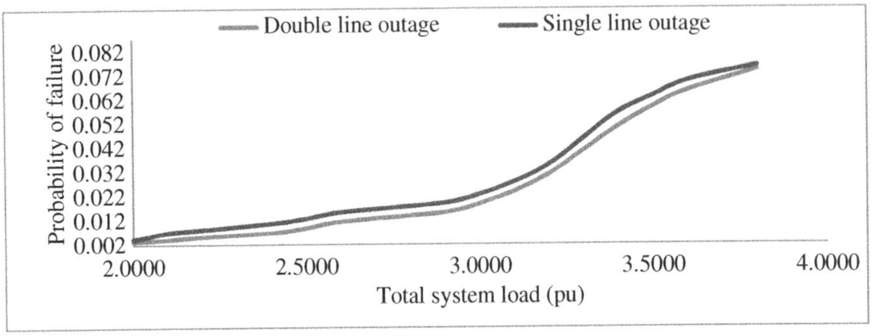

Figure 10.5 Probability of failure shown against various load levels for $\tilde{V}_1 = 1.0824$ pu, $\tilde{V}_2 = 1.0351$ pu, $\tilde{V}_3 = 0.9522$ pu, $R_T = 1.6427$ pu, $X_T = 4.6137$ pu and $Bc_T = 0.2514$ pu for 14-bus system.

0.99580437, 0.99658704, 0.99572088, 0.99545486, 0.99580437, 0.996336033, 0.99572088, 0.99580473, 0.99559204, 0.99545486, 0.99580473, 0.996336033, 0.99658704, 0.99572088, 0.99580473, 0.996336033, and 0.99658704 (Billinton and Allan 1984). Table 10.4 compares the security index-based contingency rating of lines with and without taking into account the uncertainty relating to the availability of the transmission line accounting, the voltage performance index, minimum eigen value, Schur's inequality value, and the performance index for line flow (Verma and Niazi 2012; Singh et al. 2018). Buses 7 and 5 are connected by line number 14, which is radial and whose failure causes islanding (Dester and Castro 2009). Table 10.5 lists the inputs for the RNN's input layer, values for PISI determined using the trained network and the cut-set approach while accounting for single and double-line outages, as well as the percentage error. Sixty-four neurons were added to the buried layer. The input layer's variables include line resistances ($R_1 - R_{20}$), line reactance ($X_1 - X_{20}$), line charging susceptances ($Bc_1 - Bc_{20}$), PV-bus voltages ($V_1 - V_3$), and total system load (S_d). Few testing cases, other from those used in training, have been utilized to validate trained RNN. Table 10.6 displays using a trained network and cut-set methodology, the PISI was calculated. While taking into account a single and double-line outage along with the % errors for the test instance. In order to cover a wide variety of operating circumstances for the power system, 1000 training examples were chosen for the current study. Moreover, excluding training examples, testing cases have been chosen from this range. For another 1000 test cases, the trained network has been tested. With the largest % error magnitudes, only a select number are shown in Table 10.6. The predicted chance of failure is plotted against different load levels in Figure 10.5, with $\tilde{V}_1 = 1.0824$ pu, $\tilde{V}_2 = 1.0351$ pu, $\tilde{V}_3 = 0.9522$ pu, $R_T = 1.6427$ pu, $X_T = 4.6137$ pu, and $Bc_T = 0.2514$ pu taking into account single and double-line outages. The percentage error for the failure probability as determined by trained RNN is less than 3%.

10.8.3 IEEE 25-Bus System

Twenty load buses and of five generator buses make up this system. It is assumed that lines 1 through 35 have the following availability values: 0.99559204, 0.99658704, 0.99545486, 0.99580473, 0.996336033, 0.99559204, 0.99572088, 0.99658704, 0.99545486, 0.99627723, 0.99559204, 0.99658704, 0.99545486, 0.99572088, 0.99580473, 0.996336033, 0.99559204, 0.99627723, 0.99545486, 0.99580473, 0.99658704, 0.996336033, 0.99559204, 0.99658704, 0.99627723, 0.99572088, 0.99545486, 0.99545486, 0.99580473, 0.99580473, 0.99627723, 0.99627723, 0.99559204, 0.996336033, and 0.99572088 (Billinton and Allan 1984). Table 10.7 compares rating

Table 10.7 Ranking of contingencies for the 25-bus, using the security index as the basis, 35-line system accounting uncertainty with and without taking into account the availability of transmission lines, the voltage performance index, the lowest eigen value, the value of Schur's inequality, and the line flow performance index ($S_d = 14.25$ pu).

S. no.	$\hat{p}$	Line number	SI	Line number	PI_V	Line number	λ_{min}	Line number	τ	Line number	PI_{MW}	Line number
1.	1.0000000	5	0.00091594	35	0.1442	11	0.1736	11	0.1645	5	0.910046	35
2.	0.9143000	35	0.00256579	5	0.1593	35	0.1800	5	0.1646	35	0.858553	11
3.	0.8200000	11	0.00231678	13	0.1865	5	0.2114	35	0.1946	11	0.846557	3
4.	0.8000000	13	0.00366397	11	0.1998	13	0.2416	13	0.2162	3	0.843429	5
5.	0.5645120	3	0.00064187	3	0.2211	3	0.2736	3	0.2358	4	0.804561	13
6.	0.5522361	4	0.00074016	4	0.2458	4	0.2846	10	0.2409	13	0.770274	10
7.	0.5268493	10	0.00101597	33	0.2639	33	0.3038	4	0.2418	10	0.761828	4
8.	0.4846530	33	0.00196134	10	0.2911	10	0.3091	33	0.2464	33	0.745136	33
9.	0.3281250	14	0.00361453	14	0.2966	17	0.3248	17	0.2555	17	0.744512	14
10.	0.3090909	17	0.00363611	17	0.3059	14	0.3252	14	0.2590	14	0.733536	9
11.	0.2235294	9	0.00140409	9	0.3186	9	0.3260	1	0.2658	8	0.728637	17
12.	0.2168675	8	0.00084904	1	0.3425	1	0.3261	9	0.2659	9	0.727498	8
13.	0.2125000	21	0.00062811	20	0.343	21	0.3286	8	0.2681	1	0.724276	20
14.	0.2105263	20	0.00136246	15	0.3475	15	0.3327	21	0.2698	20	0.720549	15
15.	0.2077922	1	0.00072233	26	0.3498	20	0.3341	15	0.2717	21	0.717101	1
16.	0.2023810	15	0.00076702	29	0.3531	29	0.3343	20	0.2750	15	0.716267	22
17.	0.1940299	18	0.00088321	19	0.3532	18	0.3351	29	0.2752	29	0.713124	19
18.	0.1923077	22	0.00072525	8	0.3549	8	0.3356	25	0.2770	22	0.711710	29
19.	0.1917808	29	0.00070461	21	0.3555	22	0.3370	22	0.2789	34	0.706750	21
20.	0.1891995	24	0.00064612	18	0.3557	24	0.3371	18	0.2789	19	0.705310	18
21.	0.1891892	26	0.00064573	22	0.3564	34	0.3395	34	0.2802	18	0.618581	26
22.	0.1818182	34	0.00064744	34	0.3571	26	0.3428	26	0.2832	25	0.610767	34
23.	0.1739130	25	0.00080956	25	0.3585	25	0.3438	24	0.2834	26	0.595676	25
24.	0.1714286	16	0.00060759	23	0.3613	16	0.3446	19	0.2849	24	0.582575	24
25.	0.1687556	19	0.00080457	24	0.3626	19	0.3449	23	0.2850	16	0.574432	23
26.	0.1666667	31	0.00051162	7	0.3639	23	0.3452	7	0.2861	7	0.504538	16
27.	0.1500000	7	0.00062046	16	0.3655	7	0.3555	16	0.2871	27	0.446335	7
28.	0.1465812	23	0.00041194	31	0.3671	31	0.3585	27	0.2881	23	0.415604	27
29.	0.1336782	27	0.00213633	27	0.3677	27	0.3680	32	0.2916	31	0.342200	30
30.	0.1219512	30	0.00066618	30	0.3685	32	0.3933	30	0.2936	32	0.292021	31
31.	0.1106546	32	0.0039124	32	0.3698	30	0.3966	31	0.3067	30	0.182064	32

Table 10.8 For the 25-bus test system, training instances were used to train RNN, PISI produced using the cut-set approach, and RNN.

	Total resistance of the line (pu)	Total reactance of the line (pu)	Total susceptance of the line (pu)	Voltage on the PV bus (pu)					Total system load (pu)	Single line outage			Double-line outage		
										PISI calculated using the cut-set technique	PISI calculated using the RNN	% Error	PISI calculated using the cut-set technique	PISI calculated using the RNN	% Error
Sr. no.	R_T	X_T	Bc_T	V_1	V_2	V_3	V_4	V_5	S_d						
1	3.2670	8.2441	1.8710	1.0385	1.0377	1.0246	0.9666	1.0390	16.5000	0.120877350	0.119634097	1.039213066	0.119993594	0.121999362	−1.671562965
2	3.0886	8.3500	1.7975	1.0072	0.9635	1.0868	1.0778	1.0360	16.0000	0.058184539	0.056979097	2.115587115	0.057184539	0.056102010	1.893044539
3	3.2642	8.3465	1.8327	0.9793	1.0199	1.0099	1.0020	1.0712	15.7500	0.117413828	0.116491478	0.791774879	0.112625830	0.109987978	2.342137477
4	3.2670	8.2441	1.8710	1.0385	1.0377	1.0246	0.9666	1.0390	14.7500	0.122171658	0.11904689	2.624821142	0.116697451	0.119394886	−2.311477131
5	3.2845	8.3421	1.8646	0.9508	1.0312	1.0426	1.0239	1.0369	12.5000	0.100421664	0.101478618	−1.041553954	0.101385916	0.102617698	−1.214943543
6	3.3596	8.3251	1.9022	0.9530	1.0818	0.9538	0.9954	0.9747	14.0000	0.073956824	0.074944386	−1.317727294	0.073386302	0.074901688	−2.064944416
7	3.2503	8.3769	1.8786	0.9868	1.0875	1.0350	0.9995	0.9909	14.2500	0.086792981	0.086534876	0.298266904	0.084946759	0.085880591	−1.099314605
8	3.2832	8.3042	1.8473	1.0058	1.0033	1.0865	1.0199	1.0139	16.0000	0.109911045	0.106909037	2.808002136	0.109100725	0.107460809	1.503120924
9	3.2900	8.1858	1.8506	0.9812	1.0404	0.9905	1.0971	1.0121	16.2500	0.062676858	0.06281575	−0.221110493	0.062645620	0.063584350	−1.498476958
10	3.1829	8.2850	1.8173	1.0885	1.0234	1.0109	1.0267	0.9548	15.7500	0.078953798	0.077518086	1.85209952	0.078653006	0.076337157	2.944387669
11	3.2012	8.3095	1.8615	0.9937	1.0703	1.0684	1.0514	1.0633	15.5000	0.065318686	0.067135056	−2.705546029	0.064117106	0.064917235	−1.247918051
12	3.3184	8.3020	1.8591	0.9751	1.0908	0.9682	1.0012	0.9643	14.5000	0.085620507	0.083250939	2.846295367	0.085357411	0.086357349	−1.171471263
13	3.2832	8.3042	1.8473	1.0058	1.0033	1.0865	1.0199	1.0139	15.0000	0.109708594	0.109613488	0.086764497	0.109100725	0.106248208	2.614571762
14	3.2900	8.1858	1.8506	0.9812	1.0404	0.9905	1.0971	1.0121	14.7500	0.062645620	0.062598318	0.075563533	0.060944003	0.060368534	0.944257934
15	3.2845	8.3421	1.8646	0.9508	1.0312	1.0426	1.0239	1.0369	15.2500	0.100819384	0.103226649	−2.332018836	0.098074931	0.099078070	−1.022828704

Table 10.9 Test cases created for the 25-bus system utilizing the cut-set technique to validate the trained network.

Sr. no.	Total Resistance of the line (pu)	Total Reactance of the line (pu)	Total Susceptance of the line (pu)	Voltage on the PV bus (pu)					Total system load (pu)	Single line outage			Double-line outage		
										PISI calculated using the cut-set technique	PISI calculated using the RNN	% Error	PISI calculated using the cut-set technique	PISI calculated using the RNN	% Error
	R_T	X_T	Bc_T	V_1	V_2	V_3	V_4	V_5	S_d						
1	3.2454	8.4066	2.0789	1.0073	0.9651	1.0395	1.0849	1.0827	16.50	0.120682105	0.123456209	−2.247034462	0.119993594	0.121876130	−1.568864065
2	3.0886	8.3500	1.7975	1.0072	0.9635	1.0868	1.0778	1.0360	15.50	0.058873050	0.05764253	2.134743106	0.058184539	0.059128193	−1.621830277
3	3.1536	8.2514	1.8614	1.0318	1.0711	1.0946	0.9643	0.9663	14.25	0.089710726	0.087510516	2.514224023	0.089022215	0.087322020	1.909854958
4	3.0886	8.3500	1.7975	1.0072	0.9635	1.0868	1.0778	1.0360	14.00	0.054930041	0.054130317	1.477404509	0.054241530	0.053429167	1.497677331
5	3.2544	8.2392	1.8475	0.9614	0.9523	1.0579	1.0051	1.0490	13.50	0.083563305	0.082951596	0.737428659	0.082874794	0.080902830	2.379449119
6	3.2059	8.2117	1.7948	1.0364	1.0744	0.9599	0.9906	1.0550	16.50	0.087281368	0.087421567	−0.160371195	0.086592857	0.085414580	1.360708925
7	3.2035	8.2343	1.7687	1.0357	0.9977	1.0109	0.9704	1.0295	16.25	0.126366928	0.122822889	2.885487429	0.116033292	0.113843616	1.887110493
8	3.3596	8.3251	1.9022	0.9530	1.0818	0.9538	0.9954	0.9747	14.75	0.076525784	0.075145155	1.837282319	0.073386302	0.073806304	−0.572316376
9	3.3071	8.4263	1.8431	1.0684	1.0465	1.0657	1.0529	1.0169	15.50	0.101110175	0.100538867	0.568245829	0.100421664	0.097859115	2.551789102
10	3.3596	8.3251	1.9022	0.9530	1.0818	0.9538	0.9954	0.9747	15.00	0.074074813	0.07489636	−1.096912129	0.073386302	0.073152310	0.318849495
11	3.2900	8.1858	1.8506	0.9812	1.0404	0.9905	1.0971	1.0121	13.00	0.063365369	0.062126838	1.993552128	0.062676858	0.062381957	0.470510449
12	3.1327	8.1714	1.6052	1.0362	1.0559	1.0102	0.9667	1.0846	15.25	0.104670457	0.106974157	−2.153510594	0.103981946	0.105931273	−1.874677998
13	3.3045	8.2627	1.8763	1.0044	1.0735	0.9966	1.0986	1.0657	13.50	0.073688662	0.075810501	−2.798871915	0.073000151	0.074588455	−2.175754209
14	3.1829	8.2850	1.8173	1.0885	1.0234	1.0109	1.0267	0.9548	16.00	0.079341517	0.078519208	1.047270549	0.078653006	0.078522963	0.165337993
15	3.3513	8.3514	1.8515	1.0121	1.0048	1.0153	0.9995	0.9817	14.50	0.098763442	0.099103019	−0.342650502	0.098074931	0.095188206	2.943387076

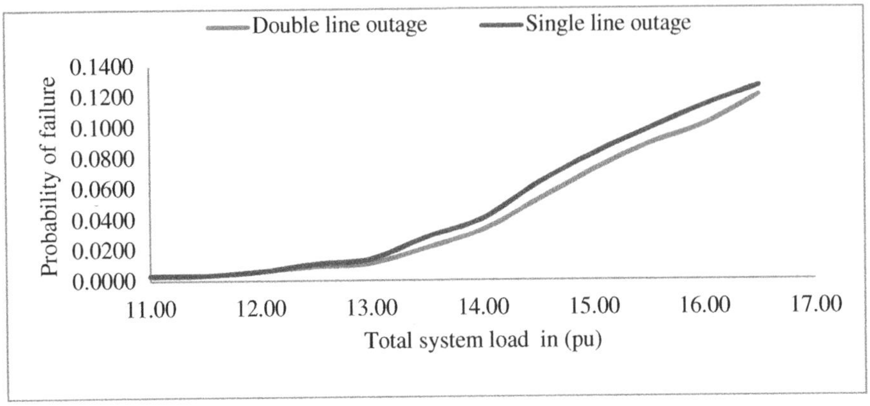

Figure 10.6 Probability of failure shown against various load levels for $\tilde{V}_1 = 0.9793$ pu, $\tilde{V}_2 = 1.0199$ pu, $\tilde{V}_3 = 1.0099$ pu, $\tilde{V}_4 = 1.0020$ pu, $\tilde{V}_5 = 1.0712$ pu, $R_T = 3.2642$ pu, $X_T = 8.3465$ pu and $Bc_T = 1.8327$ pu for 25-bus system.

of lines' contingency based on security index with and without taking into account the uncertainty relating to the availability of the trans. line accounting, the voltage performance index, minimum eigen value, Schur's inequality value, and the performance index for line flows (Verma and Niazi 2012; Singh et al. 2018). Lines 2, 6, 12, and 28 are radial lines that connect buses 1 and 16 as well as buses 1 and 25 and 15 and 16 (not included in Table 10.7), and when they are down, the buses become isolated (Dester and Castro 2009). Table 10.8 lists the inputs for the RNN's input layer, values for PISI determined using the trained network and the cut-set approach while accounting for single and double-line outages, as well as the percentage error. The buried layer's neuron count was increased to 111. Line resistance $(R_1 - R_{35})$, line reactance $(X_1 - X_{35})$, line charging susceptances $(Bc_1 - Bc_{35})$, PV-bus voltages $(V_1 - V_5)$, and total system load (S_d). are the variables of the input layer. Few testing cases, other from those used in training, have been utilized to validate trained RNN. Table 10.9 includes the PISI, % errors for test cases, and trained N/W and cut-set technique results when single and double-line outages are taken into account. 1000 training examples were chosen for the current study in order to represent a variety of power system operating scenarios. In addition, outside of training cases, testing examples have been chosen from this range. 1000 test instances have been used to test the trained network. Table 10.9 only includes a small number of items with significant % error magnitudes. In Figure 10.6, the projected likelihood of failure is shown against different load levels with the values $\tilde{V}_1 = 0.9793$ pu, $\tilde{V}_2 = 1.0199$ pu, $\tilde{V}_3 = 1.0099$ pu, $\tilde{V}_4 = 1.0020$ pu, $\tilde{V}_5 = 1.0712$ pu, $R_T = 3.2642$ pu, $X_T = 8.3465$ pu and $Bc_T = 1.8327$ pu taking into account single and double-line outages. The proportion of inaccuracy in the likelihood of failure as determined by trained ANN is less than 3%.

10.9 Conclusions

The novel approach presented in the book chapter addresses the voltage security evaluation by considering uncertainties in control variables and line parameters. It offers a comprehensive methodology that combines Monte Carlo simulation, cut-set analysis for contingency planning, and the utilization of a recurrent neural network (RNN) for efficient and accurate results. The Monte Carlo simulation is used to assess the security index, taking into account transmission line availability. This helps in identifying the most critical lines in the system. The cut-set approach is then employed to evaluate the probability of failure under different considering voltage stability, loading situations, and single- and double-line outages limits. To overcome the computational complexity associated with probabilistic approaches, the research introduces the use of RNNs. The RNN is trained using the obtained security index and PISI values for various contingencies and loading scenarios. This trained network enables fast and precise evaluations, providing prompt preventive intervention. The propose approaches effectiveness is demonstrated by comparing the contingency selection results with established approaches. The RNN-based method offers accurate and efficient results, which are crucial for ensuring the stability and reliability of complex power systems. The approach has been implemented and tested on different test systems, including the 6-bus 7-line, 14-bus 20-line, and 25-bus 35-line IEEE standard test systems, showcasing its applicability in real-world scenarios.

Appendix 10.A

10.A.1 Data for IEEE six-bus, seven-line test system (100 MVA Base)

Table 10.A.1 Line data.

Line no.	Bus no.		Resistance to lines (*R*) (pu)	Reactance to lines (*X*) (pu)	Susceptance to lines (*Bc*) (pu)	Tapping	Line flow limit (pu)
	From	To					
1	1	6	0.123	0.5180	0.0000	1.0000	3.00
2	1	4	0.0800	0.3700	0.0000	1.0000	3.00
3	4	6	0.0970	0.4070	0.0000	1.0000	3.00
4	6	5	0.0000	0.3000	0.0000	1.0000	1.00
5	5	2	0.2820	0.6400	0.0000	1.0000	2.50
6	2	3	0.7230	1.0500	0.0000	1.0000	1.50
7	4	3	0.0000	0.1330	0.0000	1.0000	2.50

Table 10.A.2 Bus data.

Bus number	Load (pu)		Generation (pu)	
	Active	Reactive	Active	Reactive
1	0.0000	0.0000	0.4500	0.0600
2	0.0000	0.0000	0.5000	0.2000
3	0.2750	0.0650	0.0000	0.0000
4	0.0000	0.0000	0.0000	0.0000
5	0.1500	0.0900	0.0000	0.0000
6	0.2500	0.0050	0.0000	0.0000

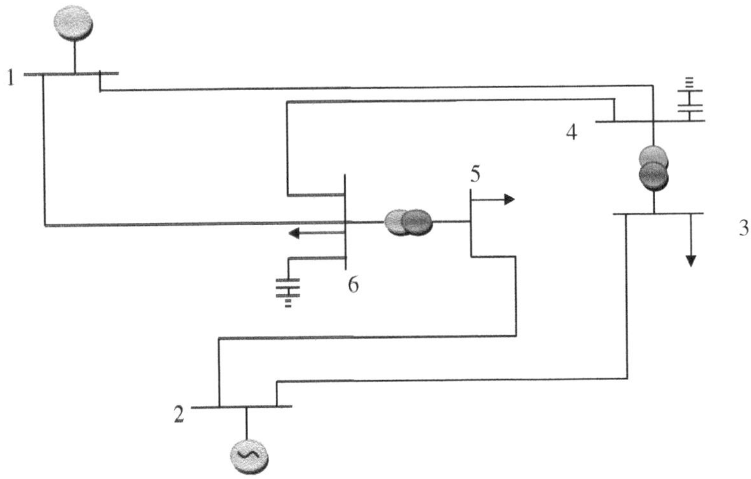

Figure 10.A.1 IEEE six-bus test system.

10.A.2 Data for IEEE 14-bus, 20-line system (100 MVA Base)

Table 10.A.3 Line data.

Line number	Bus number From	Bus number To	(*R*)	(*X*)	(*Bc*)	Tapping	Line flow limit (pu)
1	1	2	0.01938	0.05917	0.0264	1.0	0.92
2	1	8	0.05403	0.22304	0.0246	1.0	0.55
3	2	3	0.04699	0.19790	0.0219	1.0	0.41
4	2	6	0.05811	0.17632	0.0187	1.0	0.70
5	2	8	0.05695	0.17388	0.0170	1.0	0.50
6	6	3	0.06701	0.17103	0.0173	1.0	0.04
7	8	6	0.01335	0.04211	0.0064	1.0	0.42
8	6	7	0.00000	0.20912	0.0000	1.0	0.42
9	6	9	0.00000	0.55628	0.0000	1.0	0.20
10	8	4	0.00000	0.25202	0.0000	1.0	0.52
11	4	11	0.09498	0.19890	0.0000	1.0	0.09
12	4	12	0.12291	0.25581	0.0000	1.0	0.10
13	4	13	0.06613	0.13027	0.0000	1.0	0.20
14	7	5	0.00000	0.17615	0.0000	1.0	0.01
15	7	9	0.00000	0.08450	0.0000	1.0	0.42
16	9	10	0.03181	0.08450	0.0000	1.0	0.08
17	9	14	0.12711	0.27038	0.0000	1.0	0.12
18	11	10	0.08205	0.19207	0.0000	1.0	0.05
19	12	13	0.22092	0.19988	0.0000	1.0	0.025
20	13	14	0.17093	0.34802	0.0000	1.0	0.08

Table 10.A.4 Data bus.

Bus number	Load (pu)		Generation (pu)	
	Active	Reactive	Active	Reactive
1	0.0000	0.0000	1.6730	2.0000
2	0.2170	0.1270	0.7000	0.0000
3	0.9420	0.1900	0.5000	0.0000
4	0.1120	0.0750	0.0000	0.0000
5	0.0000	0.0000	0.0000	0.0000
6	0.4780	0.0390	0.0000	0.0000
7	0.0000	0.0000	0.0000	0.0000
8	0.0760	0.0180	0.0000	0.0000
9	0.2950	0.1660	0.0000	0.0000
10	0.0900	0.0580	0.0000	0.0000
11	0.0350	0.0180	0.0000	0.0000
12	0.0610	0.0160	0.0000	0.0000
13	0.1350	0.0580	0.0000	0.0000
14	0.0580	0.0580	0.0000	0.0000

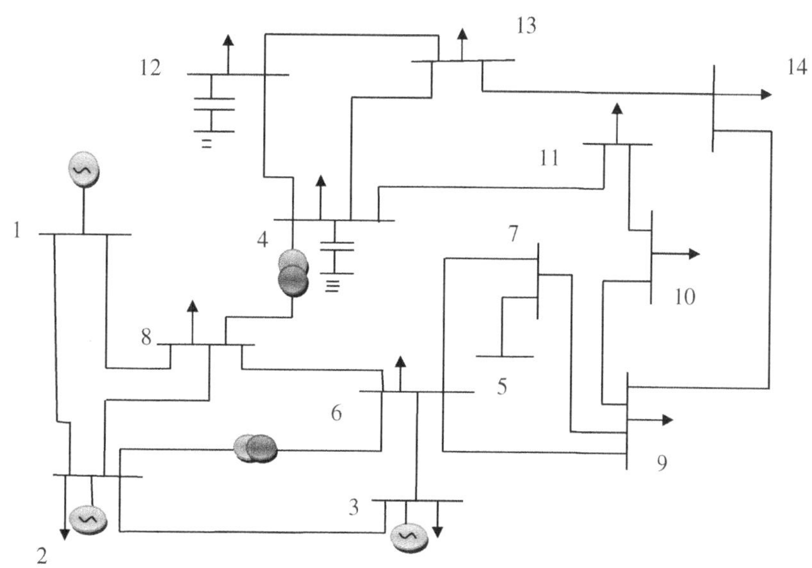

Figure 10.A.2 IEEE 14-bus test system.

10.A.3 Data for IEEE 25-bus, 35 line system (100 MVA Base)

Table 10.A.5 Line data.

Line number	Bus number		(*R*)	(*X*)	(*Bc*)	Tapping	Line flow limit (pu)
	From	To					
1	1	3	0.0720	0.2876	0.0179	1.0	0.460
2	1	16	0.0290	0.1379	0.0337	1.0	0.490
3	1	17	0.1012	0.2794	0.0148	1.0	0.170
4	1	19	0.1487	0.3897	0.0224	1.0	0.08
5	1	23	0.1085	0.2245	0.0573	1.0	0.360
6	1	25	0.0753	0.3593	0.0873	1.0	0.420
7	2	6	0.0617	0.2935	0.0186	1.0	0.300
8	2	7	0.0511	0.2442	0.0155	1.0	0.390
9	2	8	0.0597	0.2763	0.0175	1.0	0.380
10	3	13	0.0564	0.1487	0.0085	1.0	0.260
11	3	14	0.1183	0.3573	0.0185	1.0	0.490
12	4	19	0.0196	0.0514	0.0113	1.0	0.430
13	4	20	0.0382	0.1007	0.0220	1.0	0.440
14	4	21	0.0970	0.2547	0.0558	1.0	0.260
15	5	10	0.0479	0.2372	0.0577	1.0	0.370
16	5	17	0.0144	0.1269	0.1335	1.0	1.03
17	5	19	0.0929	0.2442	0.0140	1.0	0.530
18	6	13	0.0263	0.0691	0.0040	1.0	0.100
19	7	8	0.0529	0.1465	0.0078	1.0	0.075
20	7	12	0.0364	0.1736	0.0110	1.0	0.150
21	8	9	0.0387	0.1847	0.0118	1.0	0.075
22	8	17	0.0497	0.2372	0.0572	1.0	0.080
23	9	10	0.0973	0.2691	0.0085	1.0	0.110
24	10	11	0.0898	0.2359	0.0135	1.0	0.080
25	11	17	0.1068	0.2807	0.0161	1.0	0.0250
26	12	17	0.0460	0.2196	0.0135	1.0	0.026
27	14	15	0.0281	0.0764	0.0044	1.0	0.230
28	15	16	0.0256	0.0673	0.0148	1.0	0.130
29	17	18	0.0806	0.2119	0.0122	1.0	0.200
30	18	19	0.0872	0.2294	0.0132	1.0	0.050
31	20	21	0.0615	0.1613	0.0354	1.0	0.150
32	21	22	0.0141	0.1087	0.0238	1.0	0.200
33	22	23	0.2250	0.3559	0.0169	1.0	0.160
34	22	24	0.0970	0.2595	0.0567	1.0	0.095
35	24	25	0.0470	0.1458	0.0317	1.0	0.105

Table 10.A.6 Bus data.

Bus no.	Load (pu)		Generation (pu)	
	Active	Reactive	Active	Reactive
1	2.0000	0.6500	2.6730	3.0000
2	0.1000	0.0300	0.9930	1.0000
3	0.5000	0.1700	1.4719	1.0000
4	0.3000	0.1000	0.3910	1.0000
5	0.2500	0.0800	1.9300	1.0000
6	0.1500	0.0500	0.0000	0.0000
7	0.1500	0.0500	0.0000	0.0000
8	0.2500	0.0000	0.0000	0.0000
9	0.1500	0.0500	0.0000	0.0000
10	0.1500	0.0500	0.0000	0.0000
11	0.0500	0.0000	0.0000	0.0000
12	0.1000	0.0000	0.0000	0.0000
13	0.2500	0.0800	0.0000	0.0000
14	0.2000	0.0700	0.0000	0.0000
15	0.3000	0.1000	0.0000	0.0000
16	0.3000	0.1000	0.0000	0.0000
17	0.6000	0.2000	0.0000	0.0000
18	0.1500	0.0500	0.0000	0.0000
19	0.1500	0.0500	0.0000	0.0000
20	0.2500	0.0800	0.0000	0.0000
21	0.2000	0.0700	0.0000	0.0000
22	0.2000	0.0700	0.0000	0.0000
23	0.1500	0.0500	0.0000	0.0000
24	0.1500	0.0500	0.0000	0.0000
25	0.2500	0.0800	0.0000	0.0000

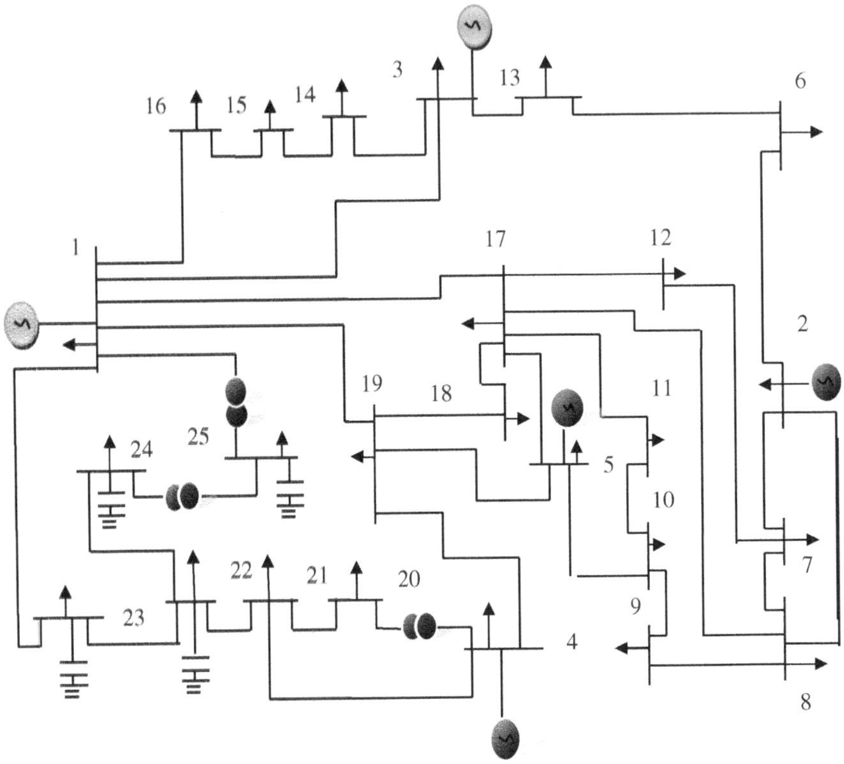

Figure 10.A.3 IEEE 25-bus test system.

References

Ajitha, A., Goel, M., Assudani, M. et al. (2022). Design and development of residential sector load prediction model during COVID-19 pandemic using LSTM based RNN. *Electric Power Systems Research* 212: 108635. https://doi.org/10.1016/j.epsr.2022.108635.

Ajjarapu, V. and Christy, C. (1992). The continuation power flow: a tool for steady state voltage stability analysis. *IEEE Transactions on Power Systems* 7 (1): 416–423. https://doi.org/10.1109/59.141737.

Amjady, N. and Esmaili, M. (2003). Voltage security assessment and vulnerable bus ranking of power systems. *Electric Power Systems Research* 64 (3): 227–237. https://doi.org/10.1016/S0378-7796(02)00196-7.

Armaghani, S., Naghshbandy, A.H., and Shahratash, S.M. (2020). Application of a new fast analytical approach for post-contingency state estimation of "N-1" transmission line outage in online vulnerability assessment of large-scale transmission networks with wind power resources. *International Journal of Electrical Power & Energy Systems* 123: 106304. https://doi.org/10.1016/j.ijepes.2020.106304.

Arya, L.D. and Verma, H.K. (1996). A method for tracing PV-curve for voltage stability analysis with voltage dependent load. *Electric Machines and Power System* 24 (6): 583–596. https://doi.org/10.1080/07313569608955696.

Arya, L.D., Pande, V.S., and Tare, R.S. (2002). Contingency selection based on vanishing eigen value concept. *Journal of the Institution of Engineers (India)* 83: 213–217.

Arya, L.D., Titare, L.S., and Kothari, D.P. (2006). Determination of probabilistic risk of voltage collapse using radial basis function (RBF) network. *Electric Power Systems Research* 76 (6-7): 426–434. https://doi.org/10.1016/j.epsr.2005.09.011.

Arya, L.D., Singh, P., and Titare, L.S. (2012). Optimum load shedding based on sensitivity to enhance static voltage stability using differential evolution. *Swarm and Evolutionary Computation* 6: 25–38. https://doi.org/10.1016/j.swevo.2012.06.002.

Bahmanyar, A.R. and Karami, A. (2014). Power system voltage stability monitoring using artificial neural networks with a reduced set of inputs. *International Journal of Electrical Power & Energy Systems* 58: 246–256. https://doi.org/10.1016/j.ijepes.2014.01.01.

Berizzi, A., Bovo, C., Cirio, D. et al. (2009). On-line fuzzy voltage collapse risk quantification. *Electric Power Systems Research* 79 (5): 740–749. https://doi.org/10.1016/j.epsr.2008.10.010.

Billinton, R. and Allan, R.N. (1984). *Reliability Evaluation of Power System*. Plenum Press, Inc. (Book).

Choube, S.C., Arya, L.D., and Sakravadia, D.K. (2001). Contingency selection based on singular value decomposition. *Journal of the Institution of Engineers (India)* 81 (4): 226–229.

Coelho, E.P.R., Alves, R.J.M., Paiva, M.H.M. et al. (2022). Topological multi-contingency screening based on current flow betweenness. *Electric Power Systems Research* 203: 10709. https://doi.org/10.1016/j.epsr.2021.107609.

Dester, M. and Castro, C.A. (2009). Multi-criteria contingency ranking method for voltage stability. *Electric Power System Research* 79 (1): 220–225. https://doi.org/10.1016/j.epsr.2008.05.023.

Hagh, M.T., Gargari, M.Z., and Pakdel, M.J.V. (2018). Sequential analysis of optimal transmission switching with contingency assessment. *IET Generation, Transmission & Distribution* 12 (6): 1390–1396. https://doi.org/10.1049/iet-gtd.2017.0435.

Haykin, S. (2003). *Neural Network: A Comprehensive Foundation*. Pearson Education, Second Edition, Inc. (Book).

Heidari, H., Hagh, M.T., and Salehpoor, P. (2022). Accurate, simultaneous and real-time screening of N-1, N-k, and N-1-1 contingencies. *International Journal of Electrical Power & Energy Systems* 136: 107592. https://doi.org/10.1016/j.ijepes.2021.107592.

Javan, D.S., Mashhadi, H.R., and Rouhani, M. (2013). A fast static security assessment method based on radial basis function neural networks using enhanced clustering. *International Journal of Electrical Power & Energy Systems* 44 (1): 988–996. https://doi.org/10.1016/j.ijepes.2012.08.014.

Kim, H. and Singh, C. (2005). Power system probabilistic security assessment using bayes classifier. *Electric Power Systems Research* 74 (1): 157–165. https://doi.org/10.1016/j.epsr.2004.10.004.

Kothari, D.P., Nagrath, I.J., and Saket, R.K. (2022). *Modern Epower System Analysis*, 5e. New Delhi, India: McGraw Hill (Book).

Kundur, P. (1994). *Power System Stability and Control*. McGraw Hill, Inc. (Book).

Lee, C.Y., Tsai, S.H., and Wu, Y.K. (2010). A new approach to the assessment of steady-state voltage stability margins using the P–Q–V curve. *International Journal of Electrical Power and Energy Systems* 32 (10): 1091–1098. https://doi.org/10.1016/j.ijepes.2010.06.005.

Naik, S.D., Khedkar, M.K., and Bhat, S.S. (2015). Effect of line contingency on static voltage stability and maximum loadability in large multi bus power system. *International Journal of Electrical Power & Energy Systems* 67: 448–452. https://doi.org/10.1016/j.ijepes.2014.12.002.

Nan, L., Liu, Y., Wu, L. et al. (2021). Graph theory based N-1 transmission contingency selection and its application in security-constrained unit commitment. *Journal of Modern Power Systems and Clean Energy* 9 (6): 1458–1467. https://doi.org/10.35833/MPCE.2019.000602.

Ni, M., McCalley, J.D., Vittal, V., and Tayyib, T. (2003). On-line risk-based security assessment. *IEEE Transactions on Power Systems* 18 (1): 258–265. https://doi.org/10.1109/TPWRS.2002.807091.

Rubinstein, R.Y. (1981). *Simulation and Monte-Carlo Methods*. John Wiley and Sons, Inc. (Book).

Sherstinsky, A. (2020). Fundamentals of recurrent neural network (RNN) and long short-term memory (LSTM) network. *Physica D: Nonlinear Phenomena* 404: 132306. https://doi.org/10.1016/j.physd.2019.132306.

Singh, P, Purey, P., Titare, L.S. et al., (2017). Enhancement of voltage security by MW-generation rescheduling based on sensitivities using black hole. *2017 International Conference on Information, Communication, Instrumentation and Control* (*ICICIC*), 1-5, https://doi.org/10.1109/ICOMICON.2017.8279045.

Singh, P., Titare, L.S., Choube, S.C., and Arya, L.D. (2018). Security assessment accounting uncertainties in line parameters and control variables with the considerations of transmission line unavailability. *Journal of Electrical Systems and Information Technology* 5 (3): 576–593. https://doi.org/10.1016/j.jesit.2017.10.002.

Singh, P., Arya, R., Titare, L.S., and Arya, L.D. (2021). Optimum value based under voltage load shedding strategies. *Australian Journal of Electrical and Electronics Engineering* 18 (3): 146–160. https://doi.org/10.1080/1448837X.2021.1945719.

Singh, P., Arya, R., Titare, L.S. et al. (2023). Value aided optimal load shedding accounting voltage stability consideration employing crow search algorithm with modification based on lampinen's criterion. *Applied Soft Computing* 143: 110391. https://doi.org/10.1016/j.asoc.2023.110391.

Srivani, J. and Swarup, K.S. (2008). Power system static assessment and evaluation using external system equivalents. *International Journal of Electrical Power & Energy Systems* 30 (2): 83–92. https://doi.org/10.1016/j.ijepes.2007.06.008.

Vaahedi, E., Fuchs, C., Xu, W. et al. (1999). Voltage stability contingency screening and ranking. *IEEE Transactions on Power Systems* 14 (1): 256–265. https://doi.org/10.1109/59.744541.

Verma, K. and Niazi, K.R. (2012). Supervised learning approach to online contingency screening and ranking in power systems. *International Journal of Electrical Power & Energy Systems* 38 (1): 94–114. https://doi.org/10.1016/j.ijepes.2011.12.025.

Verma, M.K. and Shrivastava, S.C. (2005). Approach to determine voltage control areas considering impact of contingencies. *IEE-Proceedings Generation, Transmission and Distribution* 152 (3): 342–350. https://doi.org/10.1049/ip-gtd:20041134.

Wang, L., Hu, B., Xie, K. et al. (2020). Screening model of incremental risk events for reliability analysis of transmission system. *International Journal of Electrical Power & Energy Systems* 120: 105995. https://doi.org/10.1016/j.ijepes.2020.105995.

Zarate, L.A.L., Castro, C.A., Ramos, J.L.M., and Ramos, E.R. (2006). Fast computation of voltage stability security margins using non-linear programming techniques. *IEEE Transactions on Power Systems* 21 (1): 19–27. https://doi.org/10.1109/TPWRS.2005.860945.

Zhao, S. and Singh, C. (2018). A hybrid method for reliability evaluation of line switching operations. *Electric Power Systems Research* 163, Part A: 365–374. https://doi.org/10.1016/j.epsr.2018.07.001.

Zhu, Y. and Liu, G. (2022). Application of triangle count in branch contingency screening. *International Journal of Electrical Power & Energy Systems* 135: 107392. https://doi.org/10.1016/j.ijepes.2021.107392.

11

Probabilistic Voltage Security Assessment and Enhancement Using Rescheduling of Reactive Power Control Variables

Lakhan Singh Titare[1], *Aanchal Singh S. Vardhan*[2], *Liladhar Arya*[3], *and Devkaran Sakravdia*[1]

[1]*Department of Electrical Engineering, Government Engineering College, Ujjain, Madhya Pradesh, India*
[2]*Department of Electrical Engineering, National Institute of Technology, Jamshedpur, Jharkhand, India*
[3]*Department of Electrical Engineering, Shri G.S. Institute of Technology & Science, Indore, Madhya Pradesh, India*

11.1 Introduction

Power system security studies, encompassing evaluation and regulation, are an indispensable aspect of contemporary energy management systems. Power utilities are confronting substantial obstacles in preserving the excellence and dependability of power provision due to the ever-expanding interconnections and loads in expansive power network infrastructures. Economic constraints have compelled power companies to run their generation and transmission systems at full capacity. The enhanced efficacy of protective mechanisms, Static Var Systems, rapid turbine valving, and automated voltage regulators has notably amplified the transient stability limits of the system. Consequently, the power system can now transmit greater real power over extended distances. Nevertheless, this has engendered fresh challenges in upholding the mandated bus voltages, as the system is now constrained by voltage stability (Taylor 1994). Voltage security pertains to a power system's capability to sustain stable and acceptable voltage levels, ensuring the smooth operation and functionality of electrical equipment and devices under various operating conditions, including contingencies or adverse changes in the system. It is crucial for ensuring the proper functioning and reliability of the power grid. If a power system's condition restores to the same or a state that is somewhat similar to the pre-disturbance state of operation after any little disturbance, the operating point is considered to be small disturbance stable (Kundur 1994; Concordia 1991; IEEE System Dynamic Performance Subcommitte n.d.). Additionally, it has long been accepted practice to assess reliability indicators during the planning stage using a probabilistic technique based on security evaluation including correctability (Sebastian Rios et al. 1998; Arya et al. 2009; Billinton and Wenyuan 1991). An approach for assessing a composite system's probability of failure on the basis of maximum load and system ability to repair proposed in (Arya et al. 2000). If failure probability, or, as it is sometimes referred to, probability of load shedding is lower, the system is more resistant to operating constraint violations. A large number of power flow studies under different loading conditions and outages are required for probability of load shedding evaluation. The security of the system in its operational state cannot be evaluated using this type of indicator. Such failure probability indicator may be used in online monitoring applications in power systems owing to advancements in artificial neural networks (Singh et al. 2021; Arya et al. 2007). The majority of research studies that evaluated reliability indicator

Reliability Analysis of Modern Power Systems, First Edition. Edited by R. K. Saket and P. Sanjeevikumar.

did not consider voltage security from the perspective of static voltage stability. When evaluating the composite power system's adequate suitability while taking voltage and line flow limits into account, Billinton and Khan (1992) employed the Linear Programming model. For evaluating the reliability of a transmission system, Pereira and Pinto (1992) developed Monte-Carlo (MC) sampling with a variance reduction technique. A novel method for evaluating the reliability of interconnected power systems is introduced in the research paper (Deng and Singh 1992), which includes planned outages and frequency calculations for load loss, while considering various factors impacting power supply interruptions. Billinton and Li (1993) proposed a system state transition sampling methodology to directly compute reliability of composite system. A basic framework has been designed that can be used in the probabilistic risk estimation of cascading failures (Henneaux 2015). Kirschen and Jayaweera (2007) introduced a methodology for comparing the probabilistic and deterministic assessment of transmission system security. Voltage stability constraints were taken into account while evaluating the reliability indices (Melo et al. 1997; Billinton and Aboreshaid 1998). The method for calculating voltage collapse-linked indicators and their effect on power system soundness indicators based on regaining system-solving capacity through load curtailment was probably first presented in (Melo et al. 1997). This has been accomplished by employing a univariate search technique to optimize the static voltage stability limit and a predictor-corrector-based continuation power flow technique to compute the threshold value of stable voltage (Capitaanescu and Van Cutsem 2002). A novel collapse prediction index (NCPI) is proposed in order to evaluate the power system's voltage stability circumstances as well as the critical line conditions. In addition, the sensitivity hypotheses of the currently used indices and evaluates how they affect the forecasting of voltage collapse (Mokred et al. 2023). A new methodology was developed that considers voltage stability into account while evaluating the adequate capabilities of a transmission network (Xia et al. 2021). A novel sensitivity method is employed to choose the optimal control variables for addressing the power system voltage instability. In order to account for changes in power system operation, a sensitivity analysis is carried out on a voltage stability index called Thevenin-Based Voltage Stability Margin (TVSM) (Alzaareer et al. 2020). The methodology is built up to incorporate infeed uncertainty into the evaluation of the security of the power system. It enables the evaluation of the operational security of a power system's potential state under uncertainty resulting from various contingencies and loads (de Jong et al. 2018). Benidris and Mitra (2015) presented the formulation of expressions and sensitivity evaluations of the severity index with respect to the voltage and reactive power constraints and the implications of these constraints on the reliability of the power system. This chapter introduces a novel methodology for assessing and controlling voltage security by adjusting reactive power control variables and, when required, implementing load shedding. The technique involves computing voltage stability limits for single- and double-line outage scenarios using the continuation power flow method. Based on these results, the probabilistic insecurity index (PII) is calculated using the cut-set technique. To proactively manage voltage security, the PII sensitivity is determined concerning various settings of reactive power control variables and loading conditions. This sensitivity analysis allows for preventive control measures. Developed equations in Section 11.2 are used to calculate the reactive power control variable settings in emergency situations and load-shedding conditions. The obtained results are utilized to train an Artificial Neural Network (ANN) to predict PII for any operating condition. The trained ANN is also employed to compute the sensitivity of PII with respect to reactive power control variables and load shedding at different buses. Using the ANN and the derived sensitivities, voltage security can be controlled effectively by adjusting reactive power control variables or implementing load shedding at specific buses when necessary. The advantage of employing an ANN lies in its suitability for online applications, making the proposed methodology practical and efficient for real-time voltage security assessment and control.

11.2 Computation of Probabilistic Insecurity Index (PII) Using Cut-set Technique

The computation of the PII involves assessing power system's vulnerability and security across various contingencies and operating conditions. The PII is a measure of the probability of voltage instability or collapse in the system. The cut set is a widely used and powerful technique in the field of network analysis, particularly in assessing the security and reliability of power networks. The cut-set method helps in understanding the various modes of system failure. A cut set is a set of transmission lines in the network such that their failure would lead to the disconnection of the network into two or more separate sections. By identifying these cut sets, one can determine the most vital lines or combinations of lines that, if interrupted, could cause system failure or a significant reduction in its performance. The cut-set method is particularly useful for designing robust and reliable power systems. It aids in determining redundancy requirements, assessing the impact of component failures, and designing effective preventive control measures. Moreover, the cut-set technique can be readily executed on a digital computer, allowing for efficient and practical analyses of large and complex power networks. A minimal cut set refers to a set of transmission lines that, when failed, lead to system failure. However, when any single element of the set remains operational, it does not cause the system to fail. Consequently, every constituent of a minimum cut set must be in a state of failure for the system failure to be initiated. The failure probability (F_{P}) of a power network can be expressed as follows:

$$F_{\text{P}} = P[C_{\text{s}1} + C_{\text{s}2} + C_{\text{s}3} + \ldots + C_{\text{s}n}] \tag{11.1}$$

where

$C_{\text{s}1}, C_{\text{s}2},\ldots,C_{\text{s}n}$ are minimal cut sets.

While it is theoretically possible to perform exact evaluations, doing so is often impractical and time-consuming, especially for large power networks. Therefore, in practice, failure probability assessments often resort to approximations, sacrificing some precision to achieve much faster results. Power systems with low failure rates and high repair rates for individual lines, the introduced error is typically negligible and does not significantly impact the availability data for transmission lines. Thus, the failure probability is estimated by keeping the first-order terms in the expansion of Eq. (11.1) as approximated below:

$$F_{\text{P}} = \sum_{j=1}^{n} P(C_{\text{s}i}) \tag{11.2}$$

The second approximation involves not considering cut sets with orders higher than a certain number. The number of components within a cut set corresponds to its order. In this chapter, only cut sets up to second order are taken into consideration. This decision is based on the very low probability of more than two transmission lines failing simultaneously under usual loading conditions. Additionally, a next-higher-order cut set may not necessarily be a minimal cut set, which is required for Eq. (11.1) to be valid. Therefore, Eq. (11.2) is used to calculate the probability of failure with this approximation. To model the operation of each transmission line, a two-state modal is

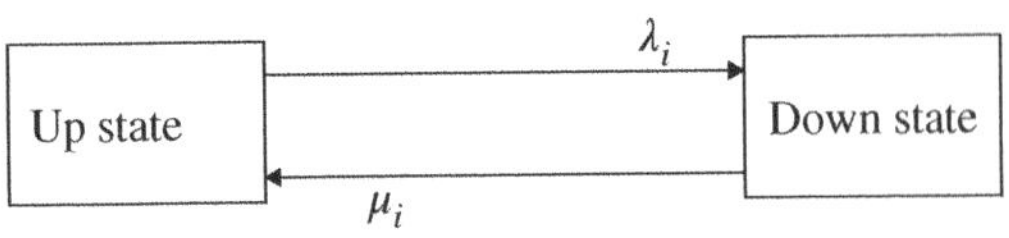

Figure 11.1 Dual-state model for individual transmission lines.

used, comprising an "up state" representing the line being in operation and a "down state" indicating that the line is either in a failed state or scheduled to be off. The failure and repair rates of the transmission lines are denoted by the symbols "λ" and "μ," respectively, as depicted in Figure 11.1. This modal is utilized to determine the long-term availability and unavailability of transmission lines. The corresponding expressions for availability and unavailability are provided in Billinton and Allan 1983:

$$A_{\mathrm{Ti}} = \frac{\mu_{\mathrm{Ti}}}{(\lambda_{\mathrm{Ti}} + \mu_{\mathrm{Ti}})} \tag{11.3}$$

$$\overline{A}_{\mathrm{Ti}} = \frac{\lambda_{\mathrm{Ti}}}{(\lambda_{\mathrm{Ti}} + \mu_{\mathrm{Ti}})} \tag{11.4}$$

where $\overline{A_{\mathrm{Ti}}}$ and A_{Ti} Indicate the unavailability and availability of the transmission line, respectively. The determination of minimal cut sets (first and second order) for a given load is achieved through a comprehensive analysis, wherein the load is compared against the static voltage stability limit in the base case and also under diverse line outage scenarios. This detailed investigation allows us to identify the critical cut sets that exert significant influence on the power system's voltage stability under various operating conditions. The static voltage stability limits under circumstances of single or double-line outage are determined by applying the continuation power flow technique (Arya and Verma 1996). For the purpose of securing proper system functioning and stability, it is of paramount importance to sustain ample stability margins. Subsequently, utilizing the outcome of the continuation power flow, prepare a table displaying capacity outages, indicating the permissible loading levels and the corresponding network conditions. By analyzing various load levels, one can identify the relevant cut sets from the capacity availability table. The PII of the transmission system is then computed using relation (11.2), representing the probability of failure. It is important to emphasize that the loadability limit is influenced by the settings of reactive power control variables. As a result, the failure probability (PII) is assessed for different configurations of reactive power control variables. The computational procedure is visually presented in Figure 11.2 using a comprehensive flow chart. This computational methodology provides a systematic and comprehensive analysis of the voltage security of the system. It aids in identifying potential stability issues, assessing their probabilities, and optimizing the control variables to enhance the overall voltage stability and security of the power transmission network.

11.3 Computation of Probabilistic Insecurity Index (PII) Sensitivity using ANN

The ANN is trained using the backpropagation algorithm, enabling it to learn from a set of instances obtained from Section 11.2 (Fu 1994). These instances cover a wide range of operating conditions, offering a comprehensive understanding of the system's behavior. The generic network architecture consists of one hidden layer, one input layer, and one output layer. The input layer's units are equal to the number of reactive power control variables and total load buses, while the output layer has a single unit that gives the PII value. Additionally, it is postulated that the neurons in the hidden layer

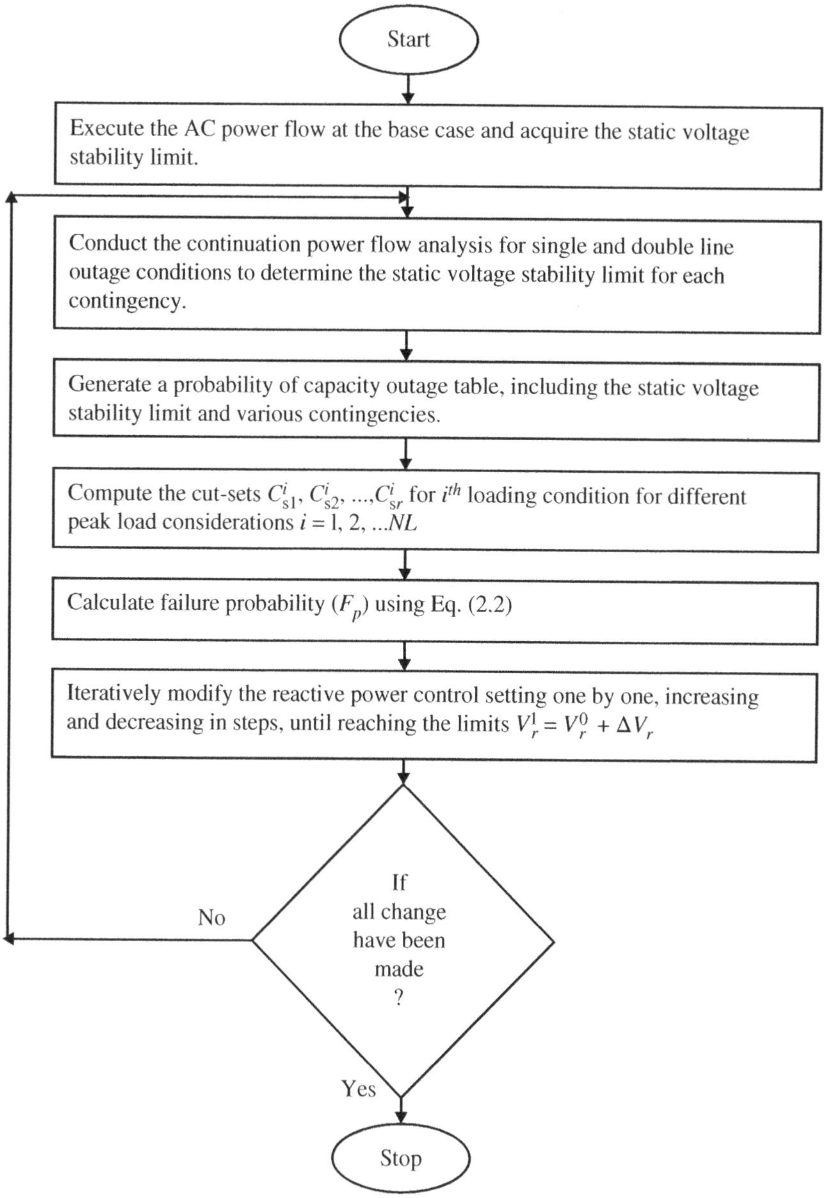

Figure 11.2 Flowchart for computing probabilistic insecurity index.

adopt a sigmoidal activation function, while the neuron in the output layer follows a nonsigmoidal (linear) activation function. Once trained, the ANN provides real-time online assessment of voltage security, enabling quick identification of potential voltage instability issues. The network equations corresponding to the architecture depicted in Figure 11.3 can be expressed as:

$$O = \sum_{j=1}^{m} (W_{jk} H_j) \tag{11.5}$$

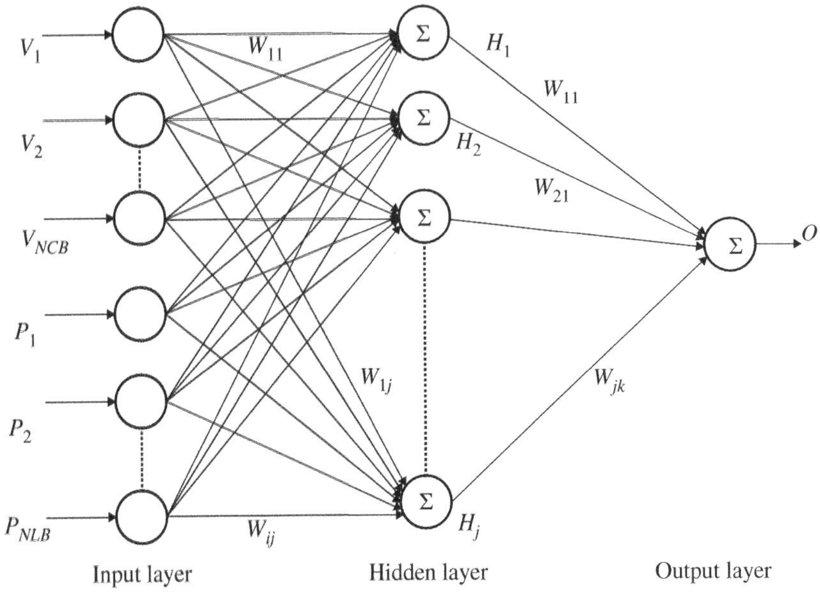

Figure 11.3 Generalized diagram of backpropagation network for the computation of probabilistic voltage security.

The output of the neuron located at the jth position in the hidden layer is as follows:

$$H_j = \frac{1}{\left(1 + e^{-\mathrm{Net}_j}\right)} \tag{11.6}$$

$j = 1,2,3,\ldots,n$

$$\mathrm{Net}_j = \sum_{j=1}^{n} W_{ij} U_i \tag{11.7}$$

where

W_{ij} is the weight that connects the ith input node to the jth hidden neuron.
U_i is the input variable at ith node (Haykin 2003). The following formulae provide the weight change W_{jk}.

$$\Delta W_{jk} = \eta \delta H_j \tag{11.8}$$

$$\delta = (T - O) \tag{11.9}$$

T – is the network's target value
O – is the network's output value,
η is the learning rate that lies between (0, 1). Using the backpropagation approach, the following weight change ΔW_{ij} expression is given for the hidden layer:

$$\Delta W_{ij} = \eta \delta U_i \tag{11.10}$$

where δ_j is the slope of error and is written as follows:

$$\delta_j = \delta \Delta W_{jk} H_j (1 - H_j) U_i \tag{11.11}$$

U_i is the component of the input vector $[U]$, where

$$[U]^{\mathrm{T}} = [V_1, V_2, \ldots, V_{\mathrm{NCB}}, P_1, P_2, \ldots, P_{\mathrm{NLB}}]^{\mathrm{T}}$$

After successfully training the ANN, the network Eqs. (11.5), (11.6), and (11.7) are employed to compute the sensitivities of the output variable "O" concerning the input variables, which can be either the reactive power control variables or load shedding. The sensitivity expression is obtained by applying the chain rule of differentiation. This mathematical technique allows us to compute how changes in the input variables occur in the context of the trained Artificial Neural Network (ANN). The chain rule enables us to analyze the complex relationship between the network's layers and efficiently determine the sensitivities of the output with respect to various input variables as follows:

$$\frac{\partial \mathrm{PII}}{\delta U_i} = \sum_{j=1}^{m} \left(\frac{\partial \mathrm{PII}}{\partial H_j}\right)\left(\frac{\partial H_j}{\partial \mathrm{Net}_j}\right)\left(\frac{\partial \mathrm{Net}_j}{\partial U_i}\right) = \mathrm{SPII}_i \tag{11.12}$$

where SPII_i is the sensitivity of PII with respect to U_i. The expression for sensitivity given by algebraic manipulation of network equations is as follows:

$$\mathrm{SPII}_i = \sum_{j=1}^{m} W_{jk} H_j (1 - H_j) W_{ij} \tag{11.13}$$

The evaluation of sensitivities based on Eq. (11.13) is characterized by its inherent simplicity and efficiency. By interpreting expression (11.13) as the sum of all path gains originating from an input node and leading to the output node. Each path gain is obtained by multiplying the weights and the gain of the nodes $\{H_j(1 - H_j)\}$ encountered along that specific path. The method for computing sensitivity is highly versatile and can be extended to accommodate multiple hidden layers within the neural network. In Section 11.4, algorithm not only ensures enhanced voltage stability but also incorporates the capability of implementing load shedding as a viable and effective corrective measure to safeguard the power system.

11.4 Voltage Security Enhancement using a Monovariable Approach

This approach focuses on controlling a single variable to improve voltage security. The variable chosen for control could be reactive power or load shedding. Trained ANN is employed for real-time voltage security calculation. Subsequently, sensitivities are assessed to facilitate corrective rescheduling. This is achieved by defining a threshold value for PII. Whenever the PII surpasses this limit, the calculated sensitivities concerning reactive power control variables are employed to perform corrective rescheduling. The reactive power control variable with the highest sensitivity is selected for the rescheduling task. The adjustment for this reactive power control variable is obtained as follows:

$$\Delta V_{\mathrm{r}} = \frac{(\mathrm{PII}_{\mathrm{o}} - \mathrm{PII}_{\mathrm{th}})}{\mathrm{SPII}_{\mathrm{r}}} \tag{11.14}$$

where

$\mathrm{PII}_{\mathrm{o}}$ is the current insecurity index obtained from the ANN,
$\mathrm{PII}_{\mathrm{th}}$ is the threshold value of the indicator, and
$\mathrm{SPII}_{\mathrm{r}}$ is the sensitivity of indicator with respect to the "r" control variable.

In the event that the control correction obtained through Eq. (11.14) exceeds the limit, it can be adjusted to the limiting value. Following that, modified sensitivities are reassessed, and an alternative control variable is chosen for corrective action. The monovariable rescheduling process continues until the PII is brought within acceptable limits. Subsequently, a continuation load flow program is executed to verify that all voltage levels are within the prescribed bounds. If it is observed that all control variables have been fully utilized and yet the PII still violates the limit, then the sensitivities of PII with respect to load shedding at the load buses are evaluated, and selection of a particular bus with the greatest sensitivity is made for load shedding, as outlined below:

$$\Delta P_i = \frac{(\mathrm{PII_o} - \mathrm{PII_{th}})}{\mathrm{SPIID}_i} \tag{11.15}$$

SPIID_i is the sensitivity of indicator (PII) with respect to load shedding at ith bus.

The limit on the sensitivity of PII with respect to load shedding at a bus should not surpass a certain maximum threshold (e.g. 80% of the total load at that bus, considering 20% as reserved for emergency services). If the total load shedding at a bus is inadequate to meet this limit, the next bus with the highest sensitivity is chosen for additional load shedding. This iterative process continues until the PII value falls within the acceptable range. In essence, the trained ANN acts as a system-wide relay, not only identifying insecure conditions but also providing recommendations for corrective measures (Zarate et al. 2006). This intelligent approach ensures the system's voltage stability and enhances overall security.

11.5 Results and Discussion

The algorithm formulated for voltage security control, employing the rescheduling of reactive power control variables or load shedding, has been applied to two IEEE standard test systems: 14-bus and 25-bus (Bijwe et al. 1993).

11.5.1 14-bus System

This system comprises five PV buses, and their voltages can be regulated between ($0.95 \leq V_i \leq 1.15$pu) to adjust the voltage of load buses. The availability of lines from line No. 1 to line No. 20 is considered as: 0.99725754, 0.99037152, 0.99027586, 0.99592631, 0.99097267, 0.99825389, 0.99772435, 0.99138266, 0.99633603, 0.99472818, 0.99927413, 0.99327504, 0.99733418, 0.99792642, 0.99078241, 0.99546628, 0.99662247, 0.99016989, 0.99906437, and 0.9908926 (Billinton and Allan 1984). Table 11.1 shows a set of samples used to train the ANN, which enabled the computation of the PII using the cut-set technique and the ANN, along with the percentage errors. The network underwent training with 150 instances, and only 18 instances with the highest error magnitudes are listed in Table 11.1. Satisfactory results were achieved by incorporating two hidden layers, each consisting of ten sigmoidal neurons. The output neuron is assumed to be linear. In Table 11.2, various weights namely $W_{j2,k}$, $W_{j1,j2}$, and $W_{i,j1}$, connect the neurons between the hidden layer-2 and the output layer, hidden layer-1 and hidden layer-2, and input node and hidden layer-1, respectively. The input layer includes variables V_1, V_2, V_3, and P_d namely PV-bus voltages and overall load of the system. Table 11.3 shows the PII obtained using the trained network and the cut-set approach, along with the percentage errors for test cases. There are no prescribed

Table 11.1 A set of samples for training of ANN. PII computed using the cut-set technique and Backpropagation Algorithm Network for a 14-bus system.

S. No	V_1(pu)	V_2(pu)	V_3(pu)	P_d(pu)	PII (by cut-set technique)	PII (by ANN)	% Error
1	0.95	0.95	0.95	6.031	0.076020248	0.078049989	−2.67
2	0.95	0.95	1	6.0271	0.067941537	0.066929208	1.49
3	0.95	1	1	5.9883	0.064533306	0.062474694	3.19
4	1	1	1	6.8166	0.072356281	0.073050901	−0.96
5	1	1	1.05	6.7161	0.060812244	0.058239886	4.23
6	1	1.05	1.05	6.6445	0.056262153	0.055018759	2.21
7	1.05	1.05	1.05	7.5022	0.067948321	0.066895122	1.55
8	1.05	1.05	1.1	7.4461	0.056266172	0.056620649	−0.63
9	1.05	1.1	1.1	6.8277	0.044472754	0.042885077	3.57
10	1.1	1.1	1.1	83,266	0.064225551	0.063531915	1.08
11	1.1	1.1	1.15	9.0376	0.071610771	0.068288031	4.64
12	1.1	1.15	1.15	7.3169	0.048195524	0.048831705	−1.32
13	1.15	1.15	1.15	9.1367	0.060812591	0.060338253	0.78

regulations governing the ratio between training and testing instances. The training instances chosen for this research have successfully encompassed a wide range of operational scenarios for the power system. Within this scope, supplementary testing instances have been chosen, alongside the training instances. The trained network was tested with 250 instances, and only 18 instances with the highest error magnitudes are listed in Table 11.3. In this range, additional testing instances have been selected in addition to the training instances. The PII values must not exceed a predefined threshold set at 0.060812590. This specific threshold is chosen to ensure that, even during outage scenarios, the distance to the saddle-node bifurcation point (measured in mega volt amperes (MVA)) remains at a minimum of 19.99% of the static voltage stability limit. This criterion is indispensable and may vary based on the operational philosophy and specific needs of the individual power utility. The threshold value holds significant importance in Eqs. (11.14) and (11.15) as they are utilized to evaluate control measures. The percentage error of PII, as computed by trained ANN, is less than ±5%. If further error reduction is desired, increasing the size of the ANN is an option. This is achieved by augmenting the number of neurons in the hidden layer and expanding the number of hidden layers. For a specific operating setting, corrective reactive power rescheduling was demonstrated with total load $P_d = 6.0271$ pu and $V_1 = 0.95$ pu, $V_2 = 0.95$ pu, $V_3 = 0.95$ pu. It was observed that the stability margin obtained was less than 20% with a PII of 0.072356281, exceeding the specified threshold value. Sensitivities of PII were computed using Eq. (11.14), and it was found that Bus No. 2 is selected for voltage correction ($\Delta V_2 = 0.1614$ pu) due to its highest sensitivity. This correction brought the PII within the desired limit of 0.060812590 as the final value. Another case was considered with values: $V_1 = 1.15$ pu, $V_2 = 1.15$ pu, $V_3 = 1.15$ pu, total load $P_d = 9.3672$ pu, and despite correction attempts using Eq. (11.14) for rescheduling of reactive power at the most sensitive buses, the PII remained above the specified limit (0.076020248). Hence, sensitivities were assessed concerning load shedding at

Table 11.2 Final weights obtained after ANN training for the 14-bus system.

W_{ij} (Weight between *i*thinput node and *j*th hidden layer-1 neuron)	Values of weights (W_{ij})	$W_{j1,j2}$ (Weight between j_1^{th}hidden layer-1 neuron and j_2^{th}hidden layer-2 neuron)	Values of weights ($W_{j1,j2}$)	$W_{j1,j2}$ (Weight between j_1^{th} hidden layer-1 neuron and j_2^{th} hidden layer-2 neuron)	Values of weights ($W_{j1,j2}$)	$W_{j2,k}$ (Weight between j_2^{th} hidden layer-2 neuron and k^{th} output node)	Values of weights ($W_{j2,k}$)
$W_{1,1}$	−0.30023216	$W_{1,1}$	−0.2776832	$W_{6,1}$	−0.64809683	$W_{1,1}$	0.7230983
$W_{1,2}$	0.27647354	$W_{1,2}$	0.19835022	$W_{6,2}$	0.07503865	$W_{2,1}$	−0.769448
$W_{1,3}$	−0.18358764	$W_{1,3}$	−0.2341812	$W_{6,3}$	0.46713774	$W_{3,1}$	−0.722583
$W_{1,4}$	−0.08670065	$W_{1,4}$	−0.6902042	$W_{6,4}$	0.35403297	$W_{4,1}$	0.6882041
$W_{1,5}$	−0.84691089	$W_{1,5}$	−0.9776295	$W_{6,5}$	0.14505076	$W_{,5,1}$	0.8545416
$W_{1,6}$	−0.11793122	$W_{1,6}$	−0.5679249	$W_{6,6}$	−0.65689846	$W_{6,1}$	0.9073756
$W_{1,7}$	0.13485305	$W_{1,7}$	−0.365493	$W_{6,7}$	0.03919319	$W_{7,1}$	0.4347044
$W_{1,8}$	−0.37024051	$W_{1,8}$	0.34264155	$W_{6,8}$	−0.12120154	$W_{8,1}$	−0.961718
$W_{1,9}$	−0.18615765	$W_{1,9}$	−0.5132074	$W_{6,9}$	−0.5650661	$W_{9,1}$	0.1031879
$W_{1,10}$	0.86567297	$W_{1,10}$	0.37565473	$W_{6,10}$	−0.01778744	$W_{10,1}$	0.2004992
$W_{2,1}$	0.66145583	$W_{2,1}$	−0.6123977	$W_{7,1}$	0.47617357		
$W_{2,2}$	0.90219146	$W_{2,2}$	0.91891559	$W_{7,2}$	−0.38991254		
$W_{2,3}$	−0.52379505	$W_{2,3}$	0.67513838	$W_{7,3}$	−0.50879259		
$W_{2,4}$	0.75761136	$W_{2,4}$	−0.4441866	$W_{7,4}$	−0.21676101		
$W_{2,5}$	−0.52157099	$W_{2,5}$	−0.362877	$W_{7,5}$	−0.4396702		
$W_{2,6}$	0.02811703	$W_{2,6}$	−0.322716	$W_{7,6}$	0.05882271		
$W_{2,7}$	−0.74248271	$W_{2,7}$	−0.736477	$W_{7,7}$	0.05713673		
$W_{2,8}$	0.44767	$W_{2,8}$	−0.2797636	$W_{7,8}$	−0.00745899		
$W_{2,9}$	0.75771368	$W_{2,9}$	0.46671175	$W_{7,9}$	0.14822149		
$W_{2,10}$	0.59574177	$W_{2,10}$	−0.37185	$W_{7,10}$	0.61642481		
$W_{3,1}$	0.69399448	$W_{3,1}$	0.32263642	$W_{8,1}$	0.86812406		
$W_{3,2}$	−0.24094788	$W_{3,2}$	0.13153567	$W_{8,2}$	−0.23425252		
$W_{3,3}$	0.138715	$W_{3,3}$	−0.9109613	$W_{8,3}$	−0.87820354		
$W_{3,4}$	0.48719812	$W_{3,4}$	0.07223889	$W_{8,4}$	0.20198286		
$W_{3,5}$	0.86200771	$W_{3,5}$	−0.6365315	$W_{8,5}$	−0.85923602		
$W_{3,6}$	0.11118879	$W_{3,6}$	−0.2011787	$W_{8,6}$	−0.54009888		
$W_{3,7}$	0.7113249	$W_{3,7}$	0.63840616	$W_{8,7}$	−0.89531795		
$W_{3,8}$	0.44375463	$W_{3,8}$	0.3039039	$W_{8,8}$	−0.67206656		
$W_{3,9}$	0.00195087	$W_{3,9}$	0.45370143	$W_{8,9}$	0.84480996		
$W_{3,10}$	−0.05467507	$W_{3,10}$	−0.7748062	$W_{8,10}$	0.17705815		
$W_{4,1}$	0.4442245	$W_{4,1}$	0.61790615	$W_{9,1}$	−0.96609938		
$W_{4,2}$	−0.02693093	$W_{4,2}$	0.23819518	$W_{9,2}$	−0.7293678		
$W_{4,3}$	−0.83992177	$W_{4,3}$	−0.8211282	$W_{9,3}$	−0.13490159		
$W_{4,4}$	−0.45336151	$W_{4,4}$	−0.5237951	$W_{9,4}$	0.47337608		

Table 11.2 (Continued)

W_{ij} (Weight between $i$thinput node and jth hidden layer-1 neuron)	Values of weights (W_{ij})	$W_{j1,j2}$ (Weight between j_1^{th}hidden layer-1 neuron and j_2^{th}hidden layer-2 neuron)	Values of weights ($W_{j1,j2}$)	$W_{j1,j2}$ (Weight between j_1^{th} hidden layer-1 neuron and j_2^{th} hidden layer-2 neuron)	Values of weights ($W_{j1,j2}$)	$W_{j2,k}$ (Weight between j_2^{th} hidden layer-2 neuron and k^{th} output node)	Values of weights ($W_{j2,k}$)
$W_{4,5}$	0.5132074	$W_{4,5}$	−0.6083042	$W_{9,5}$	−0.05384209		
$W_{4,6}$	−0.76093636	$W_{4,6}$	0.55057171	$W_{9,6}$	0.92812343		
$W_{4,7}$	0.04174126	$W_{4,7}$	−0.654054	$W_{9,7}$	0.74402351		
$W_{4,8}$	0.84932026	$W_{4,8}$	0.80304517	$W_{9,8}$	0.8716961		
$W_{4,9}$	0.69136831	$W_{4,9}$	0.63062396	$W_{9,9}$	−0.3011121		
$W_{4,10}$	0.58899559	$W_{4,10}$	0.85283625	$W_{9,10}$	0.04511855		
		$W_{5,1}$	0.14350462	$W_{10,1}$	0.50554263		
		$W_{5,2}$	−0.77981667	$W_{10,2}$	−0.9582004		
		$W_{5,3}$	0.53391204	$W_{10,3}$	−0.7336712		
		$W_{5,4}$	−0.44076614	$W_{10,4}$	−0.9653672		
		$W_{5,5}$	−0.56648787	$W_{10,5}$	−0.4090327		
		$W_{5,6}$	0.83536792	$W_{10,6}$	0.02880597		
		$W_{5,7}$	0.33216417	$W_{10,7}$	0.94920176		
		$W_{5,8}$	−0.24961992	$W_{10,8}$	−0.5693641		
		$W_{5,9}$	−0.28888735	$W_{10,9}$	0.5107654		
		$W_{5,10}$	0.78428229	$W_{10,10}$	−0.1438138		

Table 11.3 Testing of trained ANN using a set of instances produced by the cut-set technique for a 14-bus system.

S. No.	V_1(pu)	V_2(pu)	V_3(pu)	P_d(pu)	PII(by cut-set technique)	PII(by ANN)	% Error
1	0.95	0.95	0.95	5.1923	0.043650384	0.042890867	1.74
2	0.95	0.95	1	5.8459	0.056269778	0.057434562	−2.07
3	0.95	1	1	6.2883	0.072388475	0.071527052	1.19
4	1	1	1	6.5168	0.056267451	0.054872018	2.48
5	1	1	1.05	6.0432	0.036514654	0.036887103	−1.02
6	1	1.05	1.05	7.1182	0.067945921	0.065662938	3.36
7	1.05	1.05	1.05	6.6639	0.044472754	0.043191939	2.88
8	1.05	1.05	1.1	7.4461	0.044472754	0.045597915	−2.53
9	1.05	1.1	1.1	7.3312	0.051859491	0.05095195	1.75
10	1.1	1.1	1.1	7.2623	0.028822574	0.028182713	2.22
11	1.1	1.1	1.15	7.4387	0.036780674	0.03732135	−1.47
12	1.1	1.15	1.15	8.972	0.067945801	0.06739544	0.81
13	1.15	1.15	1.15	9.3672	0.076020248	0.074385813	2.15

Table 11.4 Voltage and load for the 14-bus test system at all buses, prior to and following load shedding.

	Voltage(pu)		Load(pu)	
Bus No.	Prior to load shedding	following load shedding	Prior to load shedding	following load shedding
1	1.15	1.15	3.4688	3.4688
2	1.15	1.15	0.5459	0.5459
3	1.15	1.15	2.0865	2.0865
4	0.83	0.9486	0.2927	0.2927
5	0.8652	1.0041	0	0
6	0.9966	1.0585	1.0413	1.0413
7	0.8652	1.0041	0	0
8	1.0033	1.0595	0.1696	0.1696
9	0.8204	0.9839	0.7349	0.4410*
10	0.8011	0.9606	0.2324	0.2324
11	0.8057	0.9472	0.0856	0.0856
12	0.7869	0.9151	0.1369	0.1369
13	0.7744	0.9089	0.319	0.319
14	0.7408	0.9027	0.3471	0.3471

multiple load buses employing Eq. (11.15). Bus No. 9, with the highest sensitivity of 0.0843314, was chosen for the load-shedding action. The amount of load shedding (0.3293922 pu) remained within acceptable bounds, and the computed PII following the load shedding action was 0.060812589, which is less than the threshold value. Table 11.4 provides information regarding the bus voltages and bus loads before and after the load-shedding event.

11.5.2 25-bus System

This system comprises five PV buses, and their voltages can be regulated between ($0.95 \leq V_i \leq 1.15$ pu) to adjust the voltage of load buses. The availability of lines from line No. 1 to line No. 35 is considered as: 0.99548203, 0.99659806, 0.99655488, 0.99591478, 0.996347062, 0.99557426, 0.99652046, 0.99489641, 0.99137486, 0.99857835, 0.99263634, 0.99441704, 0.99035256, 0.99095274, 0.99073713, 0.99773568, 0.99949272, 0.99227775, 0.99685410, 0.99330443, 0.99868794, 0.990636047, 0.99119283, 0.99768705, 0.99137753, 0.99122069, 0.99865404, 0.99945461, 0.99160406, 0.99070482, 0.99177757, 0.99047761, 0.99899225, 0.995236073, and 0.99932094 (Billinton and Allan 1984). Table 11.5 shows a set of samples used to train the ANN, which enabled the computation of the PII utilizing the cut-set technique and the ANN, along with the percentage errors. The network underwent training with 100 instances, and only 18 instances with the highest error magnitudes are listed in Table 11.5. In Table 11.6, various weights, namely $W_{j2,k}$, $W_{j1,j2}$, and $W_{i,j1}$, connect the neurons between the hidden layer-2 and the output layer, hidden layer-1 and hidden layer-2, and input node and hidden layer-1, respectively. The input layer includes variables V_1, V_2, V_3, V_4, V_5, and P_d namely PV-bus voltages and overall load of the system. Table 11.7 illustrates the PII acquired through the trained network and the cut-set methodology, along with the percentage

Table 11.5 Training samples and computed probabilistic insecurity index (PII) using cut-set technique and artificial neural network for a 25-bus system.

S. No	V_1(pu)	V_2(pu)	V_3(pu)	V_4(pu)	V_5(pu)	P_d(pu)	PII(by cut-set technique)	PII(by ANN)	% Error
1	0.95	0.95	0.95	0.95	0.95	16.1245	0.129134758	0.125144494	3.09
2	0.95	0.95	0.95	0.95	1	15.7275	0.077344824	0.075604565	2.25
3	0.95	0.95	1	1	1	15.385	0.073622054	0.074409809	−1.07
4	0.95	1	1	1	1	15.9937	0.069214094	0.067947476	1.83
5	1	1	1	1	1	16.9108	0.113233571	0.110221558	2.66
6	1	1	1	1	1.05	16.1893	0.081008791	0.081673063	−0.82
7	1	1	1	1.05	1.05	17.3185	0.121191671	0.119385915	1.49
8	1	1	1.05	1.05	1.05	16.3393	0.048175967	0.046417544	3.65
9	1	1.05	1.05	1.05	1.05	17.2789	0.129134758	0.130568153	−1.11
10	1.05	1.05	1.05	1.05	1.05	16.5873	0.065550127	0.06711022	−2.38
11	1.05	1.05	1.05	1.1	1.1	17.6874	0.133542718	0.13092528	1.96
12	1.05	1.05	1.1	1.1	1.1	17.2962	0.097357431	0.100112646	−2.83
13	1.1	1.1	1.1	1.1	1.1	17.6705	0.129134758	0.126732851	1.86
14	1.1	1.1	1.1	1.1	1.15	17.9886	0.085302924	0.087972905	−3.13
15	1.1	1.1	1.1	1.15	1.15	18.6302	0.119991674	0.118470404	−1.52
16	1.1	1.1	1.15	1.15	1.15	18.8887	0.116646531	0.113205458	2.95
17	1.1	1.15	1.15	1.15	1.15	19.27	0.113233571	0.114943397	−1.51

discrepancies for the test cases. The trained network was tested with 200 instances, and only 18 instances with the highest error magnitudes are listed in Table 11.7. In this range, additional instances for testing have been chosen in addition to the training instances. The values of the PII, should not be higher than a predetermined level known as the threshold value, which has been assumed as 0.109038301. The percentage error of PII, as computed by trained ANN, is less than ±5%. For a specific operating setting, corrective reactive power rescheduling was demonstrated with: total load $P_d = 24.7970$ pu and $V_1 = 1.05$ pu, $V_2 = 1.05$ pu, $V_3 = 1.05$ pu, $V_4 = 1.05$ pu. It was observed that the stability margin obtained was less than 20% with a PII of 0.152038301, exceeding the specified threshold value. Sensitivities of PII were computed using Eq. (11.14), and it was found that Bus No. 1 is selected for voltage correction ($\Delta V_1 = 0.08524$ pu) due to its highest sensitivity. This correction brought the PII within the desired limit of 0.1019038301 as the final value. Another case was considered with values: $V_1 = 1.15$ pu, $V_2 = 1.15$ pu, $V_3 = 1.15$ pu, $V_4 = 1.15$ pu, $V_5 = 1.15$ pu, total load $P_d = 28.7239$ pu and despite correction attempts using Eq. (11.14) for rescheduling of reactive power at the most sensitive buses, the PII remained above the specified limit (0.142145948). As a result, sensitivities were assessed for load shedding at different load buses using Eq. (11.15). Among these buses, Bus No. 4, with the highest sensitivity value of 0.2114285, was chosen for load shedding. The load-shedding amount (0.307905 pu) remained within acceptable bounds, and the computed PII following the load-shedding action was 0.10903828, falling below the specified threshold value. Table 11.8 provides details of all bus voltages and bus loads before and after load shedding.

Table 11.6 Final weights obtained after ANN network training for the 25-bus system.

W_{ij} (Weight between *i*thinput node and *j*th hidden layer-1 neuron)	Values of weights (W_{ij})	$W_{j1,j2}$ (Weight between j_1^{th} hidden layer-1 neuron and j_2^{th}hidden layer-2 neuron)	Values of weights ($W_{j1,j2}$)	$W_{j1,j2}$ (Weight between j_1^{th} hidden layer-1 neuron and j_2^{th} hidden layer-2 neuron)	Values of weights ($W_{j1,j2}$)	$W_{j2,k}$ (Weight between j_2^{th} hidden layer-2 neuron and *k*th output node)	Values of weights ($W_{j2,k}$)
$W_{1,1}$	0.2442573	$W_{1,1}$	0.7230983	$W_{6,1}$	−0.2776832	$W_{1,1}$	−0.19111826
$W_{1,2}$	0.7331331	$W_{1,2}$	−0.769448	$W_{6,2}$	0.19835022	$W_{2,1}$	0.216105829
$W_{1,3}$	0.0950225	$W_{1,3}$	−0.722583	$W_{6,3}$	−0.2341812	$W_{3,1}$	−0.56191311
$W_{1,4}$	−0.690432	$W_{1,4}$	0.6882041	$W_{6,4}$	−0.6902042	$W_{4,1}$	0.374601197
$W_{1,5}$	−0.773507	$W_{1,5}$	0.8545416	$W_{6,5}$	−0.9776295	$W_{5,1}$	−0.16724777
$W_{1,6}$	−0.404048	$W_{1,6}$	0.9073756	$W_{6,6}$	−0.5679249	$W_{6,1}$	0.589796739
$W_{1,7}$	−0.326991	$W_{1,7}$	0.4347044	$W_{6,7}$	−0.365493	$W_{7,1}$	0.976397132
$W_{1,8}$	−0.085284	$W_{1,8}$	−0.961718	$W_{6,8}$	0.34264155	$W_{8,1}$	−0.02965714
$W_{1,9}$	0.972212	$W_{1,9}$	0.1031879	$W_{6,9}$	−0.5132074	$W_{9,1}$	0.872369128
$W_{1,10}$	−0.654907	$W_{1,10}$	0.2004992	$W_{6,10}$	0.37565473	$W_{10,1}$	−0.39616986
$W_{2,1}$	0.5389484	$W_{2,1}$	0.0852845	$W_{7,1}$	−0.6123977		
$W_{2,2}$	−0.084517	$W_{2,2}$	0.1914532	$W_{7,2}$	0.91891559		
$W_{2,3}$	−0.381324	$W_{2,3}$	−0.609962	$W_{7,3}$	0.67513838		
$W_{2,4}$	−0.847238	$W_{2,4}$	−0.34358	$W_{7,4}$	−0.4441866		
$W_{2,5}$	−0.032479	$W_{2,5}$	−0.802622	$W_{7,5}$	−0.362877		
$W_{2,6}$	0.1945016	$W_{2,6}$	0.2634518	$W_{7,6}$	−0.322716		
$W_{2,7}$	−0.577581	$W_{2,7}$	0.2269542	$W_{7,7}$	−0.736477		
$W_{2,8}$	−0.65358	$W_{2,8}$	−0.223834	$W_{7,8}$	−0.2797636		
$W_{2,9}$	0.8746088	$W_{2,9}$	0.7524238	$W_{7,9}$	0.46671175		
$W_{2,10}$	−0.115739	$W_{2,10}$	0.2191109	$W_{7,10}$	−0.37185		
$W_{3,1}$	−0.9398377	$W_{3,1}$	0.8028368	$W_{8,1}$	0.32263642		
$W_{3,2}$	0.5577976	$W_{3,2}$	0.5600327	$W_{8,2}$	0.13153567		
$W_{3,3}$	0.8848459	$W_{3,3}$	−0.14538	$W_{8,3}$	−0.9109613		
$W_{3,4}$	0.8298412	$W_{3,4}$	0.4290769	$W_{8,4}$	0.07223889		
$W_{3,5}$	−0.9231917	$W_{3,5}$	0.4616857	$W_{8,5}$	−0.6365315		
$W_{3,6}$	−0.5588921	$W_{3,6}$	0.1024955	$W_{8,6}$	−0.2011787		
$W_{3,7}$	0.2056884	$W_{3,7}$	0.0465715	$W_{8,7}$	0.63840616		
$W_{3,8}$	0.1129604	$W_{3,8}$	−0.171932	$W_{8,8}$	0.3039039		
$W_{3,9}$	−0.0255147	$W_{3,9}$	0.5950119	$W_{8,9}$	0.45370143		
$W_{3,10}$	0.8283314	$W_{3,10}$	0.5601836	$W_{8,10}$	−0.7748062		
$W_{4,1}$	0.2134732	$W_{4,1}$	0.193971	$W_{9,1}$	0.61790615		
$W_{4,2}$	−0.311213	$W_{4,2}$	0.2719943	$W_{9,2}$	0.23819518		
$W_{4,3}$	−0.428993	$W_{4,3}$	−0.879007	$W_{9,3}$	−0.8211282		
$W_{4,4}$	0.8494294	$W_{4,4}$	0.9177006	$W_{9,4}$	−0.5237951		
$W_{4,5}$	0.3049794	$W_{4,5}$	0.2146476	$W_{9,5}$	−0.6083042		
$W_{4,6}$	−0.116272	$W_{4,6}$	−0.494274	$W_{9,6}$	0.55057171		

Table 11.6 (Continued)

W_{ij} (Weight between *i*thinput node and *j*th hidden layer-1 neuron)	Values of weights (W_{ij})	$W_{j1,j2}$ (Weight between j_1^{th} hidden layer-1 neuron and j_2^{th} hidden layer-2 neuron)	Values of weights ($W_{j1,j2}$)	$W_{j1,j2}$ (Weight between j_1^{th} hidden layer-1 neuron and j_2^{th} hidden layer-2 neuron)	Values of weights ($W_{j1,j2}$)	$W_{j2,k}$ (Weight between j_2^{th} hidden layer-2 neuron and *k*th output node)	Values of weights ($W_{j2,k}$)
$W_{4,7}$	−0.549594	$W_{4,7}$	0.6106075	$W_{9,7}$	−0.654054		
$W_{4,8}$	0.4564163	$W_{4,8}$	0.4472334	$W_{9,8}$	0.80304517		
$W_{4,9}$	0.3039145	$W_{4,9}$	−0.339636	$W_{9,9}$	0.63062396		
$W_{4,10}$	−0.335557	$W_{4,10}$	0.1843682	$W_{9,10}$	0.85283625		
$W_{5,1}$	0.1410407	$W_{5,1}$	0.5262541	$W_{10,1}$	0.14350462		
$W_{5,2}$	0.0759936	$W_{5,2}$	0.2055966	$W_{10,2}$	−0.77981667		
$W_{5,3}$	0.5457719	$W_{5,3}$	−0.724611	$W_{10,3}$	0.53391204		
$W_{5,4}$	−0.365993	$W_{5,4}$	0.5458727	$W_{10,4}$	−0.44076614		
$W_{5,5}$	0.0861291	$W_{5,5}$	−0.307851	$W_{10,5}$	−0.56648787		
$W_{5,6}$	0.2519786	$W_{5,6}$	0.3304626	$W_{10,6}$	0.83536792		
$W_{5,7}$	0.1878709	$W_{5,7}$	−0.060892	$W_{10,7}$	0.33216417		
$W_{5,8}$	−0.221088	$W_{5,8}$	0.8475661	$W_{10,8}$	−0.24961992		
$W_{5,9}$	−0.305875	$W_{5,9}$	−0.486028	$W_{10,9}$	−0.28888735		
$W_{5,10}$	0.4281549	$W_{5,10}$	−0.628961	$W_{10,10}$	0.78428229		

Table 11.7 Testing of trained ANN using a set of instances produced by the cut-set technique for a 25-bus system.

S. No	V_1(pu)	V_2(pu)	V_3(pu)	V_4(pu)	V_5(pu)	P_d(pu)	PII(by cut-set technique)	PII(by ANN)	% Error
1	0.95	0.95	0.95	0.95	0.95	16.1139	0.137950678	0.139799217	−1.34
2	0.95	0.95	0.95	0.95	1	16.2355	0.129263598	0.126872221	1.85
3	0.95	0.95	1	1	1	16.3004	0.121054491	0.117931285	2.58
4	0.95	1	1	1	1	16.3134	0.113446261	0.110655483	2.46
5	1	1	1	1	1	16.4435	0.109723491	0.111084062	−1.24
6	1	1	1	1	1.05	16.9398	0.124855638	0.12358211	1.02
7	1	1	1	1.05	1.05	17.2549	0.129134758	0.128114593	0.79
8	1	1	1.05	1.05	1.05	17.4399	0.133542718	0.138298348	−2.15
9	1	1.05	1.05	1.05	1.05	17.4789	0.137950678	0.136143524	1.31
10	1.05	1.05	1.05	1.05	1.05	16.5721	0.061271007	0.061951115	−1.11
11	1.05	1.05	1.05	1.1	1.1	17.6874	0.133542718	0.132260707	0.96
12	1.05	1.05	1.1	1.1	1.1	17.2619	0.081008791	0.079688347	1.63
13	1.1	1.1	1.1	1.1	1.1	17.495	0.121191671	0.118670884	2.08
14	1.1	1.1	1.1	1.1	1.15	18.7971	0.113233571	0.111954031	1.13
15	1.1	1.1	1.1	1.15	1.15	18.4765	0.078238604	0.077111968	1.44
16	1.1	1.1	1.15	1.15	1.15	17.2082	0.105315531	0.106316028	−0.95
17	1.1	1.15	1.15	1.15	1.15	20.07	0.125599631	0.122946157	2.11

Table 11.8 Voltage and load for the 25-bus test system at all buses, prior to and following load shedding.

	Voltage (pu)		Load (pu)	
Bus No.	**Prior to load shedding**	**following load shedding**	**Prior to load shedding**	**following load shedding**
1.	1.1500	1.1500	5.6986	5.6986
2.	1.1500	1.1500	0.2829	0.2829
3.	1.1500	1.1500	1.4311	1.4311
4.	1.1500	1.1500	0.8569	0.4285*
5.	1.1500	1.1500	0.7113	0.7113
6.	1.0490	1.0617	0.4285	0.4285
7.	1.0352	1.0499	0.4285	0.4285
8.	1.0249	1.0413	0.6775	0.6775
9.	1.0015	1.0157	0.4285	0.4285
10.	1.0355	1.0476	0.4285	0.4285
11.	1.0019	1.0237	0.1355	0.1355
12.	1.0038	1.0262	0.2710	0.2710
13.	1.0524	1.0636	0.7113	0.7113
14.	0.9653	0.9696	0.5742	0.5742
15.	0.9574	0.9620	0.8569	0.8569
16.	0.9965	1.0003	0.8569	0.8569
17.	0.9785	1.0109	1.7139	1.7139
18.	0.9952	1.0245	0.4285	0.4285
19.	1.0825	1.1059	0.4285	0.4285
20.	1.0338	1.0524	0.7113	0.7113
21.	0.9232	0.9704	0.5742	0.5742
22.	0.8037	0.8847	0.5742	0.5742
23.	0.9146	0.9747	0.4285	0.4285
24.	0.7344	0.8302	0.4285	0.4285
25.	0.7603	0.8515	0.7113	0.7113

11.6 Conclusions

A novel algorithm has been introduced, offering a new perspective has been introduced to monitor and improve voltage security based on the probability of system failure. By employing the cut-set technique and accounting for the static voltage stability limit, the algorithm computes PII under various operating conditions. These collected samples were utilized to train an ANN using the backpropagation algorithm, enabling the online application of the PII. Moreover, the ANN was employed to assess the sensitivity of the PII concerning reactive power control variables and load shedding, ensuring that the index remains at or above the threshold value. Taking into account that voltage instability is a gradual event as pointed out in reference (Zarate et al. 2006), it logically ensues that carrying out evaluations and putting into action preventive measures can be accomplished in real-time employing this ANN.

References

Alzaareer, K., Saad, M., Mehrjerdi, H. et al. (2020). A new sensitivity approach for preventive control selection in real-time voltage stability assessment. *International Journal of Electrical Power & Energy Systems* 122: 106212. https://doi.org/10.1016/j.ijepes.2020.106212.

Arya, L.D. and Verma, H.K. (1996). A method for tracing PV-curve for voltage stability analysis with voltage dependent loads. *Int Journal of Electric machines and Power Systems* 24: 583–596.

Arya, L.D., Choube, S.C., and Saket, R.K. (2000). Composite system reliability evaluation based on static voltage stability limit. *Journal of Institution of Engineers India Part El Electrical Engineering Division* 80: 133–139.

Arya, L.D., Titare, L.S., and Kothari, D.P. (2007). Probabilistic assessment and preventive control of voltage security margins using artificial neural network. *International Journal of Electrical Power & Energy Systems* 29 (2): 99–105. https://doi.org/10.1016/j.ijepes.2006.05.005.

Arya, L.D., Titare, L.S., and Kothari, D.P. (2009). An approach to mitigate the risk of voltage collapse accounting uncertainties using improved particle swarm optimization. *Applied Soft Computing* 9 (4): 1197–1207. https://doi.org/10.1016/j.asoc.2009.02.005.

Benidris, M. and Mitra, J. (2015). Reliability and sensitivity analysis of composite power systems considering voltage and reactive power constraints. *IET Generation Transmission and Distribution* 9 (12): 1245–1253.

Bijwe, P.R., Kothari, D.P., and Arya, L.D. (1993). Alleviation of line overloads and voltage violations by corrective rescheduling. *Proceedings of the IEE - Part C: Monographs* 140 (4): 249–255.

Billinton, R. and Aboreshaid, S. (1998). Voltage stability considerations in composite power system reliability evaluation. *IEEE Transactions on Power Systems* 13 (2): 655–660.

Billinton, R. and Allan, R.N. (1983). *Reliability Evaluation of Engineering Systems: Concepts and Techniques*', (Book). PITMAN Books Limited.

Billinton, R. and Allan, R. (1984). *Reliability Evaluation of Power Systems*. Plenum Press.

Billinton, R. and Khan, E. (1992). A security based approach to composite power system reliability evaluation. *IEEE Transactions on Power Systems* 7: 65–70.

Billinton, R. and Li, W. (1993). A system state transition sampling method for composite system reliability evaluation. *IEEE Transactions on Power Systems* 8 (3): 761–766.

Billinton, R. and Wenyuan, L. (1991). Hybrid Approach for reliability evaluation of composite generation and transmission systems using Monte-Carlo simulation and enumeration technique. *IEE Proceedings-C* 138 (3): 233–241.

Capitaanescu, F. and Van Cutsem, T. (2002). Preventive control of voltage security margins: a multi-contingency sensitivity based approach. *IEEE Transactions on Power Systems* 17 (2): 358–364.

Concordia, C. (1991). Voltage instability. *Journal of Electrical Power and Energy Systems* 13 (1): 14–24.

Deng, Z. and Singh, C. (1992). A new approach to reliability evaluation of interconnected power system including planned outages and frequency calculations. *IEEE Transactions on Power Systems* 7 (2): 734–743.

Fu, L.M. (1994). *Neural Network in Computer Intelligence*. Mc Graw Hill Inc.

Haykin, S. (2003). *Neural Network: A Comprehensive foundation*. Pearson Education.

Henneaux, P. (2015). Probability of failure of overloaded lines in cascading failures. *International Journal of Electrical Power & Energy Systems* 73: 141–148.

IEEE System Dynamic Performance Subcommitte (1990). Voltage stability of power systems concepts, Analytical tools and industry experience. IEEE document 90th 0358-2-PWR-1990.

de Jong, M., Papaefthymiou, G., and Palensky, P. (2018). A framework for incorporation of infeed uncertainty in power system risk-based security assessment. *IEEE Transactions on Power Systems* 33 (1): 613–621.

Kirschen, D.S. and Jayaweera, D. (2007). Comparison of risk-based and deterministic security assessments. *IET Generation Transmission and Distribution* 1 (4): 527–533. https://doi.org/10.1049/iet-gtd:20060368.

Kundur, P. (1994). *Power System Stability and Control*. New York: Mc Graw-Hill.

Melo, A.C.G., Mello, J.C., and Granville, S. (1997). Effects of voltage collapse problems in the reliability evaluation of composite systems. *IEEE Transactions on Power Systems* 12 (1): 480–487.

Mokred, S., Wang, Y., and Chen, T. (2023). A novel collapse prediction index for voltage stability analysis and contingency ranking in power systems. *Protection and Control of Modern Power Systems* 8 (7): https://doi.org/10.1186/s41601-023-00279-w.

Pereira, M.V.F. and Pinto, L.M.V.G. (1992). A new computational tool for composite reliability evaluation. *IEEE Transactions on Power Systems* 7 (1): 258–263.

Sebastian Rios, M., Victor, V.P., and Kiguel, D.L. (1998). Bus based reliability indices and associated costs in the bulk power system. *IEEE Transactions on Power Systems* 13 (3): 719–723.

Singh, P., Arya, R., Titare, L.S., and Arya, L.D. (2021). Optimal load shedding to avoid risks of voltage collapse using black hole algorithm. *Journal of The Institution of Engineers (India): Series B* 102: 261–276. https://doi.org/10.1007/s40031-021-00543-3.

Taylor, C.W. (1994). *Power System Voltage Stability*. Mc Graw Hill. Inc.

Xia, C., Zheng, X., Guan, L., and Baig, S. (2021). Probability analysis of steady-state voltage stability considering correlated stochastic variables. *International Journal of Electrical Power & Energy Systems* 131: 107105. https://doi.org/10.1016/j.ijepes.2021.107105.

Zarate, L.A.L., Castro, C.A., Ramos, J.L.M., and Ramos, E.R. (2006). Fast computation of voltage stability security margins using nonlinear programming techniques. *IEEE Transactions on Power Systems* 21 (1): 19–27.

Section 4

Reliability Analysis of Distribution Systems

12

Modern Aspects of Probabilistic Distributions for Reliability Evaluation of Engineering Systems

Aanchal Singh S. Vardhan[1], Aanchal Verma[2], Jyotsna Ogale[3], R. K. Saket[2], and Stuart Galloway[4]

[1] *Department of Electrical Engineering, National Institute of Technology, Jamshedpur, Jharkhand, India*
[2] *Department of Electrical Engineering, Indian Institute of Technology (BHU), Varanasi, Uttar Pradesh, India*
[3] *Department of Electronics Engineering, Samrat Ashok Technological Institute, Vidisha, Madhya Pradesh, India*
[4] *University of Strathclyde, Glasgow, UK*

12.1 Introduction

The reliability concept was developed in 1949 when 70% of Navy electronics equipment was not operated properly, and during the Second World War, 80% of airborne equipment was damaged during transportation. The economic values of these equipment's were so high that they immediately drew scientist's attention toward the reliability assessment and dependability engineering. The concept of reliability was primarily formulated to address situations in which nonrepairable components, equipment, devices, and systems are involved and where the initial occurrence of failure holds significant importance (Arya et al. 2000).

With development of different aspects of reliability of enormous amount of components, failure data was collected, analyzed, and published in the field of electronics engineering while at the same time, repairable system development of analytical methods for reliability assessment was at a slow pace. Nowadays, almost all branches of engineering, science, and technology studies are going on different aspects of reliability for both types of systems, which can be repaired and cannot be repaired. The first formal reliability evaluation was reported to occur when an explanation was sought for the poor performance of the missiles during the Second World War. The missiles were constructed of large number of components considered highly reliable. Over time, problems were identified related to reliability evaluation of components or systems (Arya et al. 2001). The availability of data was lacking, computational resources were limited, there were no reliability techniques of realistic nature available, the use of probabilistic techniques was averted, and understanding was lacking of the meaning, significance, and inferences drawn from probabilistic models and risk indices. As mentioned, all the problems vanished in 21st century because now we have valid and sufficient data availability and in-depth understanding of well-developed probabilistic and reliability evaluation techniques. In this chapter, discussions are made on different techniques used for evaluation of reliability, which are applicable to power system analysis and explain the significance of the various reliability indices (Bansal et al. 2002).

Whenever unreliable components, devices, and systems occupy a place in human lives, a large number of issues related to safety considerations arise, and reliability becomes the most important factor to be assessed and evaluated. In order to evaluate the system's usefulness or goodness with

Reliability Analysis of Modern Power Systems, First Edition. Edited by R. K. Saket and P. Sanjeevikumar.

the imposed constraint of cost of the system, that might be the operating cost or planning cost including initial investment, reliability consideration becomes an important part of engineering design (Bharti et al. 2019). Not only this, but the analysis of reliability also considers factors like operating environment, time of operation, and quality of service rendered by the system. Naturally, under these considerations, the usability, applicability, usefulness, or goodness of the system must be judged not only qualitatively but also quantitatively by taking numerical measures based on probability theory known as reliability indices since any qualitative judgment is meaningless for any engineer who is designing or planning a system. The operating environment must get due consideration for dependability assessment, which affects the rate of system failure (Billinton and Zhang 1998). Take an example, a transmission line installed in coastal areas and those in the region having fair weather condition will have different failure as well as repair rate, hence, will have different probability of success and failure. Therefore, any reliability modeling must account for the operating environment. The time of operation of a repairable or nonrepairable component is also of great concern. The probability of a system operating successfully decreases with an increase in the intended time of operation of a system. Quality of service requires special attention because the system may give service, but the service may not satisfy the consumer. For example, electricity supply may be available at some consumer's premises, but supply voltage and frequency may be too low, too high, or too fluctuating to use it effectively in a production process. Hence, reliability is classically defined and states that it is the probability of a component performing its own function, which is intended over a period specified under the conditions set for its operation (Kothari et al. 2022).

12.2 Life Distribution of Power Components: An Overview

In principle, any probability distribution can be employed to simulate the lifespan of the equipment. Practically, it is found that the hazard functions are monotonic and seem to be most realistic (Kumar et al. 2019). There are only a few that provide reasonable models for reliability of a device. The most commonly used models for distribution of life are given below.

12.2.1 Binomial Distribution

Let us have an experiment doing a trial, which generates only two outcomes with possibility of getting successful or failing. If $\boldsymbol{p}$ denotes the success probability, then $(\mathbf{1}-\boldsymbol{p})=\boldsymbol{q}$ will be the failure probability. If $\boldsymbol{x}$ is varying randomly and denotes the count of successful outcomes in a series of n trials. The probability of obtaining exactly x successes can be articulated as $b(x, n, p) = {}^nC_x$ $q^{(n-x)}\ p^x$, where $\boldsymbol{x} = \mathbf{0,1,2,3}\ldots\ldots \boldsymbol{n}$ and combinatorial quantities ${}^n\boldsymbol{C}_x = \frac{\boldsymbol{n}!}{(\boldsymbol{n}-\boldsymbol{x})!\boldsymbol{x}!}$ are viewed as coefficients in binomial expansions. This statistical distribution is commonly known as a binomial distribution.

Corollary: Addition of all given below

$$\sum_{x=0}^{n} f(x) = \sum_{x=0}^{n} {}^nC_x p^x q^{n-x} = q^n + {}^nC_1 p^1 q^{n-1} + \ldots p^n = (q+p)^n = 1 \tag{12.1}$$

To evaluate the number of gun firing rounds in military system for object detection by radar, in the production process calculation of number of defective samples, and in the interconnected bulk power system for estimation of reliability, the binomial distribution is usually applied (Kumar et. al 2020).

12.2.2 Exponential Distribution

The exponential function mainly uses the probability function to create models for the evaluation of reliability. Its algebraic simplicity and tractability make it a well-known model for assessment of reliability of a component and device and acting as a symbol of the operational duration within the complete lifespan of the device. Just before entering the wear-out phase, the device is typically in a state of relative stability, making the use of constant hazard rates a suitable choice for modeling device reliability as well.

If nonnegative values are assumed by a random variable of continuous nature, $\boldsymbol{X}$ then an exponential distribution will have parameter $\boldsymbol{\alpha > 0}$, and the PDF is articulated as follows:

$$f(x) = \begin{cases} \alpha e^{-\alpha x}, & x \geq 0 \\ 0, & \text{otherwise} \end{cases}$$

In integral form, this function is shown below

$$\boldsymbol{F(x)} = \int_0^x \alpha e^{-\alpha x} dx \tag{12.2}$$

12.2.3 Poisson Distribution

This distribution can be considered as a limiting case of PDF of binomial type subjected to the conditions $\boldsymbol{p \to 0}$ and $\boldsymbol{n \to \infty}$. Therefore, Poisson distribution is nothing but the relative frequency-dependent distribution of a variable $\boldsymbol{x}$:

$$f(x, m) = \frac{e^{-m} m^x}{x!} \text{and} \sum_{x=0}^{\infty} f(x, m) = \sum_{x=0}^{\infty} \frac{e^{-m} m^x}{x!} = e^{-m} \sum_{x=0}^{\infty} \frac{m^x}{x!} = e^{-m} e^m = 1. \tag{12.3}$$

It is the probability exponentially distributed with a single parameter "$\boldsymbol{m}$."

Some of the application areas of Poisson distribution are the evaluation of bomb hits spatial distribution, estimation of fragments in total number while a shell hits a target, evaluation of defective vehicles based on arrival pattern in an engineering workshop, spare parts demand pattern, and evaluation of power components reliability (Saket 2013).

Example 12.1

If 5% of MOSFET products are defectively manufactured and sold in boxes of 100 with the guarantee of 10 defective pieces per box. So find the approximate probability of failure of a box to attain the quality, which is guaranteed at the time of purchase?

Solution:

Given:- Number of products in boxes $(n) = 100$

Let p be the defective MOSFET probability $= 5\% = 0.05$

$m =$ Average of MOSFETs which are found defective in a box of n samples $=$ No. of products in boxes $\times$ probability of defective product $(n \times p) = 100 \times 0.05 = 5$

Since probability (p) is less, Poisson distribution may be used.

Within a set of 100 items, the probability of x defective products is calculated as follows:

$$P(X = x) = \frac{e^{-m} m^x}{x!} = \frac{e^{-5} 5^x}{x!}$$

where, $x = 0, 1, 2, \ldots$

The probability of getting failed to attain the quality that is guaranteed at the time of purchase,

$$P(X > 10) = 1 - P(X \leq 10) = 1 - \sum_{x=0}^{10} \frac{e^{-5} 5^x}{x!} = 1 - e^{-5} \sum_{x=0}^{10} \frac{5^x}{x!}$$

$$[1 - p(X \leq 10)] = 1 - e^{-5}\left[\frac{5^0}{0!} + \frac{5^1}{1!} + \frac{5^2}{2!} + \frac{5^3}{3!} + \ldots\ldots\ldots\ldots + \frac{5^{10}}{10!}\right]$$

$$[1 - p(X \leq 10)] = 1 - e^{-5}\ [1 + 5 + 12.5 + 20.8333 + 26.0417 + 26.0417 + 21.7014$$
$$+15.5001 + 9.6881 + 5.3823 + 2.6911]$$

$$[1 - p(X \leq 10)] = 1 - e^{-5}\ [146.3797] = 1 - 0.00673 \times 146.3797$$
$$= (1 - 0.985135) = 0.0148646$$

Example 12.2
While turning electric power switches in a factory, there is a small chance of defectiveness $\frac{1}{500}$ exists. If power switches are packed in a group of 10. With Poisson's distribution calculate number of packets approximately, which contains (i) no defective piece, (ii) single defective piece, and (iii) double defective pieces in a 1000 packets consignment.

Solution:
Given probability of defect $p = \frac{1}{500}$; $n = 10$, and $N = 1000$

$\therefore$ Mean $m = n \times p = 10 \times \frac{1}{500} = \frac{1}{50} = 0.02.$

Poisson type distribution is expressed as, $p\,(x) = \frac{e^{-m}\, m^x}{x!} = \frac{e^{-0.02}(0.02)^x}{x!}$

(i) p (probability with zero defective piece) $= P(0) = \dfrac{e^{-0.02}(0.02)^0}{0!} = 0.980198$
In a consignment of 1000 packets, total packets in number contains no defective piece $= p(0) \times N = 1000 \times 0.98019 = 9802$
(ii) p (probability with single piece defect) $= p(1) = 0.01960$ $\therefore$ In a 1000 packets consignment, total packets with single defect in piece will be $= p(1) \times N = 1000 \times 0.01960 = 20$
(iii) p (probability when double pieces of switches are in defect) $= p(2) = 0.001960$

In a consignment of 1000 packets, total number of packets having double defective pieces $= p(2) \times N = 1000 \times 0.001960 = 2$

Example 12.3
An industrial machine manufactures internal parts of an induction generator in large quantity. The mean number of defective units in a sample of 20 parts is 2. If an Indian power industry purchases 1000 such samples for installation of a wind power plant, find the chances for having atleast 3 defective parts?

Solution:
Average of defective units $(m) = 2$

Here, X is distributed in a Poisson's way whose function of probability is given as,

$$p(x) = \frac{e^{-2}2^x}{x!}$$

here, x represents total defective units.

P [of a part containing at least three defective units]

$$= p\,[X = 3] + p\,[X = 4] + p\,[X = 5] + p\,[X = 6] + p\,[X = 7]\ldots\ldots + p\,[X = n]$$

$$= 1 - \{p\,[X = 0] + p\,[X = 1] + p\,[X = 2]\}$$

$$= 1 - \left\{ \frac{e^{-2}2^0}{0!} + \frac{e^{-2}2^1}{1!} + \frac{e^{-2}2^2}{2!} \right\}$$

$$= 1 - (0.135 + 0.270 + 0.270) = 0.325$$

In 1000, the parts having at least three defective units are $0.3250 \times 1000.0 = 325$

Example 12.4

A machine produces defective items with a probability of 0.05 for modern power system applications. Such items are packed in a group of 20 for international marketing. Find the total packet number, which contains almost, exactly and minimum of two defective items in a shipment containing 1000 packets. Approximate the binomial distribution by Poisson approach.

Solution:

Given number of items in a group $(n) = 20$, and $p = 0.05$. Then $m = n \times p = 20 \times 0.05 = 1.00$

If X represents the overall count of defective items and its distribution follows poisons law. Then the probability function will be as given below

$$P(X = x) = \frac{e^{-m}m^x}{x!} = \frac{e^{-1}}{x!}$$

P [packets having at least two defective items]

$$= [X \geq 2] = 1 - p\,[X < 2]$$

$$= 1 - \{p\,[X = 0] + p\,[X = 1]\}$$

$$= 1 - \{e^{-1} + e^{-1}\} = 1 - \{0.3678 + 0.3678\} = (1 - 0.7356) = 0.2644$$

Packets containing at least 2 defective items $= 1000 \times 0.2644 = 264$

P [a packet with two defective items]

$$= p[X = 2] = \frac{e^{-1}}{2!} = 0.1839$$

The number of packet with the exactly two item is defective $= 1000 \times 0.1839 = 184$

$$P\,[X \leq 2] = p\,[X = 0] + p\,[X = 1] + p\,[X = 2]$$

$$= e^{-1} + e^{-1} + \frac{e^{-1}}{2!} = 0.9193$$

Packets having almost 2 items defective $= 1000.0 \times 0.9193 = 919$.

12.2.4 Geometric Distribution

If trial number $\boldsymbol{n}$ is not fixed in an experiment and first success achieved on the $\boldsymbol{X}^{\text{th}}$ trial with probability $\boldsymbol{p}$, preceded by $(\boldsymbol{x} - \mathbf{1})$ failures, where in $(\boldsymbol{x} - \mathbf{1})$ trials, the probability is $(\mathbf{1} - \boldsymbol{p})(\boldsymbol{x} - \mathbf{1})$. If a quantity varying randomly and assumes positive values only i.e $\boldsymbol{X}$ then it is distributed as geometric distribution and the mass function is expressed as:

$$P(X = x) = \begin{cases} pq^{x-1}, & x = 1, 2, 3, \ldots 0 < p \leq 1, q = (1 - p) \\ 0, & \text{otherwise} \end{cases} \tag{12.4}$$

Above equation shows the geometric distribution of probability.

Example 12.5

A circuit breaker operates in independent fashion in an interconnected composite electric power system. The probability of successful operation of the circuit breaker during any one fault is 0.70.

(i) What will be the probability of protection on 10th attempt?
(ii) Find the probability of breaker to attempt <4?
(iii) What will be the probability of an even number of attempts?

Solution:

Given $p = 0.70$ and $q = (1-p) = 0.30$

(i) Success probability on the 10th trial $= (0.70)\,(0.30)^{10^{-1}} = (0.70)(0.30)^9 = 0.0000139.$
(ii) The probability of taking less than four attempts is

$$P(X < 4) = \sum_{n=1}^{4-1} pq^{n-1} = \sum_{n=1}^{3} (0.7)(0.3)^{n-1}$$
$$= (0.7)[(0.3)^0 + (0.3)^1 + (0.3)^2] = (0.7)[1 + 0.3 + 0.09]$$
$$= (0.7)(1.39) = 0.973$$

(iii) The probability of taking an even number of attempts.

$$\sum_{n=1}^{\infty} (0.7)(0.3)^{2^{n-1}} = (0.7)(0.3)\sum_{n=1}^{\infty} (0.3)^{2^{n-2}} = (0.7)(0.3)\frac{1}{1-(0.3)^2} = 0.23$$

12.2.5 Weibull Distribution

The Weibull and exponential distributions are both closely related. Both distributions provide a valid model for the entire length of life for very many devices. For reliability, modeling Weibull distribution is widely used because of its flexibility of algebraic tractability and modeling different types of hazard behavior. One representation of the extreme value distribution is the Weibull distribution (Saket et al. 2006).

The formula representing the Weibull PDF for a random variable X is given as follows:

$$f(x) = \begin{cases} \alpha\beta x^{\beta-1}e^{-\alpha x^{\beta}}, & \text{for } x > 0, \alpha > 0, \beta > 0 \\ 0, & \text{elsewhere} \end{cases} \tag{12.5}$$

If a random variable follows a Weibull distribution and has a value below the integral $\int_0^a \alpha\beta x^{\beta-1}e^{-\alpha x^{\beta}}dx$, then probability is evaluated by changing the variable, $y = x^{\beta}$ we get $\int_0^{\alpha^{\beta}} \alpha e^{-\alpha y}dy = 1 - e^{-\alpha a^{\beta}}$ which shows that random variable y is also distributed exponentially.

12.2.5.1 Weibull Distribution Mean and Variance

If the Weibull distribution's parameters are represented as α and β, then the mean can be determined by calculating the integral.

$$\text{Mean } \mu = \int_0^{x} x(\alpha\beta x^{\beta-1}e^{-\alpha x^{\beta}})dx = \int_0^{\infty} \alpha\beta x^{\beta}e^{-\alpha x^{\beta}}$$

Put $\mathbf{u} = \alpha \mathbf{x}^{\beta} \Rightarrow \mathbf{du} = \alpha\beta \mathbf{x}^{\beta-1}\mathbf{dx}$

Also if $\mathbf{x} = \infty \Rightarrow \mathbf{u} = \infty$ and $\mathbf{x} = \mathbf{0} \Rightarrow \mathbf{u} = \mathbf{0}$

$$= \frac{\alpha\beta}{\alpha^{1/\beta}} \int_0^{\infty} \left(\frac{u}{\alpha}\right) e^{-u}\frac{1}{\beta}u^{(1/\beta)-1}du$$

$$= \alpha^{-1/\beta} \int_0^{\infty} u^{1/\beta} e^{-u} du = \alpha^{-1/\beta} \int_0^{\infty} e^{-u} u^{\left(\frac{1}{\beta}+1\right)-1} du$$

$$= \alpha^{-1/\beta} \Gamma\left(1 + \frac{1}{\beta}\right) \tag{12.6}$$

$$(\because \Gamma n =) \int_0^{\infty} e^{-x} x^{n-1} dx$$

Use the substitution $u = \alpha x^{\beta}$ to evaluate the variance

$$\mathbf{E}(\mathbf{X}^2) = \int_0^{\infty} \alpha \beta x^{\beta+1} e^{-\alpha x^{\beta}} dx$$

$\alpha^{-2/\beta} \Gamma^{\left(1+\frac{2}{\beta}\right)}$check carefully

Then, variance

$$\sigma^2 = E[(X-\mu)^2] = E[X^2] - \mu^2 = \alpha^{-2/\beta} \left[\Gamma\left(1 + \frac{2}{\beta}\right) - \left(\Gamma\left(1 + \frac{1}{\beta}\right)\right)^2 \right] \tag{12.7}$$

12.2.6 Normal Distribution

A continuous random variable X is distributed according to the Gaussian curve or normal curve using μ and σ, and the PDF is expressed as follows:

$$f(x) = \frac{1}{\sigma\sqrt{2\pi}} e^{-\frac{1}{2}\left(\frac{x-\mu}{\sigma}\right)^2}; -\infty < x < \infty \tag{12.8}$$

X has a standard deviation of σ and a mean of μ, and is distributed as $N(\mu, \sigma)$. When the normal curve covers the entire area under the curve and the X-axis, the area beneath the curve and between two ordinates $X = a$ and $X = b$, where $a < b$, signifies the probability of X falling between a and b. The probability is represented as $P[a < X < b]$. If $\boldsymbol{X}$ is a variable and expressed as $\left[\boldsymbol{Z} = \frac{X-\mu}{\sigma}\right]$, Eq. (12.8) is replaced as $\boldsymbol{\phi}(\boldsymbol{Z}) = \frac{1}{\sqrt{2\pi}} \boldsymbol{e}^{-\frac{z^2}{2}}, -\infty < \boldsymbol{Z} < \infty$ it is known as standard normal distribution. After substituting $\boldsymbol{\mu} = \mathbf{0}$ and $\boldsymbol{\sigma} = \mathbf{1}$ and changing $\boldsymbol{x}$ into $\boldsymbol{Z}$ this is achieved. Now x has distribution as

$\boldsymbol{N}(\boldsymbol{\mu}, \boldsymbol{\sigma})$ then $\mathbf{Z} = \left(\frac{X-\mu}{\sigma}\right)$has the distribution $\boldsymbol{N}(\mathbf{0}, \mathbf{1})$.

Given by Eq. (12.8) are a few characteristics of the normal distribution, which are as follows: Mean = $\boldsymbol{\mu}$, Variance = $\boldsymbol{\sigma}^2$, Standard deviation = $\boldsymbol{\sigma}$

$$\textbf{Average deviation about the mean} = \int_{-\infty}^{\infty} |x - \mu| f(x) dx = \sqrt{\frac{2}{\pi}} \sigma = \frac{4}{5} \sigma \text{ (approx.)}$$

$$\textbf{Quartile deviation} = \frac{2}{3} \sigma$$

Function of $N(\mu, \sigma)$ is moment generating.

$$\left(\because Z = \frac{X - \mu}{\sigma}\right)$$

$$M_x(t) = M_{\sigma Z} + \mu(t)$$
$$= e^{\mu t} M_z(\sigma t)$$
$$= e^{\mu t} \cdot e^{\sigma t t/2} \text{ (since } Mz(t) = e^{t^2/2} \text{ verify it)}$$

$$= e^{t\left(\mu+\frac{\sigma^2 t}{2}\right)} = 1 + \frac{t}{1!}\left(\mu + \frac{\sigma^2 t}{2}\right) + \frac{t^2}{2!}\left(\mu + \frac{\sigma^2 t}{2}\right)^2 + \ldots\ldots$$

$$\therefore E(X) = \mu; E(X^2) = \sigma^2 + \mu 2, \tag{12.9}$$

Thus we get $\mu_1 = 0$, $\mu_2 = \sigma^2$, $\mu_3 = 0$; $\mu_4 = 3\sigma^4$, etc.

The normal distribution approach is used to compute the probability of hitting a shot, to estimate and extract the statistical parameters in every field of engineering and technology. This approach is used to calculate the errors made unknowingly in data measurement by experiments and to evaluate the reliability of composite power systems.

Example 12.6

A particular type of randomly chosen diode is used in power converter applications in modern power system. The breakdown voltage X of the diode is normally distributed with $\mu = 40\,V$ and $\sigma = 1.50\,V$. Find the probability of break down voltage (i) lies in the range of 39 V and 42 V, and (ii) between 40 V and 43 V.

Solution:

Here X follow $N\,(40\,V, 1.50\,V)$.

(i) To determine $p[39\,V \le X \le 42\,V]$, express the event $39\,V \le X \le 42\,V$ in a standardized form:

$$39\,V \le X \le 42\,V \text{ comes to be } \frac{39-40}{1.5} \le \frac{X-40}{1.5} \le \frac{42-40}{1.5}$$

i.e. $-0.67 \le Z \le 1.33 Z = \frac{X-40}{1.5}$

$$P(39\,V \le X \le 42\,V) = p(-0.67 \le Z \le 1.33)$$
$$= \phi(Z = 1.33) - \phi(Z = -0.67)$$
$$= 0.9082 - 0.2514 = 0.6568$$

(ii) $$p(40\,V \le X \le 43\,V) = P\left[\frac{40-40}{1.5} \le \frac{X-40}{1.5} \le \frac{43-40}{1.5}\right]$$
$$= \phi\,[0 \le Z \le 2] = \phi\,[Z = 2] - \phi\,[Z = 0]$$
$$= 0.4773 - 0.0000 = 0.4773$$

Example 12.7

If the mean of lifetimes of certain kind of electronic devices is 300 hours and a standard deviation is 25 hours. Assume the distribution of lifetime follows the normal curve and is measured to the nearest hour.

(i) Determine the probability of a devices having life time > 350 hrs.
(ii) What proportion of items will exhibit lifetimes of 300 hours or less?
(iii) What percentage of items will have lifetimes ranging between 220 and 260 hours?

Solution:

Here X follows $N(\mu = 300, \sigma = 25)$

(i) $p(X > 350)$

$$= \left[\frac{X-300}{25} > \frac{350-300}{25}\right]$$

$$= \phi[Z > 2] = 1 - \phi(Z \leq 2)$$

$$= 1 - 0.9772 = 0.0228$$

The resulting percentage is 2.28%

(ii) $p(X = 300) = \left[\frac{X - 300}{25} > \frac{300 - 300}{25}\right]$

$$= \phi(Z = 0) = 0.5000$$

The resulting percentage is 50%

(iii) $p(220 \leq X \leq 260) = \phi(-3.2 \leq Z \leq -1.6)$

$$= \phi(-1.6) - \phi(-3.2)$$

$$= [1 - \phi(1.6)] - [1 - \phi(3.2)]$$

$$= [1 - 0.9452] - [1 - 0.9903]$$

$$= 0.0548 - 0.0007 = 0.0541$$

The resulting percentage is 5.41%

12.2.7 Gamma Distribution

$\boldsymbol{X}$ is a random variable of a continuous nature, and its distribution is according to the law of probability is given as:

$$f(x) = \begin{cases} \dfrac{e^{-x}x^{\alpha-1}}{\Gamma(\alpha)}; & \alpha > 0, 0 < x < \infty \\ 0, & \text{otherwise} \end{cases} \tag{12.10}$$

It is named Gamma variable with $\boldsymbol{\alpha}$ as a parameter and $\Gamma(\alpha)$ as a variant. The distribution of the function is termed as Gamma distribution.

Notes: The probability function is defined above. Since,

$$\int_0^{\infty} f(x)dx = \frac{1}{\Gamma(\alpha)} \int_0^{\infty} e^{-x}x^{\alpha-1}dx = 1 \tag{12.11}$$

The PDF of a random variable varying continuously and following gamma distribution with parameter (α, u) is given as:

$$f(x) = \begin{cases} \dfrac{\alpha^u}{\Gamma(u)} e^{-\alpha x}x^{u-1}; & \alpha, u > 0, 0 < x < \infty \\ 0, & \text{otherwise} \end{cases} \tag{12.12}$$

12.3 Failure Distribution Functions for Reliability Evaluation

Principally, to model equipment longevity, any distribution function may be used. In practice, the most realistic approach seems to have monotonic hazard functions of distribution, and in this category, only a few provide reasonable models of device reliability. The following are the main life distribution models used:

12.3.1 Evaluation of Reliability Based on Exponential Distribution

In the context of reliability modeling, the exponential distribution is widely utilized. It possesses algebraic simplicity and tractability and denotes the device life cycle's functional life interval (Saket et al. 2007). If failure times are exponentially distributed, results of great importance can be derived by connecting the mean time between failures (MTBF) of systems connected serially and parallelly. To begin with, for any component, let's initially establish a connection between reliability and service time T. Utilize the fact that

$$R(t) = 1 - F(t) = 1 - \int_0^t f(x)dx \left(\because f(t) = \frac{dF(t)}{dt} \right)$$

The PDF of the time to failure T is given if it follows exponential distributions with parameter α,

$$f(t) - \alpha.e^{\alpha t}, t \geq 0.$$

Then,

$$R(t) = 1 - \int_0^t \alpha e^{-\alpha x} dx = 1 - \alpha \left[\frac{e^{-\alpha x}}{-\alpha} \right]_0^t = 1 + [e^{-\alpha t} - 1] = e^{-\alpha t} \tag{12.13a}$$

Conversely, when

$$Z(t) = \frac{f(t)}{R(t)} = \frac{\alpha e^{\alpha t}}{e^{-\alpha t}} = \alpha, \text{constant}$$

from the equation, we get,

$$f(t) = \alpha e^{\int_0^t \alpha dx}$$

$$\Rightarrow f(t) = \alpha e^{-\alpha t}, t \geq 0 \tag{12.13b}$$

This characteristic signifies that, in the context of reliability, the exponential distribution maintains a consistent failure rate distribution.

$$\text{MTTF} = E(T) = \frac{1}{\alpha} \tag{12.14a}$$

$$\text{var}(T) = {\sigma_T}^2 = \frac{1}{\alpha^2}$$

$$R(t/T_0) = \frac{R(T_0 + t)}{R(T_0)} = \frac{e^{-\alpha(T_0+t)}}{e^{-\alpha T_0}} = e^{\alpha t} \tag{12.14a}$$

This signifies that the time taken by a component to fail does not depend on time duration when the component functions well. Reliability remains the same regardless of components, new branding, and long operation time. This is referred to as the constant failure rate (CFR) distribution's memory-less attribute.

Example 12.8
A power component has MTTR = 20.0 hours and MTBF = 100.0 hrs. If failure and repair distributions follow the exponential path, evaluate the availability and unavailability of the power component after a long time.

Solution:
Given MTBF $= = \frac{1}{\lambda} = 100 \Rightarrow \lambda = 0.01$

$$\text{MTTR} = \frac{1}{\mu} = 20 \Rightarrow \mu = 0.05$$

Availability of component

$$A(\infty) = \frac{\text{MTBF}}{\text{MTBF} + \text{MTTR}} = \frac{\frac{1}{\lambda}}{\frac{1}{\lambda} + \frac{1}{\mu}} = \frac{100}{100 + 20} = 0.83$$

The unavailability of component

$$\overline{A}(\infty) = \frac{\lambda}{\lambda + \mu} = \frac{0.01}{0.01 + 0.05} = \frac{0.01}{0.06} = 0.1666$$

Example 12.9

The failure rate of an electro-mechanical energy conversion device is 0.023 per hour. The repair time is exponential with a mean of 10 hours. Determine the reliability for 3 days and MTTF, if two devices are operating in an active redundant configuration.

Solution:

Within an active redundancy setup

$$R(t) = \frac{m_1}{m_1 - m_2} e^{m_2 t} = \frac{m_2}{m_1 - m_2} e^{m_1 t}$$

where $m_1, m_2 = \frac{1}{2}\{-(3\lambda + \mu) \pm \sqrt{\lambda^2 + 6\lambda\mu + \mu^2}\}$

Here $\lambda = 0.023$ and $\mu = \frac{1}{10} = 0.1$

$$m_1, m_2 = \frac{1}{2}\{-(0.069 + 0.1) \pm \sqrt{(0.023)^2 + 6 \times (0.023) \times (0.1) + (0.1)^2}\}$$

$$= -0.0065, -0.1625$$

Putting values in equation,

$$R(t) = \frac{-0.0065}{0.156} e^{-0.1625t} + \frac{0.1625}{0.156} e^{-0.0065t}$$

3 days =72 hours.

$$\therefore R(72) = -0.0417 \times e^{-11.7} + 1.0417 \times e^{-0.468} = 0.6524$$

$$\text{MTTF} = \frac{3\lambda + \mu}{2\lambda^2} = \frac{3 \times 0.023 + 0.1}{2 \times (0.023)^2} = 159.7 \text{ hours.}$$

Example 12.10

The failure rate of a power component is 0.02 / 1000 hrs. For a constant failure rate, determine the probability of satisfactory operation of a component for atleast 20,000 hrs? If a series carries 4 such components is connected in series, determine reliability of the power component for 5000 hr.

Solution:

The power component experiences a constant failure rate, and its failure time distribution follows an exponential distribution with a parameter of $\lambda = 0.02$.

The PDF of the power component is $f(t) = 0.02e^{-0.02xt}$, $t \geq 0$

(i) $P_r[t \geq 20]$

$$= \int_{20}^{\infty} f(t)dt$$

$$= \int_{20}^{\infty} 0.02e^{-0.02xt}dt$$

$$= -[e^{0.02xt}]_{20}^{\infty} = e^{-0.02\times 20} = 0.6703$$

(ii) The reliability function of the component is $R(t) = e^{-\lambda xt}$. There is a series of 4 components connected in series. The reliability of this combination is given below.

$$\therefore R(t = 5) = e^{-5.0\times 4.0\times 0.2} = 0.183$$

Problem 12.11

A production company makes repairs on a power transformer oil tank by employing a single welder. The welder begins working with three arc-welding units with the aim to increase welding operation reliability. The failure rate of each welding unit is 0.32/month. If the units are utilized in a standby mode by a welder with the assumption that unit failures are independent. Calculate the MTTF, and standard deviation of the time to failure. What is the probability that a welder won't be without a welding unit for the whole six-month welding work. The rule is that the broken welding equipment will not be brought back to the project after repair.

Solution:

Given rate of failure $\lambda = 0.320$/month.

T is the time to failure, and the rate of failure $= \lambda$ with exponential distribution.

Therefore the PDF is as $f(t) = xe^{-0.32t}, \geq 0$

$$= 0.32$$

and $R(t) = e^{-\lambda t} = e^{0.32t}$

Utilizing three arc-welding units, the welder begins the job.

$$MTTF = \int_{0}^{\infty} 3e^{-0.32t}dt = 9.375$$

$$\text{Var}(T) = \sigma_T{}^2 = \frac{3}{\lambda^2} = \frac{3}{(0.32)^2} = 29.297$$

$$P(T > 6) = R(t = 6) = e^{-0.32\times 6} = 0.1466.$$

12.3.2 Reliability Evaluation Based on Weibull Distribution

The strength of the tensile test specimen is observed and described by this distribution. When the rate of failure of a component increases or decreases with time, the Weibull distribution describes the failure times with parameters α and β. The distribution is given below $f(t) = \alpha\beta t^{\beta-1}e^{-\alpha t^{\beta}}$, $t > 0, \alpha > 0, \beta > 0$; the Weibull type failure-time distribution's reliability function may be stated as the following:

$$R(t) = e^{-\alpha t^{\beta}} \tag{12.15a}$$

$Z(t) = \alpha\beta t^{\beta-1\, Z(t) = t-1}$ provides the failure rate associated with the Weibull distribution.

$$\text{Then } \lambda(t) = \frac{f(t)}{R(t)} = \frac{\alpha\beta t^{-1}e^{-\alpha t\beta}}{e^{-\alpha t\beta}} \tag{12.15b}$$

$$\alpha\beta t^{\beta-1} = \frac{\beta}{\theta}\left(\frac{t}{\theta}\right)^{\beta-1} \text{ where } \theta^{\beta} = \frac{1}{\alpha}$$

Thus, we have

$$\text{MTTF} = E(T) = \theta\Gamma\left(1+\frac{1}{\beta}\right) \tag{12.16a}$$

MTTF is the Weibull model of mean time to failure.

The Weibull model variance is

$$\text{Var}(T) = \sigma_t^2 = \theta^2\left\{\Gamma\left(1+\frac{2}{\beta}\right) - \left[\Gamma\left(1+\frac{1}{\beta}\right)^2\right]\right\}$$

$$\text{Now } R\,(t/T_0) = \frac{R(t+T_0)}{R(T_0)}$$

$$= \frac{\exp\left[-\left(\frac{t+T_0}{\theta}\right)^{\beta}\right]}{\exp\left[-\left(\frac{T_0}{\theta}\right)^{\beta}\right]} = \exp\left[-\left(\frac{t+T_0}{\theta}\right)^{\beta} + \left(\frac{T_0}{\theta}\right)^{\beta}\right] \tag{12.16b}$$

The Weibull distribution is commonly employed for reliability modeling, offering the ability to represent various hazard behaviors and exhibiting algebraic simplicity. Furthermore, by utilizing two adjustable parameters, it can be effectively adapted to fit numerous real-world scenarios.

Example 12.12

If a power system has a scale parameter of 550 days and a shape parameter of 1.4 and it shows Weibull type of failure distribution, then find: (i) R (100 days); (ii) MTTF; (iii) standard deviation; and (iv) the design life assuming 90% reliability.

Solution:

PDF of the Weibull distribution is expressed as follows:

$$f(t) = \frac{\beta}{\theta}\left(\frac{t}{\theta}\right)^{\beta-1}\exp\left\{-\left(\frac{t}{\theta}\right)^{\beta}\right\}, t \geq 0$$

Now $\theta = 550$ days and $\beta = 1.4$

(i)

$$R(t) = \exp\left\{-\left(\frac{t}{\theta}\right)^{\beta}\right\}$$

$$\therefore R(100) = \exp\left\{-\left(\frac{100}{550}\right)^{1.4}\right\} = 0.9122.$$

(ii)

$$\text{MTTF} = \theta \times \Gamma\left(1+\frac{1}{\beta}\right)$$

$$= 550 \times \Gamma\left(1+\frac{1}{1.4}\right) = 550 \times 0.91057$$

$$= 500.8 \text{ days}$$

(iii)

$$\text{Var} = \theta^2\left\{\Gamma\left(1+\frac{2}{\beta}\right) - \left(\Gamma\left(1+\frac{1}{\beta}\right)\right)^2\right\}$$

$$= 550^2 \times \left\{\Gamma\left(1+\frac{2}{1.4}\right) - \left(\Gamma\left(1+\frac{1}{1.4}\right)\right)^2\right\}$$

$$= 550^2 \times \{\Gamma(2.43) - (\Gamma(1.71))^2\}$$

$$= 550^2 \times \lfloor(1.43 \times 0.8860) - (0.91057)^2\rfloor$$

S.D. $= 550 \times 0.66174 = 363.96$ days.

(iv) Let t_D be the required design life for $R = 0.90$

$$\therefore R(t_D) = \exp\left\{-\left(\frac{t_D}{550}\right)^{1.4}\right\} = 0.90$$

$$\Rightarrow \left(\frac{t_D}{550}\right)^{1.4} = 0.10536$$

$$\Rightarrow t_D = 550 \times (0.10536)^{1/1.4} = 110.2 \text{ days}$$

12.3.3 Normal Distribution-based Reliability Evaluation

Normal distribution also contributes to model device life length and popular one. For structural components of power system, it is the most appropriate model to evaluate the reliability. The PDF of T is given by, if T (time to failure) has a normal distribution with $N(\mu, \alpha)$:

$$f(t) = \frac{1}{\sigma\sqrt{2\pi}} \exp\left[\frac{-(t-\mu)^2}{2\sigma^2}\right], -\infty < t < \infty. \tag{12.17a}$$

This case is Var $(T) = \sigma_T{}^2 = \sigma^2$ and MTTF $= E(T) = \mu$.

Standard normal integrals and normal tables can be used to determine $R(t) = \int_t^\infty f(t)\mathrm{d}t$.

The conditional reliability determined by is thus called $\lambda(t) = \frac{f(t)}{R(t)}$ hazard function or instantaneous failure rate of any component:

$$R(t/T_0) = P\left\{T > \frac{T_0 + t}{T} > T_0\right\} = e^{-\int_T^{T_0} \lambda(t)\mathrm{d}t} \tag{12.17b}$$

12.4 Use of Exponential Model to Evaluate Reliability and MTBF

Very well, accepted modeling methodology for reliability assessment is the exponential distribution (Saket et al. 2009). With this model the reliability and MTBF of systems connected in parallel and series, can be evaluated as follow.

12.4.1 Components Connected in Serial

Let's suppose a system has n components connects in series, and the components have their failure rate $\alpha_1, \alpha_{2,}..\alpha_n$, respectively. Therefore, the law of product of reliability can be expressed as

$$R_s(t) = e^{-t\sum_{i=1}^{n} \alpha_i} \tag{12.18}$$

The MTBF for the series system is given as,

$$\mu_s = \frac{1}{\frac{1}{\mu_1} + \frac{1}{\mu_2} + \ldots.. + \frac{1}{\mu_n}} \tag{12.19}$$

where MTBF of the ith component is μ_i. If special case exists with CFR α for all n components and have the same μ, the system's MTBF and its failure rate $n\alpha$ is $1/\mu\alpha = \mu/n$.

12.4.2 Parallel System

Obtaining the mean time to failure for a parallel system is generally a challenging task, but if we consider the special case with the same failure rate α, of all the components are useful and interesting, it is possible to get a result. The system reliability function for this unique case can be considered as,

$$R_p(t) = 1 - (1 - e^{-\alpha t})^n = \binom{n}{1} e^{-\alpha t} - \binom{n}{2} e^{-2\alpha t} + \ldots\ldots + (-1)^{n-1} e^{-n\alpha t}$$

Utilizing that,

$$f_p(t) = R'_p(t),$$

We obtain, $$f_p(t) = \alpha \binom{n}{1} e^{-\alpha t} - 2\alpha \binom{n}{2} e^{-2\alpha t} + \ldots\ldots + (-1)^{n-1} n\alpha e^{-n\alpha t} \quad (12.20)$$

and gives the failure-time distribution's mean $\mu_p = \int_0^\infty t.f_p(t)\mathrm{d}t$

$$\alpha \binom{n}{1} \int_0^\infty te^{-\alpha t}\mathrm{d}t - 2\alpha \binom{n}{2} \int_0^\infty te^{-2\alpha t}\mathrm{d}t + \ldots\ldots + (-1)^{n-1} n\alpha \int_0^\infty te^{-n\alpha t}\mathrm{d}t$$

$$= \frac{1}{\alpha}\binom{n}{1} - \frac{1}{2\alpha}\binom{n}{2} + \ldots. + (-1)^{n-1}\frac{1}{n\alpha} \quad (12.21)$$

Then MTBF for system in which components are connected parallelly is $\mu_p = \frac{1}{\alpha}\left(1 + \frac{1}{2} + \ldots\ldots\right.$ $\left.\ldots\ldots + \frac{1}{n}\right)$. In MTBF system consists of "n" component with the same failure rate α therefore whenever the system fails, each defective component is replaced. Using two parallel components when only one rathers. Instead of being doubled, the pair's mean time to failure is 50% longer than that of the single component.

12.5 Probabilistic Methods For Reliability Evaluation

If a system performs its function adequately over an intended period of time subjected to some constraints called operating conditions then its probability is called reliability. The basic attributes of reliability are probability of success, time period, adequate function and operating conditions. The probabilistic methodologies for estimation of reliability functions have been described in this section successfully. The probabilities of success and failure and indices of reliability are evaluated by applying the Gaussian distribution method (GDA), safety factor concept (SFC), peak load consideration (PLC), and Simpson 1/3rd rule.

12.5.1 Gaussian Distribution Approach (GDA)

Over a specified time interval, load model for Gaussian distribution is given by:

$$f(P_\mathrm{d}) = \frac{1}{\sigma_\mathrm{d}\sqrt{2\pi}} e^{-0.5\left(\frac{P_\mathrm{d} - \overline{P_\mathrm{d}}}{\sigma_\mathrm{d}}\right)^2} \quad (12.22)$$

where

P_d = Demand on power generation station.
$\overline{P}_d$ = Average load on the power plant.
σ_d = Standard deviation of load demand.

The model of total generation capacity of is Gaussian distributed systems.

$$f(C) = \frac{1}{\sigma_c \sqrt{2\pi}} e^{-0.5\left[\frac{C-\overline{C}}{\sigma_c}\right]^2} \tag{12.23}$$

where

C = Generation capacity of power plant
$\overline{C}$ = Mean generation capacity of power plant
σ_c = Standard deviation for generation capacity.

The description of the probability of failure (P_F) for the load model of the developed system and the generation mentioned above is as follows:

$$P_F = (1 - P_s) \tag{12.24}$$

The probability of success (P_s) of develop the power system is using GDA can be expresses as:

$$P_s = \int_{-\infty}^{\infty} \int_{-\infty}^{C} \frac{1}{2\pi\sigma_c\sigma_d} e^{-0.5\left[\frac{C-\overline{C}}{\sigma_c}\right]^2} e^{-0.5\left(\frac{Pd-\overline{Pd}}{\sigma_d}\right)^2} dc \cdot dP_d \tag{12.25}$$

where

P_F = Failure probability of power generation station.
P_S = Success probability of power generation station.
C = Generation capacity of power plant.
$\overline{C}$ = Mean generation capacity of power plant.

Putting, $x = \frac{C-\overline{C}}{\sigma_c}$ and $y = \frac{P_d-\overline{P_d}}{\sigma_d}$

According to condition of substitution above, equation can be written in simplified way as given below.

$$P_s = \int_{-\infty}^{\infty} \left[\int_{-\infty}^{Z} e^{-0.5(x^2+y^2)} dy \right] dx \tag{12.26}$$

$$\text{where } Z = \frac{\sigma_c x + \overline{C} - \overline{P_d}}{\sigma_d}$$

Further, consider the substitutions made as follows,

$$x' = y.\sin\theta + x.\cos\theta$$
$$y' = y.\cos\theta - x.\sin\theta$$

With the above transformation, the axis rotates by an angle of θ. Equation (12.26) is modified by following substitutions.

Let $\sin\theta = \frac{\sigma_c}{\sqrt{P_d^2+\sigma_C^2}}$ and $\cos\theta = \frac{\sigma_d}{\sqrt{P_d^2+\sigma_C^2}}$

The above substitution is shown as follows in equation (12.27). The probability of success after above substitution comes:

$$P_s = \int_{-\infty}^{\infty} \int_{-\infty}^{\beta} \frac{1}{2\pi} e^{-0.5(x'^2+y'^2)} dy' dx' \tag{12.27}$$

$$\text{where } \beta = \frac{\overline{C} - \overline{P}_d}{\sqrt{\sigma_d^2 + \sigma_C^2}}$$

where β is the limit, and it becomes out to be x-independent. It is further simplified as follows:

$$P_s = \int_{-\infty}^{\beta}\left[\int_{-\infty}^{\infty}\frac{1}{\sqrt{2\pi}}e^{-0.5(x'^2)}\text{d}x'\right]\text{e}^{-0.5(y'^2)}\text{d}y'$$

$$P_s = \int_{-\infty}^{+\beta}\frac{1}{\sqrt{2\pi}}e^{-0.5(y'^2)}\text{d}y' = \varphi(\beta) \tag{12.28}$$

If we have normally distributed curve and the curve's area under it has a standard deviation $\int_{-\infty}^{\beta}[N(0,1)] = 1$ and mean of 0, it gives numerical value of $\varphi(\beta)$, which is conveniently obtained from the table of standard normal distribution. Curves illustrating the relationship between failure probability and generating capacity have been plotted for different capacity and load models.

By predicting the various load levels P_{di} daily variations in the load can be accounted for. If the relative frequency of occurrences is L_0, L_1, L_2, … , L_i and the frequency of occurrences are α_0 α_1 α_2, … , α_I for different levels, the total failure probability is provided by, and each load level failure probability may be calculated as follows:

$$P_{\text{F}} = \sum_i \alpha_i P_{fi} \tag{12.29}$$

where

α_I = Relative frequency of occurrences.
P_{fi} = Probability of failure of L_{i}^{th} load level.
From Eq. (12.29) it follows that

$$P_{fi} = 1 - \phi(\beta_N) \tag{12.30}$$

Hence, the overall failure probability is as shown below

$$P_{\text{F}} = \sum_i \alpha_i[1 - \phi(\beta_i)] \tag{12.31}$$

Equation (12.31) is used to calculate probabilities of failures. It is visible from the equation that PF depends on both the parameters α_i and β_i and these parameters are expressed as

$$\alpha_i = n_i/n$$

$$\beta_i = \frac{C_i^{-} - \overline{L}_i}{\sigma_{d_i}^2 + \sigma_{C_i}^2} \tag{12.32}$$

where

n = total study duration.
N_i = time duration for occurrence of "L_i^{th} load level."
C_i = L_i^{th} load level mean generation capacity.
L_i = Mean L_i^{th} load level mean.

12.5.2 Safety Factor Concept (SFC) and Peak Load Consideration (PLC)-based Evaluation of Reliability

The generation capacity of a plant shows the Gaussian distribution of probability therefore PDF is Gaussian. Low-level loading is dominated by peak loading of the system, whereas, at the low load level conditions, PF is almost negligible. If the system's peak load is P_{dmax}, the safety factor "S" is defined as:

$$S = C/P_{dmax} \tag{12.33}$$

If S is a random variable and the generation capacity "C" is distributed normally, then at a constant P_{dmax}, "S" is a safety factor distribution that will also be distributed normally. The function of a safety factor is related to these parameters as follows:

$$f_s = \frac{P_{dmax}}{\sqrt{2\pi}\sigma_c} e^{-0.5\left(\frac{P_{dmax}s-\overline{C}}{\sigma_c}\right)} \tag{12.34}$$

The Gaussian mean distribution of the safety factor S.

$$\overline{S} = \frac{C}{P_{dmax}} \text{ and } \sigma_s = \frac{\sigma_c}{P_{max}} \tag{12.35}$$

The system's failure probability is expressed as follows:

$$P_F = \int_{-\infty}^{1} \frac{P_{dmax}}{\sqrt{2\pi}\sigma_s} e^{-0.5\left(\frac{s-c/Pdmax\sigma_s}{\sigma_s}\right)^2} ds = \phi\left[\frac{1-\frac{\overline{C}}{P_{d\max}}}{\frac{\sigma_c}{P_{d\max}}}\right] \tag{12.36}$$

The probability of failure is obtained at different peak loads and capacity of generation using equation (12.36) as described in example 12.13. Using concept of safety factor P_Fv/s$\frac{\overline{C}}{P_{dmax}}$ plots can be drawn for the adequacy assessment and reliability evaluation of the modern power system.

12.5.3 LOLP Evaluation using Simpson's 1/3rd rule

The load duration curve is a realistic model to evaluate the failure probability of the modern power system. The daily peak loads are arranged in descending order to prepare a cumulative load model. The daily peak load variation curve (DPLVC) is drawn for the comprehensive analysis of load variations. This analysis results a load duration curve (LDC) model for reliability assessment. Normal distribution function based on this system uses the generation model. Maximum average generation capacity contributes to the evaluation of reliability (Varshney and Saket 2014). The loss of load probability (LOLP) occurs when the load probability exceeds the generation capacity. When capacity of generation exceeds the system load level, a "loss of load" occurs. To build a multi-step model, LDC is split into a number of steps. The approximation by increasing the number of approximation stages, the quantity can be decreased. For estimation of the LOLP of a composite power system, 100 steps of the stepped load duration curve (SLDC) are used. The different steps of the SLDC show the load's fluctuations over a 24-hour period. In this section, the Simpson's 1/3rd rule is used with the following approximation for evaluation of the LOLP index. The modern power system's LOLP applying SLDC is expressed as follows:

$$\text{LOLP} = \int_0^{100} \frac{t}{100}\left[\int_{-\infty}^{P_d(t)} \frac{1}{\sqrt{2\pi}\sigma_c} e^{-0.5\left(\frac{c-\overline{C}}{\sigma_c}\right)^2} dc\right] dt \tag{12.37}$$

Putting, $\frac{c-\overline{C}}{\sigma_c} = Z$

Using Eq. (12.37) for 100 steps of SLDC, the LOLP is expressed as follows:

$$\text{LOLP} = \int_0^{100} \frac{t}{100}\left[\int_{-\infty}^{P_d(t)} \frac{1}{\sqrt{2\pi}\sigma_c} e^{-0.5(z)^2} dz\right] dt \tag{12.38}$$

$$\text{LOLP} = \int_0^{100} \frac{t}{100} \varphi \left(\frac{P_d(t) - \overline{C}}{\sigma_c} \right) dt \tag{12.39}$$

Area under the normal distribution curve [N (0,1)] represents the inner integral for different value of 't', and it has been replace by $\phi\left[\frac{P_d(t)-\overline{C}}{\sigma_C}\right]$ function. The expression of Eq. (12.39) can be evaluated by any method of the area evaluation. By assuming LDC as SLDC with a maximum of 100 steps, LOLP is evaluated. It is possible only if a multilevel representation of the daily load duration curve is used. This approach is similar to the Markov load model used in calculation of the frequency and duration. The electric load varies many times in steps, if the power system operates for 24 hour per day. The SLDC is an important attribute for reliability evaluation and planning of the modern power system. The following numerical example 12.13 is taken for reliability evaluation of an interconnected power system using various probabilistic approaches.

Example 12.13

The generation capacities of an interconnected composite power system are given below:

(i) $2 \times 200 = 400$ MW Nuclear power plant
(ii) $10 \times 50 = 500$ MW Thermal power plant
(iii) $3 \times 100 = 300$ MW Hydro power plant

Daily peak load variation is from 300 MW to 1000 MW, which is measured annually.

(a) Prepare the generation capacity state table including the standard deviation data for graphical illustration and power system reliability evaluation.
(b) Calculate the failure probabilities for (i) constant mean load of 480 MW, (ii) constant generation capacity of 500 MW, and (iii) provide the graphical illustrations for reliability assessment of the power system using Gaussian distribution approach.
(c) Calculate the failure probabilities of power system considering peak loads and safety factor concept for (i) constant load of $P_d = 480$ MW, (ii) peak load variation from $P_{dmax} = 480$ MW to $P_{dmax} = 510$ MW, and (iii) provide the graphical illustration for reliability analysis of power system.
(d) Evaluate the LOLP of power system for $P_{dmax} = 1000$ MW and $P_{dmin} = 300$ MW for 10 steps of the stepped load duration curve (SLDC) and draw the LOLP v/s generation capacity curves for analysis of the modern power system.

Solution:

(a) The generation capacity, state data and data of standard deviation of power stations are as given in the table below. The subscript 0, 1, 2, 3, 4, etc., present the generation unit availability in power system due to nuclear (X), thermal (Y), and hydro (Z) power plants (Table 12.1).
(b) Failure probability evaluation using Gaussian distribution approach
   (i) Evaluation of failure probability at constant 480 MW mean load

Here, we are evaluating failure probability by keeping the mean load constant at 480MW. GDA equation and data table of standard deviation are utilized. The standard deviation for the mean load and generating capacity is assumed to be 10% of the mean loads and generating capacity, respectively. The failure probability at generation capacity $C = 490$ MW is $P_f = 0.3853$. Similarly, using this equation, the failure probability can be calculated considering available generating capacity states at 480 MW fixed mean load.

Many plots of generation capacity versus failure probability are drawn for different values of standard deviations. The failure probabilities are illustrated in curve A ($\sigma_d = 5\%$ of load and $\sigma_c = 5\%$

Table 12.1 The generation capacity state data and standard deviation data of power stations.

Sr. no.	Capacity state	Generating capacity available in MW	$\sigma_c = 5\%$	$\sigma_c = 10\%$	$\sigma_c = 15\%$	$\sigma_c = 20\%$	$\sigma_c = 25\%$
1	$X_1Y_0Z_0$	200	10	20	30	40	50
2	$X_2Y_0Z_0$	400	20	40	60	80	100
3	$X_2Y_1Z_0$	450	22.5	45	67.5	90	112.5
4	$X_2Y_2Z_0$	500	25	50	75	100	125
5	$X_2Y_3Z_0$	550	27.5	55	82.5	110	137.5
6	$X_2Y_4Z_0$	600	30	60	90	120	150
7	$X_1Y_5Z_2$	650	32.5	65	97.5	130	162.5
8	$X_1Y_6Z_2$	700	35	70	105	140	175
9	$X_2Y_7Z_0$	750	37.5	75	112.2	150	187.5
10	$X_1Y_8Z_2$	800	40	80	120	160	200
11	$X_2Y_9Z_0$	850	42.5	85	127.5	170	212.5
12	$X_1Y_{10}Z_2$	900	45	90	135	180	225
13	$X_2Y_{10}Z_1$	1000	50	100	150	200	250
14	$X_2Y_{10}Z_2$	1100	55	110	165	220	275
15	$X_2Y_{10}Z_3$	1200	60	120	180	240	300

of capacity), curve B ($\sigma_d = 15\%$ of load and $\sigma_c = 5\%$ of capacity), curve C ($\sigma_d = 5\%$ of load and $\sigma_c = 15\%$ of capacity), and curve D ($\sigma_d = 15\%$ of load and $\sigma_c = 15\%$ of capacity). The curves are labeled for different combinations of σ_c and σ_d. By comparing A and B, it is found that PF at the same generating capacity with A is less than B. This is because of the large amount of uncertainty associated with the distribution function of generating capacity. Similarly, for the same generating capacity, if we observe the curves A and C, it can be seen that C has a lower failure probability than B, and so on. The failure probability for total generating capacity is shown by curve D, the highest among all others. With an increase in both σ_c and σ_d, or both simultaneously, the chance of failure rises (Figure 12.1).

(ii) Evaluation of failure probability keeping constant generation capacity

In this section, the PF is evaluated using GDA, keeping generation capacity at a constant value i.e. 500 MW. For mean load and generation capacity, the standard deviation is assumed to be 5% of both. The failure probability at generation capacity $C = 500$ MW is $P_f = 0.4432$. Then, at a mean capacity = 500 MW (fixed), the failure probability is calculated by increasing the load steps by 5 MW. In the figure given below, different plots of PF versus capacity of generation at different values of standard deviation are shown. The failure probabilities as per the curve: There are four curves: curve A ($\sigma_d = 10\%$ of load and $\sigma_c = 10\%$ of capacity), curve B ($\sigma_d = 20\%$ of load and $\sigma_c = 10\%$ of capacity), curve C ($\sigma_d = 10\%$ of load and $\sigma_c = 20\%$ of capacity), and curve D ($\sigma_d = 20\%$ of load and $\sigma_c = 20\%$ of capacity). If the mean load and generating capacity are both equal, the PF and success are precisely equal to 50%. It is the same in all cases, and the independents of the standard deviation is shown in the figure given below. Below the point of intersections PF increases with an increase of σ_c and σ_d above the points; it varies oppositely. Using GDA for $\sigma_d = 10\%$ of $\bar{L}$ and $\sigma_C = 10\%$ of $\bar{C}$

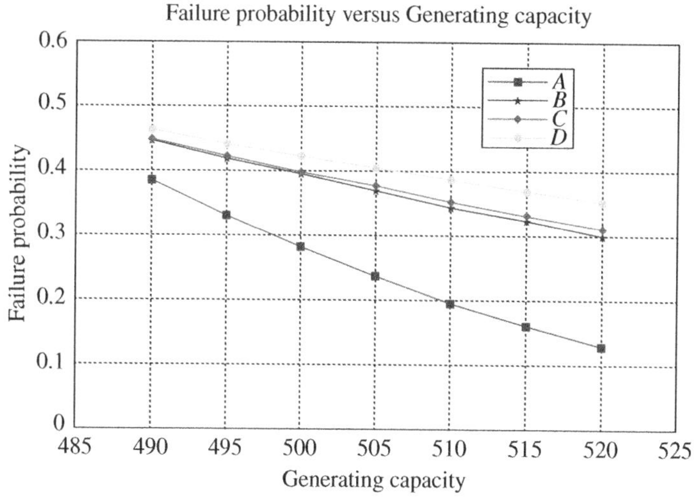

Figure 12.1 P_F versus C plot at constant load of the developed system.

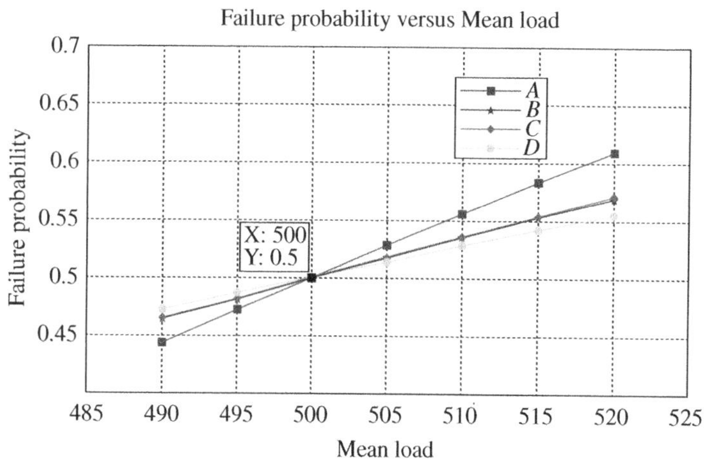

Figure 12.2 Plot of P_F v/s generation capacity at constant capacity.

the overall PF is evaluated as $P_F = 0.0594$. If σ_c and σ_d are increased from 10 to 20%, the PF overall also increases by the same amount of capacity of generation and load level at fix value of frequency of occurrence. At $\sigma_c = 20\%$ of $\bar{C}$ and $\sigma_d = 20\%$ of $\bar{L}$, the failure probability P_F is 0.0986 (Figure 12.2).

(c) Using SFC and PLC evaluate the PF

See Table 12.2.

(i) At constant load evaluation of PF

This section discusses the adequacy evaluation of the capacity of generation based on worst-case loading condition and safety factor. During evaluation, it is assumed that the maximum load on to the system is $P_{dmax} = 480$ MW. Considering the approximations, curves have been plotted at different values of standard deviations between the safety factor (C/P_{dmax}) and PF. Curve: A is for $\sigma_c = 10\%$ of C, curve B is for $\sigma_c = 12\%$ of C, curve C is for $\sigma_c = 15\%$ of C, and curve D is for

Table 12.2 Mean capacity for different load levels.

$\bar{L}_i$	1000 Mw	1200 Mw	1500 Mw	1800 Mw	2000 Mw
$\bar{C}_i$	1250 Mw	1500 Mw	1875 Mw	2250 Mw	2500 Mw
α_i	0.1	0.2	0.4	0.2	0.1

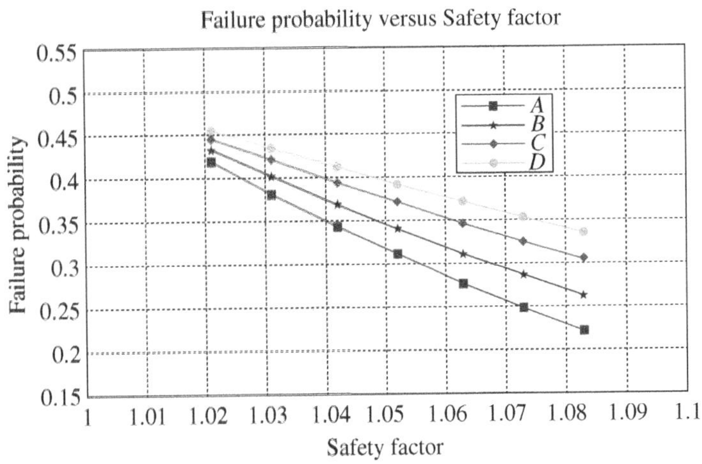

Figure 12.3 Plot of failure probability v/s safety factor at constant load.

$\sigma_c = 18\%$ of C. Observations of the curves A and B state that, the P_F with $\sigma_c = 12\%$ of C is more than $\sigma_c = 10\%$ of C for the same factor of safety and capacity of generation, if we increase σ_c for any value of safety factor and generation capacity P_f also increases. For all generation capacity values, PF decreases with an increase in safety factor. For any value of factor of safety factor of capacity of generation, the PF increases as the standard deviation rises. Given figure presents PF versus estimate of the safety factor at changing capacity of generation and constant load. For the 10.5% factor of safety, PF increases with an increase in standard deviation of capacity of generation (Figure 12.3).

(ii) Evaluation of failure probability under varying load

If load P_{dmax} is increased from 480 MW to 510 MW, keeping the safety factor's value (C/P_{dmax}), different curves are drawn as below: Curves A and B are for $\sigma_c = 10\%$ of C, Curve C is for $\sigma_c = 12\%$ of C, Curve D is for $\sigma_c = 15\%$ of C, and Curve E is for $\sigma_c = 18\%$ of C. With an increase in $\sigma_{c,}$ P_F increases at the same value of capacity of generation because of large amount of uncertainty involved in the distribution factor of capacity of generation. When the safety factor is 1, at variable load PF is constant and equal to 50% in all the cases as drawn in curves. If factor of safety is below one, the PF for $\sigma_c = 10\%$ of C is more than $\sigma_c = 12\%$ of C. If safety factor is greater than **1**, the probability of failure varies oppositely. Ratio of capacity of generation to peak load demand is called factor of safety, and it is one for equal values of peak load and capacity of generation. At this instant, both the probabilities are 50%, as shown in figure. If capacity of generation is more than load, PF of curve more than B, C, and D, A will be. Similar to this, curve A has a lower failure probability than curve B and vice versa if load exceeds generating capacity (Figure 12.4).

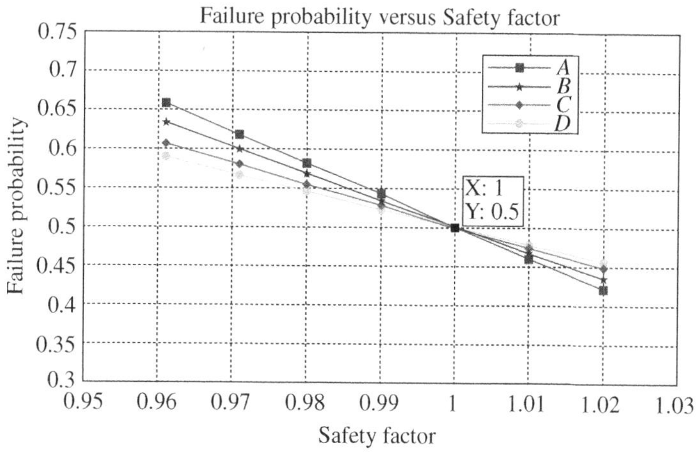

Figure 12.4 Plot of probability of failure v/s safety factor at variable load.

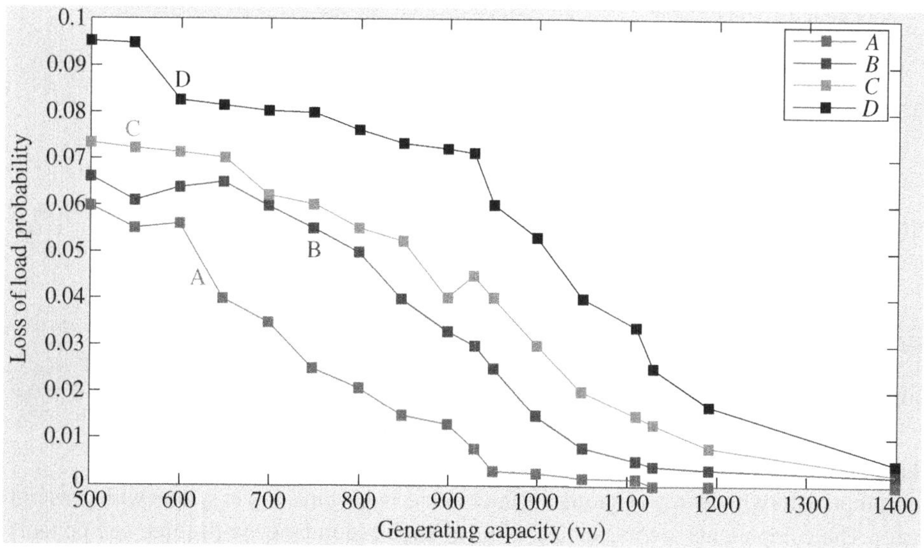

Figure 12.5 Plot of LOLP v/s capacity of generation for 10 steps.

(d) **Simpson 1/3rd rule to evaluate LOLP**

To evaluate reliability uses the Simpson 1/3rd rule $P_{dmax} = 1000$ MW and $\boldsymbol{P_{dmin}}$ **= 300 MW** is taken. LDC becomes straight line. LDC becomes a straight line. In the mentioned example of SLDC, a maximum of 100 steps are used using the Simpson 1/3rd rule approximation. During 24 hours of operation of power system load of the systems varies in many steps. Let's have a look at the following figure, which represents the relation between the capacity of generation and LOLP for 10 steps of SLDC with various σ_c. The curves A, B, C and D as shown in Figure 12.5 indicate the relationship of LOLP with capacity of generation at different values of σ_c for ten steps of the SLDC. The percentage values are 10%, 15%, 20%, and 25% respectively. The LOLP increases with an increase in σ_c because large uncertainty involved in the distribution function of capacity of generation. Also, at a particular value of capacity of generation, LOLP increases as standard deviation increases. Consider 10 steps of SLDC; during a 24-hour operation, there were a total of 10 load variations (Figure 12.5).

12.6 Additional Solved Examples

Example 12.14

The MTBF of a protective relay is 0.8 year with random failure occurrence.

(a) Calculate the probability of survival of the relay for 1 year without failure.
(b) Find the probability of more than 2 failures of the relay within a year.
(c) Find the per year expected number of failure.

Solution:

The number of failures during the interval of t follows a Poisson process since failures are random events, given by $P[N(t) = n] = \frac{e^{-\lambda t}(\lambda t)n}{n!}$, $n \geq 0$ exponential distribution with a mean of $1/\lambda$ the time between failures.

Now

$$\text{MTBF} = \frac{1}{\lambda} = 0.8\text{ year}$$

$\therefore \lambda = \frac{1}{0.8}$ per year $=1.25$ per year

(a)

$$P[N(1) = 0] = \frac{e^{-\lambda}\lambda^0}{0!} = e^{-1.25} = 0.2865$$

(b)

$$P[N(1) > 2] = 1 - e^{-\lambda}\left[\frac{\lambda^0}{0!} + \frac{\lambda^1}{1!} + \frac{\lambda^2}{2!}\right] = 1 - 0.2865\left[1 + 1.25 + \frac{(1.25)^2}{2}\right] = 0.1315$$

(c)

$$E[N(t)] = \lambda t$$

$\therefore E$ {number of failure per year} $= \lambda = 1.25$.

Example 12.15

Six power components with identical nature are interconnected in bulk power system with constant failure rates. The components are associated with (a) High redundancy with three components per sub-system, and (b) Low redundancy with two components per sub-system. In each example, determine the MTTF to necessarily get the 90% reliability in100 hours of operation.

Solution:

$R = e^{-\lambda t}$ for each component, if the failure rate for each component is constant.

(a) For high redundancy:

$$R_s(t) = 1 - [1 - \{R(t)\}^3]^2 = [1 - (1 - e^{-3\lambda t})^2]$$

$$\therefore R_s(100) = [1 - (1 - e^{-300\lambda})^2] = 0.90$$

$$\Rightarrow (1 - e^{-300\lambda})^2 = 0.10$$

$$\Rightarrow e^{-300\lambda} = 0.6837$$

$$\therefore 300\lambda = 0.3801 \Rightarrow e^{-300\lambda} = 0.3162$$

$$\therefore \text{MTTF of each component } \frac{1}{\lambda} = \frac{300}{0.3801} = 789.2\text{ hours.}$$

(b) For low redundancy:

$$R_s(t) = [1 - \{1 - [R(t)]\}^2]^3 = [1 - \{1 - e^{-\lambda t}\}^2]^3$$
$$\therefore R_s(100) = [1 - \{1 - e^{-100\lambda}\}^2]^3 = 0.90$$
$$1 - (1 - e^{-100\lambda})^2 = 0.9654, = (1 - e^{-100\lambda})^2$$
$$= 0.345, = 1 - e^{-100\lambda} = 0.1857, = e^{-100\lambda} = 0.8142,$$
$$= 100\lambda = 0.20551$$
$$\therefore \text{MTTF of each components} = \frac{1}{\lambda} = \frac{100}{0.20551} = 486.6$$

Example 12.16

If two units, standby and primary, are carried by a health monitoring system employing power transformer, and if MTTF of standby unit is 333 hours and MTTF of primary unit is 1000 hours with number of failures in standby mode of backup unit. Estimate the design life for dependability at 90% if the primary unit requires repairs at a rate of 0.01 per hour while the standby unit is in use.

Solution:

For the standby redundant system,

$$R(t) = \frac{m_1}{m_1 - m_2} e^{m_2 t} - \frac{m_2}{m_1 - m_2} e^{m_1 t} \tag{12.40}$$

where roots of the equation are $m1$ and $m2$

$$m^2 + (\lambda_1 + \lambda_2 + \mu)m + \lambda_1 \lambda_2 = 0 \tag{12.41}$$

Here $\lambda_1 = \frac{1}{1000} = 0.001/\text{hour}$; $\lambda_2 = \frac{1}{333} = 0.003/\text{hour}$ and $\mu = 0.01/\text{hour}$

Substituting the above values in Eq. (12.41), we get

$$m^2 + 0.014\, m + 0.000003 = 0$$

$$\therefore m_1, m_2 = \frac{-0.014 \pm \sqrt{(0.0 - 14)^2 - 4 \times 0.000003}}{2} = (0.00022 - 0.01378)$$

Put the values of m_1 and m_2 in Eq. (12.40)

$$R(t) = -0.01622 \times e^{-0.01378t} + 1.01622 \times e^{-0.00022t}$$

At 90% reliability, design the life of D is given by $1.01622 \times e^{-0.00022t} - 0.01622 \times e^{-0.01378D} = 0.90$, solve the equation by trial, we get $D = 550$ hours

Example 12.17

If a newly purchased distribution transformer fails with a constant rate of 0.03 per day. If we assume continuous use and constant rate of repair equal to 0.2 per day.

(a) Calculate the steady-state availability and interval availability for the first 30 days.
(b) Define the standby unit's steady-state availability. In standby, assume no failures.
(c) What is the availability at steady state if both units are active?

Solution:

$\mu = 0.2$ per day, $\lambda = 0.03$ per day,

$$A_1(T) + \frac{\mu}{\mu + \lambda} + \frac{\mu}{T \times (\lambda + \mu)^2}[1 - e^{-(\mu+\lambda)T}]$$

$$\therefore A_1(30)\frac{0.2}{0.2+0.03}+\frac{0.03}{30\times(0.03+0.2)^2}[1-e^{-(0.2\times0.03)\times30}]$$

(a)

$$A_1(T)+\frac{\mu}{\mu+\lambda}+\frac{\mu}{T\times(\lambda+\mu)^2}\left[1-e^{-(\mu+\lambda)T}\right]$$

$$\therefore A_1(30)\frac{0.2}{0.2+0.03}+\frac{0.03}{30\times(0.03+0.2)^2}\left[1-e^{-(0.2\times0.03)\times30}\right]$$

$$=0.8696+0.0031+0.8727$$

$$A(\infty)=\frac{\mu}{\lambda+\mu}=0.8696$$

(b) For the standby redundant system

$$A_s(\infty)=\frac{\lambda\mu+\mu^2}{\mu^2+\lambda\mu+\lambda^2}=\frac{(0.03\times0.2)+(0.2\times0.2)}{(0.2)^2+(0.03\times0.2)+(0.03)^2}$$

$$=>\frac{0.046}{0.0469}=0.9808$$

(c) For the active redundant system

$A_i(\infty)=1-[1-A(\infty)]^2=1-[1-0.8696]^2=0.9830$

12.7 Conclusion

The presented analysis here provides a means to relate reliability of power system to reliability of a component for different equipment. States of binary component is used in the configuration of the system. An independent component using a reductionist approach to reliability analysis, and it is component-level studies because the reliability of the system is a function of reliability of a component. This fact is very well-defined and accepted. To achieve a high level of reliability in composite systems, focus on reliability of independent and individual components is very much essential. To apply the methods few examples are taken to explain the benefits and drawbacks of different methods. In addition, case studies are taken from the experience of the author's consultancy, research, and academics. The basic concepts of theory of probability and the redundancy of the system to improve reliability are also covered comprehensively. Structure of power component for evaluation of reliability, Hazard model for analysis of component failure, different types of distribution functions of probability to illustrate reliability, and various solved, unsolved numerical examples are described in each section and sub-section of this chapter.

References

Arya, L.D., Choube, S.C., and Saket, R.K. (2000). Composite system reliability evaluation based on static voltage stability limit. *Journal of the Institution of Engineers (India): Series-B* 80: 133–140.

Arya, L.D., Choube, S.C., and Saket, R.K. (2001). Generation system adequacy evaluation using probability theory. *Journal of the Institution of Engineers (India): Series-B* 81: 170–174.

Bansal, R.C., Bhatti, T.S., and Kothari, D.P. (2002). Discussion of "Bibliography on the application of probability methods in power system reliability evaluation". *IEEE Transaction on Power System* 17 (03): 924.

Bharti, O.P., Saket, R.K., and Nagar, S.K. (2019). Reliability assessment and performance analysis of DFIG based WT for wind energy conversion system. *International Journal of Reliability and Safety, Inderscience Publishers (UK)* 13 (04): 235–266.

Billinton, R. and Zhang, W. (1998). Algorithm for failure frequency and duration assessment of composite power systems. *IEE Proceedings: Generation, Transmission and Distribution* 145 (2): 117–122.

Kothari, D.P., Nagrath, I.J., and Saket, R.K. (2022). *Modern Power System Analysis*, 5e, 01–850. New Delhi (India): McGraw Hill.

Kumar, S., Sachin Kumar, K.S., Kumar, A. et al. (2019). Probabilistic evaluation and design aspects for reliability enhancement of induction motor. *International Journal of Reliability and Safety, Inderscience Publishers (UK)* 13 (04): 267–290.

Saket, R.K. (2013). Design aspects and probabilistic approach for generation reliability evaluation of MWW based micro-hydro power plant. *Renewable and Sustainable Energy Reviews*, Elsevier 28: 917–929.

Saket, R.K., Bansal, R.C., and Singh, G. (2006). Generation capacity adequacy evaluation based on peak load considerations. *The South Pacific International Journal on Natural Sciences*, University of the South Pacific, Suva, Fiji Island 24: 38–44.

Saket, R.K., Bansal, R.C., and Singh, G. (2007, 2007). Reliability evaluation of power system considering voltage stability and continuation power flow. *Journal of Electrical Systems, Engineering and Scientific Research Groups, France* 3 (2): 48–60.

Saket, R.K., Bansal, R.C., and Singh, G. (2009). Power systems component modelling and reliability evaluation of generation capacity. *International Journal of Reliability and Safety*, Inderscience Publishers (UK) 03 (04): 427–441.

Varshney, L. and Saket, R.K. (2014). Reliability evaluation of SEIG rotor core magnetization with minimum capacitive excitation for unregulated renewable energy applications in remote areas. *Ain Shams Engineering Journal*, Elsevier 5 (03): 751–757.

13

Reliability Enhancement of Electrical Distribution Systems Considering Active Distributed Generations

Kalpesh B. Kela[1], *Bhavik N. Suthar*[2], *Smriti Singh*[3], *Rajesh Arya*[4], *and Liladhar Arya*[5]

[1] *Department of Electrical Engineering, L.D. College of Engineering, Ahmedabad, Gujarat, India*
[2] *Department of Electrical Engineering, Government Engineering College, Bhuj, Gujarat, India*
[3] *Department of Electrical Engineering, Electrical Department of Electrical Engineering, Indian Institute of Technology (BHU), Varanasi, Uttar Pradesh, India*
[4] *Department of Information Technology, Acropolis Institute of Technology and Research, Indore, Madhya Pradesh, India*
[5] *Department of Electrical Engineering, Shri G.S. Institute of Technology and Science, Indore, Madhya Pradesh, India*

13.1 Introduction

Infrastructure for delivering power and active resources are combined in active distribution systems, which also include passive infrastructure. The active distribution network is designed to create a network that is reliable, high-quality, and energy-efficient. DG, or distributed generation, is now the best option for electrical distribution firms looking to improve the reliability of their power distribution systems. DGs boost system performance by minimizing losses, improving voltage profile, and reliability. In addition, the use of DGs with low capacity at customer locations is growing for a number of reasons, such as the depletion of fossil fuel resources, increased demand, environmental concerns, and so forth.

By altering the rate of failures and time of repairs for various system segments, the reliability of the system of electrical distribution can be raised. By including DGs at the customer's end, the extra cost to accomplish this can be reduced. However, buying energy from DGs may come at a premium price. If the DG switching-over period is shorter, the interruption time of the reliability indicators can be decreased to a brief interruption.

DG placement and size have a significant impact on reliability along with other metrics in distribution systems. The distribution system's reliability has risen due to DGs placed in specified areas (Arya et al. 2012). According to numerous studies, DG implantation decreases loss. However, reliability could be more crucial for some loads than loss and profile of voltage. Many costs relating to investment, maintenance, operations, and other expenses are associated with introducing DGs into the distribution network. Therefore, the costs and advantages of deploying DGs for customers and utilities must be assessed.

In recent years, a number of academics have sought to increase the reliability of DG-based distribution networks. The best locations for DGs based on reliability indices can be found through sequential Monte Carlo simulation, according to Yousefian and Monsef (2011). The authors of Al-Muhaini and Heydt (2013) examined the reliability of the distribution system in relation to conventional and sustainable DG. For the evaluation of DG reliability, a combined Markov system simulation was used, which takes into account DG sufficiency regarding DG physical breakdown and beginning and changeover probabilities. The placement of transportable distributed electrical

Reliability Analysis of Modern Power Systems, First Edition. Edited by R. K. Saket and P. Sanjeevikumar.

power units in distribution systems was recommended by Awad et al. (2014) as a way to increase system reliability and reduce costs. In order to improve reliability in massive primary networks of distribution, it is necessary to provide the appropriate location and size for numerous dispersed generators; for that, Kumar et al. (2014) employed a method of random search known as cat swarm optimization.

In the existence of storage systems, Abbasi and Hosseini (2016) looked at how the size, shape, and DG allocation of the distribution network affected the system's reliability. Reliability is one of the features of the distribution network expansion strategy that Bagheri et al. (2015) give that integrates DGs. In order to increase reliability and take into account the whole cost of the power used by the system, Battu et al.'s (2015) proposal for where to place distributed generators in the distribution network was successful. In order to evaluate consumer- and energy-oriented reliability indices for energy distribution networks with DG, Arya (2016) established a methodology that took into consideration the impact of eliminating random allowed interruption periods at load sites. The Bootstrapping method and Monte Carlo simulation were used to complete it. With and without DGs, the ideal value for the main reliability indices in the distribution system was found by the authors of Ray et al. (2015). To do this, a differential search method was employed. DG placement in distribution systems that maximize profit and value was found by Kansal et al. (2017).

To maximize system reliability, the inventors of Rahiminejad et al. (2016) adopted DG scheduling. Here, a cost–benefit evaluation is carried out while taking the various costs related to the computations into account. In Wang (2017), Wang Yue presented a method for evaluating the analytical distribution network's reliability that takes into consideration both the constraints placed on network reconfiguration and stochastic load point disruption durations. The Roy Billinton reliability test system case studies are used in the research study to highlight the benefits of the recommended method above the conventional sector technique-based reliability evaluation method. Kela et al. (2019) suggests a strategy for boosting reliability by maximizing the entire reliability expenses associated with electrical distribution systems; both utility and consumer costs are taken into account. In order to analyze the SAIDI (system average interruption duration index) and SAIFI (system average interruption frequency index) indices based on network structure and remotely managed reclosers and switches, the authors of Lekbich et al. (2019) presented a method of analysis based on graph theory. The study Pattabiraman et al. (2019) offers a feeder reconfiguration technique for reducing reliability indices. The reconfiguration is handled by Binary Particle Swarm Optimization, which chooses the best tie switches to reduce both the duration and the number of interruptions.

A reliability assessment method for the integration of large-scale PV systems into a distribution system is provided in Su et al. (2019). In Okoye et al. (2020), researchers simulated and established a link between load increase and generator capacity using power production and demand data from the IEEE reliability test system (IEEE RTS '96 standard). An economic and reliability assessment of photovoltaic solar power systems for rural electricity supply was made by the authors in Raghuwansh and Arya (2019). The authors of Kumar et al. (2020) provide a thorough analysis of the effect of combined renewable sources on the assessment of electrical network reliability. Mahmood et al. (2020) introduces a novel multistage hybrid method based on the tie-set method and graph reduction techniques for two-terminal reliability assessment problems. The reliability of the distribution network was examined by the researchers of Idowu et al. (2021) in relation to DGs. Here, the network's reliability with and without DG units is compared. To improve distributor line reliability and voltage profile, the researchers of Agajie et al. (2021) advised placing DERs, or distributed energy resources, in the best possible locations. The article (Byk et al. 2020) demonstrates the efficiency of operating the system in a state without a power supply to it. It has shown an increase in power supply reliability as well as economic efficiency. In Ahmad and Asar (2021), authors have increased the reliability of a distribution network by injecting DGs at specific locations. The optimal locations are found with the help of the artificial neural network (ANN) technique. The researchers Mahboob et al. (2021) have found the optimal locations and sizes of DGs to improve the reliability of a distribution system. Here, power losses and the voltage profile of the system have also been assessed. The authors Xu et al. (2023) have suggested a reliability evaluation method for an active distribution system. Here, the uncertainty of the DGs has been considered, and the model has been developed. In Nasser Hussain et al. (2023), the reliability of the distribution network has been enhanced by reconfiguration of the system and optimal placement of capacitors.

According to the previous discussions, the objective of this study is to increase the distribution system's reliability by adjusting the rate of failure and times to repair its various components while taking DGs into account. An objective function created here was optimized to achieve this. However, rather than starting from any presumed positions, this optimization was carried out after finding the right places of DGs according to reliability enhancement features. Naturally, an evaluation of costs and benefits was done to support the choice. Some background knowledge on reliability indices is provided in Section 13.2. The mathematical approach to the issue is described in Section 13.3. A brief summary of the flower pollination (FP) optimization approach is given in Section 13.4. The solution to the problem is discussed in Section 13.5. The outcomes and their interpretation are covered in Section 13.6. Section 13.7 comes to a conclusion.

13.2 Electrical Distribution Reliability Indices: Customer and Energy Based

Three customer-oriented reliability indices, known as SAIFI, SAIDI, and CAIDI (index that represents the customer's average interruption duration) are commonly used by utilities (Billinton and Allan 2013):

$$\mathrm{SAIFI} = \frac{\sum \lambda_{\mathrm{sys},i} N_i}{\sum N_i} \tag{13.1}$$

$$\mathrm{SAIDI} = \frac{\sum U_{\mathrm{sys},i} N_i}{\sum N_i} \tag{13.2}$$

$$\mathrm{CAIDI} = \frac{\sum U_{\mathrm{sys},i} N_i}{\sum \lambda_{\mathrm{sys},i} N_i} \tag{13.3}$$

AENS (index that represents the average energy not supplied), one of the most essential reliability indices that is energy based, is defined as follows:

$$\mathrm{AENS} = \frac{\sum L_i U_{\mathrm{sys},i}}{\sum N_i} \tag{13.4}$$

The following are expressions for calculating $\lambda_{\mathrm{sys},i}$ and $U_{\mathrm{sys},i}$ for each load point:

$$\lambda_{\mathrm{sys},i} = \sum \lambda_k \tag{13.5}$$

$$U_{\mathrm{sys},i} = \sum \lambda_k r_k \tag{13.6}$$

13.3 Defining the Problem

Choosing the right DG placements and lowering failure rates and maintenance times in various system components are the two main objectives of this chapter, which is to increase distribution system reliability. Two objective functions were examined in this approach.

13.3.1 DG Location Determination

Enhanced reliability is the main criterion for choosing the location of DGs; the objective function examined is as follows:

$$J = \frac{\text{SAIFI}}{\text{SAIFI}_t} + \frac{\text{SAIDI}}{\text{SAIDI}_t} + \frac{\text{CAIDI}}{\text{CAIDI}_t} + \frac{\text{AENS}}{\text{AENS}_t} \tag{13.7}$$

The sum of the normalized quantities of SAIFI, SAIDI, CAIDI, and AENS is the objective function. The normalization is done in relation to the indices' respective target values. As a result, in the procedure, there will be equal weighting for each index.

It is first decided what J will be after connecting DG to load position 1. Here, J is the system's total reliability. Gradually, DGs are linked separately at each of the remaining load sites, and the value of J is determined. The system's reliability improves when DG is connected to every load point one by one. The improvement in reliability with respect to J is arranged and graded across all load locations. Choosing the top two values in the ranking, connecting the DGs at these load places, and calculating J's value are the next steps. The same methodology is then applied to the top three rankings. Until the minimum standards of reliability is attained, this procedure can be repeated. Section 13.5 goes into great length about it.

After finding the DG locations, the optimized reliability index values are found as below.

13.3.2 Connecting DGs in the System as Standby Units

The distribution firm may own the DGs, or it could encourage major consumers to become owners of the DGs. The customer-owned and operated DGs, as well as those administered by other agencies, may end up being crucial in boosting system reliability. Incorporating DGs reduces outage time as well as the cost associated with lowering failure and repair rates. It is also possible to reduce the total expense of interruption for the consumer. On the other hand, the cost of the energy borrowed through DGs could be substantial because of its higher costs (Arya et al. 2012). As a result, an objective function is written as follows:

$$F = \sum_{k=1}^{N_c} \alpha_k / \lambda_k^2 + \sum_{k=1}^{N_c} \beta_k / r_k + \text{ADCOST(EENSO-EENSD)} + \sum_{k=1}^{N_c} \text{CIC} \tag{13.8}$$

$$\lambda_k \times r_k \times L_i \times C_{p_k} = \text{CIC} \tag{13.9}$$

There are four components in the relation (13.8). The initial two terms refer to the costs of changing the rate of failure and average times for repairs for each of the distribution system components, which are, correspondingly, maintenance activity adjustment costs. The lower the value associated with these criteria, the larger the investments or expenditures utilities must make in preventative care and maintenance programes in order to meet them (Kela and Arya 2014). Both of these variables are derived from Duane's (Ebeling 2019) reliability growth model. The third term in the same relation represents the additional cost (ADCOST) to be borne by the owners of the DGs when expected energy is not supplied by them while DGs are not connected (EENSO) and when DGs are present (EENSD). Here, ADCOST is taken as Rs. 2 per kWH. The fourth component of the relation (13.8), customer interruption cost (CIC), represents the expenses incurred at the customer's end in terms of loss during a power outage. When a utility is in the occupation of providing electricity to businesses and industries, the high expenses associated with outages in electricity are of course magnified. Any load site's overall disruption expense can be estimated by adding up the expenses of all section outages, as shown in relation (13.9). The overall cost of all the customer interruptions can therefore be calculated. Here C_{p_k} in relation (13.9) is the cost of interruptions for a downtime duration r_k related to the kth distribution sector in Rs. for each kilowatt. The cost characteristics associated with the consumers at a single customer load site completely determines the customer cost at that load site. The aggregate of the expenses experienced by all types of customers as a result of the distribution outage represents the total of the consumer's costs related to any load location due to any interruption (Chowdhury and Koval 2011). The calculations are done for all the distribution sectors N_c. Table 13.3 shows the results for Roy Billinton test system bus-2 (RBTS-2) (Figure 13.1). The price of the DGs and the expense of repair plus customer interruptions are thus balanced by the function defined in (13.8). The other two costs are anticipated to decrease when the price of DGs increases. By minimizing the function specified in (13.8), this is achieved:

$$\text{EENS} = \sum L_i U_{\text{sys},i} \tag{13.10}$$

where EENS is the index that represents expected energy not supplied. $U_{\text{sys},i}$, without DG is calculated as per (13.6). The load may be shifted to DG if the power supply to load terminals from the source is disrupted while DG has been installed in the system as an on-standby unit. In this case, it is important to take into account the switch's failure rate and changeover time as it shifts the loads to DG.

By connecting the transfer switch to the system in series and the DG in parallel, the reliability simulation can be demonstrated. Assuming DG is used as a standby unit for the load sites, the estimated calculations (Arya et al. 2012) give the relevant rate of failure and maintenance time.

$$\lambda_{eq} = \lambda_s \lambda_{dg} \left(r_s + r_{dg}\right) + \lambda_{sw} \tag{13.11}$$

$$r_{eq} = \frac{\lambda_s \cdot \lambda_{dg} \cdot r_s \cdot r_{dg} + \lambda_{sw} \cdot s}{\lambda_s \cdot \lambda_{dg} \left(r_s + r_{dg}\right) + \lambda_{sw}} \tag{13.12}$$

$$U_{eq} = \lambda_{eq} r_{eq} \tag{13.13}$$

where

λ_{eq} = an equivalent rate of failure
r_{eq} = equivalent duration of interruption
λ_s = overall failure rate from the source up to the load point
r_s = average time between interruptions up to the load point, from the source
λ_{dg} = DG failure rate
r_{dg} = average time of a DG outage
λ_{sw} = rate of switch failure when shifting load to the DG
s = time for switching or service restoration with DG
U_{eq} = the load points' equivalent annual outage time when DGs are connected to the system.

The following restrictions are placed on the minimization of the objective function given in (13.8).

(i) Constraints on the decision variables

$$\lambda_{k,\min} \leq \lambda_k \leq \lambda_{k,\max} \tag{13.14}$$

$$r_{k,\min} \leq r_k \leq r_{k,\max} \quad k = 1, \ldots, N_c \tag{13.15}$$

(ii) Inequality constraints for the energy and customer-based reliability indices

$$\text{SAIFI} \leq \text{SAIFI}_t \tag{13.16}$$

$$\text{SAIDI} \leq \text{SAIDI}_t \tag{13.17}$$

$$\text{CAIDI} \leq \text{CAIDI}_t \tag{13.18}$$

$$\text{AENS} \leq \text{AENS}_t \tag{13.19}$$

The optimization procedure achieves the lowest possible failure rate and repair time for the kth sector. The reliability growth model (Ebeling 2019) is used to analyze failure and repair information as well as related expenses to create these lower limit values.

SAIFI_t, SAIDI_t, CAIDI_t, AENS_t, and SARFI_t are each of the indices' target/threshold values. They are determined by managerial or administrative decisions.

The FP optimization approach is used to solve the formulated problem (Yang 2012). Other optimization methods, such as teaching learning-based optimization (TLBO) (Rao et al. 2012) and differential evolution (DE) (Price et al. 2006), have confirmed this. The method described in this study was tested on the RBTS-2. The crew may be informed of the intended values to increase system reliability using the optimum values obtained after addressing the problem. They are determined by managerial/administrative decisions.

13.3.3 Analysis of Costs and Benefits

Cost–benefit analysis can be used to determine whether both the fixed and variable amounts put toward DGs integrated into the distribution system will generate a benefit in the long run. It supports positioning a particular number of DGs in the locations mentioned in the present research.

To evaluate the full costs and advantages of connecting a specified number of DGs at particular locations across the course of an economic cycle, the cumulative present value (CPV) technique (Raoofat and Malekpour 2011) is applied. The CPV approach, commonly referred to as the time value of money strategy, is based on the notion that someone who invests would prefer to get a specific amount of revenue now instead of the same quantity in the future. The costs and benefits of the plan are transferred in this manner to the very first year of its functioning. The concept is represented by the equations below:

$$\text{Benefit} = \text{CIC}_{\text{DGO}} - \text{CIC}_{\text{DG}} - \text{ADCOST}_{\text{DG}} - \text{CostM}_{\text{DG}} - \text{DGI} - \text{DGM} \tag{13.20}$$

where

$$\text{CIC}_{\text{DGO}} = \text{CIC}_{\text{DGO}} \times \text{CPV}_1 \tag{13.21}$$

CIC_{DGO} = Cost of customer interruption when DGs are not attached

$$\text{CIC}_{\text{DG}} = \text{CIC} \times \text{CPV}_1 \tag{13.22}$$

CIC_{DG} = Cost of customer interruption when DGs are attached

$$\text{ADCOST}_{\text{DG}} = \text{ADCOST} \times \text{CPV}_1 \tag{13.23}$$

$ADCOST_{DG}$ = Additional cost for DG operation in rupees per kWh

$$CostM_{DG} = CostM_{DG} \times CPV_2 \tag{13.24}$$

$CostM_{DG}$ = Cost of system maintenance needed to achieve the specified reliability

$$DGI = \sum_{i=1}^{n} DGI_i \times CPV_2 \tag{13.25}$$

DGI = Cost of installation for all connected DGs

$$DGM = \sum_{i=1}^{n} DGM_i \times CPV_2 \tag{13.26}$$

$$DGM = \text{The price of maintaining every DG linked to the system} \tag{13.27}$$

$$CPV_1 = \frac{(1-PV_1^{EL})}{(1-PV_1)} \tag{13.28}$$

$$PV_1 = \frac{(1 + I_{inf}) \times (1 + LG)}{(1 + I_{int})}$$

$$CPV_2 = \frac{(1-PV_2^{EL})}{(1-PV_2)} \tag{13.29}$$

$$PV_2 = \frac{(1+I_{inf})}{(1+I_{int})} \tag{13.30}$$

The economic lifespan of equipment is denoted by EL. LG stands for load growth rate. I_{inf} and I_{int} are the inflation and interest rates, respectively.

13.4 The Flower Pollination Algorithm Overview

Invented in 2012 by Xin-She Yang (2012), the FP algorithm was modeled after the flow pollination technique used by flowering. The process is governed by the following basic principles: (i) global pollination processes include biological pollination and cross-pollination; (ii) Lévy flight pattern is used by pollen carriers; (iii) pollinators, like insects, are capable of producing flower reliability, which is a measure of how closely related two flowers are, and serves as the probability of reproduction; (iv) pollination control is determined by a switching probability of [0, 1].

Because of closeness and other conditions, such as wind, local pollination accounts for a considerable fraction p of total pollination activity. The notations used to describe the FP are as follows: M, flower/pollen gamet population; D, various variables, k_{max}, maximum allowed number of generations; and p, switch probability of all pollination activities belongs to [0, 1].

Step A: Initialization: This generates a size "M" beginning population

$$S^0 = [U_1^0, U_2^0, \dots\dots, U_M^0] \tag{13.31}$$

$$U_i^0 = [U_{i1}^0, U_{i2}^0, \dots, U_{iD}^0]^T \tag{13.32}$$

where

M = number pollen gametes.

D = number of decision vector.

U_{ij}^0 = from a uniform distribution the jth parameter of the vector U_i results as follows:

$$U_{ij}^0 = U_{j,min} + (U_{j,max} - U_{j,min})\, rand_j \tag{13.33}$$

where
$U_{j,\max}$ and $U_{j,\min}$ = higher and lower limits on the variable U_j, respectively.
rand_j = random number between 0 and 1.

Step B: Vectors updated by both local and global pollinators

$$U_i^{(k+1)} = \begin{cases} U_i^k + \propto \times L\left(U_{\text{best}}^{(k)} - U_i^k\right), & \text{if rand} < p \\ U_i^{(k)} + \in \left(U_j^{(k)} - U_k^{(k)}\right), & \text{otherwise. where, } U_j^{(k)} \neq U_k^{(k)} \end{cases} \tag{13.34}$$

where
$U_{\text{best}}^{(k)}$ = current best option identified among all options.
$\propto$ = scaling factor used to manage step sizes smaller than zero.
L = Lévy distribution.
A Lévy flight can effectively reflect an insect's ability to move over longer distances with a variety of distance increments (Pavlyukevich 2007); (Yang 2010).

$$L = v \times \frac{\sigma_x(\beta)}{\sigma_y(\beta)} \tag{13.35}$$

$$v = \frac{\text{rand}_x}{\left|\text{rand}_y\right|^{1/\beta}} \tag{13.36}$$

where

$$\sigma_x(\beta) = \left[\frac{\Gamma(1+\beta) \times \left(\frac{\pi\beta}{2}\right)}{\Gamma\left(\frac{1+\beta}{2}\right) \times 2^{\left(\frac{\beta-1}{2}\right)}}\right]^{1/\beta} \tag{13.37}$$

$$\sigma_y(\beta) = 1 \tag{13.38}$$

rand_x and rand_y are two random variables that are normally distributed and have standard deviations of $\sigma_x(\beta)$ and $\sigma_y(\beta)$, β is the range of the distribution factor from 0.3 to 1.99, and $\Gamma(\cdot)$ is the function of the gamma distribution.

Step C: Assessment of the changed vectors' appropriateness in comparison to the original vectors

$$U_i^{(k+1)} = \begin{cases} U_i^{k+1}, & \text{if } f\left(U_i^{k+1}\right) < f\left(U_i^{(k)}\right) \\ U_i^{(k)}, & \text{otherwise} \end{cases} \tag{13.39}$$

The fitness of the determined best solution, $U_{\text{best}}^{(k)}$, is then found. Until the best solution is discovered or the predetermined number of iterations $(k_{\max})$ is reached, this procedure is repeated for each target vector index (i).

13.5 Solution Approach

These are the stages to complete the entire problem's solution.

13.5.1 Evaluation of the DG Locations (Steps)

Step 1: Data input λ_k, r_k, λ_{dg}, r_{dg}, λ_{sw}, s, C_{p_k}, N_i, L_i, SAIFI$_t$, SAIDI$_t$, CAIDI$_t$, and AENS$_t$.
Step 2: Failure rate and repair time should be evaluated.
Step 3: Determine ADCOST and CIC along with SAIDI, SAIFI, CAIDI, AENS, and EENS.
Step 4: Compute the value of J as per relation (13.7).
Step 5: Connect DG only at load point 1, then calculate the results for all indexes and the function J given by relation (13.7).
Step 6: At the subsequent load point, link the DG alone, and determine all the indexes and J as per (13.7). Similar to this, attach DG at all loads point one at a time and figure out what each index and J are worth.
Step 7: Considering the quantity of J in (13.7), calculate the improvement in reliability brought on by DG at each load point.
Step 8: Sort and order the increases in load location reliability (in relation to J as in (13.7)) in a decreasing sequence.
Step 9: Connect the DGs to the load locations of the first two rankings' numbers. For the DGs connected at these load sites, determine all the indices and the quantity of J.
Step 10: Pick the top three figures on the list. Obtain the system's indices in addition to J, then attach DGs to those load locations.
Step 11: By doing so, you can calculate every index and J, as well as increasing the ranking number and linking DGs at the appropriate load points.
Step 12: Repeat until you reach the desired degree of reliability.

13.5.2 Evaluating the Optimized Solution with the Help of Flower Pollination Method (Steps)

Step 1: Input data $\lambda_{k,\max}$, $r_{k,\max}$, $\lambda_{k,\min}$, $r_{k,\min}$, and cost of interruption $\left(C_{p_k}\right)$ and the target values of SAIFI, SAIDI, CAIDI, and AENS.
Step 2: By applying relation (13.33), make a set of variables of size "M" for every failure rate "λ" and repair time "r," where every vector of the pertinent population comprises the failure rate and maintenance time of every component. According to relations (13.14) and (13.15), these values are generated by picking uniformly among the smallest one and maximum boundaries.
Step 3: Compute $\lambda_{\text{sys},i}$, $r_{\text{sys},i}$, and $U_{\text{sys},i}$ at every load location.
Step 4: Compute customer and energy indexes as shown in the Eqs. (13.1)–(13.4), respectively for all the vectors of the whole population.
Step 5: Determine the value of the function F for each population vector i.e. $F\left(U_i^{(k)}\right)$, $i = 1, \ldots, M$, as mentioned in (13.8) and (13.9).
Step 6: For each vector in the population, evaluate the inequality constraints from the relations (13.16)–(13.19). Vectors that satisfy these requirements are feasible; otherwise, they are not feasible. Determine the optimal solution vector $U_{\text{best}}^{(k)}$ from the spectrum of feasible vectors depending on how much the objective function's value is.
Step 7: Set $k = 1$ as the generation counter.
Step 8: Select the vector with $i = 1$ as a target.
Step 9: Use relation (13.34), and find the latest value of the vector.
Step 10: Examine the fitness value of the modified vectors with the value of the initial vectors using relation (13.39), and keep the better ones.

Step 11: Repeat processes 3 through 6 for the revised vector.

Step 12: The target vector is increased to $i = i + 1$. Repeat from Step 9 if $i \leq M$; otherwise, raise the generation count to $k = k + 1$.

Step 13: Go to step 8 if the target optimal level is not determined or $k \leq k_{max}$.

13.5.3 Analyzing the Costs and Benefits

Use the relationships (13.20)–(13.30) to do a cost–benefit analysis for various planning horizons, such as 5, 10, and 15 years, and then make the necessary judgments.

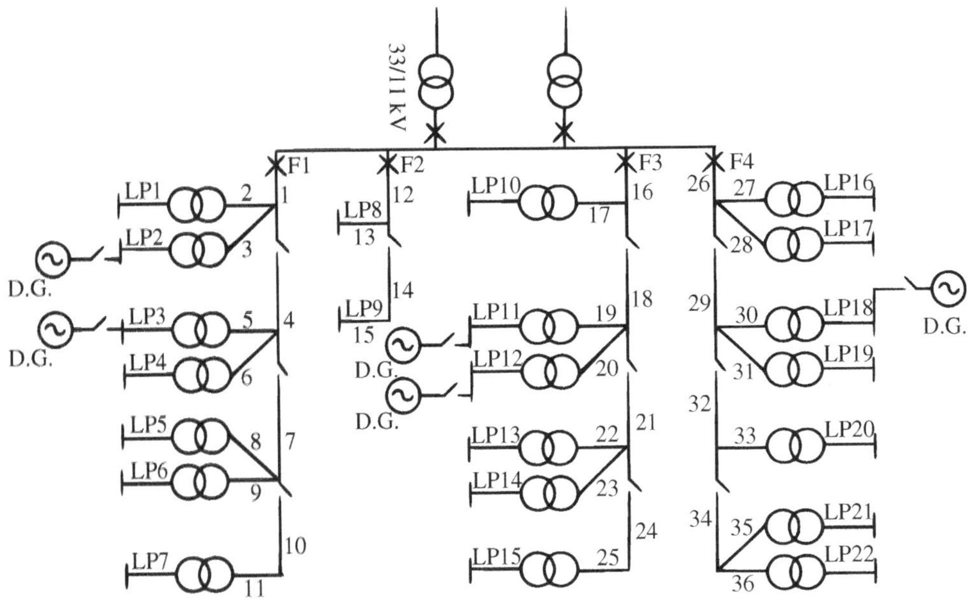

Figure 13.1 RBTS-2.

Table 13.1 Average repair times and failure rates for various RBTS-2 components.

Transformer/feeder sections/ breakers/busbars	Section number	Repair/ replacement time (h)	Rate of failure per year (λ)
Transformers	{1–22 (except 8,9)}	10	0.01500
Feeder elements/ sections (0.6 km)	{2,6,10,14,17,21,25,28,30,34}	5	0.06500
Feeder sections (0.75 km)	{1,4,7,9,12,16,19,22,24,27,29,32,35}	5	0.06500
Feeder elements/ sections (0.80 km)	{3,5,8,11,13,15,18,20,23,26,31,33,36}	5	0.06500

Table 13.2 RBTS-2 system information.

Distribution/ segment	λ_k^0(yr)	$\lambda_{k,\min}$ (yr)	r_k^0 (h)	$r_{k,\min}$ (h)	α_k (Rs.)	β_k(Rs. × 10^2)
#1	0.048750	0.036650	5.000	2.252252	2.564	22.649
#2	0.015000	0.011270	10.000	4.504504	3.205	17.589
#3	0.052000	0.039090	10.000	4.504504	1.923	4.291
#4	0.048000	0.036090	5.000	2.252252	1.282	4.291
#5	0.015000	0.011270	10.000	4.504504	2.564	13.258
#6	0.015000	0.011270	10.000	4.504504	3.045	6.647
#7	0.048750	0.036650	5.000	2.252252	3.205	5.394
#8	0.015000	0.011278	10.000	4.504504	2.564	22.649
#9	0.015000	0.011278	10.000	4.504504	3.205	17.589
#10	0.039000	0.029323	5.000	2.252252	1.923	4.291
#11	0.015000	0.011278	10.000	4.504504	1.282	4.291
#12	0.048750	0.036650	5.000	2.252250	2.564	22.649
#13	0.052000	0.039097	5.000	2.252250	3.205	17.589
#14	0.039000	0.029323	5.000	2.252250	2.564	22.649
#15	0.052000	0.039097	5.000	2.252250	3.205	17.589
#16	0.048750	0.036650	5.000	2.252250	2.564	22.649
#17	0.015000	0.011278	10.000	4.504500	3.205	17.589
#18	0.052000	0.039097	5.000	2.252250	1.923	4.291
#19	0.015000	0.011278	10.000	4.504500	1.282	4.291
#20	0.015000	0.011278	10.000	4.504500	2.564	13.258
#21	0.039000	0.029323	5.000	2.252250	3.045	6.647
#22	0.015000	0.011278	10.000	4.504500	2.564	22.649
#23	0.015000	0.011278	10.000	4.504500	3.205	17.589
#24	0.048750	0.036650	5.000	2.252250	1.923	4.291
#25	0.015000	0.011278	10.000	4.504500	1.282	4.291
#26	0.052000	0.039097	5.000	2.252250	2.564	22.649
#27	0.015000	0.011278	10.000	4.504500	3.205	17.589
#28	0.015000	0.011278	10.000	4.504500	1.923	4.291
#29	0.048750	0.036650	5.000	2.252250	1.282	4.291
#30	0.015000	0.011278	10.000	4.504500	2.564	13.258
#31	0.015000	0.011278	10.000	4.504500	3.045	6.647
#32	0.048750	0.036650	5.000	2.252250	3.205	5.394
#33	0.015000	0.011278	10.000	4.504500	2.564	22.649
#34	0.039000	0.029323	5.000	2.252250	3.205	17.589
#35	0.015000	0.011278	10.000	4.504500	1.923	4.291
#36	0.015000	0.011278	10.000	4.504500	1.282	4.291

13.6 Discussions and Outcomes

The methods described in this research was applied on RBTS-2. The FP algorithm was used to solve the problem, and the results were compared to those obtained by TLBO and DE. MATLAB R2019b software, an Intel (R), Core (TM), i5-4590S CPU running at 3.00 GHz and 4.00 GB RAM were used to run the proposed algorithm.

13.6.1 Roy Billinton Test System (RBTS-2)

The test system used in this study is the RBTS-2 (Allan et al. 1991), as shown in Figure 13.1. The average repair times and failure rates for different RBTS-2 components are shown in Table 13.1. Table 13.2 lists the current and lowest reachable rate of failure, current and lowest reachable average repair times, and cost coefficients for the rate of failure and time to repair the different RBTS-2 segments. Table 13.3 gives various customer information for RBTS-2, like the average load at different load points, the interruption costs there, etc. Table 13.4 gives information about energy-based and customer-based reliability indices, interruption costs for customers, additional costs on DGs, and the value of the function J when a single DG is connected to various load points one by one. The connected DGs are treated as backup units in this situation. The average downtime and failure rate for the DG used in this study are 13.25 hours and 0.5 failures per year, respectively. The DG changeover switch's rate of failure and the restoration of service time are 0.1 failures per year and 0.5 hours, respectively. According to the improvement in distribution system reliability, the function values J reported in Table 13.4 are arranged in Table 13.5 in the order of decrement. The rating of reliability with several DGs connected at various load sites is displayed in Table 13.6. The places of DGs are identified in Table 13.6. In this study, the function F in relation (13.8) is optimized using FP, DE, and TLBO. DGs are linked at load points 2, 3, 11, 12, and 18. Table 13.7 lists the control settings for the three optimization techniques. Table 13.8 lists the optimized values of primary reliability indices for the various RBTS-2 parts using all the optimization techniques used in this study. In Table 13.9, the optimized values for the RBTS-2's cost of maintenance, consumer interruption cost, extra expenses for buying electricity from the associated generators, and function (F) are presented. These values were obtained using the three optimization techniques. Table 13.10 shows the optimized quantities of the energy- and customer-based reliability indices created with each of the three methodologies. An evaluation of costs and benefits has been done based on Section 13.3.3. The cost of installing and maintaining each connected DG in the distribution system is estimated

Table 13.3 Customer information for the Roy Billinton test system-2.

Load point number	Load average at every load point in MW	Type of customers/ premise	Size of customers in numbers	Interruption cost (C_{p_k}) (Rs./kW)
1–3,10,11	0.5350	Residential purpose	210.0	22.290
12,17–19	0.4500	Residential purpose	200.0	6.0450
8	1.0000	Small scale user	1.0	251.420
9	1.1500	Small scale user	1.0	251.420
4,5,13,14,20,21	0.5660	Government institutes	1.0	23.980
6,7,15,16,22	0.4540	Commercial purpose	10.0	192.980

Table 13.4 The parameter values for RBTS-2's various load points with and without DG connections.

	Load points	SAIFI	SAIDI	CAIDI	AENS	EENS	Extra expenditure incurred on DG (rupees)	CIC (rupees)	Function (J) as per (13.7)
Without DG		0.09860	0.58820	5.96660	4.66410	8899.00	—	828,680.00	5.663
	1	0.1025760	0.5503740	5.3655230	4.5677580	8715.28200	367.49390	824,578.400	5.4326470
	2	0.0985090	0.5096820	5.1739540	4.4640890	8517.48200	763.09450	820,168.400	5.2011770
	3	0.0972990	0.5239790	5.3852710	4.5005130	8586.97900	624.09970	821,717.900	5.2878150
	4	0.0985720	0.5878700	5.9638380	4.4910360	8568.89800	660.26240	820,756.800	5.6038370
	5	0.0985470	0.5877420	5.9640870	4.4187840	8431.03900	935.97880	817,450.300	5.5791990
	6	0.0982620	0.5838420	5.9416680	4.4673190	8523.64500	750.76740	756,231.600	5.5767130
	7	0.0980580	0.5828210	5.9436260	4.4209550	8435.18200	927.69350	739,159.600	5.5568200
	8	0.0985780	0.5879380	5.9641800	4.4264460	8445.66000	906.73800	714,689.000	5.5826290
	9	0.0985580	0.5878360	5.9643800	4.2733290	8153.51200	1491.03400	641,237.100	5.5311470
With DG (at load points)	10	0.1025750	0.5503710	5.3655320	4.5677490	8715.26500	367.52650	827,564.200	5.4326280
	11	0.0968570	0.5217740	5.3870350	4.4948950	8576.26000	645.53770	826,723.900	5.2757400
	12	0.0969390	0.5249360	5.4150990	4.5217720	8627.54100	542.97560	827,033.900	5.3005720
	13	0.0985500	0.5877570	5.9640580	4.4272860	8447.26200	903.53280	817,839.400	5.5820980
	14	0.0985500	0.5877570	5.9640580	4.4272860	8447.26200	903.53280	817,839.400	5.5820980
	15	0.0980370	0.5827150	5.9438290	4.4161790	8426.07000	945.91660	863,237.800	5.5547720
	16	0.0987520	0.5862900	5.9370070	4.5784670	8735.71500	326.62730	795,406.900	5.6244090
	17	0.1020450	0.5504690	5.3943950	4.5792210	8737.15400	323.74950	825,945.500	5.4377430
	18	0.0969390	0.5249360	5.4150990	4.5217720	8627.54100	542.97560	825,282.900	5.3005720
	19	0.0969390	0.5249360	5.4150990	4.5217720	8627.54100	542.97560	826,898.800	5.3005720
	20	0.0985450	0.5877320	5.9641080	4.4128350	8419.68900	958.68040	815,427.100	5.5771700
	21	0.0985240	0.5876290	5.9643070	4.3550280	8309.39300	1179.27100	812,781.600	5.5574580
	22	0.0980370	0.5827150	5.9438290	4.4161790	8426.07000	945.91660	735,650.200	5.5547720

Table 13.5 The load points are shown in descending order of reliability enhancement, from largest to smallest.

Load point (LP) ranking based on reliability improvement when linked to DG	Reliability increment in (%) (in terms of *J*)
2	8.8792020
11	7.3403990
3	7.0952830
12	6.8375170
18	6.8375170
19	6.8375170
10	4.2405200
1	4.2401700
17	4.1424800
9	2.3838250
15	1.9483860
22	1.9483860
7	1.9108000
21	1.8990990
6	1.5472740
20	1.5389500
5	1.5020310
13	1.4493110
14	1.4493110
8	1.4396570
4	1.0557670
16	0.6861260

Table 13.6 Based on load point rating, reliability found for the system with multiple connected generators.

DGs linked at load points $LP)	SAIFI	SAIDI	CAIDI	AENS	EENS	Extra expenditure incurred on DG (rupees)	CIC (rupees)	Value of function (*J*) as per (7)
2,11	0.09678	0.44328	4.579908	4.294923	8194.713	1408.63	818,217.3	4.810259
2,3,11	0.09550	0.379084	3.969128	4.131375	7882.663	2032.73	811,260.1	4.427674
2,3,11,12	0.093869	0.315844	3.364737	3.989085	7611.175	2575.70	809,619	4.051987
2,3,11,12,18	0.09223	0.252604	2.738863	3.846796	7339.687	3118.68	807,977.9	3.670929
2,3,11,12,18,19	0.09059	0.189365	2.090338	3.704507	7068.199	3661.65	806,336.7	3.284208

Table 13.7 FP, TLBO, and DE control parameters for Roy Billinton test system, bus-2.

Parameter description	Parameter values
Size of population for TLBO, DE, and FP	40.0
Maximum predefined generation (k_{max}) for TLBO, DE, and FP	1000.0
Updated size of step (α) for FP	0.010
Factor of distribution for (β) FP	1.50
Switching probability for FP	0.80
Size of step for DE	0.80
Cross-over rate for DE	0.70

at 25 rupees per kW and 5.94 rupees per kW, respectively. The assumed rates for load growth, interest, and inflation are 5%, 10%, and 5%, respectively. By using all three optimization techniques, calculations for economical life span estimates of 5, 10, and 15 years have been made.

13.6.2 Comparison Research

To validate the findings acquired by FP, a comparison of the FP results with those obtained by the other two optimization techniques, TLBO and DE, was conducted. All of the strategies yield similar values for the objective function (F), but FP yields better results. The reliability indices based on customers and energy that were obtained using the three approaches are also compared. The results for FP have been proven to be superior to those of the other two approaches. The benefit gained by FP is greatest for the economic lifecycles taken into account in this work in the cost-benefit analysis as well (Table 13.11).

13.7 Conclusion

In this study, the connection of DGs boosted the reliability of a distribution system (RBTS-2). The locations of DGs have been determined using a technique that prioritizes reliability enhancement. With these locations now identified, three optimization techniques—FP, TLBO, and DE—have been used to optimize the objective function stated in this research. Of the three approaches, the FP has been found to be the best. The installation of DGs has been supported by a long-term analysis of costs and benefits, which also shows that the benefits of the values generated by FP are greater.

Table 13.8 RBTS-2's optimized rates of failure and time for repair as determined by the three optimization methodologies.

Distributor element	Rate of failure per year			Time for repair in hours		
	By DE	By TLBO	By FP	By DE	By TLBO	By FP
1	0.039839	0.036650	0.036653	2.254974	2.252253	2.252322
2	0.014873	0.015000	0.014995	4.568424	4.504504	4.504545
3	0.047409	0.052000	0.039095	4.647431	4.504504	4.508767
4	0.036107	0.036090	0.036091	2.297369	2.252256	2.252605
5	0.014789	0.015000	0.014999	7.138270	10.000000	9.997818
6	0.014874	0.015000	0.015000	8.800289	4.504504	4.513614
7	0.037194	0.036650	0.036654	2.270862	2.252252	2.252840
8	0.014694	0.015000	0.014999	4.890354	4.504542	4.516031
9	0.014840	0.015000	0.014999	4.515139	4.504507	4.505314
10	0.029675	0.029323	0.038991	2.252385	2.252252	2.252424
11	0.014999	0.015000	0.014982	4.551617	4.504504	4.505028
12	0.036650	0.036650	0.036650	2.260009	2.252250	2.252534
13	0.039132	0.039097	0.039116	2.253733	2.252250	2.257926
14	0.029329	0.029323	0.038999	2.433488	2.252250	2.255338
15	0.039671	0.039097	0.039136	2.326117	2.252250	2.252986
16	0.037130	0.036650	0.036651	2.507309	2.252250	2.268080
17	0.014994	0.015000	0.014999	5.097232	4.504500	9.987020
18	0.039415	0.039097	0.039100	2.359125	2.252250	2.262472
19	0.014222	0.015000	0.015000	5.019546	4.504500	4.510011
20	0.014976	0.015000	0.014984	9.993431	4.504500	4.504950
21	0.031561	0.029323	0.029323	2.256693	2.252250	2.252478
22	0.014541	0.015000	0.015000	5.578698	4.504500	4.508283
23	0.014421	0.015000	0.014999	4.632058	4.504500	4.504863
24	0.037447	0.036650	0.048749	2.557199	2.252276	2.256059
25	0.014903	0.015000	0.015000	4.542109	4.504500	4.506840
26	0.040029	0.039097	0.039156	2.271917	2.252250	2.252871
27	0.014648	0.015000	0.014996	4.681604	4.504500	4.505772
28	0.014113	0.015000	0.014977	8.478175	4.504500	9.982366
29	0.036795	0.036650	0.048745	2.253501	2.252250	2.253396
30	0.014993	0.015000	0.014980	9.423377	4.504500	4.505367
31	0.014868	0.015000	0.015000	4.571811	4.504500	4.538311
32	0.041654	0.036650	0.036651	2.301898	2.252250	2.252895
33	0.014998	0.015000	0.014999	9.948056	4.504500	9.979963
34	0.029505	0.030576	0.029326	2.945996	2.252250	2.252948
35	0.014890	0.015000	0.014999	4.663455	4.504500	9.938406
36	0.014664	0.015000	0.014999	5.889663	4.504500	4.505902

Table 13.9 Current and optimized objective function (J) quantities for RBTS-2 calculated by DE, TLBO, and FP.

	Current values in rupees	Values after optimization in rupees		
		DE	TLBO	FP
Cost of maintenance ($\sum_{k=1}^{N_c} \alpha_k/\lambda_k^2 + \sum_{k=1}^{N_c} \beta_k/r_k$)	232,680	258,719.8	251,732.1	254,692.6
Customer interruption cost ($\sum_{k=1}^{N_c} CIC$)	828,680	306,879.7	300,608.9	284,052.8
ADCOST (extra expenditures when generators are linked)	—	11,355.11	4230.645	12,398.98
Objective function (J)	1,064,500	576,954.6	556,571.6	551,144.3

Table 13.10 RBTS-2 reliability indices, both current and optimized.

Indices	Current values	Optimized values			Threshold/ target values
		FP	TLBO	DE	
SAIFI	0.098600	0.082604	0.085918	0.084452	0.0860
SAIDI	0.588200	0.174741	0.132302	0.111135	0.40
CAIDI	5.966600	2.163416	1.586033	1.301081	4.00
AENS	4.664100	1.688404	1.560067	1.414852	2.20

Table 13.11 Analysis of costs and benefits.

Economic lifecycle management in years	Benefit in rupees Method of optimization		
	DE	TLBO	FP
5.0	972,060.00	1,173,100.00	1,128,600.00
10.0	2,280,500.00	2,540,400.00	2,658,400.00
15.0	3,895,700.00	4,137,300.00	4,438,500.00

References

Abbasi, F. and Hosseini, S.M. (2016). Optimal DG allocation and sizing in presence of storage systems considering network configuration effects in distribution systems. *IET Generation, Transmission & Distribution* 10 (3): 617–624.

Agajie, T.F., Khan, B., Guerrero, J.M., and Mahela, O.p (2021). Reliability enhancement and voltage profile improvement of distribution network using optimal capacity allocation and placement of distributed energy resources. *Computers and Electrical Engineering* 93: 107295.

Ahmad, S. and Asar, A.u. (2021). Reliability enhancement of electric distribution network using optimal placement of distributed generation. *Sustainability* 13 (20): 11407.

Allan, R.N., Billinton, R., Sjarief, I. et al. (1991). A reliability test system for educational purposes-basic distribution system data and results. *IEEE Transactions on Power systems* 6 (2): 813–820.

Al-Muhaini, M. and Heydt, G.T. (2013). Evaluating future power distribution system reliability including distributed generation. *IEEE Transactions on Power Delivery* 28 (4): 2264–2272.

Arya, R. (2016). Determination of customer perceived reliability indices for active distribution systems including omission of tolerable interruption durations employing bootstrapping. *IET Generation, Transmission & Distribution* 10 (15): 3795–3802.

Arya, R., Choube, S.C., and Arya, L.D. (2012). Reliability evaluation and enhancement of distribution systems in the presence of distributed generation based on standby mode. *International Journal of Electrical Power and Energy Systems* 43 (1): 607–616.

Awad, A.S.A., El-Fouly, T.H.M., and Salama, M.M.A. (2014). Optimal distributed generation allocation and load shedding for improving distribution system reliability. *Electric Power Components and Systems* 42 (6): 576–584.

Bagheri, A., Monsef, H., and Lesani, H. (2015). Integrated distribution network expansion planning incorporating distributed generation considering uncertainties, reliability, and operational conditions. *International Journal of Electrical Power and Energy Systems* 73: 56–70.

Battu, N.R., Abhyankar, A.R., and Senroy, N. (2015). Dg planning with amalgamation of economic and reliability considerations. *International Journal of Electrical Power and Energy Systems* 73: 273–282.

Billinton, R. and Allan, R.N. (2013). *Reliability Evaluation of Power Systems*. Springer Science and Business Media.

Byk, F., Kakosha, Y., and Myshkina, L. (2020). Distributed power generation and power supply reliability improvement. In: *E3S Web of Conferences*, vol. 216, 01013. EDP Sciences.

Chowdhury, A. and Koval, D. (2011). *Power Distribution System Reliability: Practical Methods and Applications*. John Wiley and Sons.

Ebeling, C.E. (2019). *An Introduction to Reliability and Maintainability Engineering*. Waveland Press.

Idowu, K., Uhunmwangho, R., Okafor, E., and Big-Alabo, A. (2021). Reliability improvement study of a distribution network with distributed generation. *Applications of Modelling and Simulation* 5: 53–65.

Kansal, S., Tyagi, B., and Kumar, V. (2017). Cost–benefit analysis for optimal distributed generation placement in distribution systems. *International Journal of Ambient Energy* 38 (1): 45–54.

Kela, K.B. and Arya, L.D. (2014). Reliability optimization of radial distribution systems employing differential evolution and bare bones particle swarm optimization. *Journal of The Institution of Engineers (India): Series B* 95 (3): 231–239.

Kela, K.B., Suthar, B.N., and Arya, L.D. (2019). Reliability optimization of electrical distribution systems considering expenditures on maintenance and customer interruptions. *Indonesian Journal of Electrical Engineering and Computer Science* 14 (3): 1057–1064.

Kumar, D., Samantaray, S.R., Kamwa, I., and Sahoo, N.C. (2014). Reliability-constrained based optimal placement and sizing of multiple distributed generators in power distribution network using cat swarm optimization. *Electric Power Components and Systems* 42 (2): 149–164.

Kumar, S., Saket, R.K., Dheer, D.K. et al. (2020). Reliability enhancement of electrical power system including impacts of renewable energy sources: a comprehensive review. *IET Generation, Transmission & Distribution* 14 (10): 1799–1815.

Lekbich, A., Belfqih, A., Ouaderhman, T. et al. (2019). An analytical multicriteria model based on graph theory for reliability enhancement in distribution electrical networks. *International Journal of Electrical and Computer Engineering* 9 (6): 4625.

Mahboob, A., Mehdi, H., Noor, A., and Khan, M.A. (2021). Reliability based optimal distributed generation placement for a radial distribution network. In: *2021 International Conference on Engineering and Emerging Technologies (ICEET)*, 1–6. IEEE.

Mahmood, M.K., Uçan, O.N., Zaidan, Z., and Karim, S.M. (2020). Hybrid algorithm for two-terminal reliability evaluation in communication networks.

Nasser Hussain, A., Abed, W.R., and Yaqoob, M.M. (2023). Dual strategy of reconfiguration with capacitor placement for improvement reliability and power quality in distribution system. *Journal of Engineering* 2023: 1943535.

Okoye, M.O., Yang, J., Lei, Z. et al. (2020). Predictive reliability assessment of generation system. *Energies* 13 (17): 4350.

Pattabiraman, S., Sampath, K., Kannan, M. et al. (2019). Reliability improvement in radial distribution system through reconfiguration. In: *2019 IEEE Innovative Smart Grid Technologies-Asia (ISGT Asia)*, 1963–1967. IEEE.

Pavlyukevich, I. (2007). Lévy flights, non-local search and simulated annealing. *Journal of Computational Physics* 226 (2): 1830–1844.

Price, K., Storn, R.M., and Lampinen, J.A. (2006). *Differential Evolution: A Practical Approach to Global Optimization*. Springer Science and Business Media.

Raghuwansh, S.S. and Arya, R. (2019). Economic and reliability evaluation of hybrid photovoltaic energy systems for rural electrification. *International Journal of Renewable Energy Research (IJRER)* 9 (1): 515–524.

Rahiminejad, A., Hosseinian, S.H., Vahidi, B., and Shahrooyan, S. (2016). Simultaneous distributed generation placement, capacitor placement, and reconfiguration using a modified teaching-learning-based optimization algorithm. *Electric Power Components and Systems* 44 (14): 1631–1644.

Rao, R.V., Savsani, V.J., and Vakharia, D.P. (2012). Teaching–learning-based optimization: an optimization method for continuous non-linear large scale problems. *Information Sciences* 183 (1): 1–15.

Raoofat, M. and Malekpour, A.R. (2011). Optimal allocation of distributed generations and remote controllable switches to improve the network performance considering operation strategy of distributed generations. *Electric Power Components and Systems* 39 (16): 1809–1827.

Ray, S., Bhattacharya, A., and Bhattacharjee, S. (2015). Differential search algorithm for reliability enhancement of radial distribution system. *Electric Power Components and Systems* 44: 1–14. https://doi.org/10.1080/15325008.2015.1091864.

Su, S., Hu, Y., He, L. et al. (2019). An assessment procedure of distribution network reliability considering photovoltaic power integration. *IEEE Access* 7: 60171–60185.

Wang, Y. (2017). Analytical complex distribution system reliability evaluation considering stochastic interruption durations and network reconfigurations. *Electric Power Components and Systems* 45 (19): 2151–2163.

Xu, W., Zeng, S., Du, X. et al. (2023). Reliability of active distribution network considering uncertainty of distribution generation and load. *Electronics* 12 (6): 1363.

Yang, X.-S. (2010). *Nature-Inspired Metaheuristic Algorithms*. Luniver Press.

Yang, X.-S. (2012). Flower pollination algorithm for global optimization. In: *International Conference on Unconventional Computing and Natural Computation*, 240–249. Springer.

Yousefian, R. and Monsef, H. (2011). DG-allocation based on reliability indices by means of Monte Carlo simulation and AHP. In: *2011 10th International Conference on Environment and Electrical Engineering*, 1–4. IEEE.

14

Reliability Enhancement Strategy for Electrical Distribution Systems Considering Reward and Penalty

Kalpesh B. Kela[1], Bhavik N. Suthar[2], Liladhar Arya[3], and Rajesh Arya[4]

[1] *Department of Electrical Engineering, L.D. College of Engineering, Ahmedabad, Gujarat, India*
[2] *Department of Electrical Engineering, Government Engineering College, Bhuj, Gujarat, India*
[3] *Department of Electrical Engineering, Shri G.S. Institute of Technology & Science, Indore, Madhya Pradesh, India*
[4] *Department of Information Technology, Acropolis Institute of Technology and Research, Indore, Madhya Pradesh, India*

14.1 Introduction

In order for distribution companies to function as efficiently as possible, several countries worldwide are gradually transitioning away from traditional price-based (on the rate of return or the cost of the service) management and toward performance-oriented management/regulations. In performance-based rules, incentives are set by the companies in order to encourage high efficiency (in terms of profit). The standard of services offered to customers may suffer as a result (Fumagalli et al. 2007). Many performance-oriented rules are interwoven with quality regulations in order to satisfy these requirements, adopting either direct or indirect quality assurance measures (CEER 2011). Under indirect quality control, customers are informed about the performance quality of a distribution company, whereas under direct quality control, the output is evaluated in accordance with the financial rewards provided by regulatory bodies to ensure proper service quality (Alvehag and Awodele 2013). Three criteria are used to characterize the services quality provide by the distribution company: voltage quality, commercial quality, and reliability of operations or continuation of supply (Fumagalli et al. 2007). The regulations governing service reliability are the main topic of this chapter. Various measures put in place by regulators have regulated service reliability. For regulatory purposes in distribution networks, the rewards and penalties scheme (RPS), a direct control of quality mechanism, is a successful and popular solution. Service reliability indices like System Average Interruption Frequency Index (SAIFI), System Average Interruption Duration Index (SAIDI), and Average Energy Not Supplied (AENS) are regulated by the regulator using this method. Based on the attainment of these indices within their intended levels, a reward or penalty is agreed upon for the distribution company.

For distribution companies to keep or raise their level of quality, RPS creates financial incentives. In order to reduce the overall reliability cost to society using RPS, the regulator works to maintain a level of reliability that is socioeconomically optimal (Alvehag and Awodele 2013). In Billinton and Pan (2004), the idea of employing indices of reliability in RPS was covered. Fotuhi et al. (2006) presents various reward-penalty models with their applications and features with real system reliability information gathered by Tehran Regional Electrical Company. RPS coupled with clustering analysis is a method for increasing the operational reliability of distribution networks that

Reliability Analysis of Modern Power Systems, First Edition. Edited by R. K. Saket and P. Sanjeevikumar.

was first introduced in Mohammad Nezhad-Shourkaei et al. (2011). Here, the utilities have been categorized, and the effectiveness of the utility in one group has been evaluated in comparison to that of the other cluster members. The key conditions for developing and putting into practice a successful RPS have been examined in Mohammad Nezhad-Shourkaei and Fotuhi-Firuzabad (2011). Here, various RPS varieties have been in-depth detailed. By combining to calculate the RPS parameters for each electric business, a method is presented in Simab et al. (2012) that combines data envelopments analysis (DEA) with fuzzy C-mean grouping (FCM) based on the average system interruption time index. The essential elements of RPSs across the electric business were determined using an algorithm reported in Simab and Haghifam (2012) employing DEA that is based on reliability indices and consumers' perceived value of interruptions. Based on the idea of yardstick theory, a strategy for developing processes for reward or penalty methods has been put forward in Jooshaki et al. (2014). A new incentive program for supply continuity that was adopted in Sweden from 2016 to 2019 has been addressed in Grahn et al. (2016), where a reward or penalty is based on the DSO's performance and the financial benefit provided to the DSO. Jooshaki et al. (2017) focuses on a regulatory framework based on RPS for the construction of a distribution system, taking into account the incorporation of distributed generators. The clustering using fuzzy c-means and the analytical hierarchy process (AHP) is used in Ghasemi and Dashti (2017) to offer a model for determining the variables of the rewards and penalty program in performance-oriented management for all electricity distribution firms. Sumesh et al. (2018) has put out a performance-oriented penalty function approach that significantly affects the energy producers' state of mind to act dynamically to commit profit, resulting in predictable energy-sharing actions. A novel approach has been put forth in Wang et al. (2018) to assess load-based reliability indices in electrical distribution networks utilizing data from sophisticated metering equipment. With the aid of the recently adopted rewards and penalties regulatory framework for utility regulators, they are employed to offer reliability oversight. Jooshaki et al. (2018) discusses a novel fuzzy-based design procedure to apply reward-penalty systems in the distribution sector more successfully. A model for making decisions to ensure that the RPS for energy distribution companies (EDCs) is properly implemented is a fresh fuzzy-based design approach for the distribution sector's implementation of reward-penalty systems that work better. For EDCs, a decision-making model for the effective implementation of the RPS is described in Ghasemi and Dashti (2018). During each regulatory term, this model might modify the RPS's parameters, improving the program's efficiency. A Cubic Reward-Penalty structure is suggested in Sumesh et al. (2019) to increase the service reliability of electricity distribution firms. The employment of a reward and penalty structure that fluctuates in a cubic pattern based on performance based on reliability in the scheme's innovative characteristic. A method that utilizes relative performance to improve reliability regulation is put out in Khonakdar-Tarsi et al. (2021). Here, the feeders of the distribution system are grouped according to factors that the utilities are not in control of the reward-penalty system, which rewards feeders who perform better than the standard and penalizes feeders whose performance falls below the benchmark, is the foundation upon which the quality regulation is designed. In Janjic et al. (2021), a brand-new approach to RPS design is put out that depends on the spatial distribution of reliability performance. With this approach, the analysis provides a set of reward and penalty criteria that are more visible. An incentive-based demand response program is suggested in Ghorashi et al. (2020). Customers who minimize or reschedule their needs during the desired hours are rewarded under the initiative. If the clients' demands go above certain limits, they could also face penalties. Lu et al. (2021) explores an inventive reward-penalty mechanism created expressly to encourage net zero energy structures and communities. In order to apply the suggested reward-penalty system, two distinct strategies, the building level-based approach and the strategy based on community level, have been created. A paradigm to enhance the asset management techniques for public lighting is provided in Ghotboddini and Toossian Shandiz (2022). The reward and penalty mechanisms in the regard use the guaranteed duration and high correction time. In Amirrezai et al. (2023), the impact of distributing generations on the distribution system power factor has been studied. By improving the distribution system power factor, the penalty costs to the distribution companies have been eliminated.

By integrating rewards and penalties applied to them by the regulator, the topic of this chapter is improving distribution systems' reliability. In this case, the reliability indices' desired values can be attained using the optimal values of reward and penalty. A distribution company's financial risk and unneeded tariff changes for customers can be reduced by determining the best values for both rewards and penalties.

Briefly describing the RPS theory is Section 14.2. Section 14.3 provides a description of the issue this study aims to address. Rao algorithms are briefly described in Section 14.4. The methods used by Rao algorithms to solve the problem are discussed in Section 14.5. The results for the two sample systems are discussed in Section 14.6. In Section 14.7, conclusions are presented.

14.2 Reward and Penalty System (RPS)

To assess the system's reliability, the regulator uses system quality indicators in RPS. The utilities must comply with set targets for these initiatives. The utility will receive a reward or a penalty depending on whether or not it meets these objectives. The utility benefits increase with reliability level. Regulators thus attempt to imitate market-like conditions. The average SAIFI and SAIDI, both from the perspective of reliability, are frequently used quality indicators. Both indexes are customer-based (Billinton and Allan 1996). It is also possible to integrate these indices with energy-based reliability indicators like Expected Energy Not Supplied (EENS) (Billinton and Allan 1996).

Through RPS, the regulator aims to attain a reliability level that is socioeconomically optimal (Fumagalli et al. 2007). Network costs and customer interruption costs are related to socioeconomic reliability. High customer interruption costs result from poor reliability. Higher system reliability lowers interruption expenses for the users, but it also necessitates higher capital and maintenance costs, which are eventually covered by the users through tariffs. The level of reliability that minimizes the overall cost to society of reliability for networks and customers is known as the socioeconomic optimum reliability level, and it lies somewhere in the middle (Billinton and Li 1994).The cost-versus-reliability graph in Figure 14.1 represents the level of reliability that is socioeconomically optimal. When the rate of rise in network expenses is equal to the rate of fall in costs associated with customer disruption, the optimal level of reliability R_{opt} is attained:

$$\left.\frac{\partial C_{\text{network}}}{\partial R}\right|_{R=R_{\text{opt}}} = -\left.\frac{\partial \text{CIC}}{\partial R}\right|_{R=R_{\text{opt}}} \tag{14.1}$$

The socioeconomic optimal may or may not coincide with the utility's optimal because the utility aims to set the optimal at its minimum cost. By integrating the costs of customer interruptions into their own cost functions, optimal RPS provides incentives for achieving socioeconomic reliability (Sappington 2005). In reality, the regulator is unaware of the socioeconomic reliability level; instead, a threshold or target value is established for the level of quality index that assesses reliability. The utility sets the cost of customer interruption owing to a divergence in reliability from this target value in terms of either a reward or a penalty. When a utility achieves or falls short of its

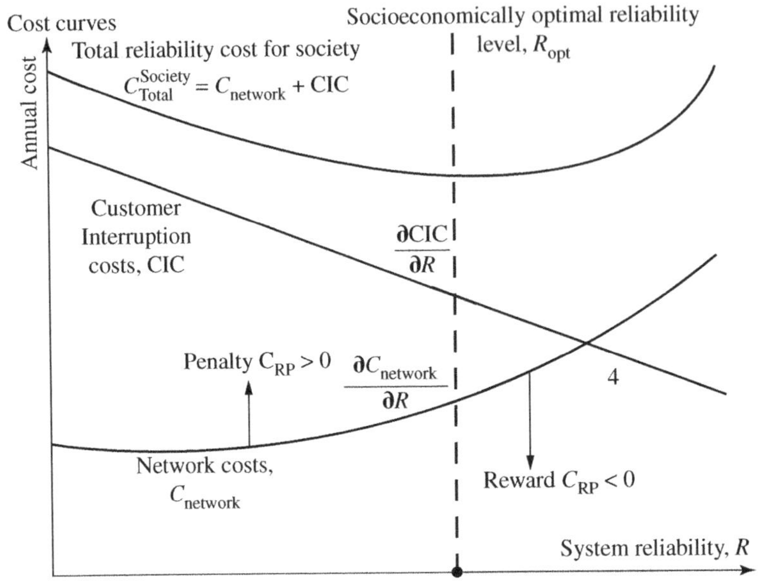

Figure 14.1 The cost versus reliability graph shows the amount of reliability that is socially and economically optimal.

target reliability level, it bears financial risk, which results in either a profit or a loss for the utility (cost of reward or penalty,C_{RP}). Here, the utility is being rewarded when $C_{RP} < 0$ and being penalized when $C_{RP} > 0$. The ultimate objective is to achieve the socioeconomic reliability level R_{opt} at the minimal value of the utility's total reliability cost curve while taking the impact of RPS into account. Figure 14.2 (Alvehag and Awodele 2013; Ajodhia 2006) depicts the various RPS schemes. This article is primarily concerned with a continuous scheme in which incentives and penalties rise as deviations from the predetermined target level do.

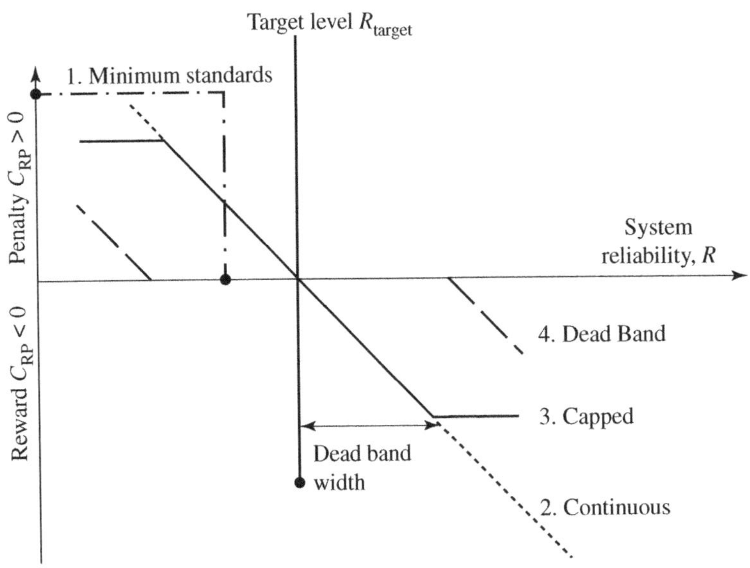

Figure 14.2 Different RPS configurations.

14.3 Problem Identification

The objective function is stated as follows with the intention of improving reliability by determining the optimal reward and penalty values:

$$F = \sum_{k=1}^{N_c} \alpha_k / \lambda_k^2 + \sum_{k=1}^{N_c} \beta_k / r_k + \text{ADCOST}(\text{EENSO} - \text{EENSD}) + \sum_{k=1}^{N_c} \text{CIC} + \sum_{k=1}^{N_c} C_{\text{RP}} \tag{14.2}$$

where

$$\lambda_k \times r_k \times L_i \times C_{\text{p}k} = \text{CIC} \tag{14.3}$$

The relation (14.2) consists of five elements. The first two terms, which are called maintenance activity adjustment costs, refer to the expenses associated with adjusting the failure rate and the average time required for repairs for every component of the distribution systems. The more investments or expenditures utilities must make in preventive care and maintenance programs in order to achieve these standards, the lower the value associated with them (Kela and Arya 2014). Both of these variables are drawn from the reliability growth model proposed by Duane (Ebeling 2004). When expected energy is not delivered by the owners of the DGs while the DGs are not connected (EENSO) or when the DGs are present (EENSD), an additional cost (ADCOST) must be incurred by them (Arya et al. 2012). This cost is represented by the third term in the same relation. ADCOST is calculated here as Rs. 2/kWH. The costs suffered at the level of the consumer in terms of loss during an outage are represented by the fourth component of the relation (14.2), known as customer interruption cost (CIC). The substantial costs linked to electricity outages are obviously amplified when a utility's business is supplying electricity to businesses and industries. By adding together the costs associated with each section outage, as indicated in relation (14.3), one may calculate the entire interruption expense for every load site. This makes it possible to estimate the total cost of all customer interruptions. Here $C_{\text{p}k}$ in (14.3) is the cost of interruptions for a downtime duration r_k related to the kth distribution sector in Rs. for each kilowatt. At a single customer load site, the cost features of the consumers are wholly responsible for determining the customer cost. The overall cost of the consumer's charges associated with any load location due to any interruptions is represented by the sum of the expenses incurred by all different types of customers as a result of the distribution outage. The calculations are done for all the distribution sectors N_c. The fifth component has to do with reward/penalty costs, which were briefly discussed in Section 14.2 (Alvehag and Awodele 2013). As a result, balances the objective functions of the costs associated with DGs, maintenance, customer interruption, and reward/penalty. By minimizing the objective function, as stated by (14.2), this is accomplished.

The following restrictions are placed on the minimization of the objective function given in (14.2).

(i) Constrained on the decision variable

$$\lambda_{k,\min} \le \lambda_k \le \lambda_{k,\max} \tag{14.4}$$

$$r_{k,\min} \le r_k \le r_{k,\max} \tag{14.5}$$

$$k = 1, \ldots, N_c$$

(ii) Constraints on inequality for the energy and customer-oriented reliability indexes

$$\text{SAIFI} \le \text{SAIFI}_t \tag{14.6}$$

$$\text{SAIDI} \leq \text{SAIDI}_t \tag{14.7}$$

$$\text{CAIDI} \leq \text{CAIDI}_t \tag{14.8}$$

$$\text{AENS} \leq \text{AENS}_t \tag{14.9}$$

The optimization procedure achieves the lowest possible failure rate and repair time for the kth sector. The reliability growth model (Ebeling 2004) is used to analyze failure and repair information as well as related expenses to create these lower limit values. SAIFI_t ,SAIDI_t, CAIDI_t, AENS_t and SARFI_t are each of the indices' target/threshold values. They are determined by managerial or administrative decisions. The reliability level (R) was created in this research, taking into account the influence of all energy and customer-oriented reliability indexes because all of these indexes are interrelated. The sum of the normalized results for SAIFI, SAIDI, CAIDI, and AENS is represented here by R. The normalization is done in relation to the various indices' target values. As a result, the method will give each index equal weight:

$$R = \frac{\text{SAIFI}}{\text{SAIFI}_t} + \frac{\text{SAIDI}}{\text{SAIDI}_t} + \frac{\text{CAIDI}}{\text{CAIDI}_t} + \frac{\text{AENS}}{\text{AENS}_t}$$

The primary aim of the study is to determine the failure rate and repairing time that will raise the distribution reliability system, with the secondary goal of selecting the best reward/penalty for the utility for the set/target reliability level. Constrained optimization of the objective function in relation (Section 14.2) is used to achieve it in this case. In this case, the socioeconomic reliability level described in Section 14.2 is the optimal reliability level. The following relations are used to determine this target/set reliability level.

For an optimal RPS, the following can be demonstrated in accordance with (Ajodhia 2006):

$$\frac{\partial C_{\text{RP}}}{\partial R} = \frac{\partial \text{CIC}}{\partial R}, \forall R \tag{14.10}$$

$$C_{\text{RP}} = \text{CIC} - K, \forall R \tag{14.11}$$

K is a randomly chosen constant.

According to (14.10), the cost of customer interruption at a certain reliability level determines the optimal reward or penalty. To obtain the optimal reward or penalty, the regulator must be able to measure and reconstruct (Ajodhia 2006). As seen in Figure 14.2, a continuous RPS scheme with a constant slope has been taken into consideration for this study. As a result, there is a linear relationship b/w reliability of the system and customer interruption costs. The amount of money for each unit of system performance indicator for reliability and, consequently, a reward rate for utility are provided by the slope of RPS in this case. The utility's and consumers' transactions are handled by the value of K. All costs associated with customer disruption will be passed on to the utility if K is set to zero. In light of this, the utility will run on overall reliability cost. In this situation, a utility that seeks to maximize profits may suffer a loss because the permitted only the stated target value generates enough income to pay all of the company's dependability expenses. Due to this, determining the value of K in a continuous RPS is equivalent to deciding what level of system quality indicator is considered to be reliable (Alvehag and Awodele 2013). Revenue only covers the overall dependability costs incurred by the organization for the specified goal value. In light of this, determining the value of K in a continuous RPS corresponds to deciding what degree of system quality indicator for dependability:

$$K = \text{CIC}\,(R = R_{\text{target}}) \tag{14.12}$$

Therefore, regardless of the reliability target value, the profit-maximizing utility can still attain the optimum level of reliability provided the rate of incentive is adjusted based on the cost of customer interruption. As a result, the utility will receive no reward or penalty (incentive rate) for meeting the target or specified reliability level. Thus, the cost of customer interruption determines the target reliability level and, consequently, the incentive or penalty.

The FP optimization method (Yang 2013) and Rao optimization techniques (Rao 2020) are used to address the formulated problem. Testing has been done on Roy Billinton Test System Bus 2 (RBTS-2) using the methodology described in this study. To increase reliability, the crew may be provided with the optimized failure rates and repair times that were so obtained as target values. While optimizing the objective function for the predetermined target level of reliability (R), the optimal reward and penalty values are found as well.

14.4 Rao Algorithms: An Overview

The Rao algorithms are three recently created, straightforward, parameter-less, metaphor-free optimization algorithms by R.V. Rao (2020) that can be used to solve both constrained and unconstrained optimization problems.

Step (a) Initialization: The following is used to produce a population of dimension "M":

$$S^0 = \left[U_1^0, U_2^0, \dots, U_{\mathrm{M}}^0\right] \tag{14.13}$$

$$U_i^0 = \left[U_{i1}^0, U_{i2}^0, \dots, U_{i\mathrm{D}}^0\right]^{\mathrm{T}} \tag{14.14}$$

U_{ij}^0 i.e. jth a uniform distribution-derived parameter of the U_i vector is obtained

$$U_{ij}^0 = U_{j,\min} + (U_{j,\max} - U_{j,\min})\mathrm{rand}_j \tag{14.15}$$

$U_{j,\min}$ and $U_{j,\max}$ are low and high limits on decision variable U_j. A random number in the [0, 1] rang is rand$_j$. D in (14.13) shows the total decision variables to be optimized. Here, decision variables are the rate of failure λ and repairing time r for different sections of the distribution systems considered in this chapter.

Step (b) Modifying/Updating vectors

$$U_i^{(k+1)} = U_i^k + r_{1,i}\left(U_{\text{best}}^k - U_{\text{worst}}^k\right) \tag{14.16}$$

$$U_i^{(k+1)} = U_i^k + r_{1,i}\left(U_{\text{best}}^k - U_{\text{worst}}^k\right) + r_{2,i}\left(\left|U_i^k \text{ or } U_l^k\right| - \left|U_l^k \text{ or } U_i^k\right|\right) \tag{14.17}$$

$$U_i^{(k+1)} = U_i^k + r_{1,i}\left(U_{\text{best}}^k - \left|U_{\text{worst}}^k\right|\right) + r_{2,i}\left(\left|U_i^k \text{ or } U_l^k\right| - (U_l^k \text{ or } U_i^k\right) \tag{14.18}$$

Relations (14.15)–(14.17) give the updated/modified vectors for $(k+1)$ iteration for there are three different algorithms: Rao 1, 2, and 3. Among all the solutions obtained in the current generation or iterations, $U_{\text{best}}(k)$ is the current best, and $U_{\text{worst}}(k)$ is current worst. U_i^k and U_l^k are the different candidate vectors of the population of M candidates.

Step (c) Evaluating the fitness between the updated and original vectors

$$U_i^{(k+1)} = \begin{cases} U_i^{k+1}, & \text{if} f\left(U_i^{k+1}\right) < f\left(U_i^{(k)}\right) \\ U_i^{(k)} & \text{otherwise} \end{cases} \tag{14.19}$$

The current best solution $U_{\text{best}}^{(k)}$ is then discovered, together with its fitness. Until the best solution is found or the predetermined number of generations ($k_{\max}$) have been carried out, this process is carried out for each target vector index i, and a new population is generated.

14.5 Steps to Solve the Problem

Data input $\lambda_{k,\max}$, $r_{k,\max}$, $\lambda_{k,\min}$, $r_{k,\min}$, λ_{dg}, r_{dg}, λ_{sw}, cost interruptions (C_{pk}), N_i, L_i, SAIFI_t, SAIDI_t, CAIDI_t, and AENS_t. λ_{dg}, r_{dg}, λ_{sw}, N_i, and L_i are the fail rate of DG, repairs the rate of DG, the failure rate of the switch connecting the standby DG to the distributions system at the time of power failure, no. of customer at load point i and total load is connected at the load point i, respectively.

1. Discover K at the indices' target values.
2. Initialization: By using relation (14.14), create "M" size is the population for each failure rate λ and repair time r, where each vector of the population contains the corresponding failure rate and repair time of each component. These values are derived via a uniform sampling procedure between the lower and higher bounds, as stated in relation (14.4).
3. In accordance with (Ghorashi et al. 2020), determine the system failure rate $\lambda_{sys,i}$, average repair time $r_{sys,i}$, and annual outage duration $U_{sys,i}$ at each load position.
4. For the vectors of the population, evaluate the energy and customer-oriented reliability indexes for the system as indicated in Ghorashi et al. (2020).
5. Determine the value of the population's vectors' objective function i.e. $F\left(U_i^{(k)}\right), i = 1, \dots, 'M'$ as stated by relations (14.2) and (14.3).
6. Determine the overall reliability R-value for each vector in the population using the formula from Section 14.3.
7. For each population vector, evaluate the inequality constraints resulting from the relations (14.5), (14.6), (14.7), and (14.8). Vectors that meet these restrictions will be feasible; otherwise, they will not be. Determine the best solution vector $U_{\text{best}}^{(k)}$ and the worst solution vector $U_{\text{worst}}^{(k)}$ value is determined by the objective functions, from each vector that can be considered viable.
8. Place $k = 1$ in the generation counter.
9. Choose the target vector with $i = 1$.
10. By using the relation (14.15)–(14.17), determine the vector's updated value.
11. Utilizing relation (14.18), compare the updated initial vectors' fitness for vectors' fitness to keep the good one.
12. Repeat steps 4 through 8 to get the modified or updated vector.
13. Target vector is increased to $i = i + 1$. Repeat from Step 11 if $i \leq M$; else, raise the generation counts $k = k + 1$.
14. If $k \leq k_{\max}$ or the required optimum value cannot be found, start over at Step 11.

14.6 A Discussion of the Findings

On RBTS-2, the methods presented in this study have been used. The FP and Rao algorithms have successfully solved the issue. The algorithms were programmed in MATLAB R2019b software, a 4.00 GB RAM and a 3.00 GHz Intel (R), Core (TM), i5-4590S CPU.

14.6.1 Test System-Bus 2 by Roy Billinton (RBTS-2)

There is a test system called the Roy Billinton Test System-Bus 2 that was employed in this study (Allan et al. 1991). It is seen in Figure 14.3. The information on average maintenance durations and failure rates of the various RBTS-2 components has been gathered from Allan et al. (1991) and Arya (2016).

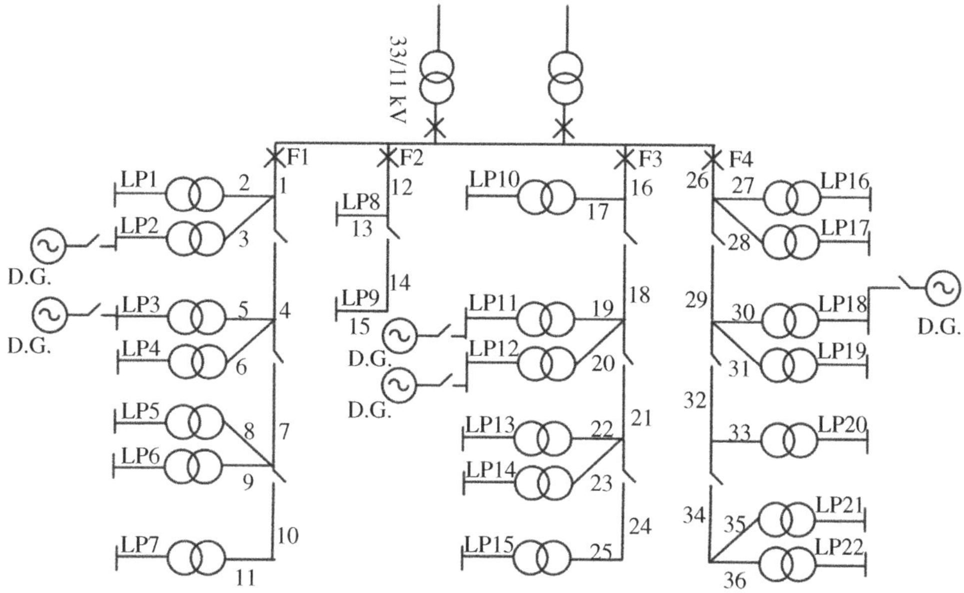

Figure 14.3 RBTS-2.

Table 14.1 provides current failure rates $(\lambda_k^0/\text{year})$ and the lowest achievable failure rates $(\lambda_{k,\min}/\text{year})$, current average repair times $(r_k^0\,(\text{h}))$ and the lowest achievable average repair times $(r_{k,\min}\,(\text{h}))$, as well as cost coefficients α_K and β_K for the various sections of the RBTS-2 (Table 14.2).

Table 14.3 shows the optimized values for maintenance expenses, additional DG costs for energy purchases, costs for customer interruption, reward/penalty costs, and objective function (F) as determined by the FP and Rao algorithms. In this study, the DGs for RBTS-2 are located at sites 2, 3, 11, 12, and 18. Table 14.4 displays the amount of customer and energy-oriented reliability indices that were optimized using the FP and Rao algorithms. Here, the comparison of the enhancements is presented in relation to the corresponding threshold or target levels.

Table 14.5 shows optimized values for maintenance costs and extra costs incurred for DGs in relation to various reliability levels attained around a target reliability level. Table 14.6 displays the network, utility, and society's optimal costs taking the effect of continuous RPS on the utility into account. The results of Tables 14.5 and 14.6 were derived using the following targets for a reliability index based on consumers and energy $\text{SAIFI}_t = 0.084746$, $\text{SAIDI}_t = 0.198559$, $\text{CAIDI}_t = 2.435000$, and $\text{AENS}_t = 1.806670$. For various reliability levels, optimized values of rewards and penalties as well as customer interruption charges are also displayed. The data from this table has been used to plot Figure 14.4.

Table 14.7 shows the optimized rates of failure and times to repair for the various RBTS-2 sections as determined through the FP and Rao algorithms. In Table 14.5, the C_{network} has been made while taking ADCOST and maintenance costs into account. The FP and Rao algorithms have been used to solve the problem with the system under examination. Following are the control parameters used: A maximum generation ($k_{\max}$) size of 1000 is defined, with a population size of 30. Updated step size ($\propto$), distribution factor (β), and switch probability for FP are 0.01, 1.5, and 0.8, respectively. The connected DGs are treated as backup or standby units in this situation. The DGs used in this study's average downtime and failure rate are 13.25 hours and 0.5 failures per year, respectively. The changeover switch at DGs has a failure rate of 0.1 failure per year and a service restoration

Table 14.1 RBTS-2 System-related data.

Distributor segment	λ_k^0/year	$\lambda_{k,min}$/year	r_k^0 (h)	$r_{k,min}$ (h)	α_K (Rs.)	β_K (Rs. × 10^2)
#1	0.048750	0.036650	5.000	2.252252	2.564	22.649
#2	0.015000	0.011270	10.000	4.504504	3.205	17.589
#3	0.052000	0.039090	10.000	4.504504	1.923	4.291
#4	0.048000	0.036090	5.000	2.252252	1.282	4.291
#5	0.015000	0.011270	10.000	4.504504	2.564	13.258
#6	0.015000	0.011270	10.000	4.504504	3.045	6.647
#7	0.048750	0.036650	5.000	2.252252	3.205	5.394
#8	0.015000	0.011278	10.000	4.504504	2.564	22.649
#9	0.015000	0.011278	10.000	4.504504	3.205	17.589
#10	0.039000	0.029323	5.000	2.252252	1.923	4.291
#11	0.015000	0.011278	10.000	4.504504	1.282	4.291
#12	0.048750	0.036650	5.000	2.252250	2.564	22.649
#13	0.052000	0.039097	5.000	2.252250	3.205	17.589
#14	0.039000	0.029323	5.000	2.252250	2.564	22.649
#15	0.052000	0.039097	5.000	2.252250	3.205	17.589
#16	0.048750	0.036650	5.000	2.252250	2.564	22.649
#17	0.015000	0.011278	10.000	4.504500	3.205	17.589
#18	0.052000	0.039097	5.000	2.252250	1.923	4.291
#19	0.015000	0.011278	10.000	4.504500	1.282	4.291
#20	0.015000	0.011278	10.000	4.504500	2.564	13.258
#21	0.039000	0.029323	5.000	2.252250	3.045	6.647
#22	0.015000	0.011278	10.000	4.504500	2.564	22.649
#23	0.015000	0.011278	10.000	4.504500	3.205	17.589
#24	0.048750	0.036650	5.000	2.252250	1.923	4.291
#25	0.015000	0.011278	10.000	4.504500	1.282	4.291
#26	0.052000	0.039097	5.000	2.252250	2.564	22.649
#27	0.015000	0.011278	10.000	4.504500	3.205	17.589
#28	0.015000	0.011278	10.000	4.504500	1.923	4.291
#29	0.048750	0.036650	5.000	2.252250	1.282	4.291
#30	0.015000	0.011278	10.000	4.504500	2.564	13.258
#31	0.015000	0.011278	10.000	4.504500	3.045	6.647
#32	0.048750	0.036650	5.000	2.252250	3.205	5.394
#33	0.015000	0.011278	10.000	4.504500	2.564	22.649
#34	0.039000	0.029323	5.000	2.252250	3.205	17.589
#35	0.015000	0.011278	10.000	4.504500	1.923	4.291
#36	0.015000	0.011278	10.000	4.504500	1.282	4.291

Table 14.2 Information regarding customers connected to Roy Billinton Test System-Bus 2.

Sr. no	Load point number	Load average at every load point in MW	Type of customers/premise	Size of customers in numbers	Interruption cost (C_{pk}) (Rs./kW)
1	1–3,10,11	0.5350	Residential purpose	210.0	22.290
2	12,17–19	0.4500	Residential purpose	200.0	6.0450
3	8	1.0000	Small scale user	1.0	251.420
4	9	1.1500	Small scale user	1.0	251.420
5	4,5,13,14,20,21	0.5660	Government institutes	1.0	23.980
6	6,7,15,16,22	0.4540	Commercial purpose	10.0	192.980

Table 14.3 Results for the FP, Rao-1, and Rao-2-derived objective function (F) for RBTS-2: both current and optimized.

Sr. no.		Current values (Rs.)	Optimized values (Rs.)	Optimized values (Rs.)	Optimized values (Rs.)
			By FP	By Rao-1	By Rao-2
1	Maintenance cost $\left(\sum_{k=1}^{N_c} \alpha_k/\lambda_k^2 + \sum_{k=1}^{N_c} \beta_k/r_k\right)$	232 680	306 348.71	266 826.3	276 386.1
2	Customer interruption cost $\left(\sum_{k=1}^{N_c} CIC\right)$	807 980	302 267.59	301 950.8	301 808.6
3	Additional cost (ADCOST) to be paid while generators are linked with the system	3118	11 434.22	11 496.36	11329.66
4	Reward/penalty $\left(\sum_{k=1}^{N_c} C_{\mathrm{RP}}\right)$	490 450	−15 264.67	−15 581.5	−15 723.6
5	Objective function (F)	1 534 228	604 785.85	564 692	573 800.7

Table 14.4 Reliability indices for RBTS-2 determined by FP, Rao-1, and Rao-2: both current and optimized.

Sr. no.	Indices	Current measures	Optimized measures	Optimized measures	Optimized measures	Threshold measures
			By FP	By Rao-1	By Rao-2	
1	SAIFI (interruptions/customer per year)	0.09860	0.080462	0.080971	0.081635	0.084746
2	SAIDI (h/customer per year)	0.58820	0.177953	0.168077	0.191913	0.198559
3	CAIDI (h/customer interruption per year)	5.96660	2.2301	2.075657	2.398312	2.435000
4	AENS (kW/customer)	4.66410	1.667670	1.651387	1.695072	1.806670

Table 14.5 Values of maintenance cost and ADCOST, as determined by FP, and related values of overall reliability R (before and after the target values) (considering $SAIFI_t = 0.084746$, $SAIDI_t = 0.198559$, $CAIDI_t = 2.435000$, $AENS_t = 1.806670$) for Roy Billinton Test System (Bus-2).

Sr. no.	Maintenance cost (Rs.)	ADCOST (Rs.)	SAIFI	SAIDI	CAIDI	AENS	*F* (Rs.)	*R*
1	292 425.50	10 128.00	0.0860	0.2250	2.6230	2.010	667 137.20	3.1320
2	290 868.10	10 385.00	0.0850	0.2230	2.660	1.9430	656 514.00	3.0940
3	295 194.80	10 630.00	0.0820	0.2260	2.7570	1.8780	658 002.00	3.060
4	264 349.60	10 650.00	0.0860	0.2040	2.4170	1.8730	617 472.50	2.9640
5	275 661.70	10 687.00	0.0860	0.1980	2.2860	1.8640	623 918.50	2.9160
6	281 435.60	10 933.70	0.08290	0.2030	2.3220	1.7980	620 153.90	2.8710
7	266 436.70	11 206.00	0.0810	0.190	2.3830	1.7280	574 835.80	2.80
8	266 249.70	11 252.00	0.0830	0.1870	2.2530	1.7150	574 703.20	2.7750
9	270 691.60	11 340.00	0.0830	0.1760	2.1560	1.6920	576 717.20	2.7120
10	2 831 670	11 400.00	0.0830	0.1780	2.140	1.6770	585 878.10	2.7070
11	295 181.60	11 375.00	0.0820	0.1750	2.1560	1.6830	597 618.40	2.6980
12	**306 348.70**	**11 434.00**	**0.08040**	**0.1780**	**2.2300**	**1.6680**	**604 785.90**	**2.6960**
13	268 835.90	11 47s1.00	0.0840	0.1720	2.1150	1.6580	560 790.90	2.6910
14	287 743.70	11 574.00	0.0830	0.170	2.0560	1.6310	585 959.70	2.6450
15	275 668.90	11 618.00	0.0820	0.1660	2.0740	1.620	571 269.70	2.6230
16	271 853.20	11 595.00	0.080	0.1660	2.1210	1.6260	562 966.30	2.6190
17	275 059.60	11 605.00	0.0810	0.1660	2.0910	1.6230	568 568.0	2.6160
18	287 907.50	11 526.00	0.080	0.1660	2.0630	1.6440	578 949.50	2.6140

time of 0.5 hours, respectively. Figure 14.4 depicts the implications of an ideal continuous RPS for the expenses associated with different parameter values for RBTS-2. The optimal cost for the n/w, utilized, social, reward or penalty, and consumer interruption are graphically illustrated here for different reliability levels.

Table 14.7 shows the optimized rates of failure and times to repair for the various RBTS-2 sections as determined through the FP and Rao algorithms. In Table 14.5, the $C_{network}$ has been made while taking ADCOST and maintenance costs into account. The FP and Rao algorithms have been used to solve the problem with the system under examination. Following are the control parameters used: A maximum generation (k_{max}) size of 1000 is defined, with a population size of 30. Updated step size ($\propto$), distribution factor (β), and switch probability for FP are 0.01, 1.5, and 0.8, respectively. The connected DGs are treated as backup or standby units in this situation. The DGs used in this study's average downtime and failure rate are 13.25 hours and 0.5 failures per year, respectively. The changeover switch at DGs has a failure rate of 0.1 failure per year and a service restoration time of 0.5 hours, respectively. Figure 14.4 depicts the implications of an ideal continuous RPS for the expenses associated with different parameter values for RBTS-2. The optimal cost for the n/w, utilized, social, reward or penalty, and consumer interruption are graphically illustrated here for different reliability levels.

Table 14.6 The effects of continuous RPS on utility (considering $SAIFI_t = 0.0847460$, $SAIDI_t = 0.1985590$, $CAIDI_t = 2.4350000$, $AENS_t = 1.8066700$) for RBTS-2 as obtained by FP.

Sr no.	CIC	C_{RP}	$C_{network}$	$C_{total}^{utility} = C_{network} + C_{RP}$	$C_{total}^{society} = C_{network} + CIC$	R
1	341 057.99	23 525.732	302 553.5096	326 079.2416	643 611.5094	3.132077
2	336 396.39	18 864.131	301 253.4363	320 117.5669	637 649.8347	3.0939145
3	334 854.53	17 322.268	305 825.2102	323 147.4785	640 679.7463	3.0603485
4	330 002.67	12 470.406	274 999.4572	287 469.8636	605 002.1314	2.963884
5	327 551.30	10 019.036	286 348.1753	296 367.2114	613 899.4791	2.9163098
6	322 658.38	5 126.11	292 369.40	297 495.5196	615 027.7874	2.8714413
7	307 362.90	−10169.36	277 642.2999	267 472.9406	585 005.2084	2.8000841
8	307 366.70	−10 165.56	277 502.0912	267 336.532	584 868.7998	2.7751445
9	306 108.78	−11 423.48	282 031.8602	270 608.3757	588 140.6435	2.7116754
10	304 421.57	−13 110.69	294 567.1973	281 456.5055	598 988.7733	2.7073760
11	304 297.26	−13 235	306 556.1356	293 321.1365	610 853.4043	2.6984982
12	**302 267.59**	**−15 264.68**	**317 782.9447**	**302 518.268**	**620 050.5358**	**2.6960579**
13	299 008.20	−18 524.07	280 306.7189	261 782.6536	579 314.9213	2.6909185
14	302 086.95	−15 445.31	299 318.1045	283 872.7911	601 405.0588	2.6446134
15	300 757.62	−16 774.65	287 286.6803	270 512.0334	588 044.3011	2.6231037
16	298 525.19	−19 007.07	283 448.1982	264 441.1262	581 973.394	2.6188200
17	299 718.06	−17 814.21	286 664.1614	268 849.9537	586 382.2215	2.6159990
18	298 523.97	−19 008.3	299 433.8125	280 425.5158	597 957.7836	2.6138074

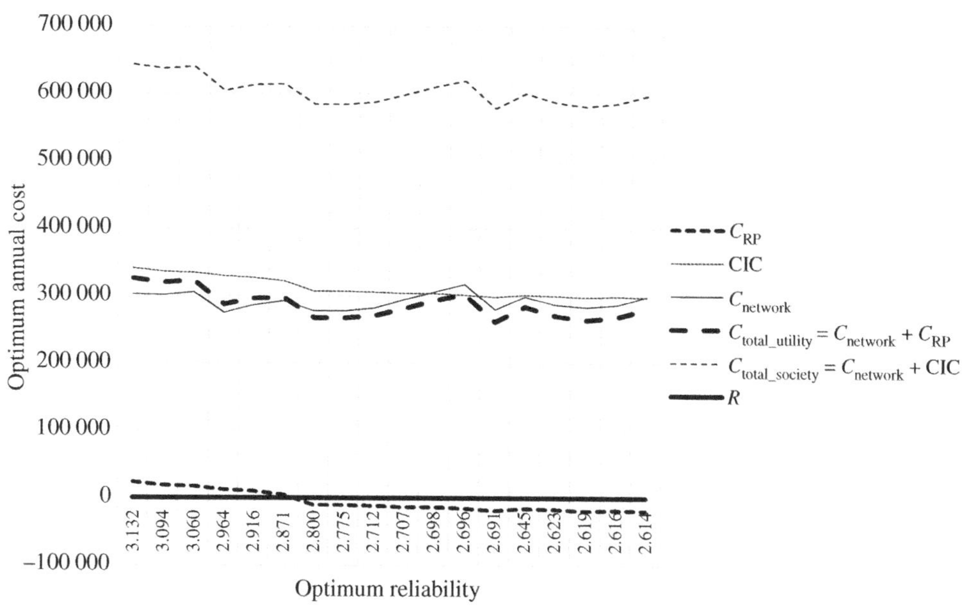

Figure 14.4 Effect of an optimal continuous RPS on the price of various RBTS-2 parameters.

Table 14.7 FP, Rao-1, and Rao-2's optimized values for rate of failure and time to repair obtained for the RBTS-2.

Distributor system segments	Failure rates (/year) With FP	Failure rates (/year) With Rao-1	Failure rates (/year) With Rao-2	Repair times (in h) With FP	Repair times (in h) With Rao-1	Repair times (in h) With Rao-2
1	0.03679	0.03666	0.03665	2.25225	2.25227	2.29738
2	0.01497	0.01499	0.01500	4.50497	4.68206	9.97628
3	0.03910	0.04089	0.03909	4.60288	4.55572	9.99998
4	0.03609	0.03666	0.03613	2.25237	2.26249	2.25225
5	0.01127	0.01468	0.01421	4.96719	9.84338	4.50450
6	0.01500	0.01478	0.01500	9.34639	4.93965	4.50609
7	0.03665	0.03759	0.03668	2.25229	2.35412	2.25344
8	0.01237	0.01461	0.01500	9.43417	9.93448	4.51630
9	0.01500	0.01473	0.01499	4.50613	4.53304	6.57998
10	0.03899	0.03880	0.02958	2.25525	2.25225	2.35627
11	0.01128	0.01500	0.01128	4.50491	4.50529	4.50487
12	0.03665	0.03732	0.03673	2.27395	2.26746	2.26344
13	0.03943	0.03922	0.03910	2.44635	2.25259	2.26420
14	0.03102	0.03261	0.02932	2.25353	2.25240	2.25227
15	0.03911	0.03910	0.03910	2.25225	2.26204	2.25225
16	0.04238	0.03704	0.03665	2.25229	2.25495	2.25225
17	0.01499	0.01454	0.01500	9.97048	6.29397	4.50451
18	0.03910	0.03944	0.03910	2.27024	2.25240	2.26192
19	0.01500	0.01495	0.01500	9.97133	4.79284	9.97537
20	0.01485	0.01468	0.01500	9.98236	8.97340	4.90722
21	0.02934	0.03895	0.02938	2.75800	2.25750	2.25662
22	0.01260	0.01224	0.01226	5.03652	4.53734	4.50469
23	0.01136	0.01500	0.01489	5.11156	9.99853	4.51759
24	0.04780	0.03692	0.03665	2.25635	2.31743	2.25255
25	0.01500	0.01129	0.01128	4.50450	4.57636	4.52306
26	0.03910	0.03919	0.04350	2.31682	2.25227	2.35457
27	0.01481	0.01447	0.01500	4.52148	4.56738	4.50450
28	0.01128	0.01478	0.01487	9.85532	4.72852	4.50450
29	0.03665	0.03874	0.03951	2.25238	2.25255	2.45385
30	0.01500	0.01476	0.01498	9.53818	4.50896	5.46092
31	0.01500	0.01500	0.01500	9.99991	4.51428	4.57499
32	0.04875	0.03665	0.04875	2.25870	2.27121	2.28080
33	0.01283	0.01464	0.01500	4.50450	5.37821	10.00000
34	0.02996	0.02963	0.02933	2.25408	2.57953	2.31643
35	0.01153	0.01494	0.01145	5.59178	4.50900	6.79013
36	0.01128	0.01499	0.01391	4.50450	4.68041	4.50546

14.7 Conclusion

By including a reward or penalty for the utilities by the regulator in this research, the distribution system reliability has improved. In light of the target values for the customer and energy-oriented reliability indexes, the best values for rewards and penalties have been identified. The results show that utility is rewarded when each of these indicators falls below the desired or target level, and it is penalized when they are above. The Flower Pollination and Rao algorithms have been used to optimize the objective function that was created. In addition to producing optimized values for all reliability indices, other components of the objective function, including costs associated with maintenance, customer disruptions, and additional expenses required to reach the level of reliability have also been produced with optimized values. All optimization approaches produced extremely similar optimized outcomes.

References

Ajodhia, V.S. (2006). Regulating beyond price. Integrated price-quality regulation for electricity distribution networks. PhD Dissertation, Delft University of Technology, Delft.

Allan, R.N., Billinton, R., Sjarief, I. et al. (1991). A reliability test system for educational purposes-basic distribution system data and results. *IEEE Transactions on Power Systems* 6 (2): 813–820.

Alvehag, K. and Awodele, K. (2013). Impact of reward and penalty scheme on the incentives for distribution system reliability. *IEEE Transactions on Power Systems* 29 (1): 386–394.

Amirrezai, M., Rezaie, H., and Goetz, S.M. (2023). Feasibility study of incorporating static compensators in distribution networks containing distributed generation considering system power factor. *Electric Power Systems Research* 219: 109253.

Arya, R. (2016). Determination of customer perceived reliability indices for active distribution systems including omission of tolerable interruption durations employing bootstrapping. *IET Generation Transmission and Distribution* 10 (15): 3795–3802.

Arya, R., Choube, S.C., and Arya, L.D. (2012). Reliability evaluation and enhancement of distribution systems in the presence of distributed generation based on standby mode. *International Journal of Electrical Power & Energy Systems* 43 (1): 607–616.

Billinton, R. and Allan, R.N. (1996). *Reliability Evaluation of Power Systems*. Springer International Edition.

Billinton, R. and Li, W. (1994). *Reliability Assessment of Electrical Power Systems using Monte Carlo Methods*. New York, NY, USA: Plenum.

Billinton, R. and Pan, Z. (2004). Historic performance-based distribution system risk assessment. *IEEE Transactions on Power Delivery* 19 (4): 1759–1765.

CEER (2011). 5th CEER benchmarking report on the quality of electricity supply, Council of European Energy Regulators, Technical Report, Brussels, Belgium.

Ebeling, C.E. (2004). *An Introduction to Reliability and Maintainability Engineering*. Tata McGraw-Hill Education.

Fotuhi, M., Shourkaei, H.M., Kharazi, M.B., and Salimi, A. (2006). Reliability assessment of utilities using an enhanced reward-penalty model in performance based regulation system. In: *2006 International Conference on Power System Technology*, 1–6. IEEE.

Fumagalli, E., Schiavo, L., and Delestre, F. (2007). *Service Quality Regulation in Electricity Distribution and Retail*. Springer Science & Business Media.

Ghasemi, M. and Dashti, R. (2017). A risk-based model for performance-based regulation of electric distribution companies. *Utilities Policy* 45: 36–44.

Ghasemi, M. and Dashti, R. (2018). Designing a decision model to assess the reward and penalty scheme of electric distribution companies. *Energy* 147: 329–336.

Ghorashi, S.M., Rastegar, M., Senemmar, S., and Seifi, A.R. (2020). Optimal design of reward-penalty demand response programs in smart power grids. *Sustainable Cities and Society* 60: 102150. https://doi.org/10.1016/j.scs.2020.102150.

Ghotboddini, R. and Toossian Shandiz, H. (2022). Reward and penalty model for the lighting of public thoroughfares contracts: an empirical study in a distribution company. *Journal of AI and Data Mining* 10 (1): 127–138.

Grahn, E., Ström, L., Alvehag, K. et al. (2016). Incentivizing continuity of supply in Sweden. In: *2016 13th International Conference on the European Energy Market (EEM)*, 1–5. IEEE.

Janjic, A., Velimirovic, L.Z., and Vranic, P. (2021). Designing an electricity distribution reward-penalty scheme based on spatial reliability statistics. *Utilities Policy* 70: 101211.

Jooshaki, M., Abbaspour, A., Fotuhi-Firuzabad, M. et al. (2014). A new reward-penalty mechanism for distribution companies based on concept of competition. In: *IEEE PES Innovative Smart Grid Technologies, Europe*, 1–5. IEEE.

Jooshaki, M., Ali Abbaspour, M.F.F., and Moeini-Aghtaie, M. (2017). Designing a regulatory framework for efficient integration of distributed generation technologies. *Turkish Journal of Electrical Engineering and Computer Sciences* 25 (5): 4117–4130.

Jooshaki, M., Abbaspour, A., Fotuhi-Firuzabad, M. et al. (2018). Designing a new procedure for reward and penalty scheme in performance-based regulation of electricity distribution companies. *International Transactions on Electrical Energy Systems* 28 (11): e2628.

Kela, K.B. and Arya, L.D. (2014). Reliability optimization of radial distribution systems employing differential evolution and bare bones particle swarm optimization. *Journal of The Institution of Engineers (India): Series B* 95 (3): 231–239.

Khonakdar-Tarsi, I., Fotuhi-Firuzabad, M., Ehsan, M. et al. (2021). Reliability incentive regulation based on reward-penalty mechanism using distribution feeders clustering. *International Transactions on Electrical Energy Systems* 31 (8): e12958.

Lu, Y., Khan, Z.A., Gunduz, H. et al. (2021). Comparison of two strategies of reward-penalty mechanism for promoting net zero energy buildings. *Sustainable Energy Technologies and Assessments* 47: 101347.

Mohammad Nezhad-Shourkaei, H. and Fotuhi-Firuzabad, M. (2011). Principal requirements of designing the reward-penalty schemes for reliability improvement in distribution systems. *21st International Conference on Electricity Distribution Measured Indices*, no. 1207, pp. 6–9.

Mohammad Nezhad-Shourkaei, H., Fotuhi-Firuzabad, M., and Billinton, R. (2011). Integration of clustering analysis and reward/penalty mechanisms for regulating service reliability in distribution systems. *IET Generation Transmission and Distribution* 5 (11): 1192–1200.

Rao, R. (2020). Rao algorithms: Three metaphor-less simple algorithms for solving optimization problems. *International Journal of Industrial Engineering Computations* 11 (1): 107–130.

Sappington, D.E.M. (2005). Regulating service quality: A survey. *Journal of Regulatory Economics* 27 (2): 123–154.

Simab, M. and Haghifam, M.R. (2012). Quality performance based regulation through designing reward and penalty scheme for electric distribution companies. *International Journal of Electrical Power & Energy Systems* 43 (1): 539–545.

Simab, M., Alvehag, K., Söder, L., and Haghifam, M.-R. (2012). Designing reward and penalty scheme in performance-based regulation for electric distribution companies. *IET Generation Transmission and Distribution* 6 (9): 893–901.

Sumesh, S., Potdar, V., and Krishna, A. (2018). Clustered prosumer penalty framework on smart grid. In: *Advances in Smart Grid and Renewable Energy*, 1–12. Singapore: Springer.

Sumesh, S., Potdar, V., and Krishna, A. (2019). Cubic reward penalty structure for power distribution companies. *International Journal of Systems Assurance Engineering and Management* 10 (3): 350–368.

Wang, B., Camacho, J.A., Pulliam, G.M. et al. (2018). New reward and penalty scheme for electric distribution utilities employing load-based reliability indices. *IET Generation Transmission and Distribution* 12 (15): 3647–3654.

Yang, X.S. (2013). Flower pollination algorithm for global optimization. In: *International Conference on Unconventional Computing and Natural Computation*, 240–241. Berlin Heidelberg: Springer.

15

Reliability Analysis of Composite Distribution System Using Frequency Duration Concept

Atul Koshti[1], *Eram Taslima*[2], *Pradeep Purey*[3], *Liladhar Arya*[4], *and Sharat C. Choube*[5]

[1] *Department of Electrical Engineering, GES'S R.H. Sapat College of Engineering, Management Studies & Research, Nashik, Maharashtra, India*
[2] *Department of Electrical Engineering, Indian Institute of Availabilities on annual basis Technology (BHU), Varanasi, Uttar Pradesh, India*
[3] *Department of Electronics Engineering, Maharaja Ranjit Singh College of Professional Sciences, Indore, Madhya Pradesh, India*
[4] *Department of Electrical Engineering, Shri Govindram Seksaria Institute of Technology and Science, Madhya Pradesh, India*
[5] *Department of Electrical Engineering, UIT, RGPV, Madhya Pradesh, India*

15.1 Introduction

Reliability indices assessment is qualitative analysis required for power system adequacy assessment. This becomes very essential in distribution system with integration of DG i.e. in composite distribution system (CDS) (Billinton and Allan 1992; Pereira and Balu 1992). DG is usually integrated at 11 KV and possesses advantages (El-Khattam and Salama 2004) such as relieving of system loading, improved voltage, etc. Also due to the uncertain environment condition i.e. stochastic nature of DG, availability of distribution system substation capacity, increased failure rate of components creates challenges for the distribution system planning in view of maintaining desired level of reliability. Adequacy review of CDS is conferred in Hegazy et al. (2003), Arya and Koshti (2008, 2012), Costa and Matos (2009), and Xu and Singh (2012). Sample-based reliability evaluation methods are discussed in Kothari et al. (2022), Probability Methods Subcommittee (1979), Billinton and Li (1993), and Wang et al. (2013). Well-being evaluation for power system is used for adequacy assessment (Wangdee and Billinton 2006; Amaral et al. 2010; Ahmadi-Khatir et al. 2009; Billinton and Lian 1994). Statistical analysis for CDS is carried out in Arya et al. (2013), Halve et al. (2023), and Koshti et al. (2023). Reliability planning is discussed in detail and advances with design (Khezri and Mahmoudi 2020; Roy et al. 2022; Ranjbar et al. 2022; Yahaya et al. 2020; Hamzeh and Vahidi 2020) for hybrid renewable energy systems (HRES). Integration of renewable energy systems (RES) for sustainable development is addressed in Hassan et al. (2022).

It is observed that there is rising interest in further segregation of reliability in three states i.e. (H), (M), and (R). A three-state model can give further insight into the reliability planning as per the deterministic criteria.

Therefore, this chapter suggests deterministic criteria-based reliability indices evaluation planning which uses a ratio of supply capacity available to the load capacity is driven as one of the criteria that segregate reliability in three state. Threshold values of this ratio are used to select CDS states H, M, or R. Moreover, the frequency and duration concept is used to evaluate frequencies of H, M, and R states along with duration indices in H, M, and risky states e.g. MHT, MRT, and MMT.

Reliability Analysis of Modern Power Systems, First Edition. Edited by R. K. Saket and P. Sanjeevikumar.

Thus Section 15.1 presents an introduction. Modeling of different components in CDS such as substation capacity, DG capacity and load is discussed in Section 15.2. Section 15.3 presents a frequency-duration concept for reliability indices evaluation. Section 15.4 presents algorithm using Monte Carlo simulation (MCS)-based reliability analysis of CDS. Section 15.5 presents result section. Section 15.6 presents illustrative examples. Section 15.7 presents conclusions.

15.2 Components Modeling in Composite Distribution System (CDS)

15.2.1 Capacity Modeling

Capacity of a CDS is given as:

$$C_{\mathrm{T}} = C_{\mathrm{S,S}} + C_{\mathrm{dgUS}} \tag{15.1}$$

where $C_{\mathrm{S,S}}$ and C_{dgUS} are the capacity available due to distribution substation (DS) and DG units and their capacity samples are obtained as follows

$$f(C_{\mathrm{S,S}}) = \frac{1}{\sqrt{2\pi\sigma_{\mathrm{S}}}} \mathrm{e}^{-0.5\left(\frac{C_{\mathrm{S,S}}-\overline{C}_{\mathrm{S,S}}}{\sigma_{\mathrm{S}}}\right)^2} \tag{15.2}$$

$$f(C_{\mathrm{dgU,i}}) = \frac{1}{\sqrt{2\pi\sigma_i}} \mathrm{e}^{-0.5\left(\frac{C_{\mathrm{dgU},i}-\overline{C}_{\mathrm{dgU},i}}{\sigma_i}\right)^2} \tag{15.3}$$

Availability and unavailability of ith DG unit are obtained in terms of repair and failure rates μ_i and λ_i follows:

$$A_i = \frac{\mu_i}{\lambda_i + \mu_i} \tag{15.4}$$

$$\overline{A}_i = \frac{\lambda_i}{\lambda_i + \mu_i} \tag{15.5}$$

And total capacity from DG is calculated as

$$C_{\mathrm{dgUS}} = \sum_{p} C_{\mathrm{dgU},p} \tag{15.6}$$

$p \epsilon$ set of available units.

15.2.2 Load Modeling

The load is treated as normally distributed of mean $\overline{\mathrm{ld}}$ and standard deviation σ_{ld}. Load at a bus in an interval is not constant. Truly speaking it is a random variable. It has been a natural choice to assume Normal distribution as probability distribution function of load at a bus:

$$f\left(l_i\right) = \left(1/(\sqrt{2\pi\sigma_{\mathrm{ld}}})\right) \mathrm{e}^{-0.5\left(\frac{\mathrm{ld}-\overline{\mathrm{ld}}}{\sigma_{\mathrm{ld}}}\right)^2} \tag{15.7}$$

15.3 Frequency-Duration Concept for Reliability Indices Evaluation

Figure 15.1 represents state transition diagram for three state model which is represented based on deterministic criteria as separation of H, M, or R states by a lines between state also it is possible

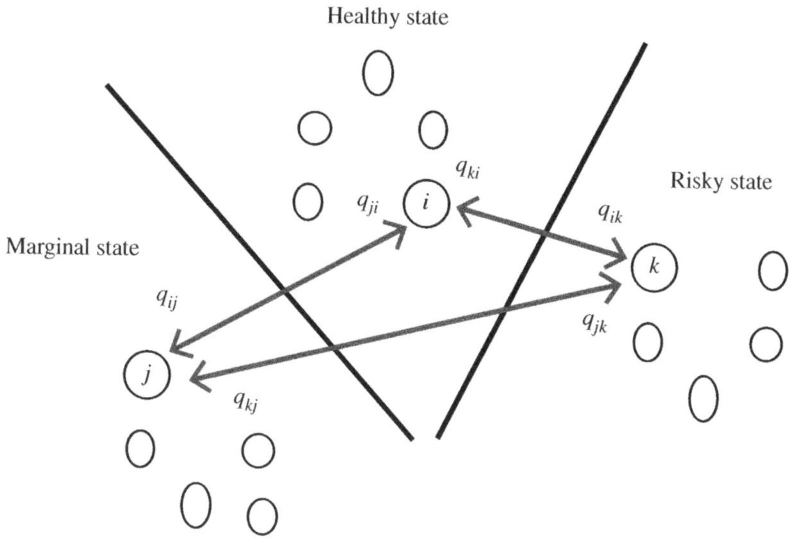

Figure 15.1 State transition diagram for three-state model.

that there may be transition from individual state to other states between each other due to failure of single component. Here if a system exists in a particular state then due to state change of the component (on/off) the state of system transition (Arya et al. 2013; Halve et al. 2023) will take place. Based on this frequency of the system and duration indices can be calculated in H, M, and R state given as follows

$$f_H = \sum_{j \in H} p_j \sum_{\substack{k \in M \\ k \in R}} (q_{jk}) \tag{15.8}$$

$$f_M = \sum_{j \in M} p_j \sum_{\substack{k \in H \\ k \in R}} (q_{jk}) \tag{15.9}$$

$$f_R = \sum_{j \in R} p_j \sum_{\substack{k \in H \\ k \in M}} (q_{jk}) \tag{15.10}$$

$$p_H = \sum_{j \in H} p_j \tag{15.11}$$

$$p_M = \sum_{j \in M} p_j \tag{15.12}$$

$$p_R = \sum_{j \in R} p_j \tag{15.13}$$

$$MHT = \frac{P_H}{f_H} \tag{15.14}$$

$$MMT = \frac{P_M}{f_M} \tag{15.15}$$

$$MRT = \frac{P_R}{f_R} \tag{15.16}$$

$$A_X = 8760 \cdot p_X \tag{15.17}$$

where X will be H, M, or R and q_{jk} are the transition rates from jth to kth state.

15.4 MCS-Based Reliability Indices Evaluation of CDS

The well-being analysis framework is used for reliability indices evaluation of CDS using MCS enlighten in the following steps:

S1: The (DS) is assumed to be available forever and capacity is evaluated using (15.2).
S2: The total DG capacity C_{dgUS} can be evaluated with (15.6).
S3: For CDS the C_T is obtained with (15.1).
S4: Sample for load (ld) is acquired using relation (15.7)
S5: Subsequently state is identified with:

System state	Deterministic criteria
Healthy state if	$(C_T/l) \geq 1.2$
Marginal state if	$1.2 > (C_T/l) \geq 1.1$
Risky state	Otherwise

S6: kth system state duration is attained as:

$$T_k = -\frac{1}{q} \cdot lnu \tag{15.18}$$

u is random numeral [0, 1] and q is total rate of transition (Arya et al. 2013).

S7: State transition method (Arya et al. 2013; Halve et al. 2023) is used for system transition due to the state change of a DG unit.

S8: Probability indices in H, M, and R states are calculated for ith sample as follows:

$$p_{H,i} = \frac{T_{H,i}}{T_D} \tag{15.19}$$

$$p_{M,i} = \frac{T_{M,i}}{T_D} \tag{15.20}$$

$$p_{R,i} = \frac{T_{R,i}}{T_D} \tag{15.21}$$

where $T_{H,i}$, $T_{M,i}$, and $T_{R,i}$ are the duration sample for H, M, and R states, respectively, and $T_D = T_{H,i} + T_{M,i} + T_{R,i}$.

S9: Convergence is ascertained with calculation of β in terms of p and NS

$$\beta = \sqrt{\frac{(1-p)}{p \cdot \text{NS}}} \tag{15.22}$$

If $\beta \leq 0.04$, then the solution is converged; otherwise, do again from steps 2 to 8 for Sample (say NS = 10,000).

S10: Frequency of H, M, and R states given as:

$$f_H = p_{H,i}(\lambda_{HM} + \lambda_{HR}) = \frac{(n_{HM} + n_{HR})}{T_D} \tag{15.23}$$

$$f_M = p_{M,i}(\lambda_{MH} + \lambda_{MR}) = \frac{(n_{MH} + n_{MR})}{T_D} \tag{15.24}$$

$$f_R = p_{R,i}(\lambda_{RH} + \lambda_{RM}) = \frac{(n_{RH} + n_{RM})}{T_D} \tag{15.25}$$

where λ_{XY} is the total transition rate and n_{XY} is the number of transitions from state X to Y where X and Y may be H, M, or R.

S11: Then, mean time for H, M, and R states evaluated correspondingly:

$$\text{MHT} = \frac{p_{\text{Hi}}}{f_{\text{H}}} \tag{15.26}$$

$$\text{MMT} = \frac{p_{\text{Mi}}}{f_{\text{M}}} \tag{15.27}$$

$$\text{MRT} = \frac{p_{\text{Ri}}}{f_{\text{R}}} \tag{15.28}$$

S12: Availabilities on annual basis in H, M, and R states are obtained as:

$$A_{\text{X}} = 8760 \cdot P_{\text{Xi}} \tag{15.29}$$

where X will be H, M, or R.

15.5 Result and Discussion

A CDS (Arya et al. 2013) is considered with (DS) data $\overline{C}_{\text{S,S}} = 2800$ KW, $\sigma_{\text{S}} = 140$ KW and shown in Figure 15.2. And reliability data of DG's are shown in Table 15.1 (Arya et al. 2013).

Load is assumed as normally distributed with mean $\overline{\text{ld}} = 2700$ KW and standard deviation $\sigma_{\text{ld}} =$ 135 KW. The simulation is run using algorithm developed in Section 15.4, and results are obtained and given in Table 15.2 with deterministic criteria explained in Section 15.4. And samples are obtained for simulation and reliability analysis is carried out to evaluate various reliability indices along with frequency and duration indices. Comparing the results shows a significant decrease in risky state probability with the addition of DG capacity is observed and also notable change

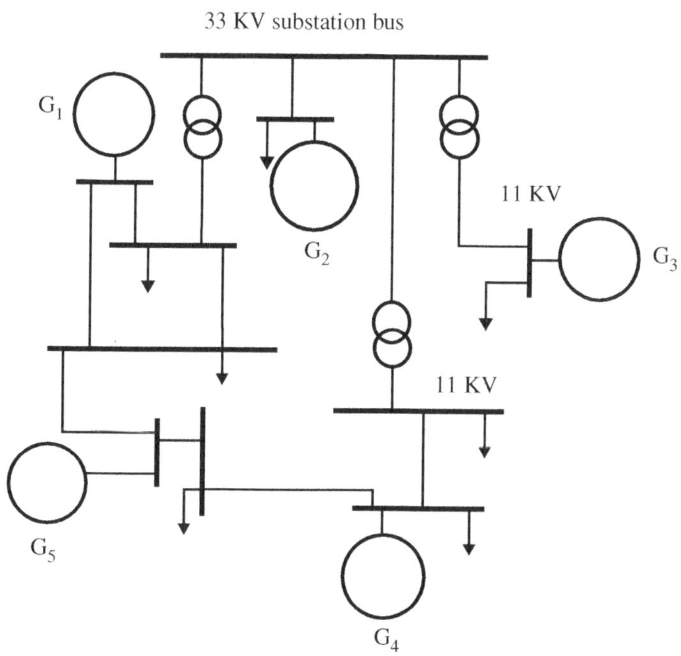

Figure 15.2 Sample composite distribution system.

Table 15.1 Reliability data of DG.

No.	I	II	III	IV	V
λ (per hour)	1×10^{-3}	2.4×10^{-3}	3×10^{-3}	4×10^{-3}	7×10^{-3}
μ (per hour)	3×10^{-3}	5×10^{-3}	6×10^{-3}	4×10^{-3}	5×10^{-3}
A	0.750	0.676	0.666	0.500	0.417
Mean capacity (KW)	100.00	100.00	100.00	100.00	100.00
Standard deviation (KW)	5.000	5.000	5.000	5.000	5.000

Source: Arya et al. (2013)/with permission of Elsevier.

Table 15.2 Reliability indices of CDS.

Sr. No.	Reliability indices	Without DG	With DG
1.	P_H	0.017400	0.616800
2.	P_M	0.188700	0.335500
3.	P_R	0.793900	0.051700
4.	f_H (occ/yr)	0.000365	0.004050
5.	f_M (occ/yr)	0.002100	0.002780
6.	f_R (occ/yr)	0.00300	0.000450
7.	MHT (h/yr)	47.67120	152.1200
8.	MMT (h/yr)	88.65000	120.5000
9.	MRT (h/yr)	264.6333	112.7200
10.	A_H (h/yr)	152.4240	5403.168
11.	A_M (h/yr)	1653.000	2938.980
12.	A_R (h/yr)	6954.600	452.8920

is observed for increase in healthy and marginal state probability which indicates that samples obtained based on deterministic criteria are shifted in H state from R state results in increase of P_H, P_M, and reduction in P_R. which gives more mean healthy time (MHT), mean marginal time (MMT), and less mean risky time (MRT). Availability A_H, A_M is observed improved is whereas significant reduction is observed in A_R.

15.6 Illustrative Examples

15.1 A distribution substation feeds a load via a distributor segment. Capacity of substation is normally distributed with mean and standard deviation $\overline{C} = 180\,\text{MW}$, $\sigma_C = 20\,\text{MW}$. And Load is also normally distributed with mean and standard deviation $\overline{l} = 100\,\text{MW}$, $\sigma_l = 10\,\text{MW}$. Then calculate reliability assuming the distributor reliability is unity.

Solution:

The reliability is given (Arya and Koshti 2008) as

$$R = \varphi(\beta) = \int_{-\infty}^{\beta} N(0,1)\,dt$$

$$\beta = \frac{\overline{C}_T - \bar{l}}{\sqrt{\sigma_c{}^2 + \sigma_l{}^2}} = \frac{80}{\sqrt{500}} = 3.577$$

From table of normal distribution [5] the reliability is obtained as

$$R = \varphi(3.577) = 0.9997674$$

15.2 Problem statement is same as above Exercise 15.1 but reliability of distributor segment is assumed to be 0.95. Obtain probability of meeting the load.

Solution:

$$R = \varphi(\beta) \cdot (\textit{Reliability}\ \text{of distributor})$$

Product law follows due to independent event in series.

$$R = 0.9997674 \times 0.95$$

$$R = 0.94977903$$

15.3 A distribution substation meets the load having distributed generation (DG) with following data. Capacity of substation is normally distributed with $\overline{C} = 100\,\text{MW}, \sigma_C = 10\,\text{MW}$. And Load is also normally distributed with $\bar{l} = 80\,\text{MW}, \sigma_l = 10\,\text{MW}$. DG is normally distributed with capacity $\overline{C}_{dg} = 20\,\text{MW}$ and $\sigma_{dg} = 5\,\text{MW}$. Obtain reliability of meeting the load.

Solution:

Total average capacity

$$\overline{C}_T = \overline{C} + \overline{C}_{dg} = 120\,\text{MW}$$

$$\bar{l} = 80\,\text{MW}$$

$$\sigma_l = 10\,\text{MW}$$

$$\sigma_T = \sqrt{\sigma_C^2 + \sigma_{dg}{}^2}$$

$$\sigma_T = 11.18\ \text{MW}$$

Safety Index is written as

$$\beta = \frac{\overline{C}_T - \bar{l}}{\sqrt{\sigma_T{}^2 + \sigma_l{}^2}}$$

$$\beta = \frac{120 - 80}{\sqrt{11.18^2 + 10^2}}$$

$$\beta = \frac{40}{14.999} = 2.67$$

Therefore

$$R = \varphi(\beta) = \varphi(2.67)$$

$$R = 0.996207$$

15.4 Now assume same capacity and load data in Exercise 6.3. Except the load is fed via a distributor with reliability 0.99. How the reliability is affected.

Solution:

$$R = \varphi(\beta) \cdot (\textit{Reliability} \text{ of distributor})$$

$$R = \varphi(2.67) \times 0.99$$

$$R = 0.98215$$

15.5 A distribution substation feeds a load via a distributor having reliability 0.98. Capacity of distribution substation and load is normally distributed with

$$\overline{C} = 100\,\text{MW}.$$

$$\sigma_C = 10\,\text{MW}$$

$$\overline{l} = 70\,\text{MW}$$

$$\sigma_l = 10\,\text{MW}$$

It is connected to a distribution generation have availability 0.9 and rated capacity 30 MW. Obtain reliability.

Solution:
Mean capacity of distribution generation

$$\overline{C}_{dg} = 0.1 \times 0 + 0.9 \times 30$$

$$\overline{C}_{dg} = 27\,\text{MW}$$

Total mean capacity

$$\overline{C}_T = \overline{C} + \overline{C}_{dg}$$

$$\overline{C}_T = 100 + 27$$

$$\overline{C}_T = 127\,\text{MW}$$

$$\sigma_{dg}{}^2 = 0.1 \times (0 - \overline{C}_{dg})^2 + 0.9 \times (30 - \overline{C}_{dg})^2$$

$$\sigma_{dg}{}^2 = 0.1 \times (0 - 27)^2 + 0.9 \times (30 - 27)^2$$

$$\sigma_{dg}{}^2 = 81$$

Therefore

$$\sigma_{dg} = 9\,\text{MW}$$

$$\sigma_T = \sqrt{\sigma_C{}^2 + \sigma_{dg}^2}$$

$$\sigma_T = \sqrt{100 + 81}$$

$$\sigma_T{}^2 = 181$$

$$\beta = \frac{\overline{C}_T - \overline{l}}{\sqrt{\sigma_T{}^2 + \sigma_l{}^2}}$$

$$\beta = \frac{127 - 70}{\sqrt{181 + 100}}$$

$$\beta = \frac{57}{16.76} = 3.400$$

Therefore, $\varphi(\beta) = \varphi(3.400)$

$$R = 0.996631 \times 0.98$$
$$R = 0.97669$$

15.6 For the deterministic criteria given below:

System state	Deterministic criteria
Healthy state if	$(\overline{C}/\overline{l}) \geq 1.2$
Marginal state if	$1.2 > (\overline{C}/\overline{l}) \geq 1.1$
Risky state	Otherwise

Determine the system state for the following cases from (a) to (c). Also determine the number of transitions from one state to another when moving from case (a) to (c) sequentially. What if determine the number of transitions from one state to another when moving from case (c) to (a) sequentially i.e. in reverse order. Where Capacity of substation is normally distributed with mean and standard deviation $\overline{C}$, σ_C. And Load is also normally distributed with mean and standard deviation $\overline{l}$, σ_l.

(a)

$$\overline{C} = 130\,\text{MW}$$
$$\sigma_C = 10\,\text{MW}$$
$$\overline{l} = 100\,\text{MW}$$
$$\sigma_l = 10\,\text{MW}$$

(b)

$$\overline{C} = 115\,\text{MW}$$
$$\sigma_C = 10\,\text{MW}$$
$$\overline{l} = 100\,\text{MW}$$
$$\sigma_l = 10\,\text{MW}$$

(c)

$$\overline{C} = 105\,\text{MW}$$
$$\sigma_C = 10\,\text{MW}$$
$$\overline{l} = 100\,\text{MW}$$
$$\sigma_l = 10\,\text{MW}$$

Solution:

(a)

$$(\overline{C}/\overline{l}) = 130/100$$
$$(\overline{C}/\overline{l}) = 1.3$$
$$\text{As } (\overline{C}/\overline{l}) \geq 1.2$$

(Therefore system state is healthy state (H).)

(b)

$$(\overline{C}/\overline{l}) = 115/100$$
$$(\overline{C}/\overline{l}) = 1.15$$
$$\text{As } 1.2 > (\overline{C}/\overline{l}) \geq 1.1$$

Therefore system state is marginal state (M).

(c)

$$(\overline{C}/\overline{l}) = 105/100$$
$$(\overline{C}/\overline{l}) = 1.05$$
$$\text{As } (\overline{C}/\overline{l}) < 1.1$$

Therefore system state is risky state (R).

As it is observed that moving from Case (a) to (b) system state is changed from H to M therefore

$$n_{\text{HM}} = 1 \text{ (number of transition from H to M)}$$

Further moving from Case (b) to (c)

$$n_{\text{MR}} = 1 \text{ (number of transition from M to R)}$$

If moves from case (c) to (a) i.e. in reverse order sequentially then it is observed that moving from Case (c) to (b) system state is changed from R to M therefore

$$n_{\text{RM}} = 1 \text{ (number of transition from R to M)}$$

Further moving from Case (b) to (a)

$$n_{\text{MH}} = 1 \text{ (number of transition from M to H)}$$

15.7 The Capacity of distribution substation and load is normally distributed with

$$\overline{C} = 100\,\text{MW}$$
$$\sigma_{\text{C}} = 10\,\text{MW}$$
$$\overline{l} = 95\,\text{MW}$$
$$\sigma_{\text{l}} = 10\,\text{MW}$$

Distribution substation is integrated with a DG unit having normally distribution with $\overline{C}_{\text{dg}} = 10\,\text{MW}$ and $\sigma_{\text{dg}} = 1\,\text{MW}$ and assuming availability as 1. Then determine the system state as per deterministic criteria given Exercise 15.6 with and without DG unit.

Solution:

Without DG unit

$$(\overline{C}/\overline{l}) = 100/95$$
$$(\overline{C}/\overline{l}) = 1.05263$$

As

$$(\overline{C}/\overline{l}) < 1.1$$

Therefore system state is risky state.

Now with single DG unit: Total system capacity

$$\overline{C}_{\mathrm{T}} = \overline{C} + \overline{C}_{\mathrm{dg}} = 110\,\mathrm{MW}$$
$$(\overline{C}_{\mathrm{T}}/\bar{l}) = 110/95$$
$$(\overline{C}_{\mathrm{T}}/\bar{l}) = 1.15789$$
$$\text{As } 1.2 > (\overline{C}_{\mathrm{T}}/\bar{l}) \geq 1.1$$

Therefore system state is marginal state.

It is observed that system state is changed from risky state to marginal state with addition of single DG unit of given capacity.

15.8 In the example Exercise 15.7 a second identical DG unit is integrated then determine the system state.

Now with single DG unit: Total system capacity

$$\overline{C}_{\mathrm{T}} = \overline{C} + \overline{C}_{\mathrm{dg}} = 120\,\mathrm{MW}$$
$$(\overline{C}_{\mathrm{T}}/\bar{l}) = 120/95$$
$$(\overline{C}_{\mathrm{T}}/\bar{l}) = 1.26315$$
$$\text{As } (\overline{C}/\bar{l}) \geq 1.2$$

Therefore system state now in healthy state.

Note: It is observed that with addition of one DG units the system state is changed from risky to marginal then further addition of DG unit system state is shifted to healthy state.

15.9 For a distribution system depending on load and supply capacity availability, the system states are represented by three-state model (H, M, R) as discussed in chapter given for 10 samples of distribution substation capacity and load distribution sequentially is given as follows:

$$S = [\mathrm{H\ M\ R\ H\ R\ M\ H\ R\ H\ M}]$$

Then determine the number of transitions from one state to another n_{HM}, n_{HR}, n_{MH}, n_{MR}, n_{RH}, and n_{RM}.

Solution:

The number of transitions from one state to another is obtained sequentially as follows:

$$n_{\mathrm{HM}} = 2$$
$$n_{\mathrm{HR}} = 2$$
$$n_{\mathrm{MH}} = 1$$
$$n_{\mathrm{MR}} = 1$$
$$n_{\mathrm{RH}} = 2$$
$$n_{\mathrm{RM}} = 1$$

15.10 A three-state model (Healthy [H], Marginal [M], and Risky state [R]) is presented for reliability analysis for a composite distribution system with the following data:

$$\lambda_{HM} = 0.5$$
$$\lambda_{HR} = 0.3$$
$$\lambda_{MR} = 0.4$$

The probabilities are given as follows for the three states:

$$p_H = 0.2$$
$$p_M = 0.7$$
$$p_R = 0.1$$

Calculate

i) Transition rates λ_{MH}, λ_{RH}, λ_{RM}
ii) Frequencies f_H, f_M, f_R
iii) Mean time in three states MHT, MMT, MRT
iv) Availabilities on annual basis in three states A_H, A_M, A_R.

Refer Figure 15.3 for three-state model.

Solution:

From Frequency balance approach (Arya et al. 2013)

$$P_H \lambda_{HM} = P_M \lambda_{MH}$$

Therefore

$$\lambda_{MH} = (P_H \lambda_{HM})/P_M$$

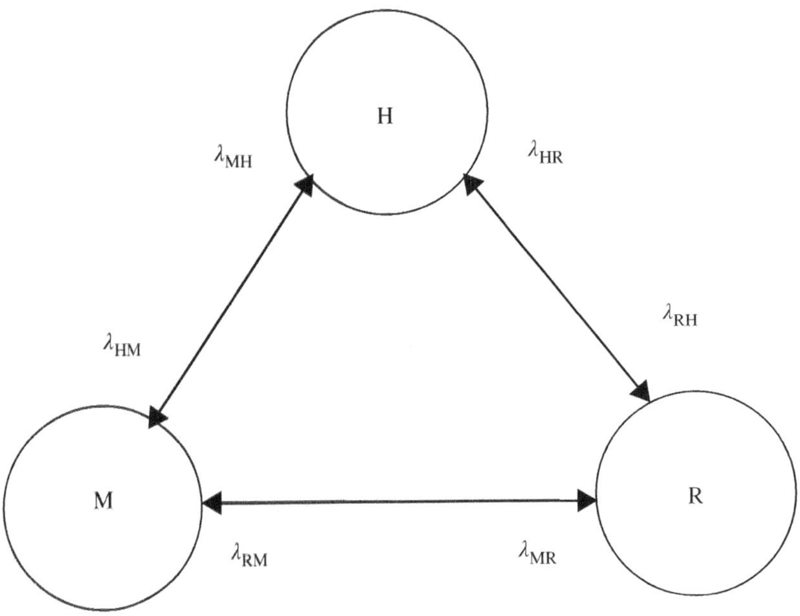

Figure 15.3 Three-state model.

$$\lambda_{\text{MH}} = \frac{0.2 X 0.5}{0.7}$$

$$\lambda_{\text{MH}} = 0.1428$$

$$P_{\text{H}}\lambda_{\text{HR}} = P_{\text{R}}\lambda_{\text{RH}}$$

Therefore

$$\lambda_{\text{RH}} = (P_{\text{H}}\lambda_{\text{HR}})/P_{\text{R}}$$

$$\lambda_{\text{RH}} = \frac{0.2 \times 0.3}{0.1}$$

$$\lambda_{\text{RH}} = 0.6$$

$$P_{\text{M}}\lambda_{\text{MR}} = P_{\text{R}}\lambda_{\text{RM}}$$

Therefore

$$\lambda_{\text{RM}} = (p_{\text{M}}\lambda_{\text{MR}})/P_{\text{R}}$$

$$\lambda_{\text{RM}} = \frac{0.7 X 0.4}{0.1}$$

$$\lambda_{\text{RM}} = 2.8$$

Now

$$\begin{aligned} f_{\text{H}} &= P_{\text{H}}(\lambda_{\text{HM}} + \lambda_{\text{HR}}) \\ &= 0.2\,(0.5 + 0.3) = 0.16 \\ f_{\text{M}} &= P_{\text{M}}(\lambda_{\text{MH}} + \lambda_{\text{MR}}) \\ &= 0.7\,(0.1428 + 0.4) = 0.37996 \\ f_{\text{R}} &= P_{\text{R}}(\lambda_{\text{RH}} + \lambda_{\text{RM}}) \\ &= 0.1\,(0.6 + 2.8) = 0.34 \end{aligned}$$

Then mean time for H, M, and R states evaluated correspondingly:

$$\text{MHT} = \frac{p_{\text{H}}}{f_{\text{H}}} = \frac{0.2}{0.16} = 1.25\,\text{h/yr}$$

$$\text{MMT} = \frac{p_{\text{M}}}{f_{\text{M}}} = \frac{0.7}{0.37996} = 1.84229\,\text{h/yr}$$

$$\text{MRT} = \frac{p_{\text{R}}}{f_{\text{R}}} = \frac{0.1}{0.34} = 0.29411\,\text{h/yr}$$

Availabilities on annual basis in H, M, and R states are obtained as:

$$AP_{\text{H}} = 8760 \cdot p_{\text{H}} = 8760 \times 0.2 = 1752\,\text{h/yr}$$

$$AP_{\text{M}} = 8760 \cdot p_{\text{M}} = 8760 \times 0.7 = 6132\,\text{h/yr}$$

$$AP_{\text{R}} = 8760 \cdot p_{\text{R}} = 8760 \times 0.1 = 876\,\text{h/yr}$$

15.11 The Capacity of distribution substation and load is normally distributed with

$$\overline{C} = 100\,\text{MW}$$

$$\sigma_{\text{C}} = 10\,\text{MW}$$

$$\overline{l} = 95\,\text{MW}$$

$$\sigma_{\text{l}} = 10\,\text{MW}.$$

Find the range of samples for distribution substation capactiy and load.

Solution:

The distribution substation capacity sample range is given as follows:

$$\overline{C} - 4\sigma_C \text{ to } \overline{C} + 4\sigma_C$$

Therefore range will be

$$((100 - 40), (100 + 40)) = (60{,}140)\,\text{MW}$$

The distribution substation load demand sample range is given as follows:

$$\bar{l} - 4\sigma l \,\text{to}\, \bar{l} + 4\sigma l$$

Therefore range will be $((95 - 40), (95 + 40)) = (55{,}135)$ MW.

15.7 Conclusions

This chapter presents a deterministic criterion-based reliability analysis of CDS for three-state model with evaluation of probabilities in H, M, and R state using MCS. The frequency-duration concept is used to assess frequencies of H, M, and R states along with duration indices in H, M, and risky states e.g. MHT, MMT, and MRT using the simulation. The results show a reliability analysis with DG to a sample composite distribution (CDS) system. Significant Improvement is observed in the performance with the addition of DG capacity such as reduction in risky state probability and mean risky state time, increase in healthy state probability and mean healthy state time based on the frequency indices. Also, illustrative examples are discussed to understand the concept.

References

Ahmadi-Khatir, A., Fotuhi-Firuzabad, M., and Goel, L. (2009). Customer choice of reliability in spinning reserve procurement and cost allocation using well-being analysis. *Electric Power Systems Research* 79 (10): 1431–1440.

Amaral, T.S., Borges, C.L.T., and Rei, A.M. (2010). Composite system well-being evaluation based on non-sequential Monte Carlo simulation. *Electric Power Systems Research* 80 (1): 37–45.

Arya, L.D. and Koshti, A. (2008). Distributed generation capacity reliability evaluation using safety index. *Journal of The Institution of Engineers (India)* 89: 3–7.

Arya, L.D. and Koshti, A. (2012). Probabilistic simulation approach for distributed generation (DG) capacity evaluation using artificial neural network representation of load duration curve. *Journal of The Institution of Engineers (India)* 93: 57–62.

Arya, L.D., Koshti, A., and Choube, S.C. (2013). Frequency-duration analysis of composite distribution system using a non-sequential Monte Carlo simulation. *International Journal of Electrical Power & Energy Systems* 46: 17–25.

Billinton, R. and Allan, R.N. (1992). *Reliability Evaluation of Engineering Systems*, vol. 792. Springer.

Billinton, R. and Li, W. (1993). A system state transition sampling method for composite system reliability evaluation. *IEEE Transactions on Power Systems* 8 (3): 761–770.

Billinton, R. and Lian, G. (1994). Composite power system health analysis using a security constrained adequacy evaluation procedure. *IEEE Transactions on Power Systems* 9 (2): 936–941.

Costa, P.M. and Matos, M.A. (2009). Assessing the contribution of microgrids to the reliability of distribution networks. *Electric Power Systems Research* 79 (2): 382–389.

El-Khattam, W. and Salama, M.M.A. (2004). Distributed generation technologies, definitions and benefits. *Electric Power Systems Research* 71 (2): 119–128.

Halve, S.S., Koshti, A., and Arya, R. (2023). A sampling method based on system state transition for distribution system adequacy assessment using distributed generation. *Journal of Operation and Automation in Power Engineering* 11 (4): 249–257.

Hamzeh, M. and Vahidi, B. (2020). Reliability evaluation of distribution transformers considering the negative and positive effects of rooftop photovoltaics. *IET Generation, Transmission & Distribution* 14 (15): 3063–3069.

Hassan, R., Das, B.K., and Hasan, M. (2022). Integrated off-grid hybrid renewable energy system optimization based on economic, environmental, and social indicators for sustainable development. *Energy* 250: 123823.

Hegazy, Y.G., Salama, M.M.A., and Chikhani, A.Y. (2003). Adequacy assessment of distributed generation systems using Monte Carlo simulation. *IEEE Transactions on Power Systems* 18 (1): 48–52.

Khezri, R. and Mahmoudi, A. (2020). Review on the state-of-the-art multi-objective optimisation of hybrid standalone/grid-connected energy systems. *IET Generation, Transmission & Distribution* 14 (20): 4285–4300.

Koshti, A.S., Verma, A., Arya, R. et al. (2023). Tolerable random interruption duration based reliability estimation of stand alone hybrid renewable energy system by network reduction and sequential Monte Carlo simulation. *IET Generation, Transmission & Distribution*.

Kothari, D.P., Nagrath, I.J., and Saket, R.K. (2022). *Modern Power System Analysis*. Tata McGraw-Hill Publishing Company.

Pereira, M.V.F. and Balu, N.J. (1992). Composite generation/transmission reliability evaluation. *Proceedings of the IEEE* 80 (4): 470–491.

Probability Methods Subcommittee (1979). IEEE reliability test system. *IEEE Transactions on Power Apparatus and Systems* 6: 2047–2054.

Ranjbar, H., Saber, H., and Sharifzadeh, M. (2022). Bi-level planning of distributed energy resources into existing transmission grids: pathway to sustainable energy systems. *IET Generation, Transmission & Distribution* 16 (24): 4963–4979.

Roy, P., He, J.B., Zhao, T., and Singh, Y.V. (2022). Recent advances of wind-solar hybrid renewable energy systems for power generation: a review. *IEEE Open Journal of the Industrial Electronics Society* 3: 81–104.

Wang, Y., Guo, C., and Wu, Q.H. (2013). A cross-entropy-based three-stage sequential importance sampling for composite power system short-term reliability evaluation. *IEEE Transactions on Power Systems* 28 (4): 4254–4263.

Wangdee, W. and Billinton, R. (2006). Bulk electric system well-being analysis using sequential Monte Carlo simulation. *IEEE Transactions on Power Systems* 21 (1): 188–193.

Xu, Y. and Singh, C. (2012). Adequacy and economy analysis of distribution systems integrated with electric energy storage and renewable energy resources. *IEEE Transactions on Power Systems* 27 (4): 2332–2341.

Yahaya, A.A., AlMuhaini, M., and Heydt, G.T. (2020). Optimal design of hybrid dg systems for microgrid reliability enhancement. *IET Generation, Transmission & Distribution* 14 (5): 816–823.

Section 5

Reliability Analysis of Distribution Systems Integrated With Renewable Energy Systems

16

Reliability Assessment of Distribution Systems Integrated with Renewable Energy Systems

Sachin Kumar[1], *Sandeep Kumar*[2], *Aanchal Singh S. Vardhan*[3], *R. K. Saket*[4], *and P. Sanjeevikumar*[5]

[1] *Department of Electrical Engineering, G.B. Pant Institute of Engineering and Technology, Pauri-Garhwal, Uttarakhand, India*
[2] *Department of Electronics and Communication Engineering, G.B. Pant Institute of Engineering and Technology, Pauri-Garhwal, Uttarakhand, India*
[3] *Department of Electrical Engineering, National Institute of Technology, Jamshedpur, Jharkhand, India*
[4] *Department of Electrical Engineering, Indian Institute of Technology (BHU), Varanasi, Uttar Pradesh, India*
[5] *Department of Electrical Engineering, IT and Cybernetics, University of South-Eastern Norway, Porsgrunn, Norway*

16.1 Introduction

Power system reliability (PSR) assessment is the process of evaluating the ability of an electrical power system to deliver electricity to consumers without interruptions or disturbances. It involves analyzing various factors that can affect the reliability of the system and identifying potential vulnerabilities or areas for improvement. Some key points to consider when conducting reliability assessments for such integrated systems are as follows.

The first step in assessing the reliability of a distribution system integrated with renewable energy systems is to model the renewable energy sources (RES)s accurately. This involves understanding the characteristics of renewable resources, such as solar irradiance or wind speed, and their variability over time. Various tools and techniques, such as weather data analysis and time series modeling, can capture the intermittent nature of renewable energy generation (Kumar et al. 2020a).

Reliability assessment for integrated distribution systems should incorporate probabilistic analysis methods. Traditional deterministic approaches may not capture the uncertainty associated with renewable energy generation and its impact on system reliability. Probabilistic methods, such as Monte Carlo simulation or stochastic optimization, can account for the variability and uncertainty in renewable energy availability and their effects on system performance.

The configuration and design of the distribution system play a crucial role in its reliability. This includes the layout of RESs, such as placing solar panels or wind turbines, and integrating energy storage systems. Proper system design can help mitigate the impacts of renewable energy variability and enhance system resilience (Kumar et al. 2020b).

Integrating renewable energy and distribution systems requires careful coordination and control mechanisms. Grid integration strategies, such as advanced power electronics, demand response, and smart grid technologies, can improve the reliability and stability of the overall system. These strategies enable efficient utilization of renewable energy, effective load management, and enhanced grid resiliency.

Several reliability metrics can be used to assess the performance of integrated distribution systems. Typical metrics include the Expected Energy Not Supplied, Average Energy Not Supplied,

Reliability Analysis of Modern Power Systems, First Edition. Edited by R. K. Saket and P. Sanjeevikumar.

System Average Interruption Duration Index, System Average Interruption Frequency Index, and Average Service Availability Index, referred to as EENS, AENS, SAIDI, SAIFI, and ASAI, respectively, which are the reliability indices taken into account for the EPDN reliability assessment. These metrics quantify the frequency and duration of power interruptions and can be used to evaluate the impact of renewable energy integration on system reliability.

Sensitivity analysis helps understand the influence of various factors on system reliability. Parameters such as renewable energy capacity, energy storage capacity, and demand patterns can significantly impact the reliability of integrated distribution systems. Sensitivity analysis identifies the critical parameters and their impact on system performance, aiding decision-making and system improvement (Kumar et al. 2021).

Reliability assessments should consider resiliency planning for integrated distribution systems. This involves identifying vulnerabilities, developing contingency plans, and implementing measures to mitigate the impact of extreme events or system disturbances. Resiliency planning ensures the system can quickly recover and restore normal operations in the face of disruptions.

Assessing the reliability of distribution systems integrated with renewable energy systems requires a comprehensive approach that accounts for the intermittent nature of RESs and their impact on system performance. Considering the factors mentioned above and employing appropriate modeling and analysis techniques, evaluating and enhancing the reliability of such integrated systems is possible.

Load analysis involves studying the patterns and characteristics of electrical demand within the power system. By analyzing historical data and conducting load flow studies, the system operator can identify peak demand periods, load growth trends, and potential areas of congestion or overload. This information helps design a reliable system to meet the expected demand.

Fault analysis examines the behavior of the power system during abnormal conditions such as short circuits or equipment failures. It involves simulating and analyzing various fault scenarios to determine their impact on system reliability. This analysis helps identify weak points in the system, assess the adequacy of protection schemes, and optimize fault clearance strategies.

The adequacy assessment evaluates the capacity of the power system to meet the load demand under normal operating conditions. It considers factors such as the availability of generation resources, transmission capacity, and the reliability of essential equipment. Adequacy assessment helps identify potential capacity shortages and determines the need for system upgrades or additional resources.

PSR assessment often involves creating a detailed power system model, including generation units, transmission lines, substations, and distribution networks. This model performs various simulations and studies to assess the system's reliability. Computer tools like power system analysis software (e.g. PSS/E, PSCAD, and DIgSILENT) are commonly used for system modeling and analysis.

Risk assessment involves evaluating the potential risks and vulnerabilities that can impact the reliability of the power system. This includes analyzing factors like extreme weather events, equipment aging, cyber threats, and human errors. System operators can implement preventive measures and develop contingency plans to minimize disruptions and improve system reliability by assessing these risks.

Based on the findings of the reliability assessment, various measures can be implemented to enhance PSR. These measures may include infrastructure upgrades, system reinforcement, equipment maintenance and replacement strategies, grid modernization initiatives, and the implementation of advanced monitoring and control systems. PSR assessment is ongoing as power systems evolve and new challenges arise. It helps system operators and planners make informed decisions to ensure a reliable and resilient consumer electricity supply. For reliability assessment, in continuation, the chapter deals with essential reliability functions, various RESs, optimization, and control strategies, and some practical case studies with challenges and future directions. The chapter's flow can be easily depicted in Figure 16.1.

16.2 Reliability Functions

By analyzing the reliability or survival function ($\text{Re}(t)$), cumulative distribution function ($f_c u(t)$), failure density function ($f_f(t)$), and hazard rate function ($\text{Hz}(t)$), one could evaluate the reliability of a system. The reliability of a component, or the likelihood that a system or piece of equipment won't malfunction within a certain amount of time $(0, t)$, is a method for describing a component's failure probability. This relationship has been expressed using Eq. (16.1):

$$\text{Re}(t) = \text{Probability Pr}\{T \geq t\} \tag{16.1}$$

The $f_{cu}(t)$ is defined in Eq. (16.2), which is the failure probability that occurs before time "t":

$$f_{cu}(t) = 1 - \text{Re}(t) = \text{Pr}\{T < t\} \tag{16.2}$$

The derivative of $f_c(t)$ is termed $f_f(t)$, as given in Eq. (16.3), and the area under the entire $f(t)$ is calculated to be 1:

$$f_f(t) = \frac{\text{d}f_c(t)}{\text{d}t} = -\frac{\text{dRe}(t)}{\text{d}t} \tag{16.3}$$

Equation (16.4) states that $R(t)$ depends on the term failure rate function or hazard rate function ($H(t)$) and that a higher $H(t)$ will result in a lower $\text{Re}(t)$:

$$\text{Re}(t) = \text{e}^{-\int_0^t \text{Hz}(t)\text{d}t} \tag{16.4}$$

where

- $\text{Hz}(t) = \text{Pr}\{t \leq T \leq t + \Delta t | T \geq t\} = \dfrac{\text{Re}(t) - \text{Re}(t + \Delta t)}{\text{Re}(t)}$

$$\text{Hz}(t) = \frac{f_f(t)}{\text{Re}(t)} = -\frac{\text{dRe}(t)}{\text{d}t} \times \frac{1}{\text{Re}(t)} \tag{16.5}$$

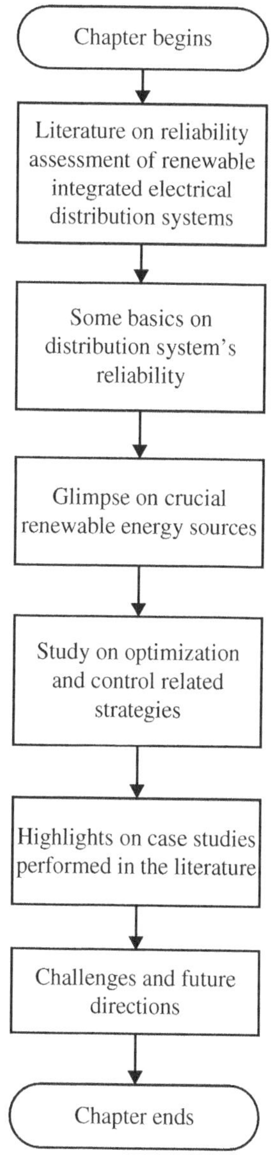

Figure 16.1 The flow of the chapter.

From Eq. (16.2), Infant mortality is an example of an early life failure, and $\text{Hz}(t)$ describes these failures as well as wear-out failures, in which the failure rate significantly rises. A bathtub curve may be used to demonstrate this. $\text{Hz}(t)$ is a different method of expressing a failure distribution. When $\text{Hz}(t)$ is rising, decreasing, or constant, failure rates may be described as increasing, decreasing, or constant.

To simulate equipment lifetime, any distribution may be utilized. In reality, distribution functions with monotonic hazard functions seem to be the most realistic, and a handful within that class are widely considered to offer the most plausible system reliability models. It is important to note that the following distributions can be used to obtain the life distribution of an electrical power system or component: Binomial, Exponential, Poisson, Geometric, normal, Gamma, and Weibull Distributions.

The basic parameter used in the reliability evaluation is 'forced outage' at some distant time in the future. This probability parameter is defined as the system or component unavailability. It is known as the forced outage rate in power systems, as mentioned in Eqs. (16.6) and (16.7):

$$\text{Unavailability} = \frac{\lambda_f}{\lambda_f + \mu_r} \tag{16.6}$$

$$\text{Availability} = \frac{\mu_r}{\lambda_f + \mu_r} \tag{16.7}$$

where

- λ_f = expected failure rate = $\dfrac{1}{\text{Mean-Time-To-Failure (MTTF)}}$
- μ_r = expected repair rate = $\dfrac{1}{\text{Mean-Time-To-Repair (MTTR)}}$

The aforementioned reliability parameters operate as reliability indices. At load point "*L*," the reliability parameters can be estimated.

The reliability assessment of a distribution network is as crucial as other components of the EPDN. The IEEE guide for EPDN reliability is based on the IEEE guide, referenced in standard 1366-2012. The reliability of an EPDN can be evaluated using a few reliability indices under the provided specification. According to Equations, these indexes are divided into two categories (16.8)–(16.14):

$$\text{EENS} = \sum[(\text{load at } L\text{th load}) \times (\text{Annual } D_l \text{ at } L\text{th load})] \tag{16.8}$$

$$\text{AENS} = \frac{\sum(\text{EENS at } L\text{th load})}{\text{Total no. of customers at all loads}} \tag{16.9}$$

$$\text{SAIDI} = \frac{\sum[(\text{annual } D_L \text{ at } L\text{th load}) \times (\text{No. of customers at } L\text{th load})]}{\text{Total no. of customers at all loads}} \tag{16.10}$$

$$\text{SAIFI} = \frac{\sum[(\text{average } \lambda_f \text{ at } L\text{th load}) \times (\text{No. of customers at } L\text{th load})]}{\text{Total no. of customers at all loads}} \tag{16.11}$$

$$\text{ASAI} = 1 - \frac{\text{SAIDI}}{8760} \tag{16.12}$$

$$\text{CAIDI} = \frac{\sum_{L=1}^{n_L} U_L N_L}{\sum_{L=1}^{n_L} \lambda_f N_L} \text{ hour/customer/interruption} \tag{16.13}$$

$$\text{ASUI} = 1 - \text{ASAI p.u.} \tag{16.14}$$

where

- λ_f = failure/year
- U_L = hour/year
- D_L = repair rate in hour

16.3 Renewable Energy Sources

Renewable energy is derived from naturally replenished sources and can be sustained over time without depleting finite resources. These energy sources are considered environmentally friendly because they impact the planet less than fossil fuels can contribute to climate change by releasing greenhouse gases. RESs include the following sources.

16.3.1 Solar Energy

This is generated by capturing sunlight using photovoltaic cells (solar panels) to convert it directly into electricity or using solar thermal systems to produce heat for various applications. IEEE 1562-2021 recommends practice for sizing stand-alone photovoltaic (PV) systems. Solar PV systems convert sunlight directly into electricity using semiconductor materials in solar cells, as demonstrated in Figure 16.2.

The basic operating principle involves the creation of electron-hole pairs when photons from sunlight strike the semiconductor material. These electron-hole pairs generate an electric current, which can be harnessed for various applications.

The fundamental equation that describes the current–voltage relationship in a solar cell is the diode equation, which combines the current components due to the light-generated current and the diode-like behavior of the semiconductor material, as depicted in Figure 16.3 and the related equation is given as:

$$I = I_L - I_D \left(\exp\left(\frac{V}{V_T} \right) - 1 \right) \tag{16.15}$$

where

- I = the solar cell output current.
- I_L = the light-generated current $\propto$ intensity of incident light.
- I_D = the diode current, dependent on the voltage across the solar cell.
- V = the voltage across the solar cell terminals.
- V_T = the thermal voltage, a characteristic temperature-related voltage.

An equivalent circuit can represent a PV cell and model its behavior under different conditions. The equivalent circuit is an electrical representation that simplifies the complex physical processes

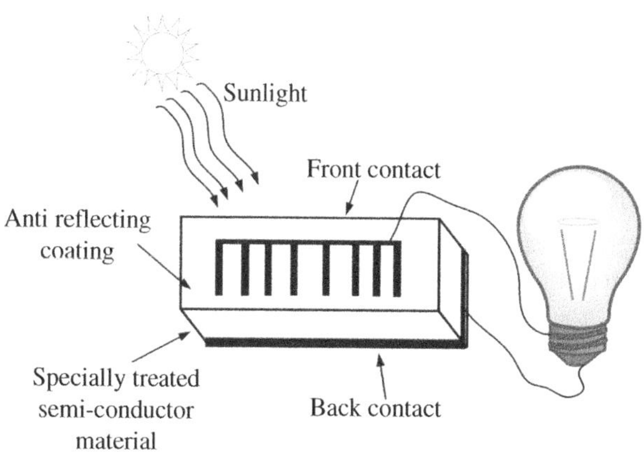

Figure 16.2 PV cell: basic operation.

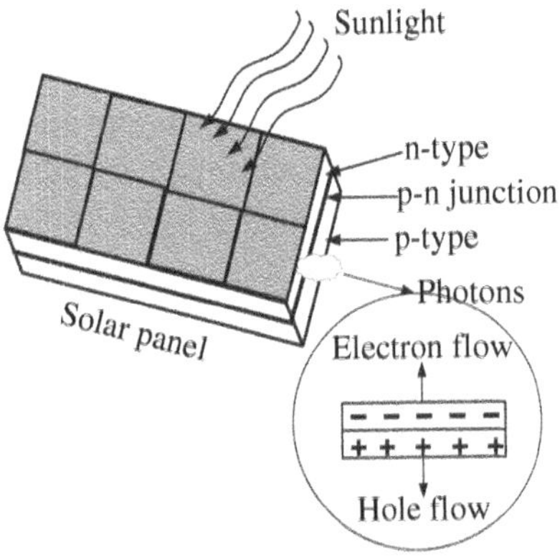

Figure 16.3 PV effect: a schematic diagram.

occurring within the PV cell. However, the most popular comparable circuit model for a PV cell is the one-diode model; double- and triple-diode models have also been implemented.

By multiplying the current and voltage at each point along the curve, one can determine the power output of a solar cell, also known as the power-voltage ($P - V$) curve:

$$P = I \times V \tag{16.16}$$

According to Figure 16.4, the $P - V$ curve of a solar photovoltaic (PV) module represents the relationship between the power output and the voltage across the device's terminals. This curve is essential for understanding the behavior and performance of a solar cell or module under different operating conditions.

The shape of the $P - V$ curve can vary based on factors such as temperature, shading, and the characteristics of the solar cell or module. In most cases, the curve has a characteristic "hump" shape, with the MPP located on the descending side of the curve, typically at around 70–80% of the open-circuit voltage.

Solar PV systems are designed to ensure that they operate near the MPP under varying conditions (e.g. changing sunlight intensity and temperature) to maximize energy production and efficiency. MPP Tracking algorithms in solar inverters or charge controllers often achieve this.

Engineers use more complex models that consider factors such as semiconductor material properties, shading effects, and system efficiency for detailed analysis and design of solar PV systems. In addition, reliability and availability are crucial factors in designing, operating, and maintaining

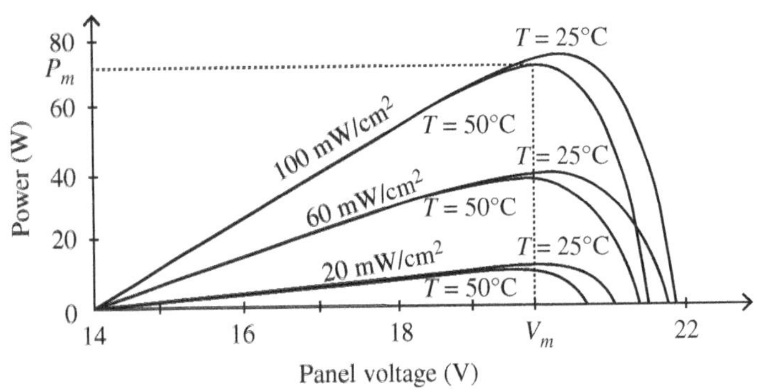

Figure 16.4 Power-voltage curve of a solar PV.

grid-connected solar PV systems. These systems aim to generate renewable energy consistently and contribute to the overall energy supply.

16.3.2 Wind Energy

Wind turbines capture the kinetic energy of moving air and turn it into electricity. When exposed to the wind, turbine blades rotate, producing wind power. The IEEE 1834-2019 standard is available for wind turbine rotor systems.

A wind energy conversion system (WECS) is designed to harness the wind's kinetic energy and convert it into electrical energy. Several components and technologies are involved in a typical WECS: wind turbine, generator, transmission system, power electronics, yaw and pitch control systems, tower and foundations, nacelle, anemometer, wind vane, and monitoring and control systems. A typical DFIG-based WECS is illustrated in Figure 16.5.

The air density (kg/m^3), the WT rotor radius (m), the wind speed (m/s), and the power coefficient all affect the mechanical and electrical power outputs. The coefficient, which depends on the blade pitch angle and tip speed ratio, is calculated as the WT power's ratio to the wind stream's power, as represented by Eq. (16.17):

$$P_W = \begin{cases} 0; & 0 \leq V \leq V_{\text{c,in}} \\ & \text{or} \\ & V \geq V_{\text{c,out}} \\ P_{W,\text{r}} \times \left(\dfrac{V - V_{\text{c,in}}}{V_{\text{r}} - V_{\text{c,in}}} \right); & V_{\text{c,in}} \leq V \leq V_{\text{r}} \\ P_{W,\text{r}}; & V_{\text{r}} \leq V \leq V_{\text{c,out}} \end{cases} \tag{16.17}$$

where P_W = output power, $P_{W,\text{r}}$ = rated output power, V_{r} = rated wind speed, $V_{\text{c,in}}$, and $V_{\text{c,out}}$ = cut-in and cut-out speeds, respectively.

Wind energy conversion systems come in various sizes, from small residential turbines to large utility-scale wind farms with multiple turbines. The technology has advanced significantly over the years, improving efficiency, reducing costs, and increasing reliability, making wind energy an increasingly important contributor to the global energy mix.

A WECS, also commonly referred to as a wind turbine or wind power system, is a technology that converts the wind's kinetic energy into electrical energy. Wind turbines are a critical component

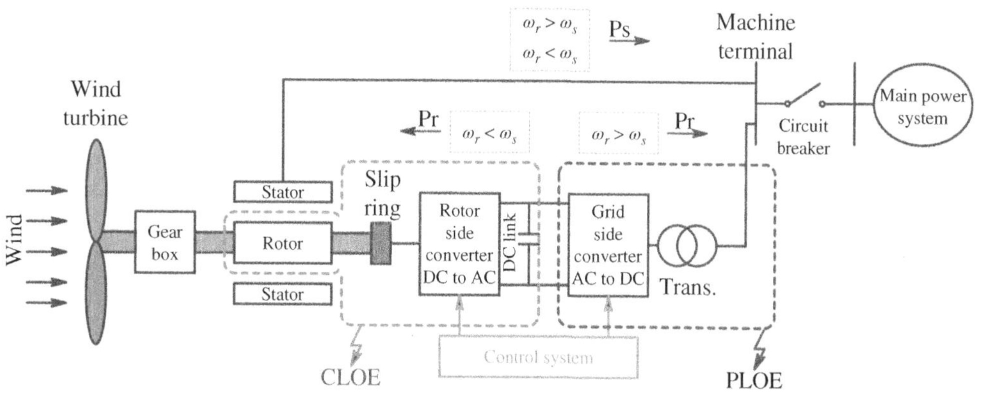

Figure 16.5 DFIG-based WECS structure. ⇒ω_r and ω_s are the rotor and synchronous speeds, P_r and P_s are rotor and stator active powers. CLOE and PLOE are the complete and partial loss of excitations.

of WECS and are used to capture the energy from the wind and transform it into a usable form of electricity. Here are the key components and aspects of a typical Wind Energy Conversion System:

1. **Wind Turbine**: The wind turbine is the primary component capturing wind energy. There are two main types of wind turbines: horizontal-axis wind turbines (HAWT) and vertical-axis wind turbines (VAWT). HAWTs are the most common and resemble traditional windmills, while VAWTs have a vertical rotor axis and are less common.
2. **Rotor Blades**: In HAWTs, the rotor blades are attached to the rotor hub and capture the wind's kinetic energy. The blades, such as fiberglass or composite materials, are typically lightweight and durable.
3. **Rotor Hub**: The rotor hub connects the blades to the main shaft, allowing the blades to rotate. It also houses various control systems and sensors.
4. **Main Shaft**: The main shaft is connected to the rotor hub and transfers the rotational energy from the blades to the generator.
5. **Generator**: The generator converts mechanical energy from the rotating shaft into electrical energy. Most wind turbines use synchronous, asynchronous, or permanent magnet generators.
6. **Tower**: The tower supports the wind turbine and elevates it to a height where it can capture stronger and more consistent winds. Taller towers are often used to access higher wind speeds at greater heights.
7. **Nacelle**: The nacelle is a protective housing that encloses the generator and other critical components. It is usually mounted on the tower and contains control systems, gearboxes (if needed), and other necessary equipment.
8. **Anemometer and Wind Vane**: These instruments are used to measure wind speed and direction to optimize the orientation of the wind turbine blades for maximum energy capture.
9. **Control System**: A control system regulates the wind turbine's operation. It adjusts the pitch angle of the blades and the rotor speed to ensure efficient energy capture and protect the turbine in varying wind conditions.
10. **Power Electronics**: Power electronics components, such as inverters and converters, convert the generator's variable voltage and frequency output into grid-compatible electricity.
11. **Grid Connection**: The electricity generated by the wind turbine is typically connected to the electrical grid. This allows for the distribution of wind-generated power to homes, businesses, and industries.
12. **Energy Storage (Optional)**: Some WECS may include energy storage systems, such as batteries, to store excess energy when wind conditions are favorable and release it when the wind is not blowing.

Further, increasing the availability and reliability of wind turbine generator systems is crucial for optimizing their performance and reducing downtime. Here are some strategies and approaches that can help achieve this goal: regular maintenance and inspections, condition monitoring, remote monitoring and control, redundancy and backup systems, lifecycle planning, grid integration, and energy storage (Kumar et al. 2021).

16.3.3 Vehicle Charging-Discharging

Integrating electric vehicle (EV) charging and discharging into the distribution system can enhance its reliability, but it also introduces new challenges that must be carefully managed. Here are some ways in which EV charging and discharging, as represented in Figures 16.6 and 16.7, respectively, can impact distribution system reliability (DSR).

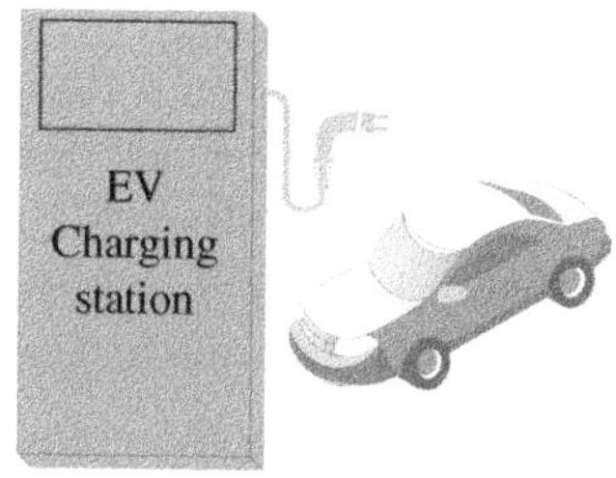

Figure 16.6 EV charging structure.

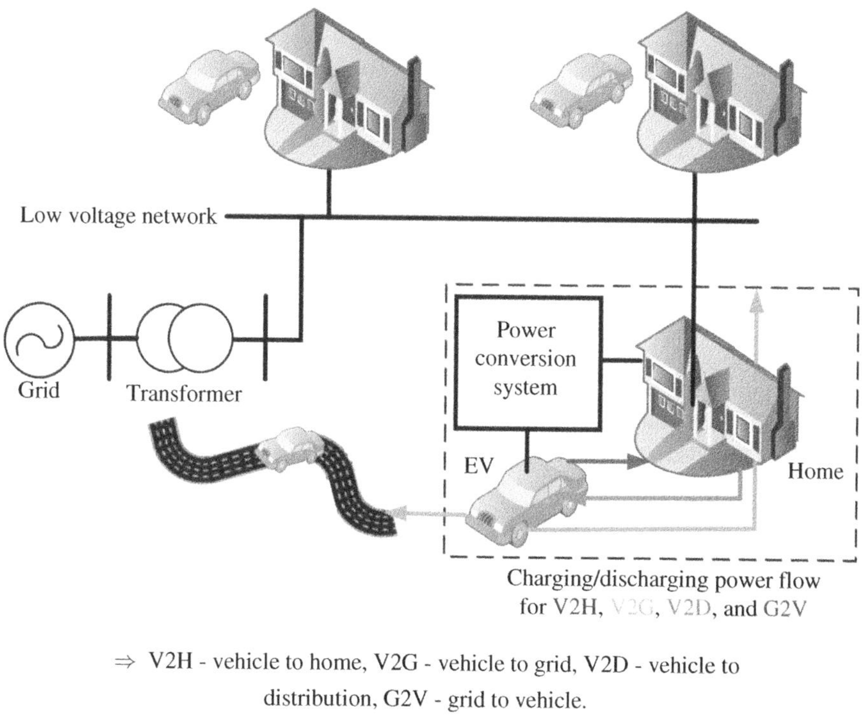

Figure 16.7 Strategies related to of EV-to-everything. ⇒ V2H – vehicle to home, V2G – vehicle to grid, V2D – vehicle to distribution, G2V – grid to vehicle.

1. **Demand Response and Peak Shaving:** EVs can be used for demand response, which involves adjusting their charging or discharging patterns to help balance electricity demand with supply. During times of high demand, EV charging can be temporarily reduced or shifted to off-peak hours, alleviating stress on the distribution grid and preventing grid overload during peak periods.
2. **Energy Storage:** Some EVs have substantial battery capacities, particularly electric buses and larger vehicles. These batteries can be utilized for energy storage, absorbing excess energy from the grid during periods of low demand and injecting it back into the grid during peak demand, effectively acting as distributed energy storage systems.
3. **Grid Ancillary Services:** EVs can provide ancillary services to the grid, such as frequency regulation and voltage support. These services help maintain the stability and reliability of the grid by quickly responding to fluctuations in supply and demand.

4. **Voltage Regulation:** Strategically placing EV charging infrastructure across the distribution network can help regulate voltage levels, ensuring that they remain within acceptable limits and minimizing voltage fluctuations.
5. **Islanded Operation:** In the case of a localized power outage, EVs with bidirectional charging capability can potentially power critical infrastructure or homes during islanded operation, enhancing resiliency and reliability.

Some primary EV charging and discharging equations are depicted in Eqs. (16.18)–(16.22). The modeling and other descriptions can be observed as follows.

1. **Charging Power (P_c):** The power at which an EV is being charged is given by:

$$P_c = V_c \times I_c \tag{16.18}$$

   where
   - P_c = charging power (W)
   - V_c = charging voltage (V)
   - I_c = charging current (A)
2. **Battery Charging Time (T_c):** The time it takes to charge an EV's battery is calculated using:

$$T_c = E_{\text{cap}}/P_c \tag{16.19}$$

   where
   - T_c = charging time (h)
   - E_{cap} = energy capacity of the battery (Wh)
3. **State of Charge (SoC):** The State of Charge represents the current energy level of the battery as a percentage. It can be calculated using:

$$\text{SoC} = (E_{\text{cur}}/E_m) \times 100 \tag{16.20}$$

   where
   - E_{cur} = current energy level of the battery (Wh)
   - E_m = maximum energy capacity of the battery (Wh)
4. **Discharging Power (P_{dis}):** The power drawn from the battery during discharging is given by the same equation as the charging power:

$$P_{\text{dis}} = V_{\text{dis}} \times I_{\text{dis}} \tag{16.21}$$

   where
   - P_{dis} = discharging power (W)
   - V_{dis} = discharging voltage (V)
   - I_{dis} = discharging current (A)
5. **Battery Discharging Time (T_{dis}):** The time for which the battery can discharge at a certain power before reaching empty is calculated using:

$$T_{\text{dis}} = E_{\text{cel}}/P_{\text{dis}} \tag{16.22}$$

   where
   - T_{dis} = discharging time (h)
   - E_{cel} = current energy level of the battery (Wh)

These equations explain power, energy, and time relationships during EV charging and discharging. In practice, efficiency losses, charging profiles, and other factors might affect the actual charging and discharging times. Modern charging systems might incorporate advanced features like fast charging, variable charging rates, and smart charging algorithms.

However, there are challenges associated with integrating EVs into the distribution system as well (Kumar et al. 2020a):

1. **Load Management:** Sudden and widespread EV charging during peak hours can lead to localized grid congestion and overloads. Proper load management strategies are needed to ensure that charging is distributed intelligently across time and locations.
2. **Infrastructure Upgrades:** Integrating many EVs might necessitate infrastructure upgrades in some areas to accommodate increased demand. This can be expensive and requires careful planning.
3. **Bidirectional Charging:** While bidirectional charging (vehicle-to-grid, or V2G) can offer benefits, it requires sophisticated control systems to manage the bidirectional power flow safely and effectively.
4. **Communication and Control:** To coordinate EV charging and discharging to benefit the grid while minimizing interruptions to car owners, reliable communication and control systems are also essential.
5. **User Behavior:** Predicting and influencing user behavior regarding EV charging patterns can be complex, and incentives may be required to encourage the desired behavior.

In summary, while integrating EVs into the distribution system can enhance reliability and provide grid support, careful planning, advanced technologies, and effective policies are necessary to manage the associated challenges and optimize the benefits.

16.3.4 Other Renewable Energy Sources

It is mentioned in the literature that several RESs, such as hydropower, biomass, geothermal, and ocean energy, have the potential to significantly contribute to our global energy needs while minimizing environmental impact. These RESs offer several advantages, including reduced greenhouse gas emissions, minimal air pollution, long-term sustainability, and improved PSR. However, each source also has its own set of challenges.

The critical advantage of RESs is that they are sustainable and do not produce harmful emissions during energy generation. They help mitigate climate change, reduce greenhouse gas emissions, and rely less on limited fossil resources. However, challenges like intermittent availability (solar and wind) and the need for advanced technology and infrastructure remain to be addressed for the broader adoption of these energy sources. Simultaneously, dealing with these uncertain RESs is crucial by implementing various optimization and control techniques.

16.4 Optimization and Control

Optimization and control techniques enhance DSR by effectively managing grid operations, mitigating faults, and minimizing downtime. Optimizing DSR involves employing various techniques and methodologies to improve the stability and resilience of the electrical grid (Kumar et al. 2020b). Here are some optimization techniques specifically focused on enhancing DSR:

1. **Reliability-Centered Maintenance (RCM):** RCM is a methodical technique that identifies crucial assets, evaluates potential failure modes, and chooses the most efficient maintenance strategies. Utilities can minimize downtime and reduce the risk of outages by prioritizing maintenance efforts on the most critical components.

2. **Volt-VAR Optimization (VVO):** VVO adjusts voltage levels and reactive power flows to optimize system performance and minimize energy losses. This technique maintains voltage within desired limits, improving equipment efficiency and reducing the risk of voltage-related outages.
3. **Fault Location, Isolation, and Service Restoration (FLISR):** FLISR techniques use real-time data and automation to quickly detect faults, isolate affected areas, and restore power to unaffected zones. These systems minimize outage duration and improve service reliability.
4. **Optimal Placement of Distributed Energy Resources:** Using optimization algorithms, determine the best locations for integrating RESs, energy storage systems, and microgrids. This approach can enhance grid resilience by reducing dependency on centralized generation and improving local supply during outages.
   Advanced control techniques manage the operation of DERs, such as solar panels, wind turbines, and energy storage systems. These techniques optimize the utilization of renewable resources and improve grid resilience.
5. **Load Balancing and Load Shedding:** Employ optimization algorithms to balance loads across distribution feeders and substations. In overloads or emergencies, load-shedding algorithms can identify and disconnect non-critical loads to maintain grid stability.
6. **Demand Response Optimization:** Develop algorithms that optimize demand response programs, considering customer preferences, load profiles, and grid constraints. This ensures efficient load reduction during peak periods.
7. **Microgrid Operation Optimization:** Microgrid optimization techniques maximize the use of local resources (generation, storage, and demand) to maintain critical loads during grid outages. These algorithms balance supply and demand within the microgrid.
8. **Stochastic Optimization:** Stochastic optimization accounts for uncertainties in load forecasts, equipment failures, and renewable energy generation. It helps utilities make robust decisions in the face of varying conditions.
9. **Dynamic Voltage and Reactive Power Optimization:** These techniques adjust real-time voltage levels and reactive power settings to maintain optimal voltage profiles and minimize energy losses. Proper voltage control enhances equipment lifespan and reduces the risk of voltage-related outages.
10. **Risk Assessment and Mitigation:** Use risk assessment techniques to identify vulnerabilities in the distribution system and prioritize mitigation strategies. This can involve identifying weak points, single points of failure, and potential sources of disruption.
11. **Data Analytics and Predictive Maintenance:** Implement machine learning and data analytics to predict equipment failures based on historical data. Predictive maintenance can schedule interventions before issues lead to outages.
12. **Optimal Switching and Reconfiguration:** Algorithms for optimal switching and reconfiguration help reroute power during faults or outages, minimizing the affected customers and reducing the duration of outages.
13. **Wide-Area Monitoring and Control (WAMC):** WAMC systems use synchronized data from multiple points in the distribution network to make real-time decisions that improve grid stability and reliability.
14. **Scenario Analysis and Contingency Planning:** Conduct scenario analyses to model various failure scenarios and their impacts on the distribution system. Develop contingency plans that outline actions to be taken for specific events.

15. **Communication and Control Networks:** Implement robust communication systems to ensure quick and reliable information exchange between control centers, substations, and field devices.
16. **Cybersecurity Measures:** Utilize control techniques to secure communication networks and protect control systems from cyber threats that could compromise grid reliability.

The effectiveness of these techniques often depends on integrating them into an overarching control and optimization framework that considers the dynamic nature of distribution networks and the diverse range of operating conditions. This approach enables utilities to manage, adapt, and respond to various challenges while maintaining a reliable distribution system.

Each optimization technique requires a combination of advanced data analytics, modeling, simulation, and control strategies. The choice of technique depends on the specific goals and challenges of the distribution system, and a holistic approach often involves integrating multiple techniques for comprehensive reliability improvement.

16.5 Case Study

Any nation's ability to flourish economically and socially depends on the reliability of its power distribution infrastructure. Industrial, commercial, residential, and other institutions providing services are expanding quickly. All of these organizations require reliable power sources. As a result, utilities must make a solid effort to guarantee that customers receive a reliable power supply. However, the essential issues with any power distribution system are higher power interruption frequency and longer interruption duration, which impact customers' everyday lives.

Indeed, let's delve into a case study demonstrating how DSR can be enhanced using various strategies. A suburban region with a growing population has been experiencing frequent power outages due to aging infrastructure, adverse weather conditions, increasing electricity demand, etc. The local utility company aims to improve DSR to ensure consistent and high-quality power supply to its customers. Therefore, the following objectives must be guaranteed while assessing the reliability improvement of the power distribution system.

1. Minimize the frequency and duration of power outages.
2. Improve response and restoration times after outages.
3. Enhance the overall resilience of the distribution network.

Overall, the Roy Billinton Test System (RBTS) and IEEE Reliability Test System (RTS) significantly advance the understanding of PSR and develop and test new reliability assessment methods. Both are widely used in academic research, industrial applications, and educational settings within the power engineering community. Some case studies have been tabulated in Table 16.1 and discussed in Sections 16.5.1–16.5.3.

16.5.1 IEEE Reliability Test System-79

The IEEE RTS is another commonly used benchmark system for studying the reliability and performance of power systems. The IEEE RTS provides a standardized platform for researchers and practitioners to evaluate the effectiveness of various reliability analysis methods and techniques.

The IEEE RTS is a simplified power system network representation consisting of generators, transmission lines, loads, and other components. The IEEE Reliability Test System Task Force

Table 16.1 Case studies for assessing the various distribution systems' reliability.

S. No.	References	Case study
1	Ranjbar et al. (2022)	Practical power system – installed capacity = 40,270 MW, maximum load = 36,872 MW, and installed capacity of RESs = 3010 MW with ratio of WT to PV = 2:3
2	Taheri Otaghsara and Asghari Gharakheili (2022)	IEEE RTS 24 buses
3	Alizadeh et al. (2022)	RBTS4 and Finnish Distribution System
4	Karimi et al. (2021)	34 feeders in Ahvaz city, Iran
5	Sinishaw et al. (2021)	Bata 34 bus feeder of Bahir Dar distribution system, Ethiopia
6	Afzal et al. (2020)	IEEE 69-bus system
7	Bagen et al. (2019)	29-buses actual utility substation in the Province of Manitoba, Canada
8	Zare et al. (2018)	IEEE 69-bus system
9	Chen et al. (2016)	RBTS and IEEE 15-bus system
10	Benidris and Mitra (2015)	IEEE RTS 24 buses

created it to make testing and contrasting various reliability assessment approaches easier. The system has undergone several updates and versions to reflect power system technology and practice changes.

It's worth mentioning that the IEEE RTS also comes in various versions, as described later in this section, each representing different levels of complexity and realism.

The application of probability methods (APM) Subcommittee of the Power System Engineering Committee created and released the initial IEEE Reliability Test System (RTS-79) in 1979. To meet the need for a uniform model to test and compare outcomes from various PSR evaluation approaches, RTS-79 was created. The model was created primarily to study the effects of transmission and generation outages on the system and is represented in Figure 16.8.

In 1986, the RTS-79 noticed a minor update (RTS-86). The improvement of the information about the creation of feet received special consideration in this version. Unit derates (partial outages), load forecast uncertainty and unit scheduled maintenance are essential features incorporated in this release. The model was increased to allow users to examine multi-area interconnection effects and improvements relating to reliability indices, including loss of load probability, loss of load expectation, and loss of energy expectation. The RTS-86 connected two identical RTS-79 versions using a 300 MW-rated 230 kV tie line. The RTS-86 also provides precise system reliability indices for benchmarking and case study purposes.

The RTS received its most recent upgrade in 1996 (RTS-96). This update significantly increased the size and complexity of the model; the RTS-96 is made up of three associated RTS-79 models. The industry at the time saw a sharp rise in interregional transactions and advancements in computing capacity. Thus, a tractable analysis of larger multi-area models might be performed. The most significant change to the RTS-96 was adding information on operating costs and restrictions for the generating units: unit start-up heat input, net plant incremental heat rates, unit cycling restrictions, ramping rates, and unit emissions data.

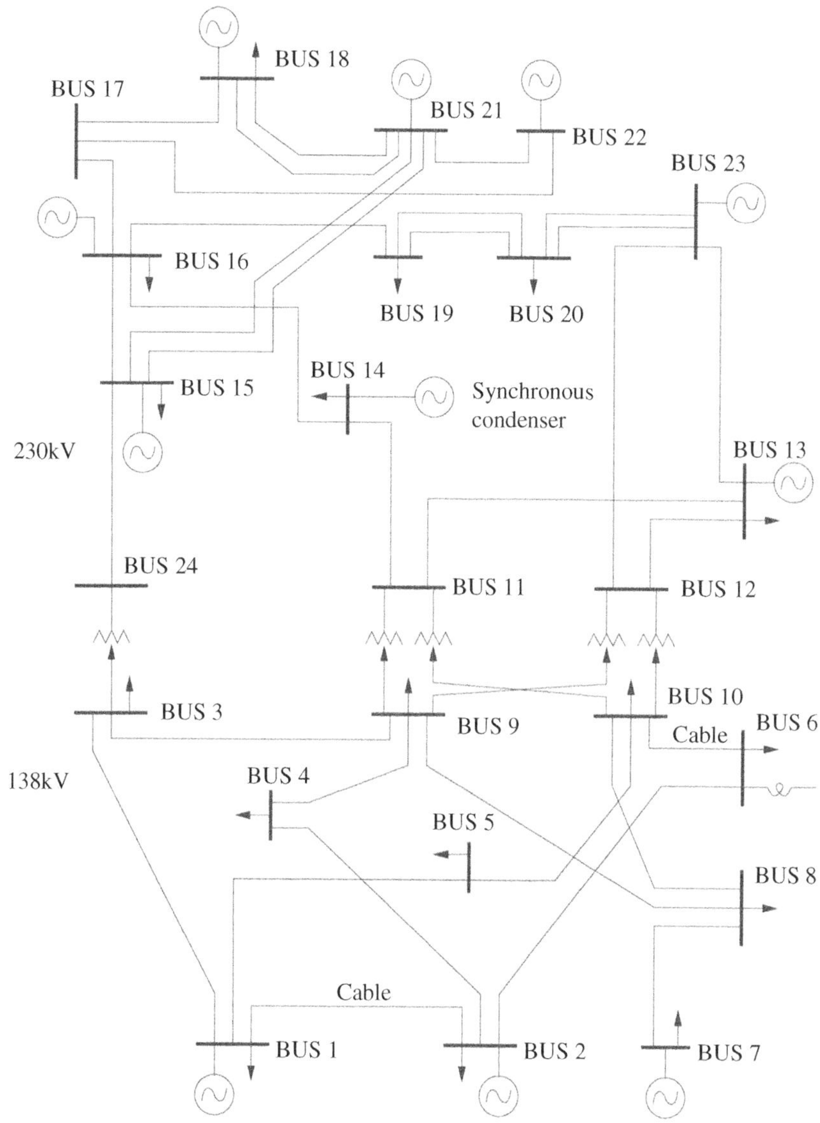

Figure 16.8 Reliability case study structure IEEE RTS-79. Source: Benidris and Mitra (2015)/The Institution of Engineering and Technology.

The IEEE RTS-79 system is a classical PSR analysis case consisting of 24 buses, 32 generating units, and 42 transmission lines. Four more lines expand the original system. Figure 16.8 depicts the redesigned network topology of the IEEE RTS-79 system. The MTTF and MTTR of units for various capacities are provided via the modified IEEE RTS-79 system. In this scenario, the settings are slightly changed to improve the reliability of the larger-capacity units. It's a widely used case study system for evaluating the reliability and performance of power systems.

It's important to note that while the IEEE RTS 24-bus version provides a more detailed representation than smaller versions, it's still a simplified power system model. This allows for controlled testing and comparison of methods while avoiding the complexity of real-world networks.

Researchers and engineers have widely adopted the IEEE RTS-79 version in the power industry to develop and validate new reliability assessment techniques and explore the impact of various factors on PSR.

New sensitivity criteria that can be helpful for site selection are discussed. The technique may integrate the available DGs by determining the optimal sites in a power system.

16.5.2 Roy Billinton Test System

The RBTS is a widely used case study for evaluating power systems' reliability and risk assessment. Roy Billinton and his colleagues developed it to provide a standardized platform for researchers and practitioners to test and compare reliability assessment methods and techniques, represented as Figure 16.9. The RBTS consists of a simplified representation of a power system network, including generators, transmission lines, and loads. It's designed to represent various aspects of power system behavior, such as generation, transmission, and distribution, which are crucial for assessing reliability and risk. The RBTS includes different scenarios with varying levels of complexity to determine the performance of other reliability analysis techniques.

Using a Monte Carlo simulation-based method, examined the effects of the stochastic nature of distributed and renewable resources, including solar PV, battery energy storage, and electric vehicles, on the reliability of distribution networks. The results are illustrated, and customer-side

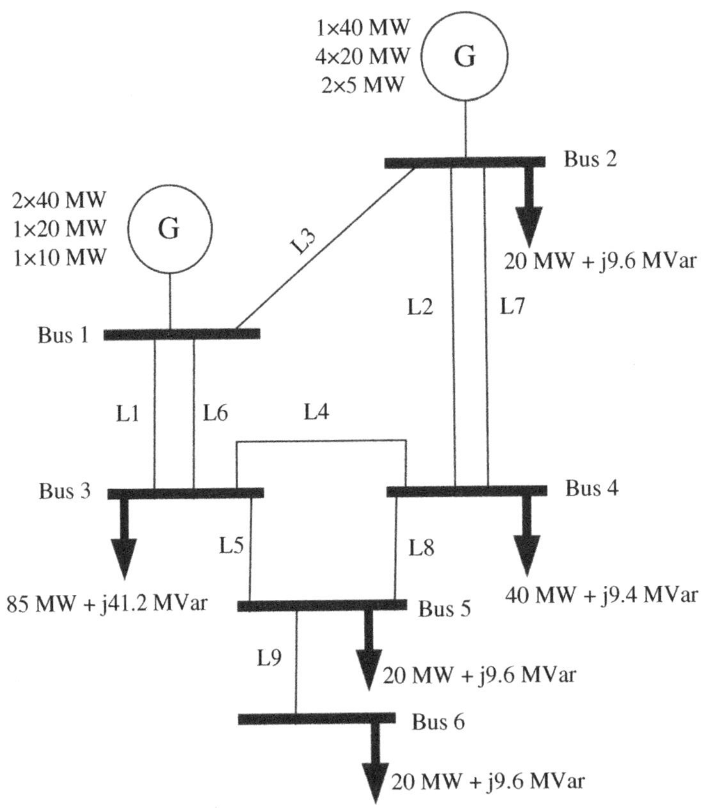

Figure 16.9 Reliability case study structure RBTS. Source: Adapted from Chen et al. (2016).

dependability indices, such as SAIDI, SAIFI, and CAIDI, are computed using a modified RBTS, considering capacity considerations.

16.5.3 Bahir Dar's Power Distribution Test System

Substation-1 66/45/15 kV and Substation-2, including 230/66/15 kV and 230/132/15 kV substations, supply the electrical loads for Bahir Dar city. The electricity supply system for Bahir Dar Town includes the Bata, Ghion, Papyearus, Air Force, Sematate, and Industry feeders. The chapter focuses on the Bata feeder, which has a more significant yearly interruption duration and annual interruption frequency. A single-line diagram of the Bata feeder distribution network is displayed in Figure 16.10. The figure shows that the network contains 34 buses and is configured in a radial pattern. Node 1 is the slack bus out of the entire number of buses. The final four buses or nodes are common coupling nodes, while the remaining 29 are coupled to loads through step-down distribution transformers. According to the single-line diagram of the Bata feeder, the distribution line nodes, which correspond to the numbers up to 34, are represented in dark gray.

In contrast, the sectional lines of the feeder, which correspond to the numbers up to 33, are represented by black color. The operational glitches are at their peak and need to be reduced. The Bata feeder's distribution system is modeled, and the system's reliability indicators are computed.

Using smart grid technology, the reliability is improved. In recent years, generation and transmission networks have principally conceived about and implemented smart grid applications. Three smart grid technology applications enhance the distribution system's reliability, although the conceptual framework of smart grid technology is quite broad. These technologies include quick power restoration, effective network reconfiguration, and efficient switching device location.

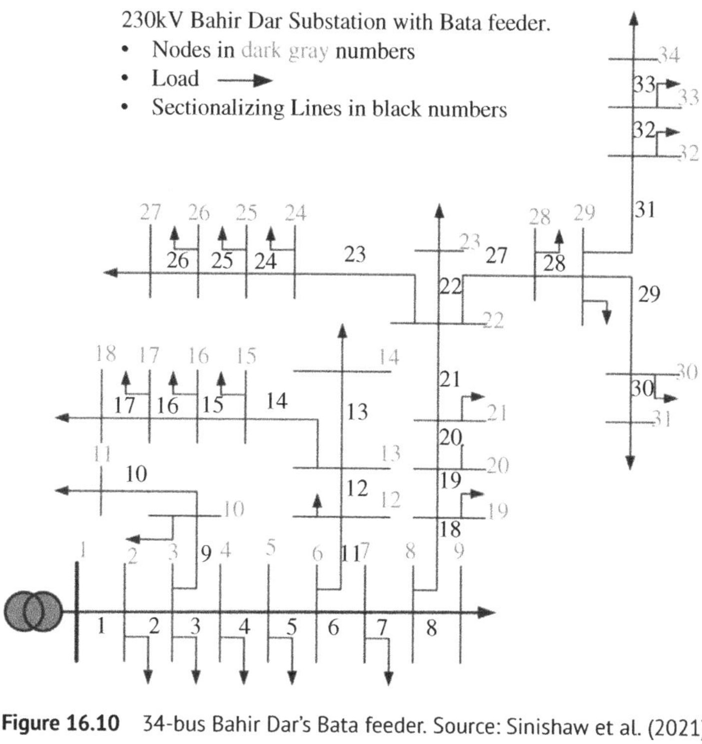

Figure 16.10 34-bus Bahir Dar's Bata feeder. Source: Sinishaw et al. (2021)/with permission of Elsevier.

Rapid power restoration reduces the annual frequency and length of disruptions for consumers by placing switching devices optimally and reconfiguring the network optimally. So, improving Ethiopia's electricity distribution systems' reliability through smart grid technology is preferable.

A Bata feeder from the Bahir Dar power distribution system with 34 buses is considered to analyze and evaluate the applicability of smart grid technology. The duration and frequency of interruption statistics from Ethiopian Electric Utility were gathered and examined during 2017–2019. The Bata feeder experiences 712.43 hours of average interruption each year, with an average frequency of 604 interruptions per year, according to the interruption data studied. The reliability indices derived are SAIFI = 131.15 int/yr, SAIDI = 245.84 h/yr, CAIDI = 1.88 h/cust.int, and ASAI = 0.972 pu.

16.6 Challenges and Future Directions

Ensuring DSR in the presence of RESs presents challenges and opportunities for the energy industry. Here are some of the key challenges and future directions in this area.

16.6.1 Challenges

1. **Intermittency and Variability:** RESs like solar and wind are intermittent and variable. This can lead to fluctuations in power generation, making it challenging to maintain a stable supply-demand balance in the distribution grid.
2. **Grid Stability:** The intermittent nature of RESs can affect grid stability by causing voltage and frequency fluctuations. Sudden changes in power output can lead to voltage violations and potential blackouts if not appropriately managed.
3. **Integration and Penetration Limits:** Integrating a high penetration of renewable energy into the distribution system requires careful planning to avoid overloading the grid. The existing infrastructure might not be designed to handle the bidirectional power flow and variable generation associated with RES.
4. **Limited Control:** Operators have limited control over renewable energy generation, unlike traditional power plants. This lack of control can make managing and optimizing the distribution system's performance challenging.
5. **Storage and Energy Management:** Effective energy storage solutions are essential to store excess renewable energy during periods of high generation and release it during periods of low generation. Developing cost-effective and efficient storage technologies is crucial.
6. **Grid Planning and Resilience:** Integrating renewable sources requires rethinking grid planning and resilience strategies. The intermittent nature of these sources makes traditional planning approaches less effective, necessitating new methods to ensure reliable operation.

16.6.2 Future Directions

1. **Advanced Forecasting:** Improving forecasting models for renewable energy generation can help grid operators anticipate fluctuations and proactively balance supply and demand.
2. **Energy Storage Development:** Research and development of better energy storage technologies, such as advanced batteries, compressed air energy storage, and pumped hydro storage, can efficiently utilize surplus renewable energy.

3. **Demand Response and Flexibility:** Implementing demand response programs and flexible load management can help match energy consumption with intermittent renewable energy generation patterns.
4. **Microgrids and Decentralization:** Creating microgrids and decentralized energy systems can enhance reliability by allowing localized generation, consumption, and storage, reducing the dependency on central grids.
5. **Smart Grid Solutions:** Integrating smart grid technologies, including advanced sensors, communication networks, and automation, can enable real-time monitoring and control of the distribution system, improving its resilience.
6. **Hybrid Energy Systems:** Combining multiple RESs (solar, wind, hydro, etc.) can mitigate intermittency, providing a more consistent power output.
7. **Grid-Forming Inverters:** Developing grid-forming inverter technologies that allow RESs to contribute to grid stability, even during grid disturbances, can enhance system resilience.
8. **Policy and Regulation:** Governments and regulatory bodies need to create supportive policies that encourage the integration of RESs and provide incentives for investing in grid reliability enhancement.
9. **Advanced Control Algorithms:** Implementing advanced control algorithms, such as predictive control and machine learning, can help manage the complex interactions between renewable generation, storage, and demand.
10. **Collaborative Research:** Continued collaboration between research institutions, academia, industry, and government agencies is essential to address the technical and operational challenges of integrating RESs while maintaining the distribution system's reliability.

While integrating RESs into distribution systems poses challenges, innovative technologies, strategies, and supportive policies can pave the way for a more reliable, resilient, and sustainable energy future.

Problems

16.1 What is the AENS if SAIDI is given as A hour/year and all customers have the same annual consumption, B kWhr/year/customer?

16.2 CAIDI is given as A hour/failure, and SAIDI is B hour/year/customer. What is SAIFI?

16.3 If ASAI is 0.9999, and all failures always last exactly one hour and affect the whole system, what are SAIDI, SAIFI, and ASUI?

16.4 Why does CAIDI sometimes increase when a redundancy eliminating some failures is introduced?

16.5 How is SAIFI affected if the disconnecting switches are removed?

16.6 A transformer and distributor are part of a radial system that is 400 m long. The transformer and distributor have annual failure rates of 0.80 and 8.2 failures per kilometer, respectively. Transformer and distributor repairs take six and five hours, respectively. Could you find the system's failure rate, the duration of each outage, and the total annual outage time?

16.7 The 132/11 kV transformer, 11 kV transformer, 11 kV feeder, and 11/0.44 kV distribution transformer constitute a distribution system. It functions in a two-state fluctuation environment where normal and stormy weather durations are anticipated to last 120 and 3 hours, respectively. The following table displays the outage rates and repair times:

Equipment	λ	λ'	λ''	r	r''
132/11 kV transformer	0.1	0.1	0.5	6	8
11 kV transformer	0.15	0.15	0.4	4	5
11 kV feeder	1.2	5.2	2	5	3
11/0.44 kV distribution transformer	0.3	0.3	0.6	5	4

Find the system's yearly outage rate, downtime per outage, and annual total outage time. The following variables are provided:
λ = Normal weather rate per year, λ' = Stormy weather rate per year, λ'' = component outage rate per year, r = expected repair time forced outage hours/year, r'' = expected maintenance time hours/year, N = expected value for normal hours/year, S = expected value for stormy hours/year.

16.8 A double circuit 220 kV line operates in a two-state fluctuating mode environment having expected normal weather and stormy weather durations of 110 and 4 hours, respectively. Both circuits are similar. The outage data for each circuit is as follows:
$\lambda = 3$ outages/year of normal weather, $\lambda' = 9$ outages/year of stormy weather, $\lambda'' = 1$ outage/year, $r = 6$ hours, $r'' = 8$ hours.
Find annual outage rate, downtime per outage, and total outage time per year.

16.9 Explain the difference between primary and secondary failures in a power system.

Solution

16.1 A × B; kWh per year per customer.

16.2 B/A; failure per year per customer.

16.3 SAIDI = 0.876 hours per year = SAIFI, ASUI = 0.0001.

16.4 CAIDI is a measure of the average length of interruption. Suppose an interruption that lasts less than the average interruption time of the system is eliminated. In that case, SAIFI and SAIDI decreases, but the average interruption length, CAIDI, increases: CAIDI = SAIDI/SAIFI, so when SAIFI decreases more than SAIDI, CAIDI increases.

16.5 SAIFI is unaffected since the same customers will have the same failures in both cases. Disconnecting switchers will only shorten the average failure length, affecting, for example, SAIDI, CAIDI, and AENS, but not SAIFI.

16.6 failure rate = 4.1 per year, downtime = 5.1 hours per outage, total outage time = 21.2 hours per year.

16.7 annual outage rate = 5.35, outage per year = 5.86 hours, total outage time per year = 31.3 hours.

16.8 annual outage rate = 0.023, downtime per outage = 3.1 hours, total outage time per year = 0.07 hours.

16.9 A primary failure is when a component fails independently of the other components in the system, for example, due to wear or external effects such as falling trees. Secondary failures are those caused by other failures somewhere else in the system, for example, caused by overload or by a circuit breaker that does not open in a situation where it could stop a failure from causing further effects on the system.

16.7 Conclusion

In conclusion, the reliability assessment of distribution systems integrated with renewable energy systems is critical in ensuring modern power grids' stability and resilience. The chapter has shed light on the complex interplay between conventional and renewable energy sources through comprehensive analysis and evaluation, highlighting challenges and opportunities for enhancing system reliability. Integrating renewable energy resources brings significant benefits, including reduced carbon emissions, enhanced energy independence, and potential cost savings. However, it also introduces unique operational and reliability considerations due to the intermittency and variability of renewable generation. Advanced modeling techniques, innovative control strategies, and robust grid infrastructure are essential to address these challenges. Moreover, effective coordination among stakeholders, including utilities, regulators, and technology providers, is crucial for maximizing the reliability and performance of distribution systems with renewable energy integration. Three case studies on the 24-bus IEEE Reliability Test System-79, the 6-bus Roy Billinton Test System, and the 34-bus Bahir Dar's Power Distribution Test System highlight the importance of power system reliability's significance. Ongoing research and development efforts are needed to refine reliability assessment methodologies further, incorporate real-time data analytics, and optimize system planning and operation. By embracing these advancements, we can pave the way for a more sustainable and resilient energy future, where renewable energy plays a central role in powering our communities while ensuring the reliability and stability of our electrical power grids.

References

Afzal, S., Mokhlis, H., Illias, H.A. et al. (2020). State-of-the-art review on power system resilience and assessment techniques. *IET Generation, Transmission & Distribution* 14(25): 6107–6121.

Alizadeh, A., Kamwa, I., Moeini, A., and Mohseni-Bonab, S.M. (2022). An exact MILP model for joint switch placement and preventive maintenance scheduling considering incentive regulation. *IET Generation, Transmission & Distribution* 16(23): 4672–4688.

Bagen, B., Huang, D., and Fattal, K. (2019). Enhanced probabilistic approach for substation reliability assessment. *IET Generation, Transmission & Distribution* 13(12): 2488–2495.

Benidris, M. and Mitra, J. (2015). Reliability and sensitivity analysis of composite power systems considering voltage and reactive power constraints. *IET Generation, Transmission & Distribution* 9(12): 1245–1253.

Chen, Y., Zheng, Y., Luo, F. et al. (2016). Reliability evaluation of distribution systems with mobile energy storage systems. *IET Renewable Power Generation* 10(10): 1562–1569.

Karimi, H., Niknam, T., Dehghani, M. et al. (2021). Automated distribution networks reliability optimization in the presence of DG units considering probability customer interruption: a practical case study. *IEEE Access* 9: 98490–98505.

Kumar, S., Saket, R.K., Dheer, D.K. et al. (2020a). Reliability enhancement of electrical power system including impacts of renewable energy sources: a comprehensive review. *IET Generation, Transmission & Distribution* 14(10): 1799–1815.

Kumar, S., Sarita, K., Vardhan, A.S.S. et al. (2020b). Reliability assessment of wind-solar PV integrated distribution system using electrical loss minimization technique. *Energies* 13(21): 5631.

Kumar, S., Sarita, K., Saket, R.K. et al. (2021). Reliability assessment for DFIG-based WECS considering the impact of 3-phase fault and lightning impulse voltage. *International Transactions on Electrical Energy Systems* 31(8): e12952.

Ranjbar, H., Saber, H., and Sharifzadeh, M. (2022). Bi-level planning of distributed energy resources into existing transmission grids: pathway to sustainable energy systems. *IET Generation, Transmission & Distribution* 16(24): 4963–4979.

Sinishaw, G.Y., Bantyirga, B., and Abebe, K. (2021). Analysis of smart grid technology application for power distribution system reliability enhancement: a case study on Bahir Dar power distribution. *Scientific African* 12: e00840.

Taheri Otaghsara, S.S. and Asghari Gharakheili, M. (2022). An optimum budget-based development in maintenance for electric power transmission networks. *IET Generation, Transmission & Distribution* 16(22): 4520–4527.

Zare, M., Azizipanah-Abarghooee, R., Hooshmand, R.-A., and Malekpour, M. (2018). Optimal reconfigurattion of distribution systems by considering switch and wind turbine placements to enhance reliability and efficiency. *IET Generation, Transmission & Distribution* 12(6): 1271–1284.

17

Reliability Evaluation and Performance of Hybrid Photovoltaic Energy Systems for Rural Electrification Using Markov Process

Santosh S. Raghuwanshi[1], Smriti Singh[2], Akanksha Singh S. Vardhan[3], Rajesh Arya[4], and R. K. Saket[2]

[1]*Department of Electrical Engineering, Medi-Caps University, Indore, Madhya Pradesh, India*
[2]*Department of Electrical Engineering, Indian Institute of Technology (BHU), Varanasi, Uttar Pradesh, India*
[3]*Department of Electrical Engineering, Shri GS Institute of Technology and Science, Indore, Madhya Pradesh, India*
[4]*Department of Information Technology, Acropolis Institute of Technology and Research, Indore, Madhya Pradesh, India*

17.1 Introduction

Energy plays a crucial role in propelling and nurturing the progress and development of society. Every sector of the economy requires energy input in some manner. Unfortunately, the generation and utilization of energy give rise to various environmental issues. Different nations and the international community confront a range of problems related to demand and environmental concerns. These encompass local and regional environmental consequences, potential global climate change and sea level rise due to the greenhouse effect, and threats to international relations under oil supply conditions leading to nuclear proliferation. Consequently, each country ought to take actions to limit environmental harm and energy shortages (Akhtar et al. 2021; Raghuwanshi and Arya 2020b).

So required alternative energy resources called renewable energy resources. The alternative energy resources like sun, wind, biomass, ocean, etc., promise clean and abundant sources of energy for all times to cane. Many believe that alternative energy sources will form the foundation of energy supplies in the next century. Biomass has dictionary been used very widely for cooking in developing countries. Expansion of alternative sources faces three obstacles viz., technical, economic, and reliability. Moreover, the available energy fluctuates with natural cycles. In view of these, the contribution from these sources will remain marginal in the foreseeable future. Nevertheless, the global prospects for photovoltaic (PV) cells and wind energy utilization can increase sustainability if cost are reduced. The cost of PV system would have to reduce by more than 50% before they can penetrate more into the energy scene. Similarly, if the cost of wind energy utilization can bc brought down by about 30%, the contribution of wind energy can be sustainable (Alanazi et al. 2022; Ostovar et al. 2021).

Solar energy is the most abundant and constant stream of energy. It is available in both direct and indirect forms of energy. The design process for the solar PV system encompasses several key steps, such as calculating the load requirements, estimating losses, choosing an array size that aligns with the load, and determines the appropriate storage battery capacity. Additionally, a controller is essential to prevent both battery overcharging and excessive discharging. PV systems offer several benefits, including prolonged lifespan, minimal maintenance requirements, a high degree

Reliability Analysis of Modern Power Systems, First Edition. Edited by R. K. Saket and P. Sanjeevikumar.

of reliability, preservation of conventional resources, and pollution reduction, leading to savings. The uses of the solar PV system encompass a wide range of applications, such as illuminating village streets, powering railways, supporting offshore rooftops and oil platforms, facilitating battery charging, providing domestic lighting, enabling community center illumination, supporting TV and radio systems, serving defense needs, powering microwaves and repeater stations, facilitating water pumping, running solar refrigerators, and supporting satellite operations, among others. For efficient functioning and continue energy supplies used renewable hybrid PV energy system (RHPVES). RHPVES is includes wind/biomass/fuel cell/solar batteries and conventional diesel engine (Raghuwanshi and Arya 2019a; Kundur and Malik 2022).

The effectiveness of a system is determined by its ability to perform the intended task and utilize resources efficiently. The system's ability to perform its task is primarily dependent on its reliability and quality. Reliability refers to the probability of failure occurring in a given time interval (Saket 2011). Product reliability is the probability that a unit will adequately perform its intended function for a specified duration, under specific environmental conditions (Niu et al. 2021).

The reliability of systems is often compromised due to poor design and manufacturing. To improve reliability, it is important to plan and design new systems with reliability as a key parameter. With increasing automation and complexity, subsystems become more integral, leading to potential conflicts and decreased reliability. This highlights the importance of stringent reliability standards to ensure the success of these systems in performing critical tasks. Additionally, variations in temperature and irradiance can impact panel performance, which must be considered by the International Commission on Accreditation (ICA) (Abunima and Teh 2020; Tawfiq et al. 2021).

The reliability of a system is dependent on the reliability of its individual parts and components. The failure of a single component can cause the entire system to fail, making it crucial to consider the reliability of each component when assessing system reliability. To achieve this, it is necessary to analyze the functional relationships between components and their failure models (Saket 2011). While it may be difficult to establish failure rates for all working conditions, it is possible to predict reliability under specific stress conditions through extrapolation or interpolation of available data (Kebede et al. 2021).

This chapter presents the Markov process and calculates important reliability indices, such as LOLP, availability, failure and repair time, and EENS. The hybrid renewable energy system is considered for the evaluation of these reliability indices.

The chapter's structure unfolds as follows: Section 17.1 serves as the chapter's Introduction. Section 17.2 provides an explanation of Reliability Indices. Section 17.3 delves into the Markov process. Section 17.4 explores the Reliability of Series, Parallel, and Complex Systems, along with an explanation of the components of the Renewable Hybrid PV System. This section also includes illustrative numerical examples with solutions to enhance comprehension. Lastly, Section 17.5 brings the chapter to a Conclusion.

17.2 Reliability Indices

- **Availability**: The performance of maintained equipment can be measured by its availability, which takes into account both reliability and maintainability factors. Availability is determined by the frequency of failures and the speed at which faults are corrected. Steady-state availability refers to the duration of time that equipment is available for use (Postnikov and Mednikova 2022).

- **Repairability**: Repairability is the likelihood that a failed equipment can be repaired and restored to operable condition within a specified downtime period, assuming maintenance is performed under optimal conditions.
- **Failure**: A device or system experiences a failure when its properties are altered to the extent that it cannot function properly, either partially or completely. While some components have clearly defined failure modes, others do not.
- **EENS Index**: It is a measure of the amount of energy that is not delivered to customers due to outages in a power system. It is calculated by multiplying the probability of an outage by the amount of energy that would have been delivered if the outage had not occurred. The EENS index is typically expressed in megawatt hours (MWh) (Kuo et al. 2022).
- **LOLP**: LOLP is a measure of the probability that a power system will not be able to meet the demand for electricity. It is calculated by multiplying the probability of each component in the system failing by the amount of load that each component can supply. The LOLP is typically expressed as a percentage. A lower LOLP indicates a more reliable power system. A high LOLP could indicate that there are problems with the system's infrastructure, such as aging equipment or inadequate maintenance. It could also indicate that the system is not operating at its full capacity (Hou et al. 2023).

These reliability indices help engineers, analysts, and decision-makers understand and improve the performance, maintenance, and design of systems by identifying potential weaknesses or areas for improvement in terms of reliability and dependability.

17.3 Markov Process

Markov analysis examines a system's various states, including when all components are functioning and when one component has failed while others continue to operate. The fundamental concept of Markov analysis is that the likelihood of a system moving from one state to another is determined exclusively by its present state, with no consideration of past states. This property, similar to exponential distribution in finance, makes Markov analysis ideal for predicting failure times. To analyze a system's reliability, Markov process can be used, with the system's N components falling into either an operating or failed state. The system's state can be defined as one of the $2n$ possible combinations of functioning and nonfunctioning components (Kuo et al. 2022).

In a Markov process, there are two key components:

1. **States**: A set of distinct states that the system can be in. These states can represent various situations, conditions, or configurations of the system being modeled. For example, in weather forecasting, states could be "sunny," "cloudy," or "rainy" (Kalantar-Neyestanaki and Cherkaoui 2021).
2. **Transition Probabilities**: The probabilities linked to transitions between different states are usually depicted using a transition matrix or transition diagram. Within this matrix, each individual element signifies the likelihood of moving from one state to another within a single time interval (Rayati et al. 2023).

A Markov process can be employed to represent the unpredictable behavior of a PV system, and a Markov state diagram can be employed for its depiction. The main advantage of this approach is that it offers a comprehensive representation of all possible states and the transitions between them. The occurrences of PV string(s), module(s), and inverter(s) failures and subsequent repairs are responsible for the shifts between different Markov states. Consequently, this strategy can be especially advantageous for managing individual components (Singh et al. 2023). Examine a repairable single-component system as illustrated in the state-transition diagram. In Figure 17.1, the state-transition diagram for the Markovian model can be seen. Each component in this system can exist in one of two operational states. When the component is in state 1, it functions as intended, while in state 0, it is considered failed. The component's failure rate is denoted as λ, and its repair rate is represented by μ. Thus, the condition of the component, whether it is operational or faulty, can be depicted using a Markov chain featuring two states: λ and μ (as shown in Figure 17.2).

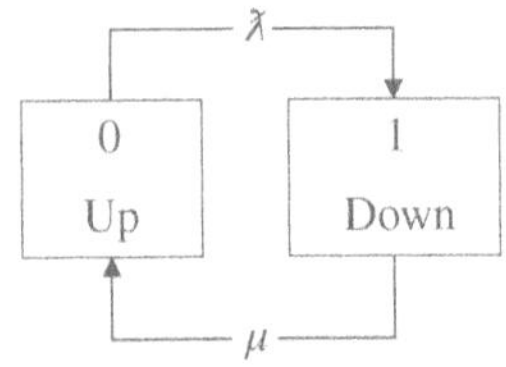

Figure 17.1 State-transition diagram for single repairable component system.

The steady-state probabilities is given by Eqs. (17.1) and (17.2):

$$P_{\text{up}} = \frac{\mu}{\lambda + \mu} \tag{17.1}$$

$$P_{\text{down}} = \frac{\lambda}{\lambda + \mu} \tag{17.2}$$

Similarly, for a two repairable component system the staedy state probabilities is given by Eqs. (17.3) and (17.4):

$$P_{\text{up}} = \frac{\mu_1 \mu_2}{(\lambda_1 + \mu_1)(\lambda_2 + \mu_2)} \tag{17.3}$$

$$P_{\text{down}} = \frac{\lambda_1 \lambda_2}{(\lambda_1 + \mu_1)(\lambda_2 + \mu_2)} \tag{17.4}$$

The key assumption in a Markov process is that the future behavior of the system depends only on its current state and not on how it arrived at that state. This property is known as the Markov property.

Markov processes have several important properties:

1. **Homogeneity**: The transition probabilities remain constant over time. This means that the probability of transitioning from one state to another does not change as time progresses.
2. **Irreducibility**: In a finite number of steps, it's feasible to reach any state from any other state with a nonzero probability. In other words, there are no isolated subsets of states that cannot be reached from other states (Kou et al. 2022).
3. **Stationarity**: The long-term behavior of a Markov process reaches an equilibrium distribution called the stationary distribution or steady-state distribution. In this distribution, each state has a fixed probability of being occupied.

Markov processes find applications in various fields such as finance (stock market modeling), biology (population dynamics), physics (particle movements), and computer science (algorithm analysis). They are also used in machine learning algorithms like Hidden Markov Models (HMMs) for speech recognition and natural language processing (Spertino et al. 2020).

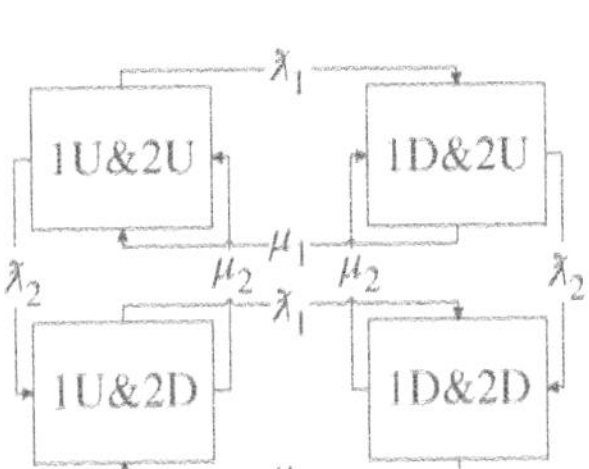

Figure 17.2 State-transition diagram for two repairable component system.

The steps of a Markov process for calculating reliability indices are as follows:

1. **Step 1**: Define the system: Identify the components and their states that make up the system. Determine the possible states of each component, such as functioning or failing.
2. **Step 2**: Construct the state-transition diagram: Create a diagram that represents the possible transitions between states for each component. This diagram shows how the system can move from one state to another based on component failures or repairs.
3. **Step 3**: Assign probabilities: Assign probabilities to each transition in the state-transition diagram. These probabilities represent the likelihood of moving from one state to another, such as the probability of a component failing or being repaired.
4. **Step 4**: Calculate steady-state probabilities: Solve for the steady-state probabilities of each state in the system using mathematical equations or numerical methods, such as matrix operations or iterative algorithms. These steady-state probabilities represent the long-term behavior of the system and indicate its reliability (Spertino et al. 2020).
5. **Step 5**: Compute reliability indices: Once you have obtained the steady-state probabilities, you can calculate various reliability indices, such as availability, LOLP, EENS, and system reliability.
6. **Step 6**: Analyze results and make decisions. Analyze the calculated reliability indices to assess the performance and dependability of the system. Use these results to make informed decisions (Samy et al. 2020).

17.4 Reliability of the System

The system components may be connected in three ways: series, parallel, and complex (series–parallel) configuration. In series, all components must be working for the system to work. In a parallel configuration, at least one component must work for system to work.

17.4.1 System with Series Components

In series system, all the components are necessary to operate for system to work. The block diagram in Figure 17.3 illustrates the representation of two series components within the system. If any of the n components fail, the information will not reach the OUT end in the system.

The reliability of series components system is (Samy et al. 2020; Xu et al. 2022):

$$R(t)_{\text{series}} = \prod_{j=1}^{n} e^{-\lambda_j t} = e^{-\sum_{j=1}^{n} \lambda_j * t} \tag{17.5}$$

The unavailability, failure rate, and interruption duration are as follows (Samy et al. 2020; Xu et al. 2022):

$$U_{\text{series}} = \sum_{j=1}^{n} \lambda_j * r_j \tag{17.6}$$

$$\lambda_{\text{series}} = \sum_{j=1}^{n} \lambda_j \tag{17.7}$$

Now,

$$r_{\text{series}} = \frac{\sum_{j=1}^{n} \lambda_j * r_j}{\sum_{j=1}^{N} \lambda_j} \tag{17.8}$$

where n is number of system components.

Figure 17.3 Series system.

17.4.2 System with Parallel Components

Two or more components are in parallel configuration if all units must fail for the system fail. If one components or units operate system will operate. Two parallel components system are represented by the block diagram shown in Figure 17.4 shows that there are n parallel paths between the IN and OUT ends, with each component connected across each other.

The reliability of parallel components system is as follows (Billinton et al. 1992):

$$R(t)_{\text{parallel}} = 1 - \prod_{j=1}^{n} [1 - e^{-\lambda_j t}] \tag{17.9}$$

To calculate the unavailability, failure rate, and interruption duration of a parallel system with two components, the following equations can be utilized (Billinton et al. 1992; Abunima and Teh 2020).

$$U_{\text{parallel}} = \frac{\lambda_1 \lambda_2}{(\lambda_1 + \mu_1)(\lambda_2 + \mu_2)} \tag{17.10}$$

Under the assumptions that $\lambda_1 \ll \mu_1, \lambda_2 \ll \mu_2$, approximate relations can be derived for the indices:

$$U_{\text{parellel}} = (\lambda_1 r_1) * (\lambda_2 r_2) \tag{17.11}$$

$$r_{\text{parellel}} = \frac{r_1 r_2}{r_1 + r_2} \tag{17.12}$$

$$\lambda_{\text{parallel}} = \lambda_1 \lambda_2 (r_1 + r_2) \tag{17.13}$$

17.4.2.1 Complex (Series–Parallel) System

Systems are not always series or parallel configuration. A complex system is the combination may be series or parallel or series–parallel configuration shown in Figure 17.5. For the complex system successful operation of the system, there should be at least one continuous path between IN and OUT terminals.

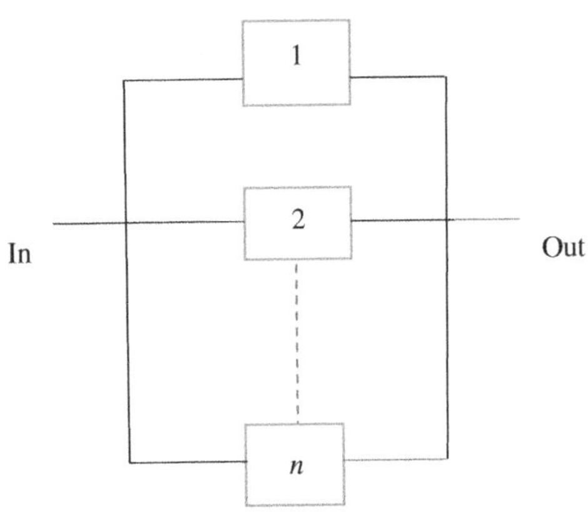

Figure 17.4 Parallel system.

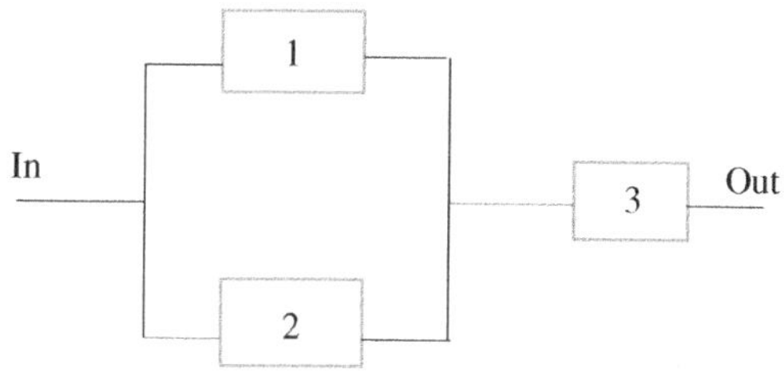

Figure 17.5 Series–parallel system.

Series–parallel system indices are as follows (Khan et al. 2022):

$$\lambda_{\text{comp}} = (\lambda_1\lambda_2) * (r_1 + r_2) * \lambda_3 \tag{17.14}$$

$$U_{\text{comp}} = (\lambda_{\text{parellel}} \cdot r_{\text{parallel}} + \lambda_3 \cdot r_3) \tag{17.15}$$

$$r_{\text{comp}} = \frac{(\lambda_{\text{parallel}} \cdot r_{\text{parallel}}) + \lambda_3 \cdot r_3}{\lambda_1 \cdot \lambda_2 \cdot \lambda_3 \cdot (r_1 + r_2)} \tag{17.16}$$

EENS is as follows Raghuwanshi and Arya (2019b) and Ghania et al. (2022):

$$\text{EENS} = L_{\text{avg}} \cdot U_{\text{system}} \text{ kWh/yr} \tag{17.17}$$

where L_{avg} is average load in kW. LOLP is Raghuwanshi et al. 2023 and Raghuwanshi and Arya (2020a):

$$\text{LOLP} = \frac{U_{\text{system}}}{8760} \tag{17.18}$$

17.4.3 Renewable Hybrid PV System

Figure 17.6 depicts the RHPVES, which is comprised of PV arrays, a converter, an inverter, a charge controller, a solar battery, a DG set, and a load. The PV arrays and solar batteries are connected to the direct current (DC) bus via the charge controller, while the inverter is connected between the DC bus and alternating current (AC) bus. The AC bus is connected to the DG set via a changeover switch (COS). The farm house has two types of loads: household and irrigation.

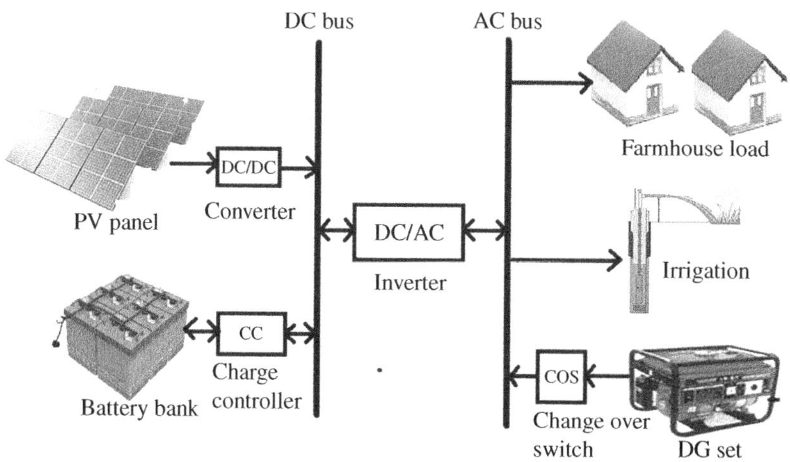

Figure 17.6 Renewable hybrid PV energy system.

17.4.3.1 PV Arrays

PV effect is the direct conversion of sun energy into electrical energy. Solar cells, which are devices that employ this effect to convert sunlight into electricity, can be made up of a single converter cell, also known as a photovoltaic cell, or a combination of cells called a solar array. These cells contain semiconductors that generate voltage proportional to the amount of light or radiation they receive. Iron selenium cells are a type of PV cell commonly employed, despite being very rare (Alfawaz et al. 2023).

PV arrays generator power is as follows (Alfawaz et al. 2023):

$$P_{\text{avg}} = \left[I_{Lg} - I_{\text{r}}\left(\exp\left(\frac{eV}{kT}\right) - 1\right)\right] \times v \tag{17.19}$$

In this context, I_{Lg} and I_{r} correspond to the light-generated current and reverse saturation current, respectively. Additionally, e represents the electron charge, k denotes Boltzmann's constant, T stands for temperature, and v signifies the output voltage of the cells.

17.4.3.2 Converter

A converter is an electronic device that can transform a fixed voltage or DC voltage from a PV panel into a variable, average voltage that is suitable for a specific load. This is achieved by inserting a high-speed switch, such as an IGBT, GTO, SCR, MOSFET, or another similar component, between the DC source and the load. In a solar PV energy system, the converter is usually installed between the PV panels and the DC bus. Buck, boost, and buck–boost converters are link between PV and inverter or PV-battery-inverter. Generally, buck–boost converter topology is used.

The output voltage of the buck–boost converter is as outlined below Raghuwanshi and Khare (2018),

$$V_{\text{o}} = \frac{d}{1-d} V_{\text{PV}} \tag{17.20}$$

where d and V_{PV} are the duty cycle and PV voltage, respectively.

17.4.3.3 Inverter

An inverter is an electronic circuit designed to convert DC power into AC power. This conversion process can be achieved using either transistors or silicon-controlled rectifiers (SCRs). Transistorized inverters are ideal for low and medium power outputs, while SCRs are more suitable for high-power outputs. This is because SCRs are capable of handling significantly higher load currents than transistors. As a result, SCRs have become the preferred choice for high-power output applications. In solar energy systems, an inverter is typically installed between the DC bus and the AC bus. This enables the converted AC power generated by the solar panels to be supplied to the AC loads in the building or fed back to the utility grid.

The inverter power is as follows Raghuwanshi and Khare (2018),

$$P_{\text{inverter}} = \frac{P_{\text{Load}}}{\eta_{\text{inverter}}} \tag{17.21}$$

where P_{Load} and inverter are the load power and inverter efficiency, respectively.

17.4.3.4 Charge Controller and Solar Batteries

Charge controller is circuit that connects the DC bus to solar batteries. This charge controller is controlling the charging and discharging of the solar batteries. The charge controller also prevents the overcharging and deep charging of the battery (Billinton 1970). Electrical Storage batteries are devices that can store and release electricity and are comprised of individual cells that work together. Each cell is made up of a material and an electrolyte that works together to produce electrochemical energy storage. When a battery is connected to an electric load, separation of charge occurs at the interface between the electrodes and the electrolyte, creating free electrons and ions. The electrons flow through the external load, while the ions flow through the electrolyte and combine with the other electrode to create a circuit. This process allows the battery to convert chemical energy into electrical energy that can be used to power various devices (Raghuwanshi and Khare 2018).

Energy storage ability in battery is as follows,

$$E_{\text{battery}} = \frac{(E_{\text{avg}} \times \text{DA})}{\eta_{\text{inverter}} \times \eta_{\text{battery}} \times \text{DOD}} \tag{17.22}$$

where E_{avg} and DA are average demand energy and autonomy, respectively, DOD is the depth of discharge, η_{inverter} and η_{battery} are inverter and battery efficiency, respectively.

17.4.3.5 DG Set

A DG is a device that consists of a diesel engine and an alternator, working together to convert mechanical energy into electrical energy. Diesel engine operates by repeating a cycle of four stages or strokes during which the Piston move up and down. Diesel engine is connected to synchronous generator with couple via shaft. A diesel engine generates mechanical power, which is then transformed into electrical power through an electrical generator. The DG is linked to an AC load.

DG-generated power is as follows (Raghuwanshi et al. 2022),

$$P_{\text{DG}} = P \times L \times A \times N \times \frac{\pi}{30} \tag{17.23}$$

where P and L are the operative pressure and length of stroke, respectively, A and N are the area of cylinder bore and RPM, respectively.

Problem 17.1

Design a Reliability Block Diagram (RBD) of the hybrid energy system is shown in Figure 17.6. Calculate the following reliability indices of PV-only and DG-only system: failure rate, unavailability, EENS and LOLP. Table 17.1 provides the failure rate and repair time data for each individual component of the energy system. Consider L_{avg} as 5 kW.

Table 17.1 Failure rate and repair time information for each component within the system

S. No.	System apparatuses	Failure rate (λ) in yr^{-1}	Repair time (r) in h
1	PVG	0.055	40
2	Converter	0.0357	70
3	Solar inverter	0.0945	65
4	Charge controller	0.0312	55
5	Solar batteries	0.012	25
6	DG set	0.499	90
7	Switch	0.011	15

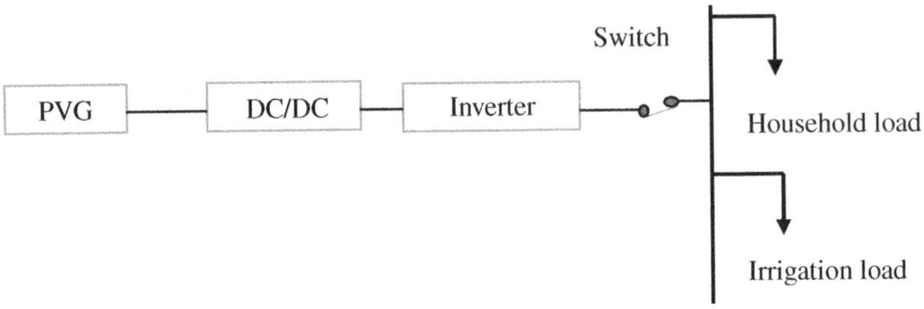

Figure 17.7 RBD of PV-only system.

Solution:

1. The PV-only system includes PV arrays, a DC–DC converter, a solar inverter, a switch, and a load, all connected in series. The RBD of the PV-only system is depicted in Figure 17.7.
   The reliability indices are calculated using Eqs. (17.5)–(17.18). The failure rate, unavailability, EENS, and LOLP are as follows:

$$\lambda_{sys} = 0.1955\,\text{yr}^{-1}$$
$$U_{sys} = 10.995\,\text{h/yr}$$
$$r_{sys} = 56.24\,\text{h}$$
$$\text{EENS} = 54.97\,\text{kWh/yr}$$
$$\text{LOLP} = 0.001255\,\text{yr}^{-1}$$

   In series system, if any one component is not working, the complete system is shutdown. The electricity is not supplied to the load. This system is not reliable system.
2. **DG Only System**: DG only system consists diesel-generator set, switch, and load. The RBD of DG only system is shown in Figure 17.8. The system components are connected in series.
   The failure rate, unavailability, EENS, and LOLP are as follows:

$$\lambda_{sys} = 0.51\,\text{yr}^{-1}$$
$$U_{sys} = 45.075\,\text{h/yr}$$
$$r_{sys} = 88.38\,\text{h}$$

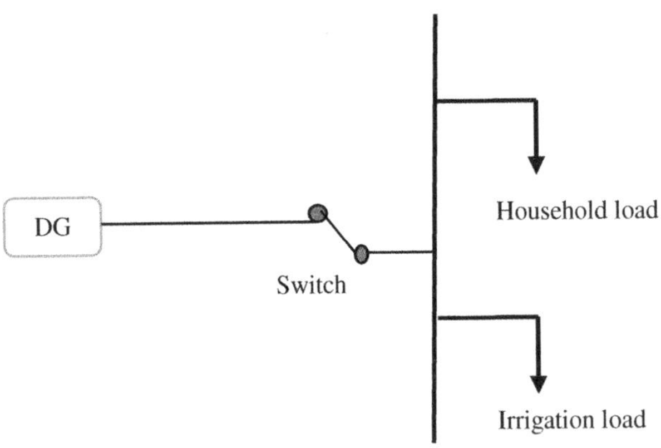

Figure 17.8 RBD of DG only.

$$\text{EENS} = 225.375\,\text{kWh/yr}$$

$$\text{LOLP} = 0.005145\,\text{yr}^{-1}$$

Reliability indices shows the system failure rate is high and system is not reliable. DG is failed, energy not supplied the load. This system configuration is not adequate system in remote areas.

Problem 17.2

Design a Reliability Block Diagram (RBD) of the hybrid energy system shown in Figure 17.6. Calculate the following indices of PV/battery, DG/battery, and PV/DG energy system. Table 17.1 provides the failure rate and repair time data for each individual component of the energy system.

Solution:

1. **PV with Battery Backup System**: The PV/battery system comprises PV arrays, a DC–DC converter, a charge controller, a solar batteries, a solar inverter, a switch, and a load. These components are interconnected in both series and parallel configurations, as shown in Figure 17.9. The PV arrays generate electricity which is converted into DC power by the DC–DC converter. This power is then sent to the charge controller which regulates the charging of the solar batteries. The solar batteries stores the energy for later use. When energy is needed, the solar inverter converts the DC power stored in the solar batteries into AC power for the load to use. The switch is used to control the power flow between the PV arrays, the solar batteries, and the load.
   The failure rate, unavailability, EENS, and LOLP are calculated using Eqs. (17.5)–(17.18).

   $$\lambda_{sys} = 0.21\,\text{yr}^{-1}$$

   $$U_{sys} = 7.86\,\text{h/yr}$$

   $$r_{sys} = 39\,\text{h}$$

   $$\text{EENS} = 39.3\,\text{kWh/yr}$$

   $$\text{LOLP} = 0.000897\,\text{yr}^{-1}$$

   This system is not feasible in terms of load demand. Inverter is failed the overall system is shut down and not supplied the electricity.

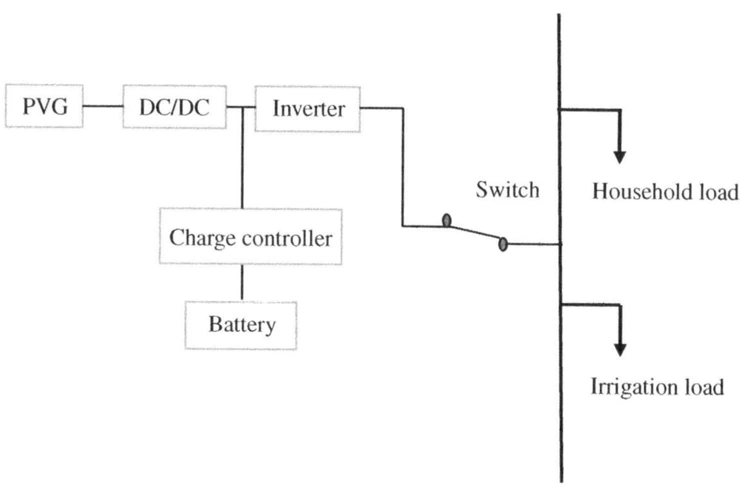

Figure 17.9 RBD of PV system with battery.

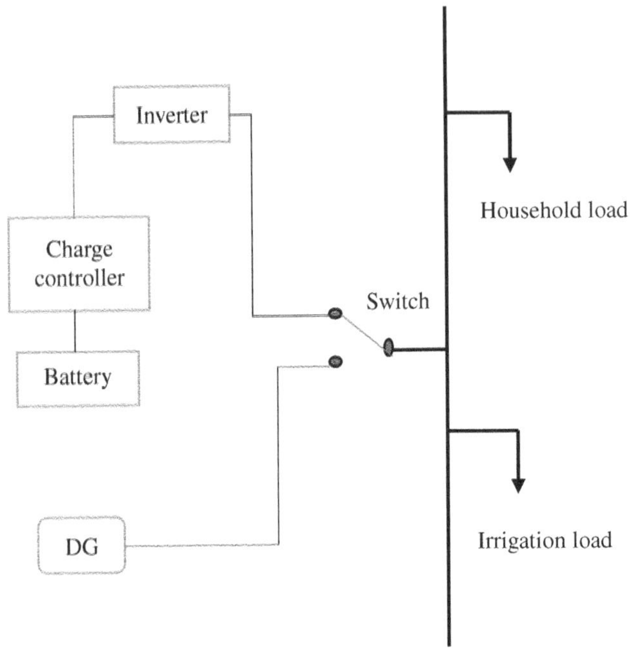

Figure 17.10 RBD of DG/battery.

2. **Diesel with Battery Backup System**: The DG/battery system is composed of a DG generator, a charge controller, a solar-batteries, a solar inverter, a switch, and a load. The block diagram of the system can be seen in Figure 17.10. The DG generator is connected in parallel with the solar batteries in order to ensure that the power generated can be efficiently stored. The charge controller regulates the charging of the solar batteries to ensure that it is fully charged and ready for use. When energy is required, the solar inverter converts the DC energy stored in the solar batteries into AC power for the load to use. The switch is used to control the power flow between the DG generator, the solar batteries, and the load. This system provides reliable backup power, especially in areas where there are frequent power outages.
   The failure rate, unavailability, EENS, and LOLP are calculated using Eqs. (17.5)–(17.18).

$$\lambda_{sys} = 0.0318\,\text{yr}^{-1}$$
$$U_{sys} = 0.0992\,\text{h/yr}$$
$$r_{sys} = 9.88\,\text{h}$$
$$\text{EENS} = 0.496\,\text{kWh/yr}$$
$$\text{LOLP} = 0.0000113\,\text{yr}^{-1}$$

3. **PV with DG System**: The PV/DG system comprises of PV arrays, a DC–DC converter, a solar inverter, a DG generator, a switch, and a load. As seen in Figure 17.11, the PV arrays are connected in series to ensure effective power generation. The PV system is connected in parallel with the DG generator to ensure optimal utilization of the power generated. The DC–DC converter is responsible for regulating the voltage levels of the DC power generated by the PV arrays. The solar inverter then converts the DC power into AC power, which can be used directly by the load. The switch is used to control the flow of power between the PV arrays, the DG generator, and

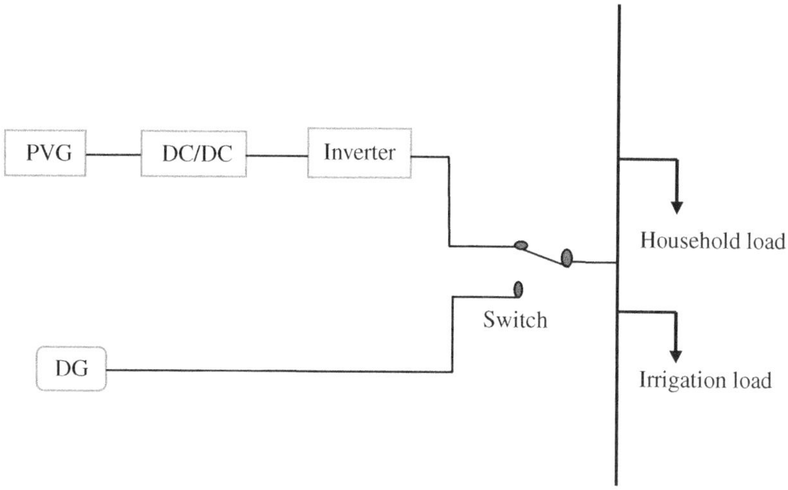

Figure 17.11 RBD of PV/diesel energy system.

the load. This system is ideal for areas with ample sunlight and unreliable grid power, providing a stable and consistent source of electricity.

PV/diesel energy system reliability indices are calculated using Eqs. (17.5)–(17.18):

$$\lambda_{sys} = 0.0245\,\text{yr}^{-1}$$

$$U_{sys} = 1.023\,\text{h/yr}$$

$$r_{sys} = 10.20\,\text{h}$$

$$\text{EENS} = 5.105\,\text{kWh/yr}$$

$$\text{LOLP} = 0.0001167\,\text{yr}^{-1}$$

This system is adequate system, if one system is repair on that time second system is supply the electricity.

Problem 17.3

Design a Reliability Block Diagram (RBD) of the hybrid energy system shown in Figure 17.12. Calculate the following Reliability Indices of PV/DG/battery: failure rate, unavailability, EENS, and LOLP. Table 17.1 provides the failure rate and repair time data for each individual component of the energy system.

Solution:

PV/DG with battery backup: The PV/DG/battery system is made up of several essential components, including PV arrays, a DC–DC converter, a charge controller, a solar inverter, a DG generator, a switch, and a load. This system is illustrated in the block diagram shown in Figure 17.12.

The PV arrays are connected in series, whereas the PV/battery system is connected parallel with the DG set. Moreover, the PV system is connected parallel with solar-batteries, ensuring that any excess energy generated is stored for future use. The charge controller is responsible for regulating the charging of the battery, ensuring it does not overcharge or undercharge. Overall, this system is ideal for remote areas where there is no access to the power grid, providing efficient and reliable electricity for various applications. The indices are:

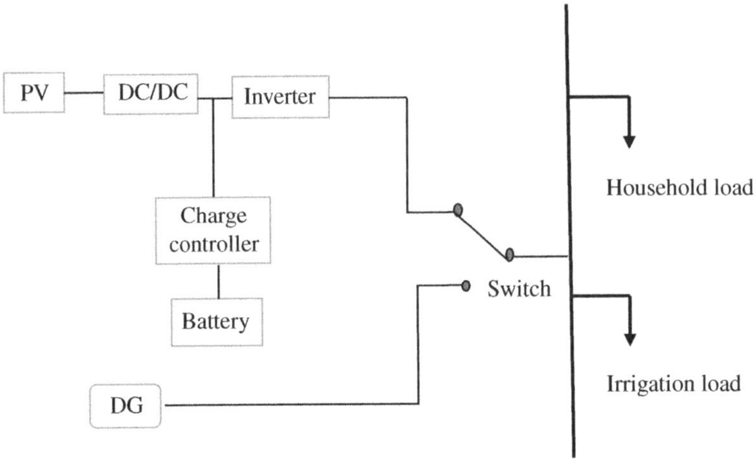

Figure 17.12 RBD of PV/DG/battery energy system.

$$\lambda_{hybrid} = 0.0821\,yr^{-1}$$

$$U_{hybrid} = 0.0603\,h/yr$$

$$r_{hybrid} = 0.8923\,h$$

$$EENS = 0.3015\,kWh/yr$$

$$LOLP = \frac{U_{hybrid}}{8760} = 0.000006883$$

The indices results show that RHPVES is suitable and more feasible energy system in remote areas.

17.5 Conclusion

This chapter's primary focus was on assessing the reliability of standalone solar PV energy systems. To achieve this objective, six different system configurations were analyzed, including PV-only, diesel-only, PV/diesel, PV/battery, PV/diesel/battery, and diesel/battery. The development of the system's operational strategy was based on various indices, such as failure rate, repair time, unavailability, LOLP, and EENS. By utilizing the Markov process method, a reliability evaluation was conducted, which allowed for a thorough analysis of the system's suitability for the intended application. The system configurations were analyzed based on their unavailability, failure rate, EENS, and LOLP. Among all configurations considered, the hybrid PV/battery/DG system demonstrated the lowest values across all indices and proved to be the most reliable and sustainable option. The reliability performance of a system can be better understood through these indices, and as such, the PV/battery/DG system showcases superior reliability and sustainability compared to other configurations. Therefore, it can be concluded that PV/battery/DG system is the most reliable and sustainable option among all system configurations.

References

Abunima, H. and Teh, J. (2020). Reliability modeling of PV systems based on time-varying failure rates. *IEEE Access* 8: 14367–14376.

Akhtar, I., Kirmani, S., and Jameel, M. (2021). Reliability assessment of power system considering the impact of renewable energy sources integration into grid with advanced intelligent strategies. *IEEE Access* 9: 32485–32497.

Alanazi, A., Alanazi, M., Nowdeh, S.A. et al. (2022). An optimal sizing framework for autonomous photovoltaic/hydrokinetic/hydrogen energy system considering cost, reliability and forced outage rate using horse herd optimization. *Energy Reports* 8: 7154–7175.

Alfawaz, O., Khedr, A.M., Alwasel, B., and Osamy, W. (2023). Reliability evaluation for chain routing protocols in wireless sensor networks using reliability block diagram. *Journal of Sensor and Actuator Networks* 12 (2): 34.

Billinton, R. (1970). *Power System Reliability Evaluation*. Taylor & Francis.

Billinton, R., Allan, R.N., Billinton, R., and Allan, R.N. (1992). Discrete Markov chains. In: *Reliability Evaluation of Engineering Systems: Concepts and Techniques*, 260–279. Boston, MA: Springer.

Ghania, S.M., Mahmoud, K.R.M., and Hashmi, A.M. (2022). A reliability study of renewable energy resources and their integration with utility grids. *Engineering, Technology & Applied Science Research* 12 (5): 9078–9086.

Hou, K., Tang, P., Liu, Z., and Dong, Z. (2023). An efficient reliability assessment approach for multiple energy systems based on shadow price. *Energy Reports* 9: 829–836.

Kalantar-Neyestanaki, M. and Cherkaoui, R. (2021). Coordinating distributed energy resources and utility-scale battery energy storage system for power flexibility provision under uncertainty. *IEEE Transactions on Sustainable Energy* 12 (4): 1853–1863.

Kebede, F.S., Olivier, J.-C., Bourguet, S., and Machmoum, M. (2021). Reliability evaluation of renewable power systems through distribution network power outage modelling. *Energies* 14 (11): 3225.

Khan, A.A., Minai, A.F., Pachauri, R.K., and Malik, H. (2022). Optimal sizing, control, and management strategies for hybrid renewable energy systems: a comprehensive review. *Energies* 15 (17): 6249.

Kou, L., Chu, B., Chen, Y., and Qin, Y. (2022). An automatic partition time-varying Markov model for reliability evaluation. *Applied Sciences* 12 (12): 5933.

Kundur, P.S. and Malik, O.P. (2022). *Power System Stability and Control*. McGraw-Hill Education.

Kuo, T.-C., Pham, T.T., Bui, D.M. et al. (2022). Reliability evaluation of an aggregate power conversion unit in the off-grid PV-battery-based DC microgrid from local energy communities under dynamic and transient operation. *Energy Reports* 8: 5688–5726.

Niu, M., Xu, N.Z., Kong, X. et al. (2021). Reliability importance of renewable energy sources to overall generating systems. *IEEE Access* 9: 20450–20459.

Ostovar, S., Esmaeili-Nezhad, A., Moeini-Aghtaie, M., and Fotuhi-Firuzabad, M. (2021). Reliability assessment of distribution system with the integration of photovoltaic and energy storage systems. *Sustainable Energy, Grids and Networks* 28: 100554.

Postnikov, I. and Mednikova, E. (2022). A reliability analysis of fuel supply for district heating systems based on statistical test method. *Energy Reports* 8: 304–311.

Raghuwanshi, S.S. and Arya, R. (2019a). Renewable energy potential in India and future agenda of research. *International Journal of Sustainable Engineering* 12 (5): 291–302.

Raghuwanshi, S.S. and Arya, R. (2019b). Reliability evaluation of stand-alone solar PV energy system for irrigation. *Proceedings of Recent Advances in Interdisciplinary Trends in Engineering & Applications (RAITEA)*.

Raghuwanshi, S.S. and Arya, R. (2020a). Design and economic analysis of a stand-alone hybrid photovoltaic energy system for remote healthcare centre. *International Journal of Sustainable Engineering* 13 (5): 360–372.

Raghuwanshi, S.S. and Arya, R. (2020b). Reliability evaluation of stand-alone hybrid photovoltaic energy system for rural healthcare centre. *Sustainable Energy Technologies and Assessments* 37: 100624.

Raghuwanshi, S.S. and Khare, V. (2018). Sizing and implementation of photovoltaic water pumping system for irrigation. *IAES International Journal of Artificial Intelligence* 7 (1): 54.

Raghuwanshi, S.S., Masih, A., and Raghuwanshi, P. (2022). Design and optimisation of stand-alone hybrid energy system for rural areas. *International Journal of Ambient Energy* 43 (1): 8707–8721.

Raghuwanshi, S.S., Raghuwanshi, P., Masih, A., and Singh, P. (2023). Modeling and optimization of hybrid renewable energy with storage system using flamingo swarm intelligence algorithms. *Energy Storage* 5 (7): e470.

Rayati, M., Bozorg, M., Carpita, M., and Cherkaoui, R. (2023). Stochastic optimization and Markov chain-based scenario generation for exploiting the underlying flexibilities of an active distribution network. *Sustainable Energy, Grids and Networks* 34: 100999.

Saket, R.K. (2011). Reliability evaluation of defence support systems. In: *Innovations in Defence Support Systems-2: Socio-Technical Systems*, vol. 338 (ed. L.C. Jain, E.V. Aidman, and C. Abeynayake), 241–286. Berlin, Heidelberg: Springer-Verlag.

Samy, M.M., Mosaad, M.I., El-Naggar, M.F., and Barakat, S. (2020). Reliability support of undependable grid using green energy systems: economic study. *IEEE Access* 9: 14528–14539.

Singh, S., Saket, R.K., and Khan, B. (2023). A comprehensive review of reliability assessment methodologies for grid-connected photovoltaic systems. *IET Renewable Power Generation* 17 (7): 1859–1880.

Spertino, F., Chiodo, E., Ciocia, A. et al. (2020). Maintenance activity, reliability, availability, and related energy losses in ten operating photovoltaic systems up to 1.8 MW. *IEEE Transactions on Industry Applications* 57 (1): 83–93.

Tawfiq, A.A.E., El-Raouf, M.O.A., Mosaad, M.I. et al. (2021). Optimal reliability study of grid-connected PV systems using evolutionary computing techniques. *IEEE Access* 9: 42125–42139.

Xu, X., Tang, J., and Xiang, H. (2022). Data transmission reliability analysis of wireless sensor networks for social network optimization. *Journal of Sensors* 2022: 1–12.

18

Probabilistic Distribution and Monte Carlo Approach for Reliability Evaluation of SEIG-Based Micro Hydro Power Generation System

Lokesh Varshney[1], Kanhaiya Kumar[2], Gautam Singh Dohare[3], Udaya M. Bhaskara Rao[4], and Jitendra Singh Shakya[5]

[1]*Department of Electrical, Electronics & Communication Engineering, Galgotias University, Greater Noida, Uttar Pradesh, India*
[2]*Electrical & Electronics Engineering Department, Galgotias College of Engineering & Technology, Greater Noida, Uttar Pradesh, India*
[3]*Department of Electrical, Electronics & Communication Engineering, Galgotias University, Greater Noida, Uttar Pradesh, India*
[4]*Engineering Division, CSIR National Metallurgical Laboratory, Jamshedpur, Jharkhand, India*
[5]*Department of Electrical Engineering, Samrat Ashok Technical Institute, Vidisha, Madhya Pradesh, India*

18.1 Introduction

Compared to an alternator, a self-energized induction generator (SEIG) has various benefits. However, it has certain issues, like reactive energy use and weak voltage regulation. The SEIG works either as the SEIG was constructed by the engineer or as an appropriate level of capacitor is attached in shunt with the induction motor (IM) contacts. In comparison to the IM, the accessibility of developed SEIG is not very common. But in highlands or isolated places, uncontrolled renewable energy (RE) sources like micro hydro and wind are fairly common. SEIG has a direct link to a turbine that has a renewable energy source connected to it. Dams, gear-pulleys, and various other mechanical attachments are not required. An IM's effectiveness when functioning as a SEIG with an appropriate capacitance level is unsatisfactory. The choice of the parameter values made throughout the design stage, which optimized the efficacy of the device as a motor instead of as a power source, is what causes the poor performance. Understanding that an IM is a magnetic circuit and the way hysteresis will affect it is crucial for illustrating this notion. In light of this hysteresis, an IM has two operational areas. In particular, an IM will function in either a linear or nonlinear saturation zone. In the majority of instances, it is not advisable to run an IM at saturation because doing so lowers the iron's relative permeability and raises the motor's magneto motive force (MMF) need. However, working in the linear zone only partially makes use of the iron's abilities, making this strategy uneconomical. As a consequence, the knee of the curve of saturation represents the ideal point of operation for an IM. These ideas maximize iron use while reducing saturation. A SEIG's functioning is steady when its magnetic circuit reaches saturation, though. As a result, an IM's voltage at the terminal is raised till the magnetic circuit becomes saturated with the goal to use the IM as a generator (Nakorn et al. 2021; Silva et al. 2018). As a consequence, IM work prior to the knee of the saturate curve, whereas SEIG works after the

Reliability Analysis of Modern Power Systems, First Edition. Edited by R. K. Saket and P. Sanjeevikumar.

knee. The IM's rotor core is designed with less hysteresis loss in mind; hysteresis loss is dependent on the size of the hysteresis loop. Because of this circumstance, an IM's rotors core is made of soft magnetic material, resulting in a smaller hysteresis loop. On the opposite hand, the rotating core of the planned SEIG, wherein hysteresis loop is wide, uses hard magnetic materials. As a result, the rotor of this machine has a greater hysteresis impact. As a result, the rotating core possesses enough remaining magnetism to cause the SEIG's first activation. When using an IM as a SEIG under these circumstances, the issue of excitation loss may arise. But as compared to how SEIG was designed, IM is more accessible and affordable. Consequently, utilizing IM as a SEIG is fairly common. It is necessary to assess the machine's rotor's dependability because it will function as SEIG.

The structure of this research is as follows: Section 18.2 has information on residual magnetism losses and recovery. These three techniques for restoring residual magnetism are briefly described in this section. The idea, the causes, and variables that contribute to failure functioning of SEIG are described in Section 18.3. The measured minimal capacitance value for SEIG excitation is evaluated in Section 18.4. The reliability of the SEIG's rotors core magnetization is assessed in Section 18.5 along the lines of the curve, hazard rate, failure density function, survivor function, cumulative failure distribution, and utilizing Monte Carlo simulation (MCS), the odds of either succeeding or failing have been assessed. The work's summary is presented in Section 18.6.

18.2 Residual Magnetism in SEIG: Restoration and Loss

18.2.1 Residual Magnetism Losses

The short-circuiting state and coupling of an additional reactive loading are to blame for the depletion of residual magnetism in SEIG. These circumstances are what lead to the SEIG's abrupt endpoint voltage reduction and the reduction in remaining magnetism in the rotating core. Figure 18.1 depicts the tested configuration with reactive loads. In order to temporarily excite the

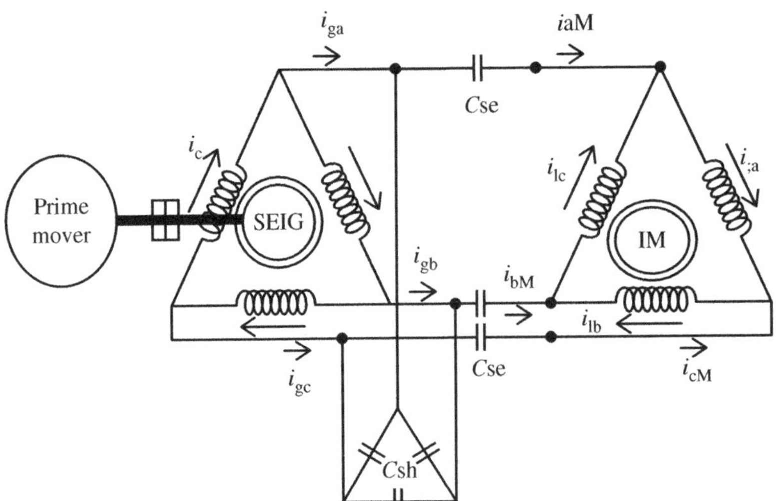

Figure 18.1 Reactive loading configuration for a study.

core of iron and restore the remaining magnetism, one may employ (any one of) the methods described in Section 2.2.

18.2.2 Residual Magnetism Restoration

In broadly, three techniques are effective for restoring residual magnetism.

1. For a period of 10–15 minutes, use the machine as a motor powered by an exciting ac unit.
2. By bringing in capacitors with charge terminals. In most cases, even with a degaussed rotor, the current at the discharge point will be enough to trigger self-excitation if the capacitors are charged to a very high voltage, such as the rated machine voltage.
3. By raising the machine velocity over its certified value, which exceeds the resonant frequency at weak magnetization and starts self-excitation (keep in mind that the rotors and bearing of the unit must be certified for the increased velocity).
4. Attach six-volt batteries across two of the equipment's connections when it is at resting and leave it there for 10–15 minutes. Or, enough of residual magnetism may be ensured by running a DC current across the equipment before it is brought up to speed. With the power grid being available, the initial option is superior to the other strategies. The third option is helpful when the grid is unavailable (Chaturvedi and Goel 2019).

18.3 Problems with SEIG Excitation Failure in RE Systems

In this study, an IM serving as a SEIG with an appropriate capacitor size is utilized. Shaft velocity, residual magnetism, lower permeability at lower magnetization, and the capacity of the linked capacitor all affect SEIG production. By raising the rate, the rating of the capacitor, or both of them, the self-excitation reliability has to be extremely high (Bu et al. 2020; Ion 2020). While the SEIG produces active energy; capacitors supply reactive energy to the reactive load (Q_L) and to the SEIG for excitation. With a proper size for the capacitor, SEIG is initiated without any load. During reliability assessments, this simulated SEIG condition has been taken into account (Saket et al. 2022; Saket 2011; Varshney et al. 2013a, 2021; Varshney and Saket 2014). The minimum required terminal capacitor, which is necessary for SEIG to withstand voltage deposits, was used. It is advised to compare the reliability of various SEIGs using a small amount of capacitor.

18.4 SEIG Tests with Lowest Capacitive Excitation

The SEIG IM has the following specifications: 12.3 A, 3-phase, 50 Hz, 4-pole, 230 V, and 5.5 hp. In terms of units, the IM's per-phase equivalent variables are Rs = 0.04960, Rr = 0.03500, Xs = 0.13440, and Xr = 0.13440. Figure 18.2 depicts the estimated magnetization gradient of the IM. The capacitor's estimated lowest possible value is 24.48 μF (Chaturvedi 2018).

The use of Vg/F for the evaluation machine is shown in Figure 18.3. IM working as a SEIG is assessed through the examination of junction voltage at various primary mover speeds. Figure 18.4 displays the differences in produced delta linked no load terminals voltages with varied prime mover speeds at 25 μF and 36 μF capacitors delta linked. As can be observed, the junction voltage's match between observed and calculated values is enough to support the reliability of the Khan and Khan (2022) assessment.

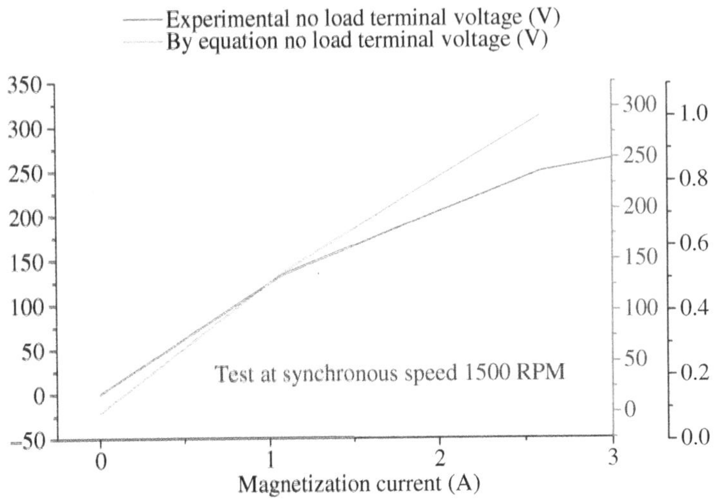

Figure 18.2 IM magnetization graph without load.

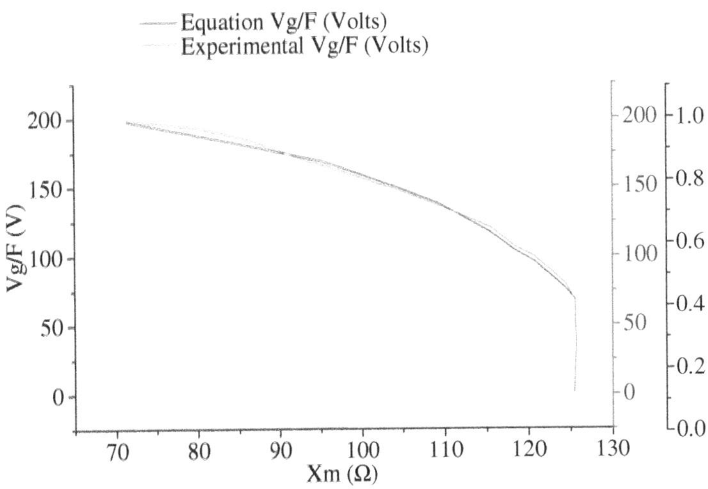

Figure 18.3 Vg/F variation with Xm.

18.5 Rotor Core Magnetization of SEIG Reliability Assessment Using Least Capacitor Score

Degradation of excitation was only taken into account in this work when the SEIG's production failed. The primary reason for a breakdown of machine performance is thought to be the lack of excitement. Throughout the testing phase, other SEIG system breakdown factors, such as generating failure caused by design/manufacturing flaws and operational factors, are not taken into account (Saket and Varshney 2012; Varshney et al. 2016; Varshney and Saket 2013). Utilizing the smallest size for the capacitor, the reliability of SEIG excitation was evaluated. Figure 18.5 lists the evaluated dependability indices.

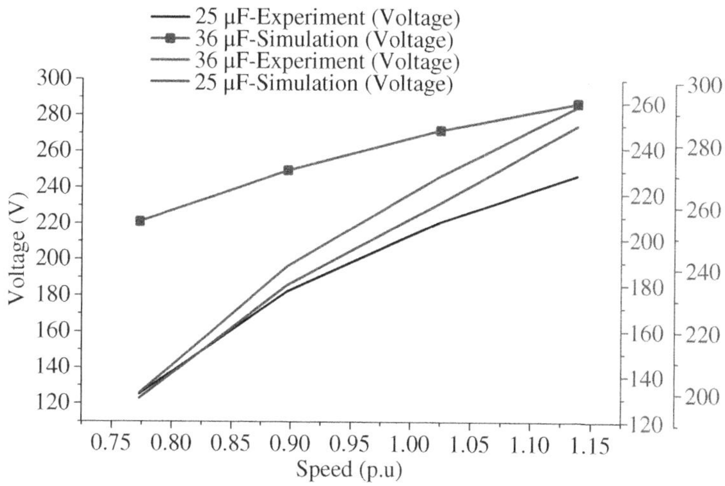

Figure 18.4 Change of the voltage at the terminal for without a load with respect to time as well as capacitance.

18.5.1 Assessment of Reliability-Related Functions

Probability distribution curves are used here to characterize the reliability assessment of SEIG excitation. In the beginning, SEIG (Varshney et al. 2013b) was the subject of over 32 tests on the first day. It indicates a 32-fold rise in reactive burden. The overall number of tests was determined the following day by adding up all successful tests and all unsuccessful ones from the preceding time period (Kumar et al. 2021; Saket 2011). In tests, the total number of fails is tallied across seven periods. Figure 18.5 describes the probability indices such as the hazard function, failure density function, survival function and cumulative failure distribution. The following is the process for evaluating dependability indices:

1. Computation for each interval's overall failures. The time span is one day.
2. Assessment of reliability metrics include the risk rate, failed density function, and cumulative failed distribution. Figures 18.6–18.9 show a graphical representation of each dependability index, accordingly.

18.5.2 Probabilities of Failure and Success

In this work, the probability of failure and success have been assessed utilizing the MCS technique. The method used to assess the dependability indicators differs among the analytic and simulation-based approaches (Kumar et al. 2021). Analytic approaches use a mathematical framework for expressing the data, which is frequently streamlined and employ direct mathematical approaches to assess the model's reliability scores. However, MCS calculates the reliability indexes by mimicking the system's actual procedure and random characteristics. As a result, the technique approaches the issue as a set of genuine tests carried out over a period of modeled time. By recording the instances of an occurrence, it may calculate probability along with other indices. The setting up for lab tests began with a least capacitance of 25 μF. Utilizing MCS, it is being practically tested if the IM will successfully excite when used as a SEIG. In entirety, 71 experiments were run over the course of 8 consecutive days. Figure 18.10 displays different test findings (success/failure excitation) from a study on the SEIG.

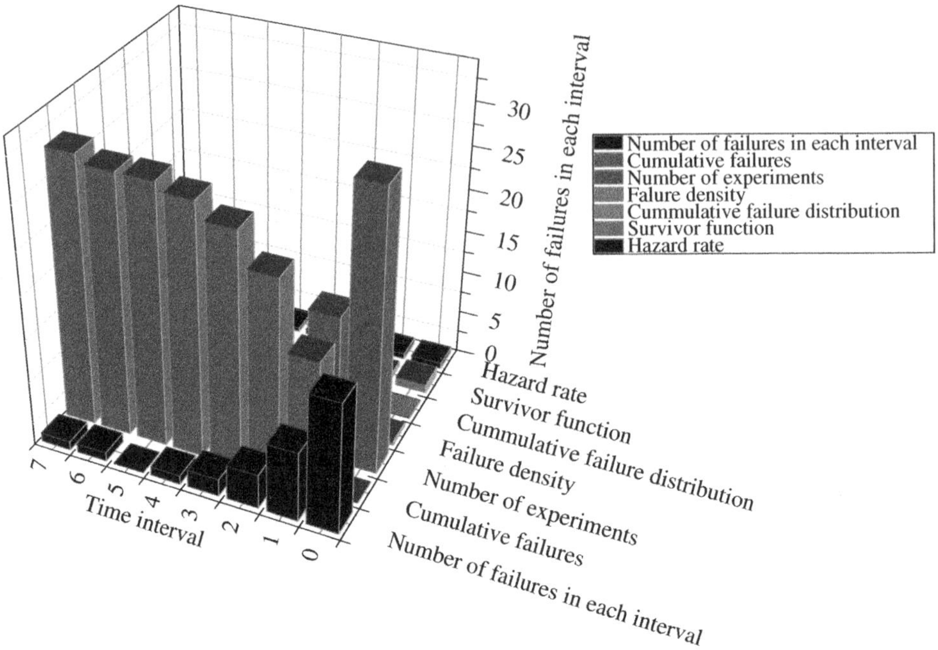

Figure 18.5 Data on actual failures and an assessment of reliability indexes.

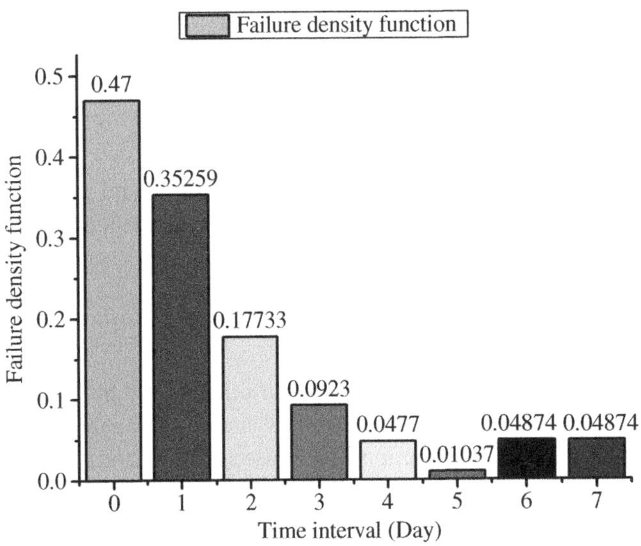

Figure 18.6 Graph of a failed density function.

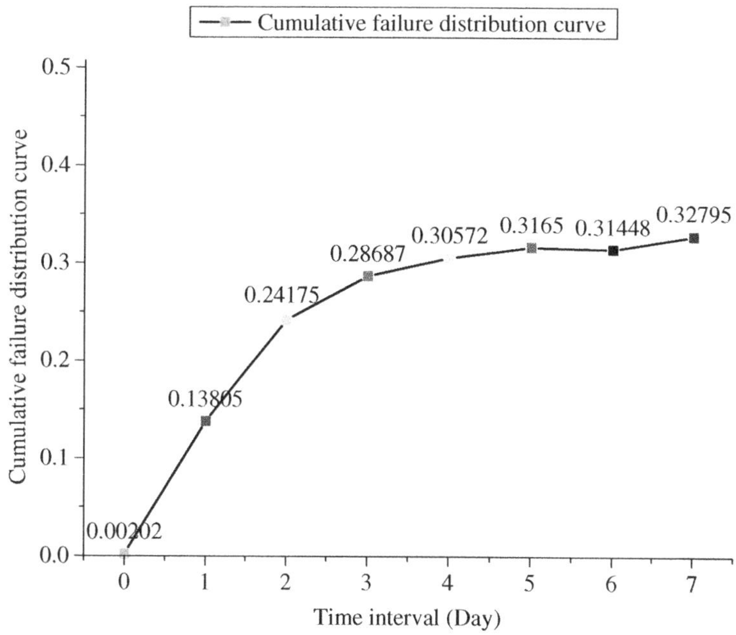

Figure 18.7 Distribution curve for cumulative failures.

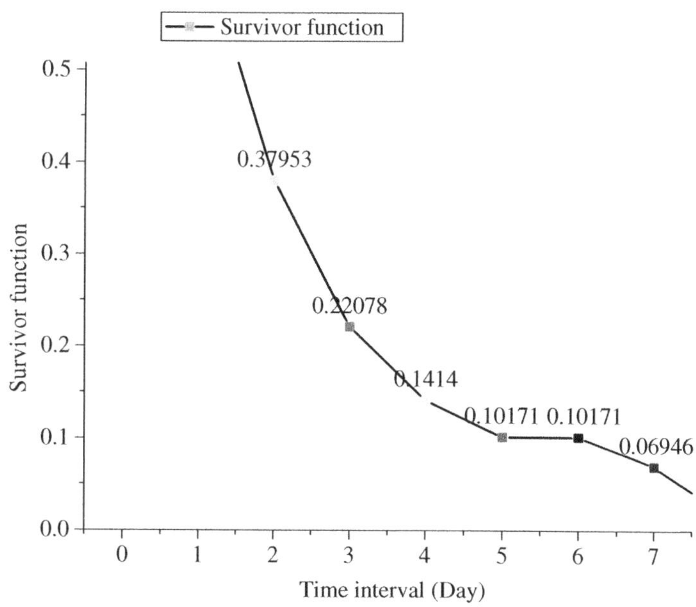

Figure 18.8 Curve for the survivor function.

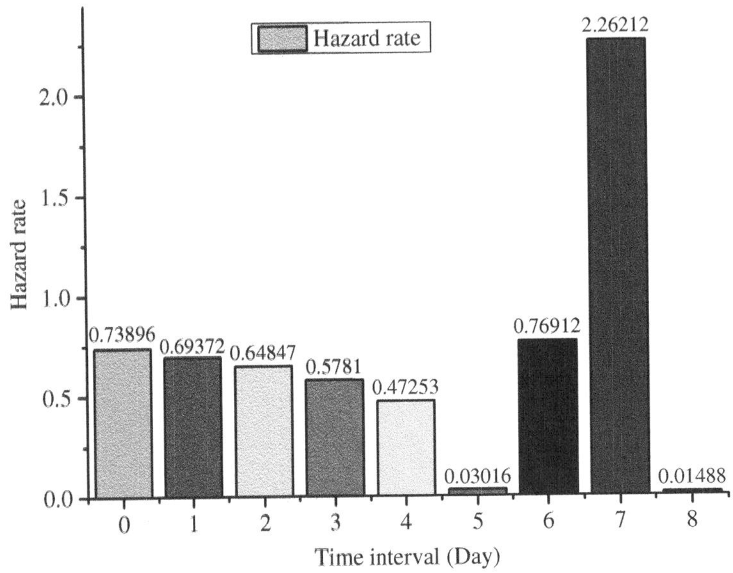

Figure 18.9 Hazard graph.

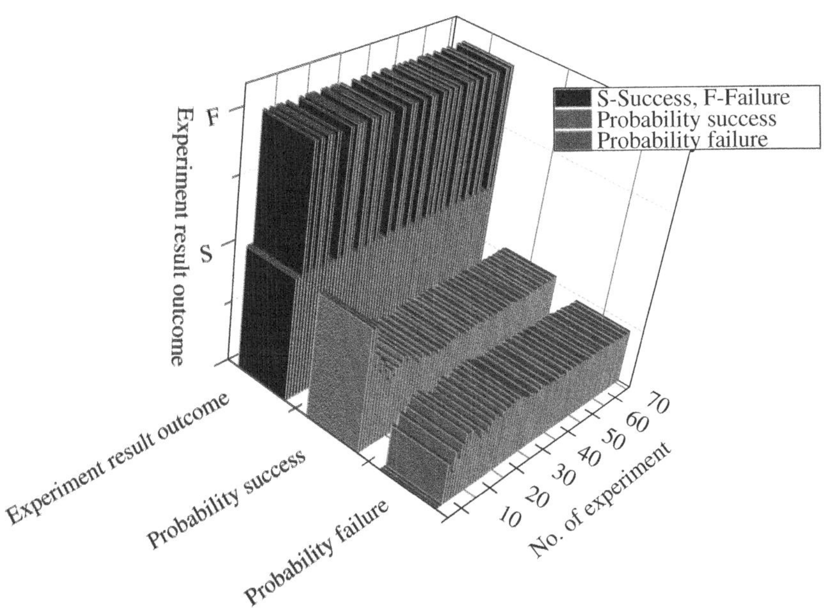

Figure 18.10 Results of the SEIG's development of failures and successes.

18.6 Discussion and Outcomes

The SEIG's chances of failure and success with the lowest capacitor size have been assessed utilizing Monte Carlo modeling. The simulation parameters are displayed in Figures 18.11 and 18.12. The simulation findings lead to the following conclusions:

1. SEIG experiments are run frequently, which improves the results in terms of failure and success probabilities.
2. After enough testing, the chance of failure and success fluctuates about its true value. The true value cannot be accurately estimated using the mean of the failure and success probabilities.
3. As the quantity of experiments rises, the estimation of the likelihood of failure and success tends toward the true value.
4. Based on the outcomes of the simulations, the probabilities of failure and success were inclined toward 0.55 and 0.45, correspondingly.

Employing a statistical approach, the probability of success (Ks) and failure (Kf) has been determined. The proportion of the quantity of successful/failed assessments to the total number of results is the probability of success (Ks)/failure (Kf). The ratings are listed below.

$$\text{Success probability Ks} = \text{s/J}$$

$$\text{Ks} = 39/71 = 0.549$$

Similarly,

$$\text{Failure probability Kf} = \text{f/J}$$

$$\text{Kf} = 32/71 = 0.451$$

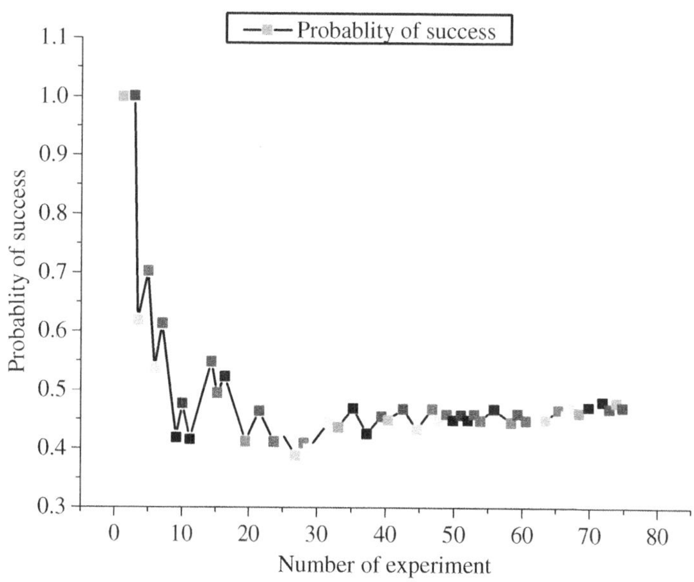

Figure 18.11 Probability for a successful simulation using the Monte Carlo method.

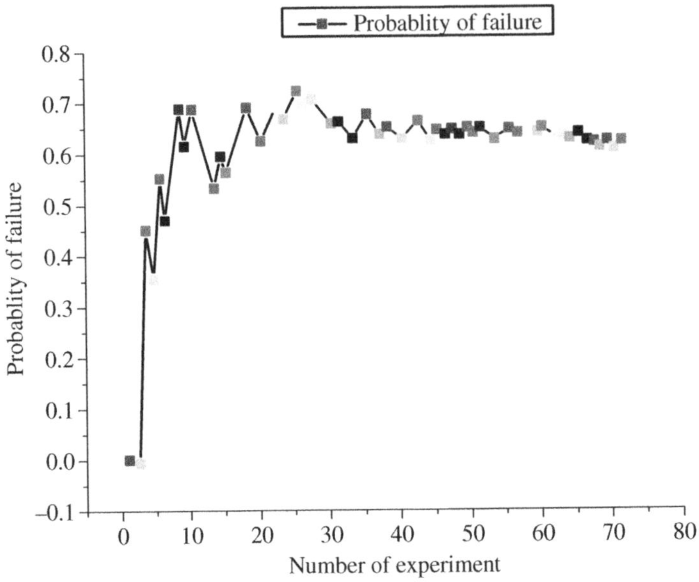

Figure 18.12 Chance of fail simulation using the Monte Carlo method.

18.7 Conclusion

Despite the least amount of capacitive stimulation, the SEIG's performance dependability is noted. This study correctly assesses the possibilities of failure and success of 3-phase, 5.5 hp, and 25 μF. The analytical outcomes are Kf = 0.452 and Ks = 0.549. The results go toward Ks = 0.55 and Kf = 0.45 and are centered on MCS. The findings of the simulations as well as the analysis are nearly identical. The outcomes of experiments have been used to derive the failure density slope, survival graph, cumulative graph, and hazard graph. The area under the failure density graph is one, and also production failure diminishes as the quantity of experiments is reduced.

References

Bu, F., Liu, H., Huang, W. et al. (2020). Induction-machine-based starter/generator systems: techniques, developments, and advances. *IEEE Industrial Electronics Magazine* 14 (1): 4–19.

Chaturvedi, Y.K. (2018). A dissertation report on modelling of self excited induction generator for minimum capacitance requirement.

Chaturvedi, Y. and Goel, A. (2019). Wind powered electricity generation through self excited induction generator. *World Scientific News* 121: 90–100.

Ion, C.P. (2020). A comprehensive overview of single-phase self-excited induction generators. *IEEE Access* 8: 197420–197430.

Khan, M.F. and Khan, M.R. (2022). Analysis of a nine-phase self excited induction generator equipped with optimum excitation capacitances. In: *2022 International Conference for Advancement in Technology (ICONAT)*, 1–6. IEEE.

Kumar, K., Varshney, L., Ambikapathy, A. et al. (2021). Solar tracker transcript—a review. *International Transactions on Electrical Energy Systems* 31 (12): e13250.

Nakorn, P., Machot, P., Kinnares, V., and Manop, C. (2021). Study of three-phase self-excited induction generator operating as single-phase induction generator supplying non-linear load. In: *2021 18th International Conference on Electrical Engineering/Electronics, Computer, Telecommunications and Information Technology (ECTI-CON)*, 806–809. IEEE.

Saket, R.K. (2011). Reliability evaluation of defence support systems. In: *Innovations in Defence Support Systems-2: Socio-Technical Systems* (ed. L.C. Jain, E.V. Aidman, and C. Abeynayake), 241–286. Berlin, Heidelberg: Springer-Verlag.

Saket, R.K. and Varshney, L. (2012). Self excited induction generator and municipal waste water based micro hydro power generation system. *International Journal of Engineering and Technology* 4 (3): 282.

Saket, R.K., Kothari, D.P., and Nagrath, I.J. (2022). *Modern Power System Analysis*. Tata McGraw-Hill Publishing Company.

Silva, F.B., da Silva Gonçalves, F.A., Vanço, W.E. et al. (2018). Application of bidirectional switches in the development of a voltage regulator for self-excited induction generators. *International Journal of Electrical Power & Energy Systems* 98: 419–429.

Varshney, L. and Saket, R.K. (2013). Reliability evaluation of residual magnetism in rotor of SEIG. In: *2013 Students Conference on Engineering and Systems (SCES)*, 1–4. IEEE.

Varshney, L. and Saket, R.K. (2014). Reliability evaluation of SEIG rotor core magnetization with minimum capacitive excitation for unregulated renewable energy applications in remote areas. *Ain Shams Engineering Journal* 5 (3): 751–757.

Varshney, L., Saket, R.K., and Eslamian, S. (2013a). Power estimation and reliability evaluation of municipal waste water and self-excited induction generator-based micro hydropower generation system. *International Journal of Hydrology Science and Technology* 3 (2): 176–191.

Varshney, L., Varshney, V., Newwel, A., and Saket, R.K. (2013b). Squirrel cage rotor design for safety and reliability improvement of a three phase induction machine. In: *Proceedings of the 3rd International Conference on Trends in Information, Telecommunication and Computing*, 267–273. Springer.

Varshney, L., Gupta, C., and Debnath, D. (2016). Performance improvement of squirrel cage rotor by design aspects. In: *2016 International Conference on Recent Advances and Innovations in Engineering (ICRAIE)*, 1–4. IEEE.

Varshney, L., Vardhan, A.S.S., Vardhan, A.S.S. et al. (2021). Performance characteristics and reliability assessment of self-excited induction generator for wind power generation. *IET Renewable Power Generation* 15 (9): 1927–1942.

19

Reliability and Mean Life Assessment of Solar Panel by Cooling

Rahul Agrawal[1], Jyotsna Ogale[2], Nga T. T. Nguyen[3], R. K. Saket[4], and Joydeep Mitra[5]

[1] *Department of Electrical Engineering, Indian Institute of Technology (BHU) Varanasi, Uttar Pradesh, India*
[2] *Department of Electronics Engineering, Samrat Ashok Technological Institute, Vidisha Madhya Pradesh, India*
[3] *Department of Electrical Engineering and Computer Science, University of Wyoming, Wyoming, USA*
[4] *Department of Electrical Engineering, Indian Institute of Technology (BHU), Varanasi, Uttar Pradesh, India*
[5] *Department of Electrical Engineering, Michigan State University, Michigan, USA*

19.1 Introduction

The world is widely harnessing solar power for its energy needs and decreasing its dependence on diminishing fossil fuels. Solar PV panel works in the sunlight the higher the exposure to the sun, the more energy the solar cell will produce. But as a consequence, the temperature of the Solar cell rises due to rising in the ambient temperature and irradiance (Markvart 2000); due to the rise in the temperature of the solar cell, its lifetime gets reduced (Hacke et al. 2015). So, solar panels reliability and mean life are essential parameters for the more extensive acceptability of solar power generation.

This work focuses on the environmental effects (temperature and humidity) on the reliability of solar panels and how they affect the mean life of solar panels (Park et al. 2013; Luque and Hegedus 2011). Mean life and reliability were studied and compared with and without cooling of the panel (Sharaf et al. 2022; Teo et al. 2012; Sohani et al. 2021; Mawoli et al. 2020). The environmental impact on the solar cell tends to reduce the lifetime of the panel, which will decrease its reliability. As a part of environmental impact, temperature and relative humidity are the prime factors that cause the degradation of solar panels (Park et al. 2013; Chikate et al. 2015). To reduce the temperature impact, various cooling methods are being introduced to improve the lifetime of solar panels (Hacke et al. 2015; Sohani et al. 2021). This research aims to compare the improvement in mean life and compare the reliability of solar panels by cooling and without cooling (Park et al. 2013). The proposed work studied the two degradation models, namely, Erying and Peck, to calculate the degradation after cooling. The cooling methods for which the reliability assessment is done are water cooling (Moharram et al. 2013), Air cooling using extended fin area (Ravi and Porpatham 2017), Wick structure with nanofluids using CuO/water nanofluids (Kusiak et al. 2013), and passive cooling by heat pipe (Akbarzadeh and Wadowski 1996). Corresponding to the mean life obtained from these models, a reliability study of solar panels is done, and the reliability comparison for without cooling (Teo et al. 2012) and cooling (considering the Erying and Peck model) (Park et al. 2013; Firoozzadeh et al. 2019) is done.

Reliability Analysis of Modern Power Systems, First Edition. Edited by R. K. Saket and P. Sanjeevikumar.

It is observed that the output will be maximum if the irradiance falling on the PV surface is maximum. As a consequence of high irradiance, cell temperature also increases as irradiance increases, reducing the lifetime of the solar cell. It is observed that the efficiency improvement of solar PV systems by applying cooling techniques (Irwan et al. 2015; Cuce et al. 2011) used to be the prime focus of research. At the same time, reliability is also a prime factor that should be considered after using the cooling methods (Firoozzadeh et al. 2019; Nižetić et al. 2017). The reliability of subsystems of solar cells without cooling is calculated in (Makrides et al. 2018) using statistical techniques Ordinary Least Squares (OLS), Classical Seasonal Decomposition, and the Year-on-Year (YOY) comparative technique to obtain the degradation rates (Meeker et al. 2022; Aminisharifabad et al. 2021). As observed from the literature survey, some of the cooling methods and the parameters they found are tabulated in Table 19.1.

In (Moharram et al. 2013), it is calculated that using the water-cooling method for a five-minute operation of the cooling system is capable of reducing the solar cell temperature by 10 °C and increasing the efficiency by 12.5%. The wet cotton wick formation that is placed on the back surface of the panel is estimated to cool the solar panel due to the capillary action of the cotton wick (Kusiak et al. 2013).

For the work to be carried out, data for the location in central India, longitude 23.181 and latitude 79.986, that is for the city Jabalpur in Madhya Pradesh, is taken. This data contains atmospheric temperature, relative humidity, and solar irradiance in one-hour intervals taken from March 3, 2021, to April 27, 2022. The plots for atmospheric temperature and relative humidity are shown in Figures 19.1 and 19.2.

Table 19.1 Summary of improvements by cooling.

Researcher	Cooling methods	Temperature reduction	Parameters calculated
Moharram et al. (2013)	Water cooling	Significant	Efficiency improved by 12.5% for a temperature reduction of 10 °C
Ravi and Porpatham (2017)	Air cooling with extended fins area	16 °C	Adding fins increases thermal efficiency to 15.5% and electrical efficiencies to 10.5%
Kusiak et al. (2013)	Wick structure was placed in the back panel with nanofluids	Up to 11% using CuO/water and 17% by Al_2O_3/water	Maximum efficiency attained is 10.4%
Akbarzadeh and Wadowski (1996)	Passive cooling with heat pipe	Significant	Passive cooling with a heat pipe can reduce the temperature by 38 °C
Hasan et al. (2017)	PCM-RT42	Annual avg. by 10.5 °C	Electrical output incremented by 5.9%
Karthick et al. (2020)	PCM-eutectic	Maximum 9 °C	Average energy gains 103.6 Wh
Salem et al. (2019)	PCM – a mixture of water and Al_2O_3	Significant	Efficiency increments by 10.9% to 22.7%
Tana et al. (2017)	Organic paraffin wax	By 15 °C	Efficiency improved by 5.39%

Source: Sohani et al. 2021; Mawoli et al. 2020; Moharram et al. 2013; Karthick et al. 2020; Dwivedi et al. 2020; Salem et al. 2019.

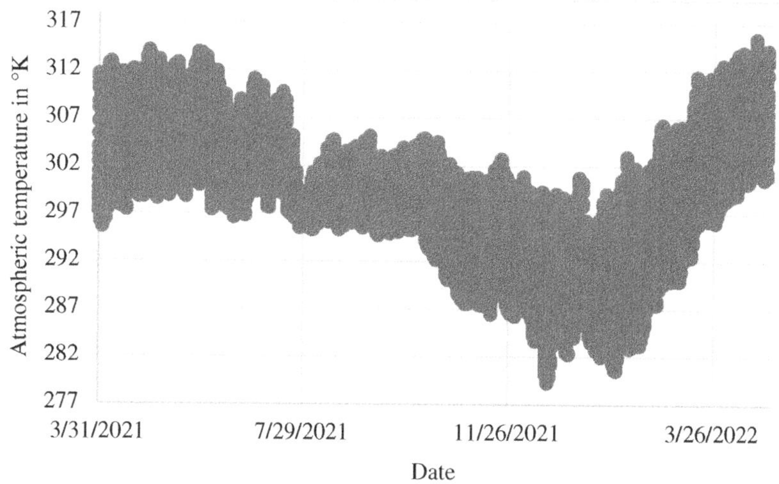

Figure 19.1 Atmospheric temperature in °K at longitude 23.181 and latitude 79.986. Source: Anonymous (n.d.). Adapted from Live and Forecast API.

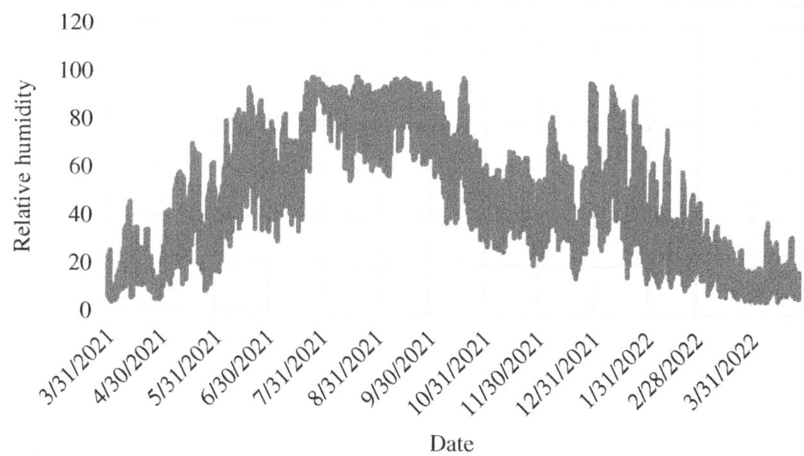

Figure 19.2 Relative humidity at longitude 23.181 and latitude 79.986. Source: Anonymous (n.d.). Adapted from Live and Forecast API.

Only 4–17% of incoming solar radiation is typically converted into electricity by PV modules. The type of solar cells used, as well as the working conditions, influence conversion efficiency. As a result, more than half of the incoming solar power is transformed to heat. Short term losses are caused by an increase in cell temperature. A reduction in electrical rate of returns, and reduces reliability and solar panel efficiency too (Moharram et al. 2013).

19.2 Methodology

The workflow of the proposed work starts with defining a location to work upon, followed by the atmospheric condition in that location, which affects the change in solar panel parameters, and

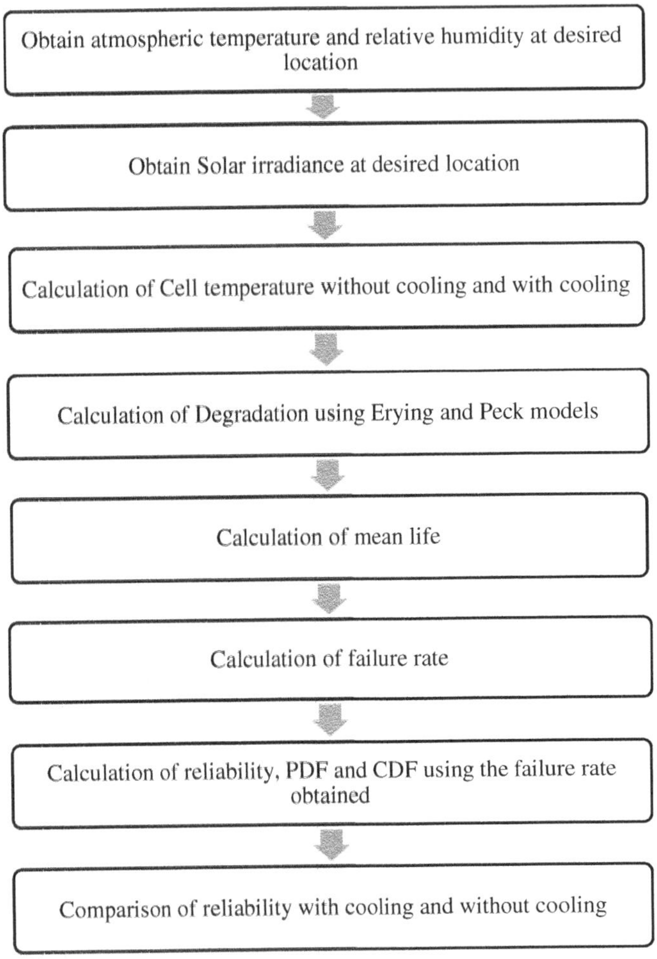

Figure 19.3 Flowchart of proposed work.

the degradation through those parameters is calculated, followed by the calculation of failure rates, mean life and reliability (Meeker et al., 2022). The flowchart of the work is shown in Figure 19.3.

19.2.1 Cell Temperature

The nominal operating conditions temperature (NOCT) and standard temperature conditions (STC), which depend upon the atmospheric temperature, can be estimated by using the equation for the temperature of the PV cell. The equation of cell temperature is shown below (Markvart 2000)

$$T_{\text{cel}} = T_{\text{amb}} + \frac{\text{NOCT} - 20}{800} * S$$

- T_{cel} – Temperature of solar cell
- T_{amb} – Ambient temperature of the atmosphere
- NOCT – nominal operating conditions temperature

where at STC, the temperature is 20 °C, and the irradiance is 800 W/m^2, while at NOCT, NOCT is given by the solar cell manufacturer, and it lies in the range 45 ± 2 °C (Sun et al. 2020).

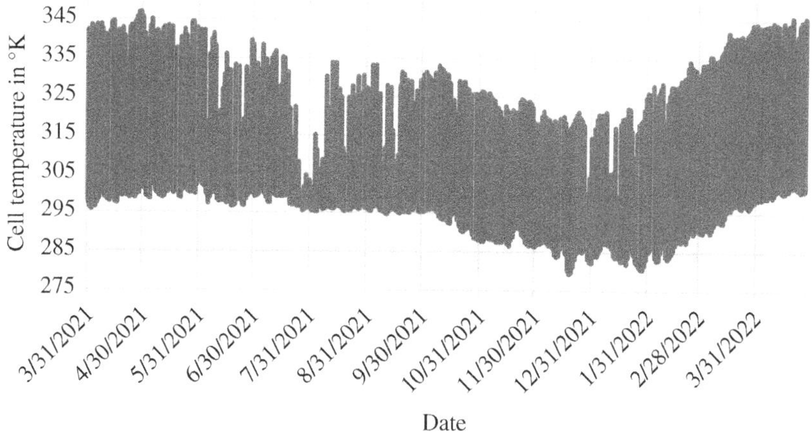

Figure 19.4 Cell temperatures in °K without cooling.

The temperature is taken 46 °C according to requirement of the experimental setup. In the ideal condition, the maximum irradiance that can hit the PV surface is 1000 W/m^2. So, from the atmospheric temperature and irradiance from March 31, 2021, to April 27, 2022, the temperature of the solar cell is calculated by putting the values in equation A. The cell temperature comes out to be as shown in Figure 19.4.

19.2.2 Degradation Equations

This work is performed considering two degradation models:

i. Erying
ii. Peck

Erying degradation model (Park et al. 2013; David 2002)

$$R_{\text{D Erying}} = A \exp\left(-\frac{E_{\text{act}}}{k_B T} - \frac{b}{\text{rh}}\right) \tag{19.1}$$

where

S – Solar irradiance in W/m^2
E_{act} – Thermal activation energy of the degradation Process (eV)
K_B – Boltzmann constant (8.62* 10-5 eV/K)
T – Temperature (°K)
rh – Relative humidity (%)
A and b – Two constants dependent on the failure mode
R_D (%/h) – Inverse to the meantime to failure (MTTF).

According to the Peck model (Park et al. 2013),
The degradation rate is given as,

$$R_{\text{D Peck}} = Be^{\left(\frac{-E_{\text{act}}}{K_B T}\right)(\text{rh}^n)} \tag{19.2}$$

where

B and n – Constants dependent on failure mode

19.2.3 Mean Life Without Cooling

The mean life of solar panels is 25 years without cooling (Sharaf et al. 2022; Kim et al. 2021; Guérin et al. 2006). Therefore for the without cooling temperature in this work, the mean life equation is optimized such that to find constants. Since, the solar data from March 31, 2021, to April 27, 2022, contains values per hour, so there are 9430 of the total values.

By the Erying method using Eq. (19.2), mean life is

$$\sum_{i=1}^{9430} R_{\text{D Erying}_i} = R_{\text{D Erying}_1} + R_{\text{D Erying}_2} + R_{\text{D Erying}_3} + \ldots + R_{\text{D Erying}_{9430}}$$

$$\text{Mean life} = \frac{9430}{\sum_{i=1}^{9430} R_{\text{D}_i}}$$

This equation is optimized using the SciPy library method, which contains the SLSQP optimization technique.

To maximize:

$$\text{Mean life} = \frac{9430}{\sum_{i=1}^{9430} R_{\text{D Erying}_i}}$$

Constraints:

$$\frac{9430}{\sum_{i=1}^{9430} R_{\text{D Erying}_i}} \leq 25$$

$$\text{A} = (2 * 10^6, 4.5 * 10^6)$$

$$b = (1, 4)$$

Such that the constants A and b obtained are:

$$\text{A} = 4.00^{*}106$$

$$\text{b} = 3.02793706$$

Similarly, for Peck equation, as mentioned in Eq. (19.3),

$$\sum_{i=1}^{9430} R_{\text{D Peck}_i} = R_{\text{D Peck}_1} + R_{\text{D Peck}_2} + R_{\text{D Peck}_3} + \ldots + R_{\text{D Peck}_{9430}}$$

$$\text{mean life} = \frac{9430}{\sum_{i=1}^{9430} R_{\text{D Peck}_i}}$$

This equation is optimized using the SciPy library method, which contains the SLSQP optimization technique.

To maximize:

$$\text{mean life} = \frac{9430}{\sum_{i=1}^{9430} R_{\text{D Peck}_i}}$$

Constraints:

$$\frac{9430}{\sum_{i=1}^{9430} R_{\text{D Peck}_i}} \leq 25$$

$$B = (1 * 10^6, 10 * 10^6)$$

$$n = (10^{-6}, 10)$$

Such that the constants B and n obtained are:

$$B = 3.4789 * 10^6$$

$$n = 1.9 * 10^{-6}$$

19.2.4 Mean Life After Cooling

For the calculation of mean life using four cooling techniques that are water cooling (Moharram et al. 2013), Air cooling using extended fin area (Ravi and Porpatham 2017), Wick structure with nanofluids Using Al_2O_3/water and CuO/ water nanofluids (Kusiak et al. 2013), Passive cooling by heat pipe (Akbarzadeh and Wadowski 1996), the mean life is calculated in three different cases: when cooling is performed to cool cell temperature not less than 25 °C (298 °K), cooling is performed to cool cell temperature not less than 35 °C (308 °K), cooling is performed to cool cell temperature not less than 45 °C (318 °K). Which implies that with the constraints of reducing the temperature for the cooling methods, the temperature above 25, 35, and 45 °C, the cooling will be done as per the capability of technique, but for the cell temperature less than these temperatures naturally, then the cooling method will not reduce the temperature further.

19.2.4.1 Mean Life with Cooling up to 25 °C

When water cooling (Moharram et al. 2013), air cooling using extended fin area (Ravi and Porpatham 2017), Wick structure with nanofluids Using Al_2O_3/water and CuO/ water nanofluids (Kusiak et al. 2013), passive cooling by heat pipe (Akbarzadeh and Wadowski 1996), cooling methods are applied such that to bring down the temperature not less than 25 °C that is 298 °K.

For different cooling techniques taken into consideration and applying cooled temperature according to constraints of those cooling techniques, the following Figures 19.5–19.8 are obtained. These figures compare the cell temperature before and after applying the cooling techniques.

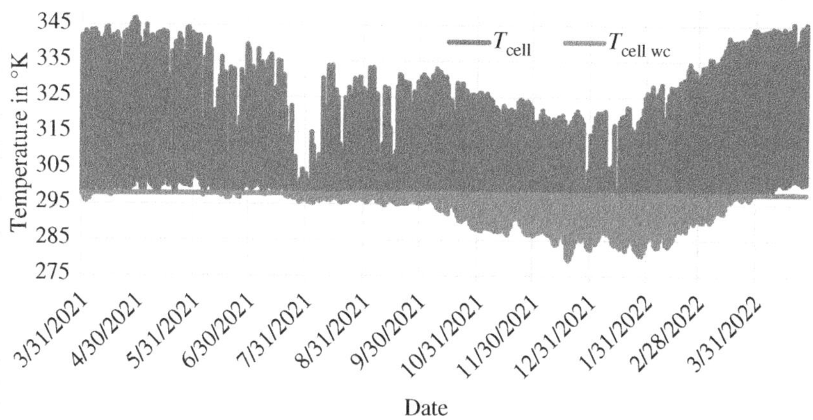

Figure 19.5 Cell temperature after water cooling.

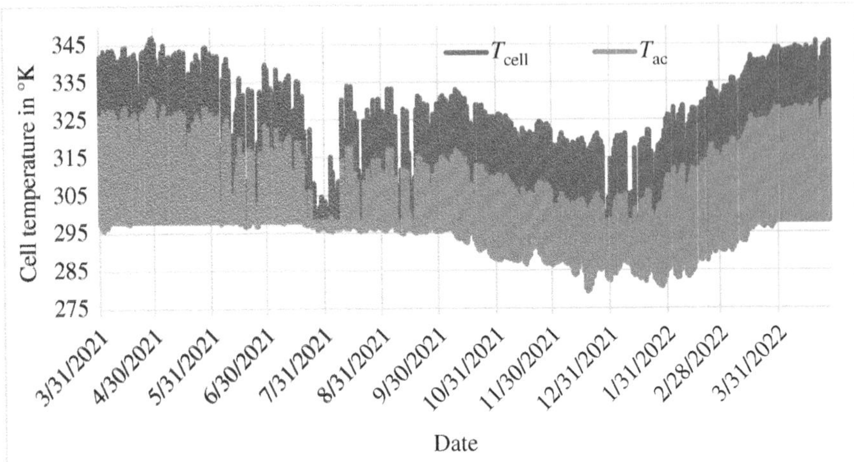

Figure 19.6 Cell temperature after air cooling with extended fins.

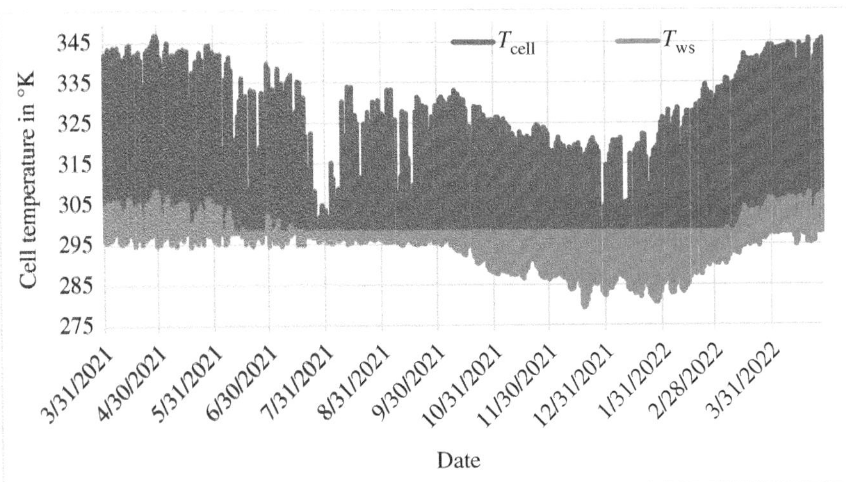

Figure 19.7 Cell temperature after cooling by wick structure.

Using the cooled temperature and putting it in Eq. (19.1) of Eyring and Eq. (19.2) of Peck, to obtain the mean life after cooling and also the failure rate is obtained, the failure rate is reciprocal of mean life for a constant failure rate function (Ebeling 2004), as shown in Table 19.2.

19.2.4.2 Mean Life with Cooling up to 35 °C

For different cooling techniques taken into consideration and applying cooled temperature up to 35 °C and according to constraints of those cooling techniques, the following Figures 19.9–19.12 are obtained. The estimated data are shown in Table 19.3 for graphical illustration. These figures compare the cell temperature before and after applying the cooling techniques.

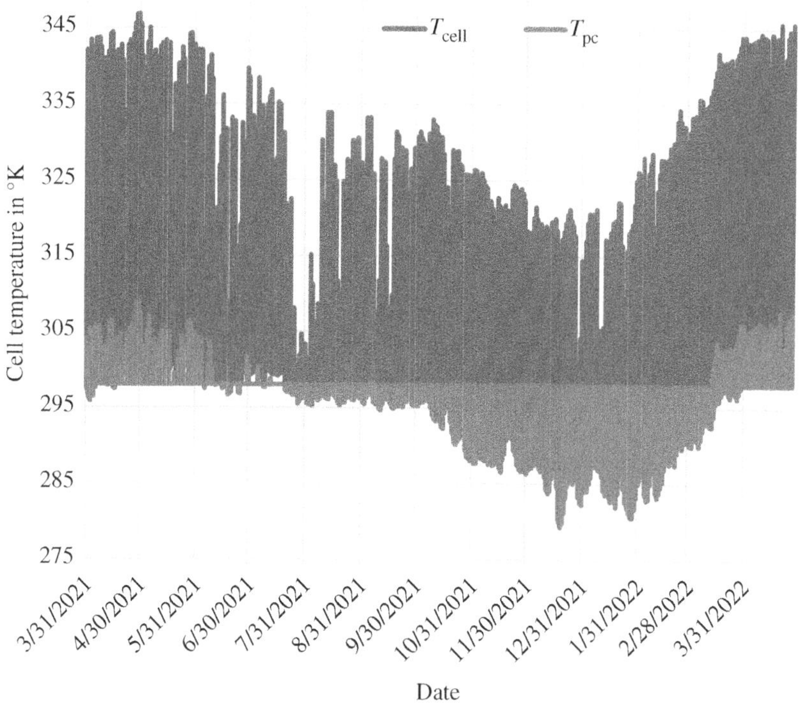

Figure 19.8 Cell temperature after cooling with passive cooling heat pipe.

Table 19.2 Mean life and failure rate calculated when cooled up to 25 °C.

Cooling technique	Degradation equation	Mean life	Failure rate (λ) failures per year
Without cooling		25	0.04
Water cooling	Erying	59.6170	0.01677
	Peck	61.5025	0.01625
Air cooling extended fin	Erying	44.6956	0.02237
	Peck	44.5812	0.02243
Wick surface	Erying	58.9043	.01698
	Peck	60.5570	.01651
Passive cooling heat pipe	Erying	58.8606	0.01698
	Peck	60.5322	0.01652

19.2.4.3 Mean Life with Cooling up to 45 °C

For different cooling techniques taken into consideration and applying cooled temperature up to 45 °C and according to constraints of those cooling techniques, the following Figures 19.13–19.16

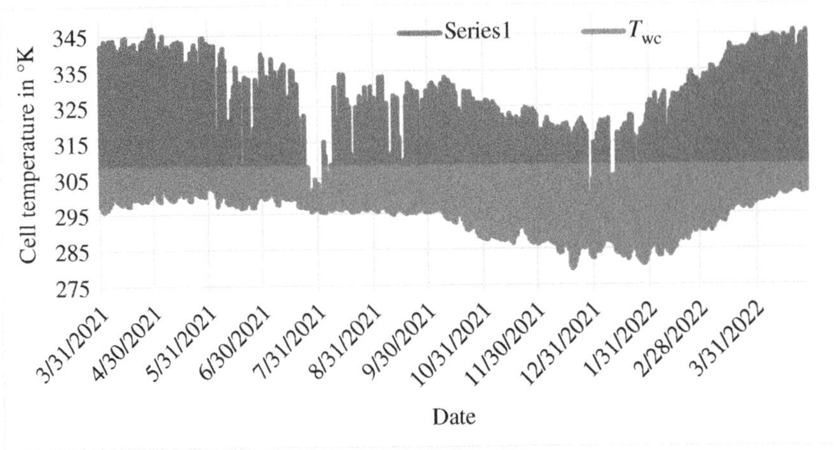

Figure 19.9 Cell temperature after water cooling.

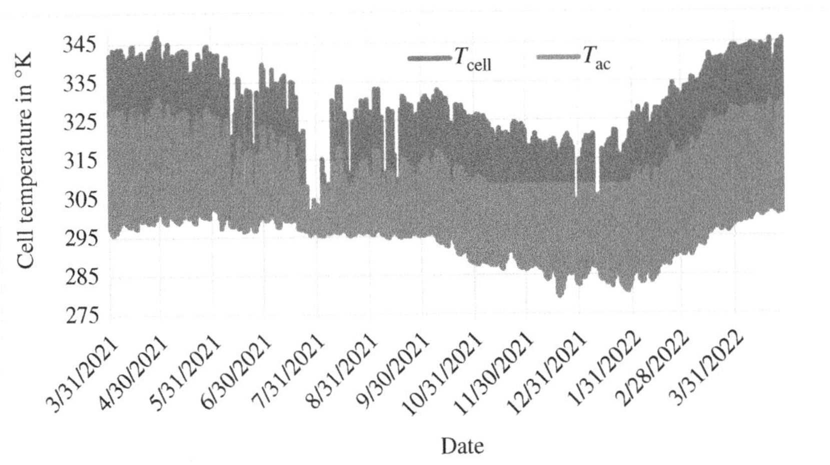

Figure 19.10 Cell temperature after air cooling with extended fin.

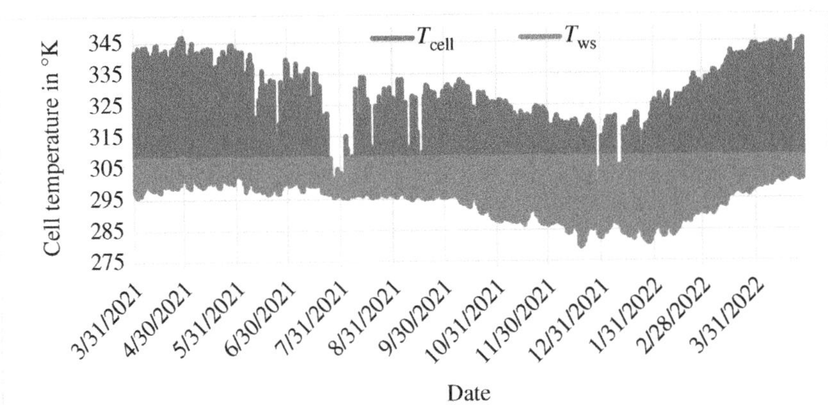

Figure 19.11 Cell temperature after cooling by wick structure.

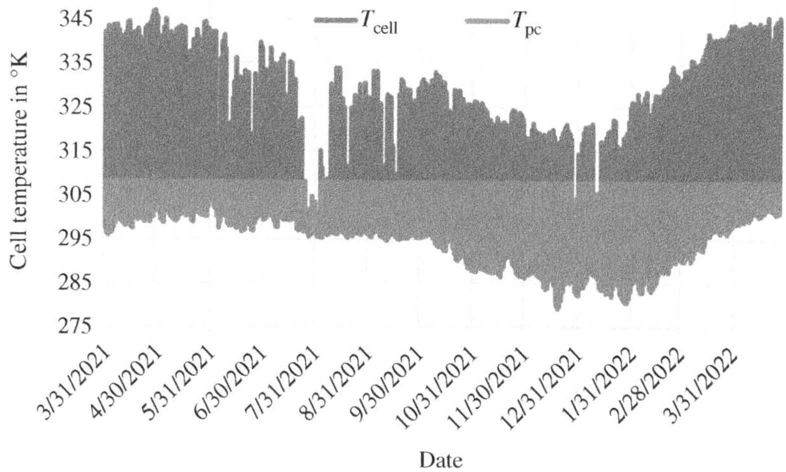

Figure 19.12 Cell temperature after cooling by passive cooling heat pipe.

Table 19.3 Mean life and failure rate calculated when cooled up to 35 °C.

Cooling technique	Degradation equation	Mean life	Failure rate (λ)
Without cooling		25	0.04
Water cooling	Erying	42.2947	0.02364
	Peck	43.1613	0.02317
Air cooling extended fin	Erying	37.5643	0.02662
	Peck	37.6175	0.02658
Wick surface	Erying	42.2947	0.02346
	Peck	43.1613	0.02317
Passive cooling heat pipe	Erying	42.2943	0.02364
	Peck	43.1605	0.02317

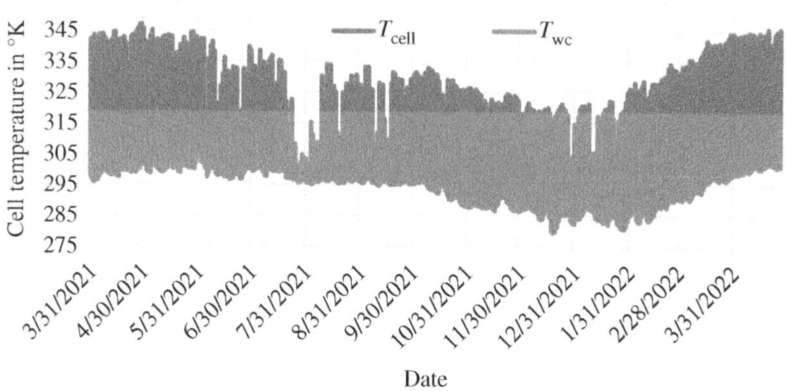

Figure 19.13 Cell temperature after water cooling.

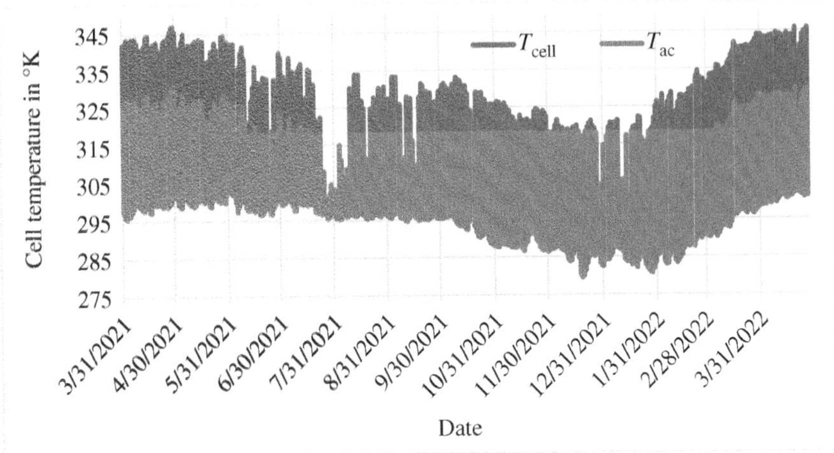

Figure 19.14 Cell temperature after air cooling with extended fins.

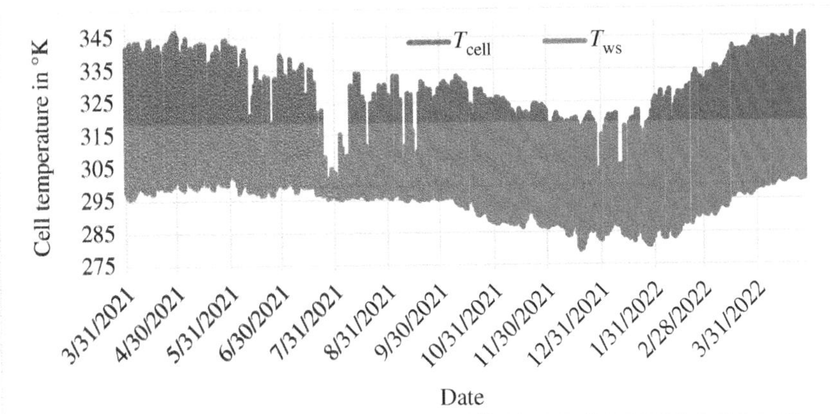

Figure 19.15 Reliability comparison of reliabilities of cooling method calculated by Erying and without cooling.

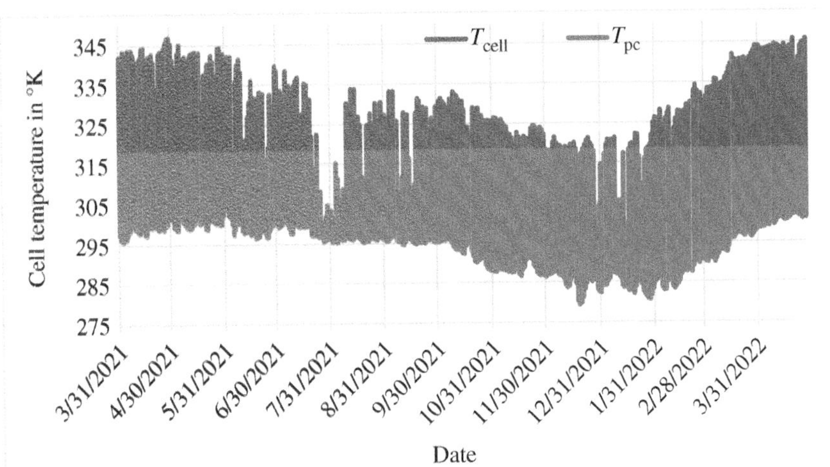

Figure 19.16 Reliability comparison of reliabilities of cooling method calculated by Peck and without cooling.

Table 19.4 Mean life and failure rate calculated when cooled up to 45 °C.

Cooling technique	Degradation equation	Mean life (years)	Failure rate (λ) (failures per year)
Without cooling		25	0.04
Water cooling	Erying	33.2692	0.03006
	Peck	33.5838	0.02978
Air cooling extended fin	Erying	32.2318	0.03102
	Peck	32.3162	0.03094
Wick surface	Erying	33.2692	0.03006
	Peck	33.5838	.02978
Passive cooling heat pipe	Erying	33.2692	0.3006
	Peck	33.5838	0.02978

are obtained. The estimated data are shown in Table 19.4 for graphical illustration. These figures compare the cell temperature before and after applying the cooling techniques.

19.3 Reliability Assessment

The probability of a system or the components of a system to perform a desired task or function for a provided period of time when used under specified operating conditions. It is the probability of not failing for a given period of time (Ebeling 2004). Mathematically, it can be expressed as for any random time T is the time of failure; $T \geq 0$. Then, the reliability is given as:

$$R(t) = Py\,(T \geq t)$$

where

$R(t)$ – reliability function
Py – probability
$R(t) \geq 0$
$R(0) = 1$

Considering that system is not failing at $T = 0$

$$\lim_{t \to \infty} R(t) = 0$$

and reliability will lie in the range $R(t) \in [0, 1] R(t)$ is the probability that the time to fail is greater than or equal to t.

Considering the constant failure rate (Ebeling 2004), therefore taking the exponential distribution curve, Therefore,

$$R(t) = \mathrm{e}^{-\lambda t}$$

where λ is the failure rate (Ebeling 2004), the reliability equations calculated are shown in Table 19.5.

Table 19.5 Reliability $R(t)$ calculations.

Cooling	Degradation equation	Reliability for cooling up to 25 °C	Reliability for cooling up to 35 °C	Reliability for cooling up to 45 °C
Without cooling		$e^{-0.04t}$	$e^{-0.04t}$	$e^{-0.04t}$
Water cooling	Erying	$e^{-0.01677t}$	$e^{-0.02364t}$	$e^{-0.03006t}$
	Peck	$e^{-0.01625t}$	$e^{-0.02317t}$	$e^{-0.02978t}$
Air cooling with extended fins	Erying	$e^{-0.02237t}$	$e^{-0.02662t}$	$e^{-0.03102t}$
	Peck	$e^{-0.02243t}$	$e^{-0.02658t}$	$e^{-0.03094t}$
Wick surface	Erying	$e^{-0.01698t}$	$e^{-0.02364t}$	$e^{-0.03006t}$
	Peck	$e^{-0.01651t}$	$e^{-0.02317t}$	$e^{-0.02978t}$
Passive cooling heat pipe	Erying	$e^{-0.01698t}$	$e^{-0.02364t}$	$e^{-0.03006t}$
	Peck	$e^{-0.01652t}$	$e^{-0.02317t}$	$e^{-0.02978t}$

19.3.1 Cooling up to 25 °C

As observed from Figure 19.17 by the Erying degradation equation, the reliability with cooling improves significantly with all four cooling techniques; reliability improvement with water cooling, wick structure, and passive cooling with heat pipe are similar, but with air cooling extended fins cooling reliability, curve lies below from the curve of other cooling which implies that the reliability improvement using air cooling with extended fins is less than the others.

As observed from Figure 19.18, the reliability with cooling improves significantly; reliability improvement with water cooling, wick structure, and passive cooling with heat pipe are similar, but with air cooling extended fins cooling reliability, improvement is less than the others.

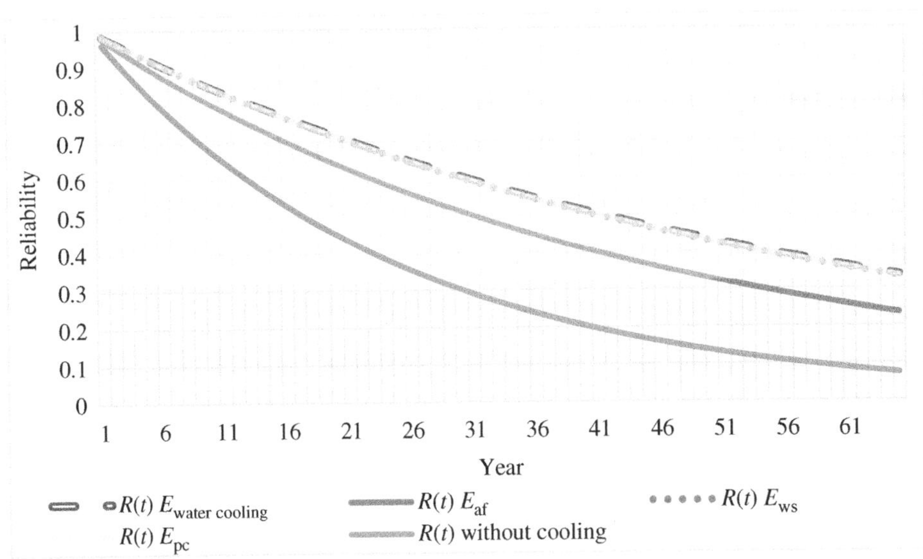

Figure 19.17 Cell temperature after cooling by wick structure.

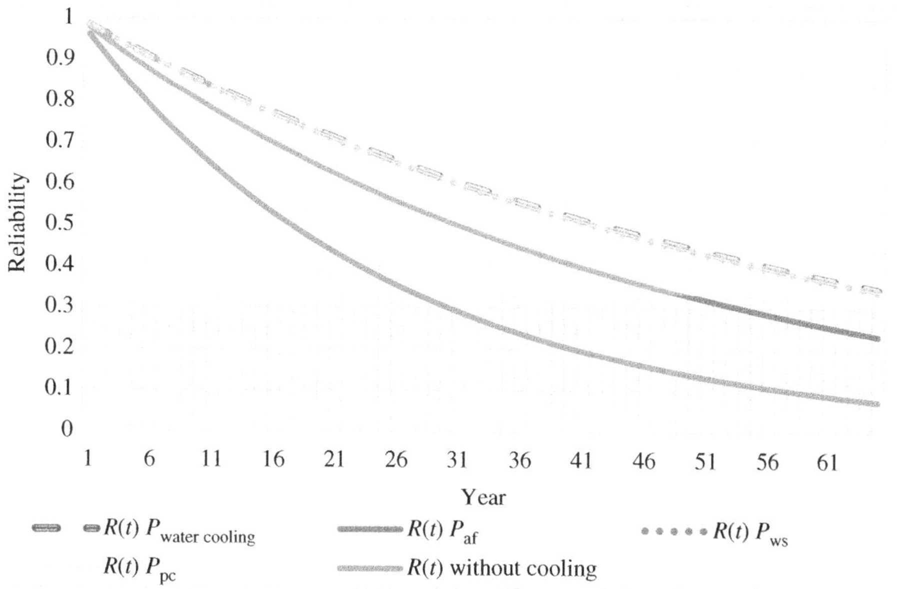

Figure 19.18 Cell temperature after cooling by passive cooling heat pipe.

19.3.2 Cooling up to 35 °C

As observed in Figure 19.19, the reliability with cooling improves significantly. Reliability improvement with water cooling, wick structure, and passive cooling with heat pipe are similar, but with air cooling, extended fins cooling improvement is less than the others. Up to 35 °C cooling, the reliability difference becomes lesser than 25 °C cooling. For both Erying and Peck equations (by Figure 19.20), a similar observation is made.

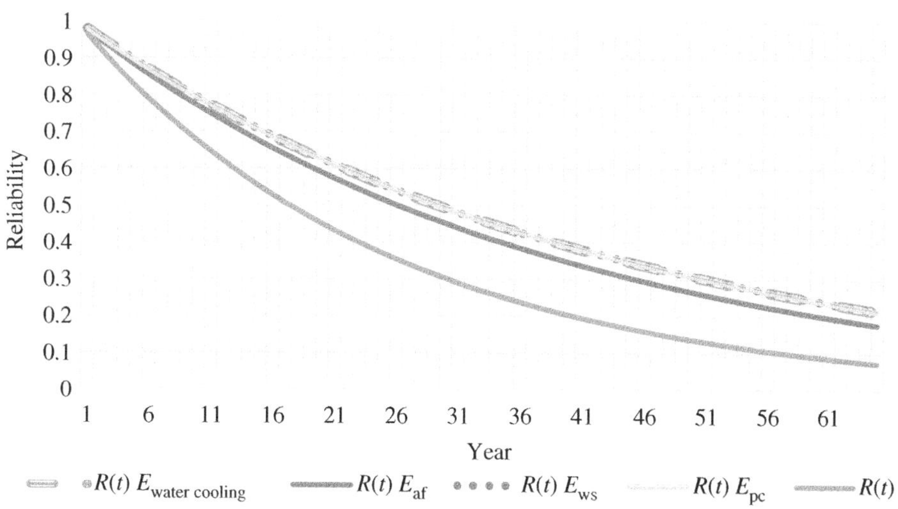

Figure 19.19 Reliability comparison of reliabilities of cooling method calculated by Erying and without cooling.

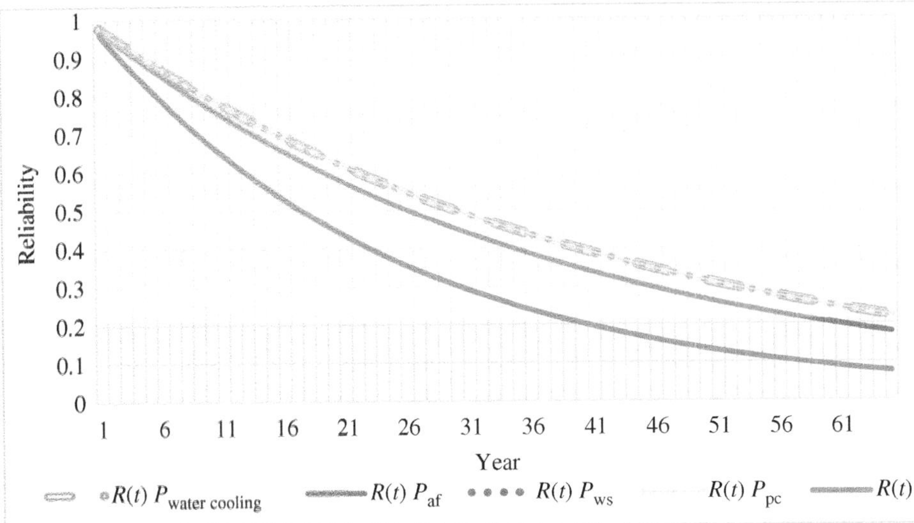

Figure 19.20 Reliability comparison of reliabilities of cooling method calculated by Peck and without cooling.

19.3.3 Cooling up to 45 °C

Up to 45 °C cooling, the reliability improvement is lesser than 25 and 35 °C cooling but still significant, as seen by the Erying equation in Figure 19.21. And in this case, all four cooling methods' reliability curves overlap.

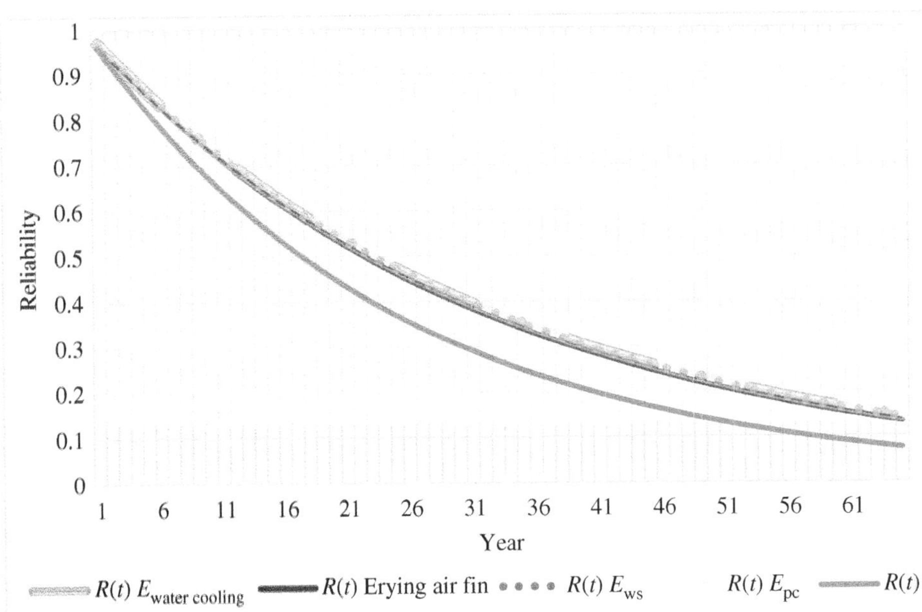

Figure 19.21 Reliability comparison of reliabilities of cooling method calculated by Erying and without cooling.

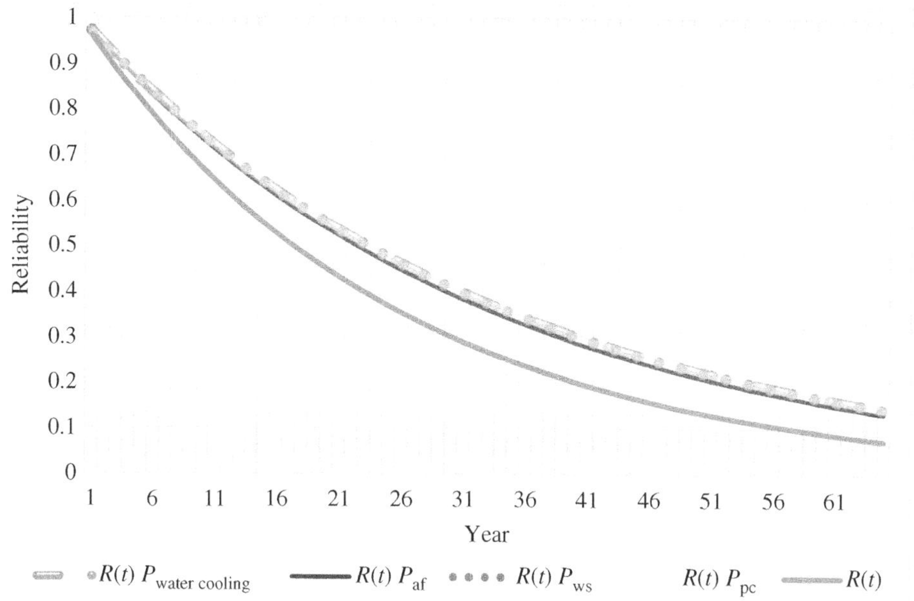

Figure 19.22 Reliability comparison of reliabilities of cooling method calculated by Peck and without cooling.

Observation from the Peck equation is also similar to Erying, as observed in Figure 19.22. The distance between cooling and without curves reduces but is still significant. And in this case, all four cooling methods' reliability curves overlap.

19.4 Probability Density Function

This PDF function describes the shape of the failure distribution (Ebeling 2004).

Where $f_{PDF}(t)$ – Probability density function (PDF).

$$f_{PDF}(t) = \frac{dF_{CDF}(t)}{dt} = -\frac{dR(t)}{dt}$$

The range of PDF is $f(t) \in [0, \infty)$

$$\int_0^\infty f_{PDF}(t)dt = 1$$

For constant failure rate

$$f_{PDF}(t) = \lambda e^{-\lambda t}$$

The equations formed are shown in Table 19.6.

19.4.1 Cooling up to 25 °C

The area under the probability distribution curve till time t shows the probability of failure until time "t" (Ebeling 2004). As seen in Figures 19.23 and 19.24, up to any time "t," the PDF curve

Table 19.6 Probability density function $f(t)$ calculations.

Cooling	Degradation equation	PDF for cooling up to 25 °C	PDF for cooling up to 35 °C	PDF for cooling up to 45 °C
Without cooling		$0.04 * e^{-0.04t}$	$0.04000 * e^{-0.04000t}$	$0.04 * e^{-0.04t}$
Water cooling	Erying	$0.01677 * e^{-0.01677t}$	$0.02364 * e^{-0.02364t}$	$0.03006 * e^{-0.03006t}$
	Peck	$0.01625 * e^{-0.01625t}$	$0.02317 * e^{-0.02317t}$	$0.02978 * e^{-0.02978t}$
Air cooling with extended fins	Erying	$0.02237 * e^{-0.02237t}$	$0.02662 * e^{-0.02662t}$	$0.03102 * e^{-0.03102t}$
	Peck	$0.02243 * e^{-0.02243t}$	$0.02658 * e^{-0.02658t}$	$0.03094 * e^{-0.03094t}$
Wick surface	Erying	$0.01698 * e^{-0.01698t}$	$0.02364 * e^{-0.02364t}$	$0.03006 * e^{-0.03006t}$
	Peck	$0.01651 * e^{-0.01651t}$	$0.02317 * e^{-0.02317t}$	$0.02978 * e^{-0.02978t}$
Passive cooling heat pipe	Erying	$0.01698 * e^{-0.01698t}$	$0.02364 * e^{-0.02364t}$	$0.03006 * e^{-0.03006t}$
	Peck	$0.01652 * e^{-0.01652t}$	$0.02317 * e^{-0.02317t}$	$0.02978 * e^{-0.02978t}$

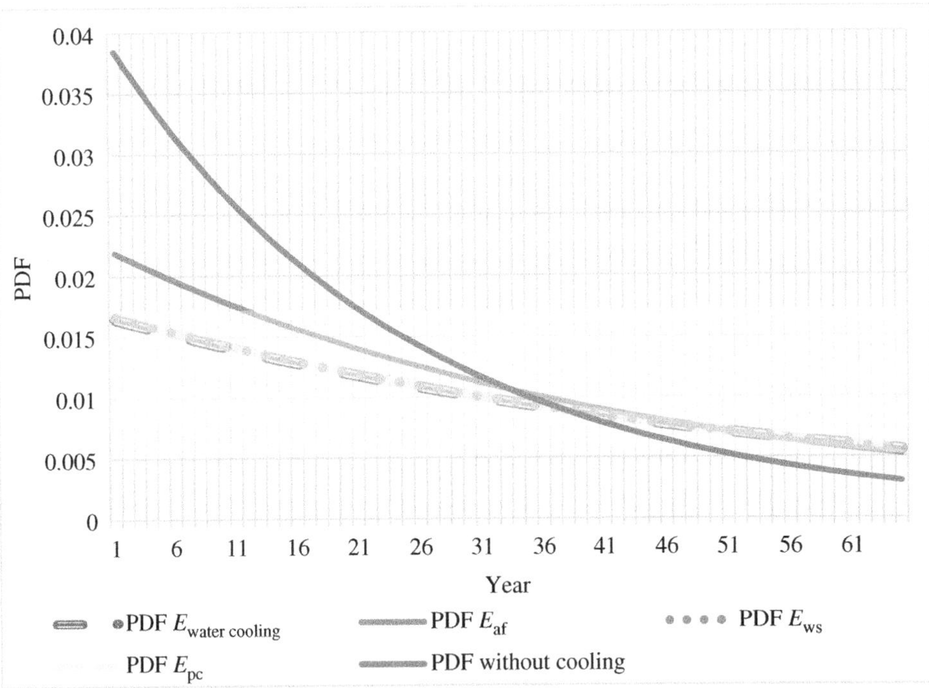

Figure 19.23 PDF plot for the various cooling methods by Erying equation for cooling up to 25 °C.

without cooling will have an area more than that of any of the four cooling methods, which implies that the probability of failure is higher without cooling, which le it is also seen for air cooling with fins cooling method has a higher area; therefore, it will have a higher chance of failure than the other three cooling techniques.

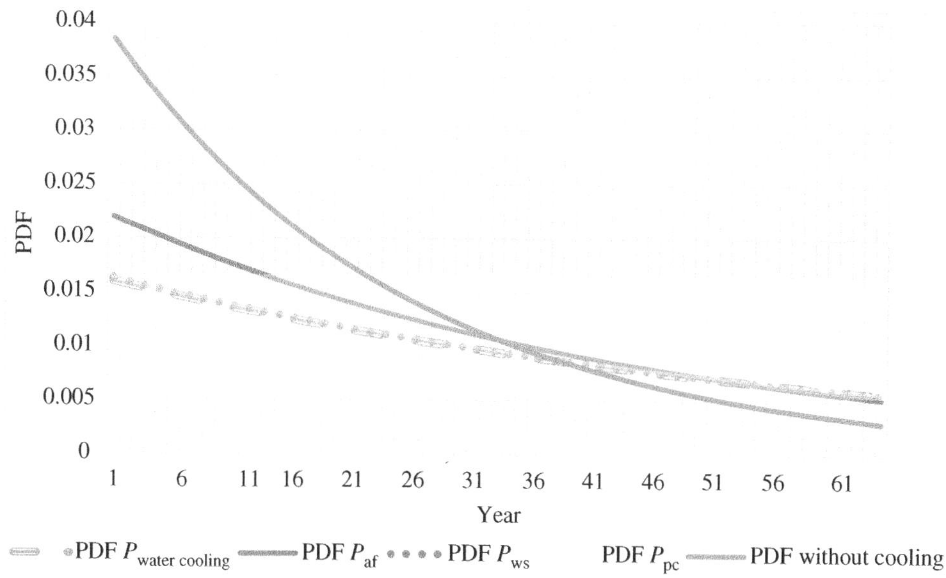

Figure 19.24 PDF plot for the various cooling methods by Peck equation for cooling up to 25 °C.

19.4.2 Cooling up to 35 °C

From Figures 19.25 and 19.26, as the cooled temperature increased to 35 °C from 25 °C, the cooling curves overlapping each other, as well as they became closer to the PDF curve without cooling, implying that with failure probability of 35 °C cooling is more than that of 25 °C cooling.

19.4.3 Cooling up to 45 °C

From Figures 19.27 and 19.28, it is observed that all the PDF curves of cooling overlap each other, which implies that the failure probability with all four types of cooling will be approximately the same, as well as the failure probability of 45 °C, will be higher than both 25 and 35 °C cooling. From time 0 to any time "t," the area under the curve of PDF gives the failure probability, while the area under the curve from time t to infinity defines the reliability of the system or component. So, if the intersection of two curves is considered as without cooling and with air cooling, then the area under the curve before the intersection time shows that the failure probability without a cooling system is higher than that with cooling techniques and after the intersection time it is seen that area under the curve of without cooling is less which implies that the reliability of with cooling is lower. Similar results are seen in Figures 19.23–19.28.

19.5 Cumulative Distribution Function

The cumulative distribution function (CDF) is the probability up to any time T, which is in between the zero time and the maximum time. CDF does signify that the failure occurs before time t (Ebeling 2004):

$$F_{\text{CDF}}(t) = 1 - R(t) = Py(T < t)$$

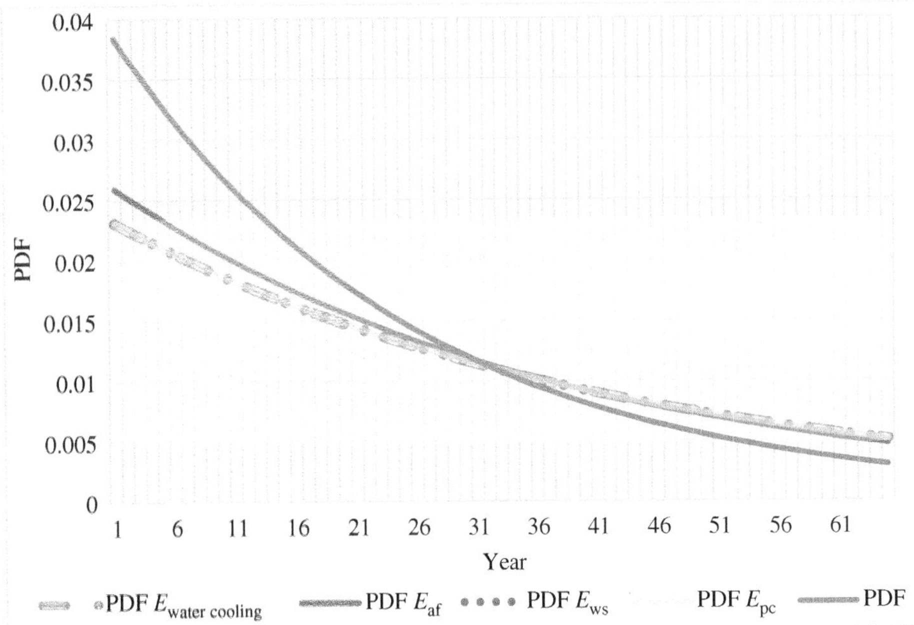

Figure 19.25 PDF plot for the various cooling methods by Erying equation for cooling up to 35 °C.

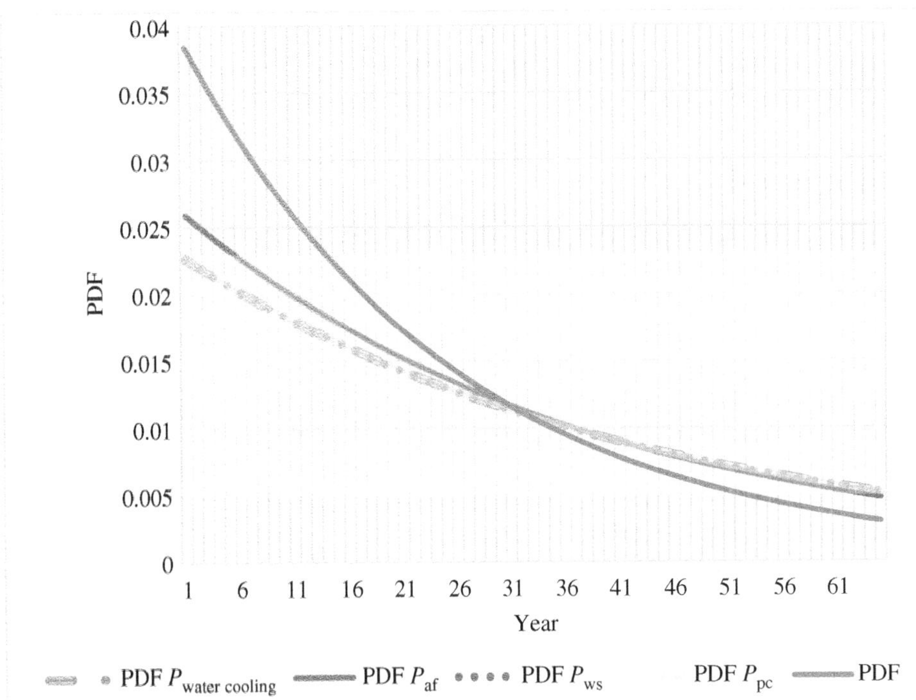

Figure 19.26 PDF plot for the various cooling methods by Peck equation for cooling up to 35 °C.

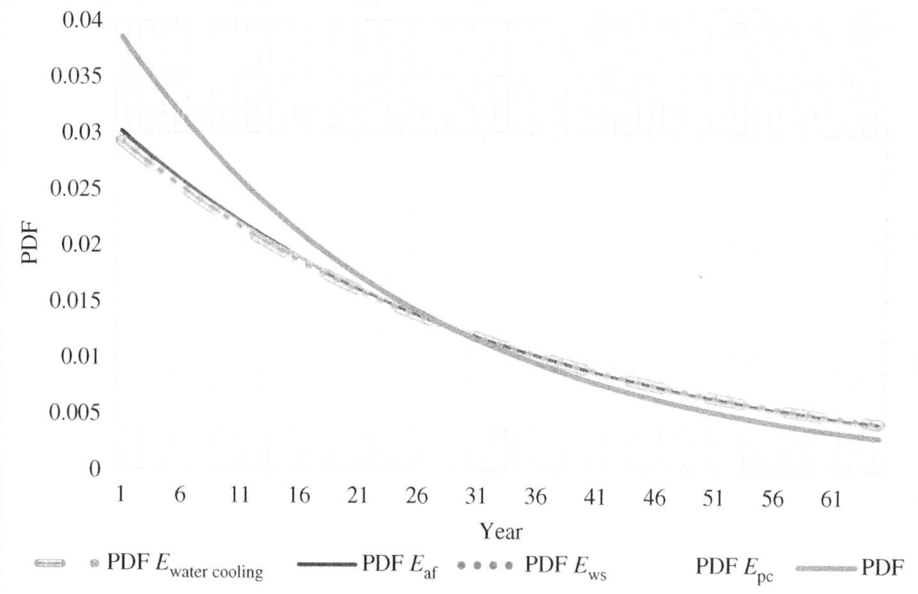

Figure 19.27 PDF plot for the various cooling methods by Erying equation for cooling up to 45 °C.

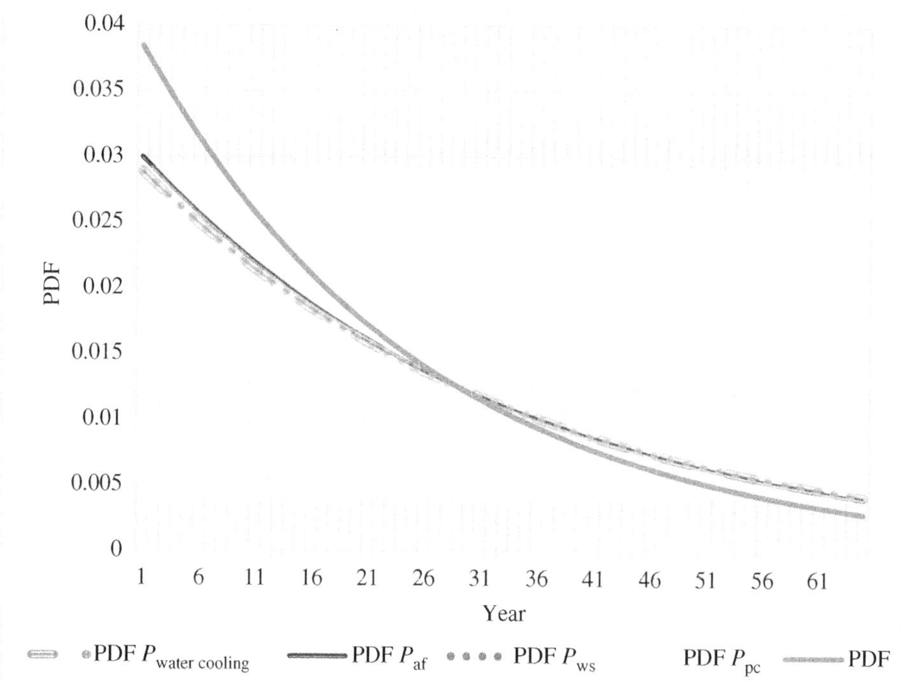

Figure 19.28 PDF plot for the various cooling methods by Peck equation for cooling up to 45 °C.

Table 19.7 Cumulative distribution function F_CDF(t) calculations.

Cooling	Degradation equation	CDF for cooling up to 25 °C	CDF for cooling up to 35 °C	CDF for cooling up to 45 °C
Without cooling		$1-e^{-0.04t}$	$1-e^{-0.04t}$	$1-e^{-0.04t}$
Water cooling	Erying	$1-e^{-0.01677t}$	$1-e^{-0.02364t}$	$1-e^{-0.03006t}$
	Peck	$1-e^{-0.01625t}$	$1-e^{-0.02317t}$	$1-e^{-0.02978t}$
Air cooling with extended fins	Erying	$1-e^{-0.02237t}$	$1-e^{-0.02662t}$	$1-e^{-0.03102t}$
	Peck	$1-e^{-0.02243t}$	$1-e^{-0.02658t}$	$1-e^{-0.03094t}$
Wick surface	Erying	$1-e^{-0.01698t}$	$1-e^{-0.02364t}$	$1-e^{-0.03006t}$
	Peck	$1-e^{-0.01651t}$	$1-e^{-0.02317t}$	$1-e^{-0.02978t}$
Passive cooling heat pipe	Erying	$1-e^{-0.01698t}$	$1-e^{-0.02364t}$	$1-e^{-0.03006t}$
	Peck	$1-e^{-0.01652t}$	$1-e^{-0.02317t}$	$1-e^{-0.02978t}$

where

$$F_{\mathrm{CDF}}(t) - \mathrm{CDF}$$

$$F_{\mathrm{CDF}}(0) = 0$$

$$\lim_{t\to\infty} F_{\mathrm{CDF}}(t) = 1$$

Implies the range of CDF $F(t) \in [0, 1]$
$F(t)$ is the probability that a failure occurs before time t.
For constant failure rate,

$$F_{\mathrm{CDF}}(t) = 1 - e^{-\lambda t}$$

The equations for PDF are shown in Table 19.7.

19.5.1 Cooling up to 25 °C

From Figures 19.29 and 19.30, it seems that the chance of occurrence of failure before any time is greater without cooling compared to cooling. The chance of failure of air cooling with fins also has a higher chance of failure than the other three cooling methods.

All the other three methods have their curve overlapping, implying that the chance of failure for water cooling, wick structure cooling, and passive cooling with heat pipe is nearly the same.

19.5.2 Cooling up to 35 °C

From Figures 19.31 and 19.32, it is observed that the difference between the chance of failure without cooling and with cooling is less than 25 °C cooling but still significant. Also, the chance of failure of air cooling is higher than in other cooling methods.

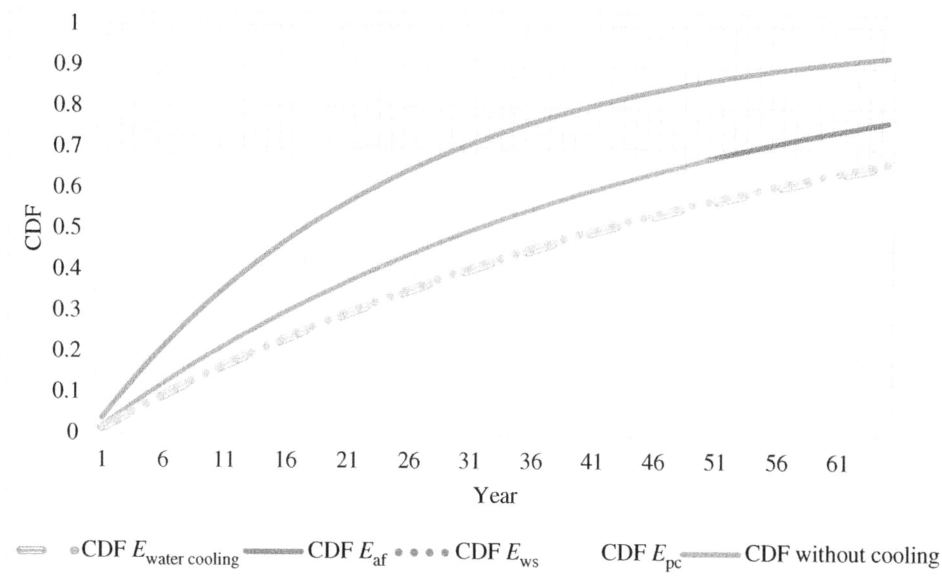

Figure 19.29 CDF plot for the various cooling methods by Erying equation for cooling up to 25 °C.

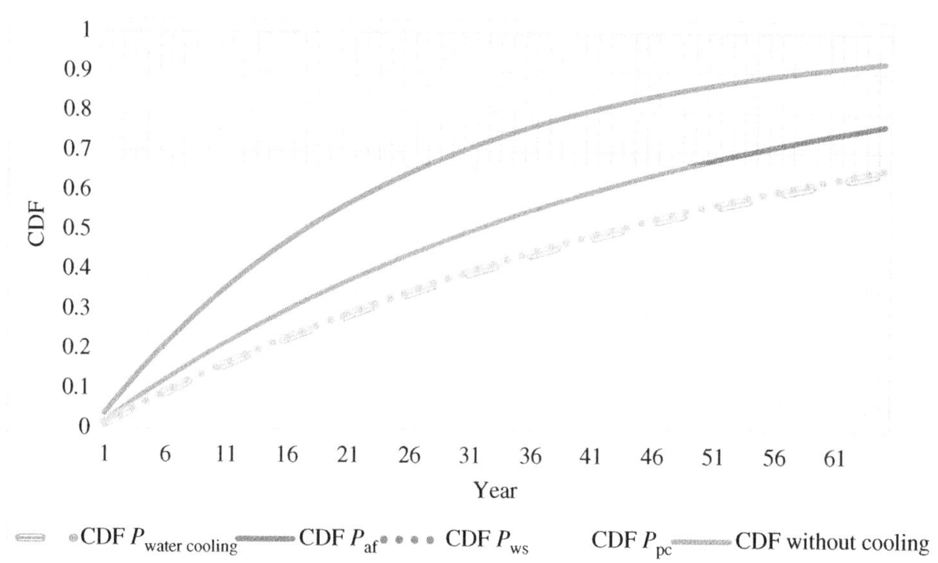

Figure 19.30 CDF plot for the various cooling methods by Peck equation for cooling up to 25 °C.

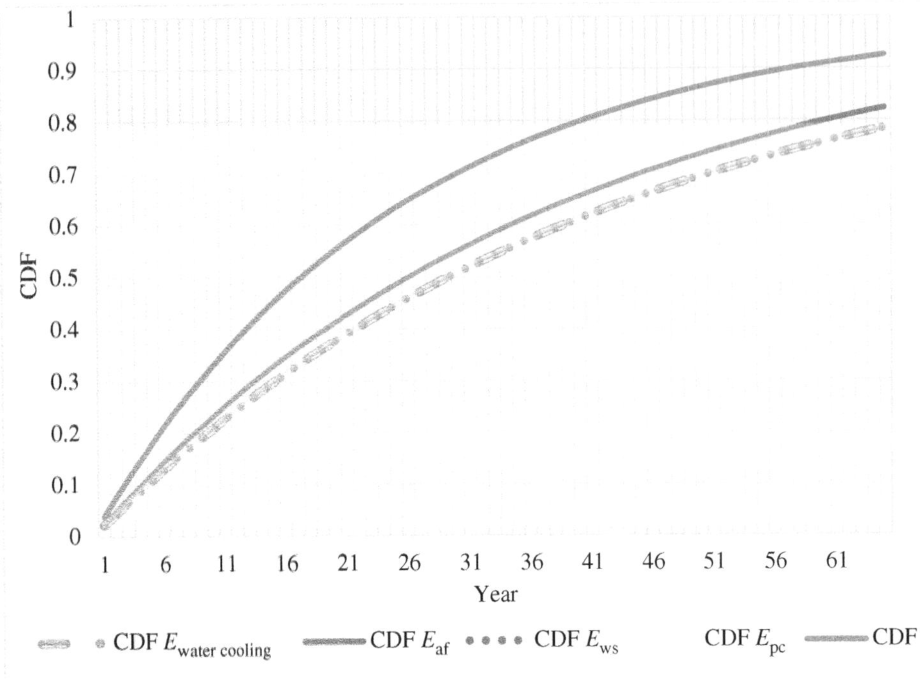

Figure 19.31 CDF plot for the various cooling methods by Erying equation for cooling up to 35 °C.

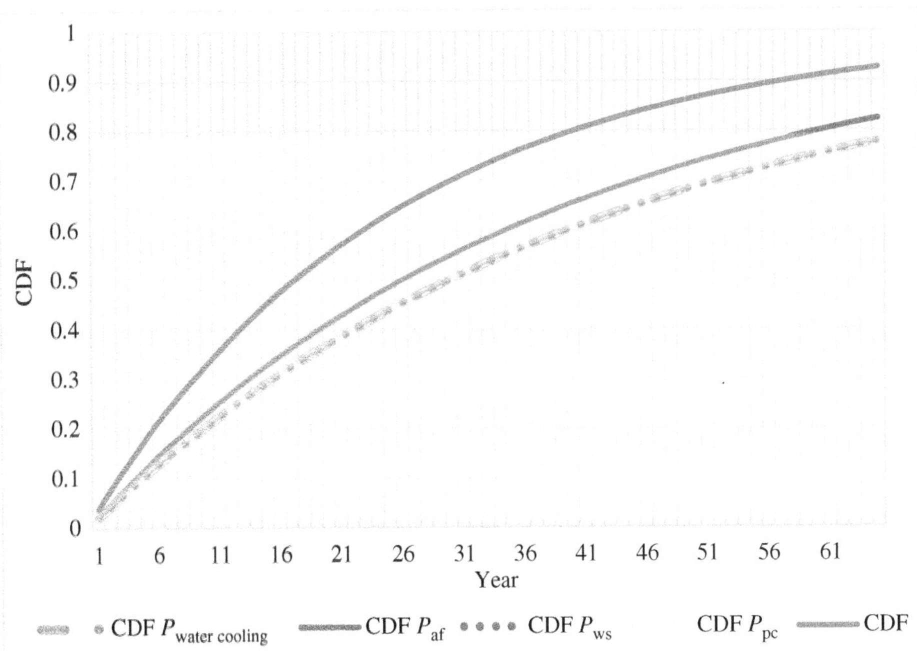

Figure 19.32 CDF plot for the various cooling methods by Peck equation for cooling up to 35 °C.

19.5.3 Cooling up to 45°C

From Figures 19.33 and 19.34, it can be seen that the chance of failure for all the cooling methods is nearly similar and is less than that of a chance of failure without cooling, but the chance of failure for 45 °C is more than that of 25 and 35 °C. So, the results obtained from CDF are just another representation of the probability of failure observed by the area of the PDF till time t. By all the three entities measured, that is, reliability, PDF, and CDF, the more cooling is done, the more the lifetime and the probability of failure is high if less cooling is done.

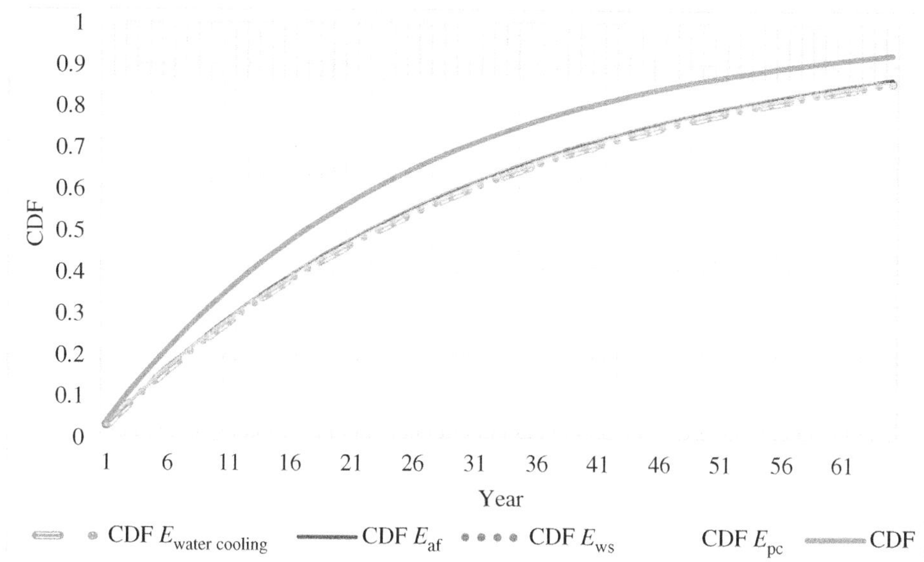

Figure 19.33 CDF plot for the various cooling methods by Erying equation for cooling up to 45 °C.

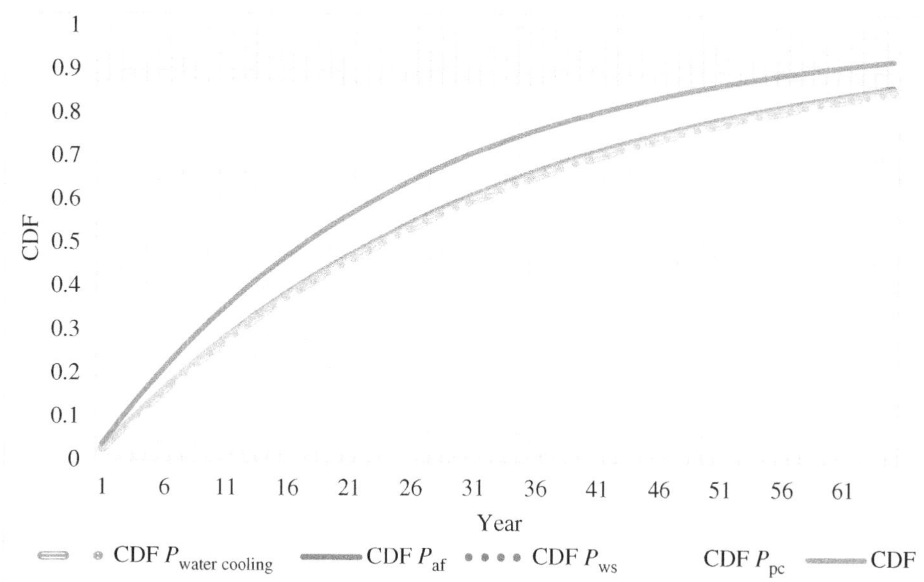

Figure 19.34 CDF plot for the various cooling methods by Peck equation for cooling up to 45.

19.6 Results

The calculations of the rate of degradation using Eq. (19.1), and Erying degradation using equation Eq. (19.2) have been illustrated in details in Section 19.5. The Peck degradation equation is used for evaluation of the mean life of the system. Mean life as calculated for various cases is shown in Table 19.8. The mean life of solar panels is 25 years without cooling (Kim et al. 2021; Guérin et al. 2006); therefore, the mean life after cooling is shown in Table 19.9.

19.7 Conclusion

From both the degradation equations, Erying and Peck, the mean life obtained is nearly equal in every case. It is observed that less the cooling temperature of the cell, higher mean life, higher reliability, and less probability of failure, as observed in Table 19.9. Therefore, incorporating cooling methods increase the mean life and reliability of solar panels, and as Table 19.1 seen, that increases

Table 19.8 Mean life by different cooling methods.

Cooling technique	Degradation equation	Mean life by cooling up to 25 °C	Mean life by cooling up to 35 °C	Mean life by cooling up to 45 °C
Water cooling	Erying	59.6170	42.2947	33.2692
	Peck	61.5025	43.1613	33.5838
Air cooling extended fin	Erying	44.6956	37.5643	32.2318
	Peck	44.5812	37.6175	32.3162
Wick surface	Erying	58.9043	42.2947	33.2692
	Peck	60.5570	43.1613	33.5838
Passive cooling heat pipe	Erying	58.8606	42.2943	33.2692
	Peck	60.5322	43.1605	33.5838

Table 19.9 Increment in mean life under various conditions.

Cooling technique	Degradation equation	Increment in mean life by cooling up to 25 °C	Increment in mean life by cooling up to 35 °C	Increment in mean life by cooling up to 45 °C
Water cooling	Erying	34.6170	17.2947	8.2692
	Peck	36.5025	18.1613	8.5838
Air cooling extended fin	Erying	19.6956	12.5643	7.2318
	Peck	19.5812	12.6175	7.3162
Wick surface	Erying	33.9043	17.2947	8.2692
	Peck	35.5570	18.1613	8.5838
Passive cooling heat pipe	Erying	33.8606	17.2943	8.2692
	Peck	35.5322	18.1605	8.5838

the efficiency, too. As seen from the above study, cooling increases the mean life and reliability of solar panels significantly, so cooling techniques should be implemented in solar power plants. Mean life and reliability for some of the different cooling methods are compared, and it can be extended to many other best cooling techniques to obtain the best cooling method in terms of reliability improvement. Reliability should also be taken into consideration while calculating the reliability of solar power systems along with the efficiency and performance of PV systems. Economic aspects of the cooling system should also be incorporated to find out if installing the solar panel with the cooling system is economically viable or not.

References

Akbarzadeh, A. and Wadowski, T. (1996). Heat pipe-based cooling systems for photovoltaic cells under concentrated solar radiation. *Applied Thermal Engineering* 16 (1): 81–87.

Aminisharifabad, M., Yang, Q., and Wu, X. (2021). A deep learning-based reliability model for complex survival data. *IEEE Transactions on Reliability* 70 (1): 73–81.

Chikate, B.V., Sadawarte, Y., and Sewagram, B.D.C.O.E. (2015). The factors affecting the performance of solar cell. *International Journal of Computer Applications* 1 (1): 0975–8887.

Cuce, E., Bali, T., and Sekucoglu, S.A. (2011). Effects of passive cooling on performance of silicon photovoltaic cells. *International Journal of Low Carbon Technologies* 6 (4): 299–308.

David, L. (2002). *Handbook of Batteries*. New York: McGraw-Hill.

Dwivedi, P., Sudhakar, K., Soni, A. et al. (2020). Advanced cooling techniques of PV modules: a state of art. *Case studies in thermal engineering* 21: 100674.

Ebeling, C.E. (2004). *An Introduction to Reliability and Maintainability Engineering*. Tata McGraw-Hill Education.

Firoozzadeh, M., Shiravi, A., and Shafiee, M. (2019). An experimental study on cooling the photovoltaic modules by fins to improve power generation: economic assessment. *Iranian (Iranica) Journal of Energy & Environment* 10 (2): 80–84.

Guérin, F., Lantieri, P., and Dumon, B. (2006). Applying accelerated life models to halt testing. *Reliability Modeling, Analysis And Optimization* 9: 107.

Hacke, P., Spataru, S., Terwilliger, K. et al. (2015). Accelerated testing and modeling of potential-induced degradation as a function of temperature and relative humidity. *IEEE Journal of Photovoltaics* 5 (6): 1549–1553.

Hasan, A., Sarwar, J., Alnoman, H., and Abdelbaqi, E.S. (2017). Yearly energy performance of a photovoltaic-phase change material (PV-PCM) system in hot climate. *Solar Energy* 146: 417–429.

Irwan, Y.M., Leow, W.Z., Irwanto, M. et al. (2015, May). Comparison of solar panel cooling system by using dc brushless fan and dc water. In: *Journal of Physics: Conference Series*, vol. 622, No. 1, 012001. IOP Publishing.

Karthick, A., Ramanan, P., Ghosh, A. et al. (2020). Performance enhancement of copper indium diselenide photovoltaic module using inorganic phase change material. *Asia-Pacific Journal of Chemical Engineering* 15 (5): e2480.

Kim, J., Rabelo, M., Padi, S.P. et al. (2021). A review of the degradation of photovoltaic modules for life expectancy. *Energies* 14 (14): 4278.

Kusiak, A., Zhang, Z., and Verma, A. (2013). Prediction, operations, and condition monitoring in wind energy. *Energy* 60: 1–12.

Luque, A. and Hegedus, S. (ed.) (2011). *Handbook of Photovoltaic Science and Engineering*. Wiley.

Makrides, G., Theristis, M., Bratcher, J. et al. (2018). Five-year performance and reliability analysis of monocrystalline photovoltaic modules with different backsheet materials. *Solar Energy* 171: 491–499.

Markvart, T. (ed.) (2000). *Solar Electricity*, vol. 6. Wiley.

Mawoli, M., Yayha, H.N., Danshehu, B.G. et al. (2020). Development and performance evaluation of solar photovoltaic module's surface-to-rear temperature controlled valve for cooling application. *Nigerian Journal of Technological Development* 17 (1): 20–27.

Meeker, W.Q., Escobar, L.A., and Pascual, F.G. (2022). *Statistical Methods for Reliability Data*. Wiley.

Moharram, K.A., Abd-Elhady, M.S., Kandil, H.A., and El-Sherif, H. (2013). Enhancing the performance of photovoltaic panels by water cooling. *Ain Shams Engineering Journal* 4: 869–877.

Nižetić, S., Papadopoulos, A.M., and Giama, E. (2017). Comprehensive analysis and general economic-environmental evaluation of cooling techniques for photovoltaic panels, part I: passive cooling techniques. *Energy Conversion and Management* 149: 334–354.

Park, N.C., Oh, W.W., and Kim, D.H. (2013). Effect of temperature and humidity on the degradation rate of multicrystalline silicon photovoltaic module. *International Journal of Photoenergy* 2013.

Ravi, K. and Porpatham, E. (2017). Effect of piston geometry on performance and emission characteristics of an LPG fuelled lean burn SI engine at full throttle condition. *Applied Thermal Engineering* 110: 1051–1060.

Salem, M.R., Elsayed, M.M., Abd-Elaziz, A.A., and Elshazly, K.M. (2019). Performance enhancement of the photovoltaic cells using Al_2O_3/PCM mixture and/or water cooling-techniques. *Renewable Energy* 138: 876–890.

Sharaf, M., Yousef, M.S., and Huzayyin, A.S. (2022). Review of cooling techniques used to enhance the efficiency of photovoltaic power systems. *Environmental Science and Pollution Research* 29: 26131–26159.

Sohani, A., Shahverdian, M.H., Sayyaadi, H. et al. (2021). Selecting the best nanofluid type for a photovoltaic thermal (PV/T) system based on reliability, efficiency, energy, economic, and environmental criteria. *Journal of the Taiwan Institute of Chemical Engineers* 124: 351–358.

Sun, V., Asanakham, A., Deethayat, T., and Kiatsiriroat, T. (2020). A new method for evaluating nominal operating cell temperature (NOCT) of unglazed photovoltaic thermal module. *Energy Reports* 6: 1029–1042.

Tan, L., Date, A., Fernandes, G. et al. (2017). Efficiency gains of photovoltaic system using latent heat thermal energy storage. *Energy Procedia* 110: 83–88.

Teo, H.G., Lee, P.S., and Hawlader, M.N.A. (2012). An active cooling system for photovoltaic modules. *Applied Energy* 90 (1): 309–315.

20

Reliability Assessment of Different Topologies in Photovoltaic System

Laxman Chaudhary[1], *Aanchal Verma*[1], *Ramesh C. Bansal*[2], *and R. K. Saket*[1]

[1]*Department of Electrical Engineering, Indian Institute of Technology (BHU), Varanasi, Uttar Pradesh, India*
[2]*Department of Electrical Engineering, University of Sharjah, Sharjah, United Arab Emirates*

20.1 Introduction

The fundamental issue is global climate change, which is caused by the detrimental impact of carbon emissions produced by traditional electrical energy generation methods. Many governments and organizations around the world are financially supporting the implementation of renewable energy such as wind and solar (Patel 2005). PV system is an environmentally favorable electrical energy generation technique because it produces no carbon dioxide and does not need fossil fuels. It is currently the world's most rapidly expanding product. Total global installed PV capacity expanded fast from 41.5 to 1177 GW in 2010–2022 and is expected to hit 5400 GW by 2030 (Solar Power Europe 2021, IRENA report).

An electric power distribution is a massive, intricate web of electrical components that delivers electricity from power plants to users. Its major goal is to constantly meet load demand at the lowest possible cost while maintaining the level of reliability and environmental compliance. Therefore, it is preparing for future electrical demands using environmentally friendly energy supplies, guaranteeing that load requirements are met. Moreover, To reduce the risk of an electrical service disruption, it is important to evaluate the power systems reliability. For developing and running reliable power systems, the creation of energy supply reliability models is crucial. The capacity of an electrical power network to provide a consistent and reliable electricity supply is assessed through power system reliability assessments (Billinton and Allan 1992). Because of its size, number of components, variables, interrelations, and various purposes and aims of the systems, evaluating the reliability of the complete power system is intrinsically difficult. Hence, The reliability studies regarding this topic are divided into three hierarchical levels (HL): HL-1 (generation), HL-2 (transmission), and HL-3 (distribution). Generation management is critical for electric power companies because generation facilities determine the economy of power systems (Billinton and Allan 1984, 1996). The study presented in this paper focuses on generation systems at the HL-I level, incorporating different solar energy methods.

With rising electrical consumption, the use of fossil fuels has grown. Fossil fuels are in a limited amount, lasting in the next few years (Li and Billinton 1994)—so they are looking for alternate sources to generate electricity. Solar power demands are increasing to generate electricity because

Reliability Analysis of Modern Power Systems, First Edition. Edited by R. K. Saket and P. Sanjeevikumar.

it is freely available and easily accessible. The power system needs for PV systems rise, so their dependency will also grow. Their involvement in the generating system leads to an increase in their failure mechanism, so they focus on how to minimize their failure rates and try to make the system more reliable. In most studies, PV systems are a core component, according to the reliability assessment of PV systems, which does not focus on sub-components. The focus of this research is on a sub-component of PV systems (Peyghami et al. 2021).

The PV system topology can be classified based on component arrangement are central-inverter, string-inverter, and micro-inverter (Zhang et al. 2013; Jana et al. 2017). The structural parts of central and string structures are identical from electrical point of view, but their connections are different. The central PV systems are composed of many PV arrays linked to only one inverter, and one inverter is linked to each string in a string PV system. In micro-inverter, there is one inverter per solar panel. Developing detailed reliability models is required to incorporate these components to find the overall system reliability. The different topologies in PV system with their pros and cons comparison as shown in Table 20.1.

20.1.1 Basic Reliability Estimation Concepts

The probability that system will fulfill a defined function under typical operating conditions for a given time is referred to as reliability (Denson 1998; Ebeling 1997). The continuous random variable is used to mathematically express this relationship. The term "reliability" can be defined as in Eq. (20.1):

$$\mathrm{Re}(t) = \mathrm{Pr}\,(T \geq t) \tag{20.1}$$

$$0 \leq \mathrm{Re}(t) \leq 1$$

where T is time taken for component to fail.

Cumulative distribution function is chance that failure occurs before time (Ebeling 1997). Mathematically shown in Eq. (20.2),

$$\mathrm{Fa}(t) = 1 - \mathrm{Re}(t) = \mathrm{Pr}\,(T < t) \tag{20.2}$$

$$0 \leq \mathrm{Fa}(t) \leq 1$$

The probability density function describes the failure distribution's shape. Mathematically shown in Eq. (20.3),

$$\mathrm{fp}(t) = \frac{\mathrm{dFa}(t)}{\mathrm{d}t} = -\frac{\mathrm{dRe}(t)}{\mathrm{d}t} \tag{20.3}$$

When computing reliability, failure probability, or a visual representation of the failure distribution, the function Re(t), Fa(t), and fp(t) are utilized (Ebeling 1997).

20.1.2 Hazard Rate Function

Failure rate function often knows as hazard rate function and represented by $\lambda(t)$. It delivers an instantaneous failure rate. The hazard rate function is used to define the reliability function in Eq. (20.4):

$$\mathrm{Re}(t) = \exp^{(-\int_0^t \lambda(u)\,\mathrm{d}u)} \tag{20.4}$$

Table 20.1 Summary of different topologies.

Topology	Pros	Cons
Multi-stage inverter	Make smoother DC input	High cost
with HFT (Bose et al. 1985)	Better power quality and decoupling	Less reliability
	Less harmonic	
Multi-stage boost	Better reliability and efficiency	Lower power decoupling
inverter by GEC (Bose et al. 1985)	Lower cost and harmonic	
Multi-stage boost inverter	Lower cost and size	Poor power decoupling
Pseudo-dc-link	Good reliability	More harmonic
Half bridge topology	Less conversion stage	Poor power decoupling
with dc–dc-boost	Good reliability	and quality
	Easily handle power levels	
Fly-back inverter with	Improve flexibility of PV	Small life-span
active power decoupling (Shimizu et al. 2006)	generation system	Lower efficiency
	Less harmonic and cost	problem of leakage energy
Cascaded H-bridge	Improve flexibility of PV modules	Extra power loss
multilevel inverter	Lower switching frequency	Unbalanced power supply
(Xiao et al. 2014; Villanueva et al. 2009)	Increase efficiency	Higher cost
Two-state power	Enhance voltage gain,	Large size, cost
conversion (Dogga and Pathak 2019)	design flexibility	and weight
	Lesser conversion stage	Unbalanced power supply
	Increase energy	More losses
Traditional buck inverter	Less input volt.	Larger size
and line frequency	Lesser conversion stage	High cost
transformer (Kjaer et al. 2005; Xue et al. 2004)	Large input electrolytic capacitor	Low efficiency
Central-inverter	Lower cost	High replacement cost
(Solar Mango 2022)	More credibility	High-risk factor
	Produce high power	
String-inverter	Robustness	Shorter lifespan
(Sino Voltaics 2022)	Durability	Less efficient
	Quick troubleshooting	Complex system
Micro-inverter	Longer lifespan	High installation and
(Baker Home Energy 2022)	High Quality, Small Size	replacement cost

Suppose that the hazard rate function value is time-independent. In that case, that rate is called the constant failure such that $\lambda(t) = \lambda$ for $t \geq 0$, then the reliability functions will be evaluated with the help of exponential distribution (Ebeling 1997).

$$\text{Re}(t) = \exp^{(-\lambda \times t)} \tag{20.5}$$

$$\text{Fa}(t) = 1 - \exp^{(-\lambda \times t)} \tag{20.6}$$

$$\text{fp}(t) = \lambda \exp^{(-\lambda \times t)} \tag{20.7}$$

20.1.3 Reliability Block Diagram

It is a graphical representation of a system's components that allows the systems reliability to be calculated based on the components reliability. The reliability performance structure is reflected in the system model, which includes series, parallel, and mixed combinations. Each block in a reliability block diagram (RBD) represents a system component or subsystem. The RBD of the system is indicated by the connecting lines, which may or may not correlate to the functional block diagram (Ebeling 1997; Catelani et al. 2019).

All the components must be operated to perform the whole system in series. If one of the systems fails, the entire systems fail during the operation. The series relationship is represented in Figure 20.1.

The series equivalent Reliability equation is:

$$\text{Re}_s(t) = \text{Re}_A(t) * \text{Re}_B(t) * \ldots * \text{Re}_n(t) \tag{20.8}$$

In a parallel configuration, at least one component must be operational. If all components must fail for whole system fail. The parallel relationship is represented in Figure 20.2.

The parallel equivalent Reliability equation is:

$$\text{Re}_p(t) = 1 - (1 - \text{Re}_A(t)) * (1 - \text{Re}_B(t)) * \ldots * (1 - \text{Re}_n(t)) \tag{20.9}$$

Systems are arranged in the series and parallel combination and use the above results of series and parallel to get complete system reliability (Ebeling 1997). The combined RBD is in Figure 20.3.

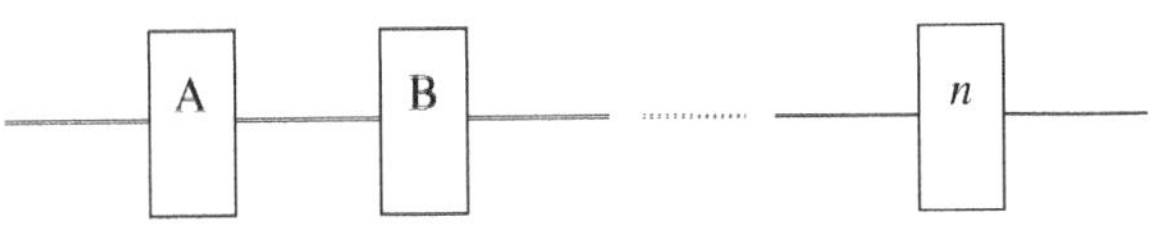

Figure 20.1 Reliability block for elements in series.

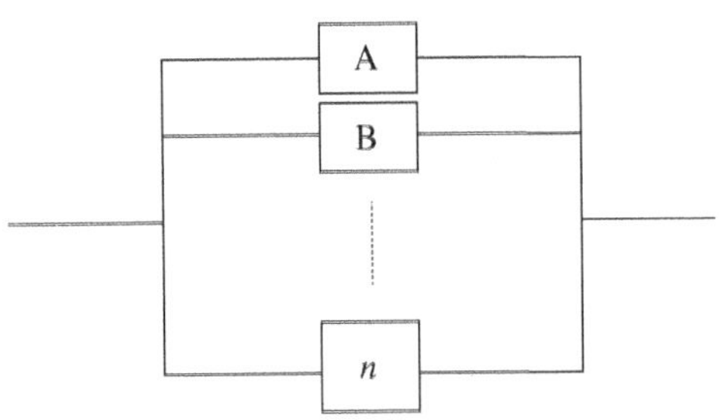

Figure 20.2 Reliability block for elements in parallel.

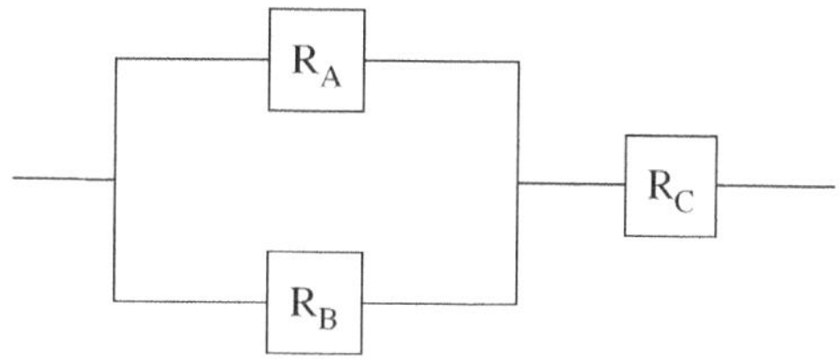

Figure 20.3 Reliability block for components in combination.

The equivalent Reliability equation is:

$$Re_{eq.}(t) = [1 - (1 - Re_A(t)) * (1 - Re_B(t))] * Re_C(t)$$

This chapter applies the above block diagram technique to evaluate the overall PV system reliability of different solar topologies.

20.2 Reliability Modeling of PV Topology

It is necessary to create a generation model for the entire system, including the PV system power output as well as the traditional generation method. In this study, the failure rates of the critical components of the PV system are capacitors, diodes, switches, and converters. PV systems are modeled using various topologies like central, string and micro-inverter (Ristow et al. 2008). A probabilistic methodology is used to construct PV system reliability model. The reliability is evaluating these different PV topologies by considering three assumptions, and these assumptions are (Anand et al. 2014; Billinton et al. 2001; Allan et al. 1994):

1. The entire systems will fail, if any one of them fails.
2. During operation, the failure rates of each component stay constant.
3. The entire system is modeled as a series and parallel RBD.

This study gives a summary of the reliability models for central, string, and micro-inverter PV systems. Reliability is calculated by considering individual component failure rate, component RBD and their arrangement in the combination of series and parallel. The different PV topologies reliability modeling and their RBDs are shown in Figures 20.4–20.6.

20.2.1 Reliability Modeling of Central PV Topology

Every part of the central PV system is represented by two possibilities in which it is either operational or fails with a specific probability. The key components of a central PV system are solar array, Bulk DC-Link capacitor, inverter, and filter (Alferidi and Karki 2017; Song and Wang 2013; Wikstrom et al. 2000). The central PV systems are shown in Figure 20.4.

20.2.2 Reliability Modeling of String PV Topology

Modeling of string PV system topology is almost the same as central but different in the connection of the power electronics components, and one extra component is used, which is a DC–DC converter. The key component of a string PV system is solar array, two parallel DC–DC converters,

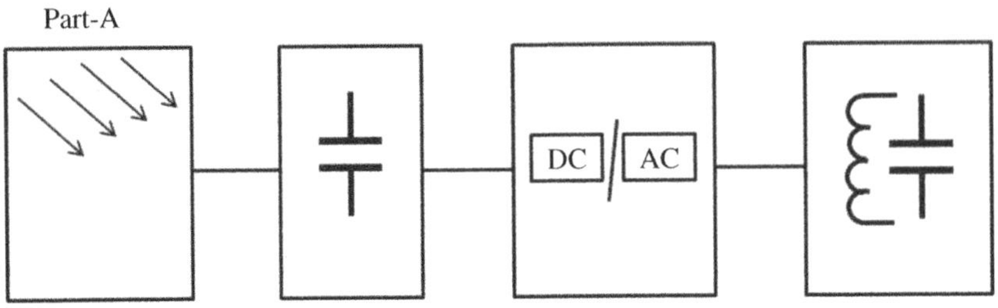

Figure 20.4 Schematic diagram of central-inverter.

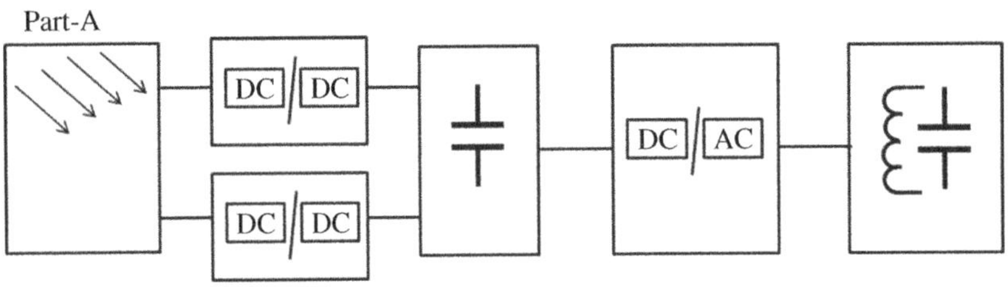

Figure 20.5 Schematic diagram of string-inverter.

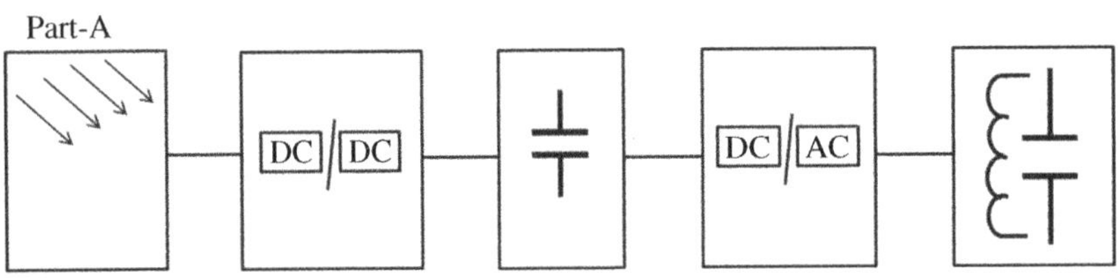

Figure 20.6 Schematic diagram of micro-inverter.

Bulk DC-Link capacitor, an inverter and filter (Alferidi and Karki 2017; Atasoy and Tezcan 2021). A string-inverter PV system is shown in Figure 20.5.

20.2.3 Reliability Modeling of Micro PV Topology

The key component of a micro-inverter PV system topology is solar array, DC–DC converters, Bulk DC-Link capacitor, inverter, and filter. Its modeling arrangement is shown in Figure 20.6 (Alferidi and Karki 2017; Tariq et al. 2018; Li and Wolfs 2008).

The above shows the reliability modeling of different topologies. As earlier mentioned, the most critical components are capacitors, diodes, switches, and converters, so focus on these component's failure rates and how the failure rates of these components depend on several factors. The MIL-HDBK-217F second edition of the military handbook provides a failure database for the power electronic elements. This information is used in this research to estimate the failure rate of power electronic components, and the results of this stage are used to investigate component's reliability (Military Handbook, 217 F Notice 2 1995).

20.3 Estimation of Failure Rate

20.3.1 DC-Link Capacitor

Its goal is to provide a stable DC voltage and limit the fluctuation of the system. DC-link capacitors can fail for a variety of reasons, including design flaws, material wear, operational temperature, voltage, current, moisture, and mechanical stress (Wang and Blaabjerg 2014). As for the consideration of the system requirements electrolytic capacitor has a voltage rating of 250 V and a peak temperature of 50°C (Harb and Balog 2012). Capacitor failure rate equation is:

$$\lambda_{\text{cap}} = n * \lambda_{\text{base}} * \pi_{\text{CF}} * \pi_{\text{QF}} * \pi_{\text{ES}} * \pi_{\text{TF}} \tag{20.10}$$

$$\pi_{\text{CF}} = 0.34 * C^{0.18} \tag{20.11}$$

$$\pi_{\text{TF}} = \exp^{-4061.74*(\frac{1}{T_j+273} - \frac{1}{298})} \tag{20.12}$$

where

n = Number of capacitor in DC-Link
λ_{base} = Capacitor basic failure rate
λ_{cap} = Capacitor failure rate
π_{ES} = Effect of environment stress
π_{QF} = Quality factor
π_{CF} = Capacitance factor
π_{TF} = Temperature factor
C = Capacitor value in micro Faraday
T_j = Junction temperature in degree Celsius

20.3.2 Diodes

The equation of failure rates of diodes as show (Military Handbook, 217 F Notice 2 1995):

$$\lambda_{\text{diode}} = n * \lambda_{\text{base}} * \pi_{\text{CCF}} * \pi_{\text{QF}} * \pi_{\text{ES}} * \pi_{\text{TF}} * \pi_{\text{ESF}} \tag{20.13}$$

$$\pi_{\text{TF}} = \exp^{-3091*(\frac{1}{T_j+273} - \frac{1}{298})} \tag{20.14}$$

$$\pi_{\text{ESF}} = \begin{cases} 0.054, & \text{for } V_{\text{Sr}} \leq 0.3 \\ V_{\text{Sr}}^{2.43}, & \text{for } 0.3 \leq V_{\text{Sr}} \leq 1 \end{cases} \tag{20.15}$$

$$V_{\text{Sr}} = \frac{\text{operated voltage}}{\text{rated voltage}} \tag{20.16}$$

where

n = Number of diodes
λ_{base} = Diode basic failure rate
λ_{diode} = Diode failure rate
π_{ES} = Effect of environment stress
π_{QF} = Quality factor
π_{CCF} = Contact Construction factor

π_{TF} = Temperature factor
V_{Sr} = Voltage stress factor
T_j = Junction temperature in degree Celsius

20.3.3 Switch

The equation of failure rates of switch as show (Military Handbook, 217 F Notice 2 1995):

$$\lambda_{switch} = n * \lambda_{base} * \pi_{QF} * \pi_{ES} * \pi_{TF} \tag{20.17}$$

$$\pi_{TF} = \exp^{-1925*\left(\frac{1}{T_j+273}-\frac{1}{298}\right)} \tag{20.18}$$

where
n = Number of switch
λ_{base} = switch basic failure rate
λ_{switch} = switch failure rate
π_{ES} = Effect of environment stress
π_{QF} = Quality factor
π_{TF} = Temperature factor
T_j = Junction temperature in degree Celsius

20.3.4 Filter

LC filter is used for flattening the AC signal in PV system. As for the consideration of system requirement MIL-F-18327 LC filter is used (Military Handbook, 217 F Notice 2 1995). The equation of failure rate of filter as show:

$$\lambda_{filter} = n * \lambda_{base} * \pi_{QF} * \pi_{ES} \tag{20.19}$$

where
n = Number of filter
λ_{base} = filter basic failure rate
λ_{Filter} = filter failure rate
π_{ES} = Effect of environment stress
π_{QF} = Quality factor

The component failure rate is calculated using these failure rate equations. The exponential distribution is used to find the reliability of component (Military Handbook, 217 F Notice 2 1995). The calculation of failure rates needs some fundamental value of failure rate, it is shown in Table 20.2.

20.4 Reliability Estimation Using RBD

Equations (20.10)–(20.19) are used to calculate the failure rate. The RBD and their series-parallel combinations are used to find the overall system reliability. This chapter first evaluates the single block of reliability and similarly finds all other block reliability, and finally uses a parallel system to get the reliability of the parallel blocks after that apply series system to estimate the overall system reliability.

Table 20.2 Basic failure rate of components.

Components	Failure rate (occur/yr)
Switches	0.012
Capacitor	0.0314
Diodes	0.025
Circuit-breaker	0.003
Filter	0.0011
Transformer	0.006

20.4.1 Central Topology

In Figure 20.4 of the central inverter, the reliability estimation of Block-A consists of DC-Link capacitance, inverter, and filter. Use Eqs. (20.10)–(20.20) to get their failure rates. Similarly, the second and third blocks are arranged. After that, all these blocks are parallel connected to the series arrangement of circuit breaker and transformer (Obeidat and Shuttleworth 2017; Lindquist et al. 2008). The whole arrangement of the Central topology is shown in Figure 20.7.

DC-Link capacitor failure rate is estimated from Eqs. (20.10)–(20.12), where the capacitor value is 330 μF (Harb and Balog 2012), the junction temperature is 50°C (Harb and Balog 2012), $n = 3$, $\lambda_{\text{base}} = 0.0314$, $\pi_{\text{ES}} = 1$, $\pi_{\text{QF}} = 1$, $\pi_{\text{CF}} = 0.9656$, and $\pi_{\text{TF}} = 2.8718$. These values are considered as per the system requirement (Military Handbook, 217 F Notice 2 1995):

$$\lambda_{\text{cap}} = 3 * 0.0314 * 0.9656 * 1 * 1 * 2.8718$$

$$\lambda_{\text{cap}} = 0.2613\ \text{occur/yr}$$

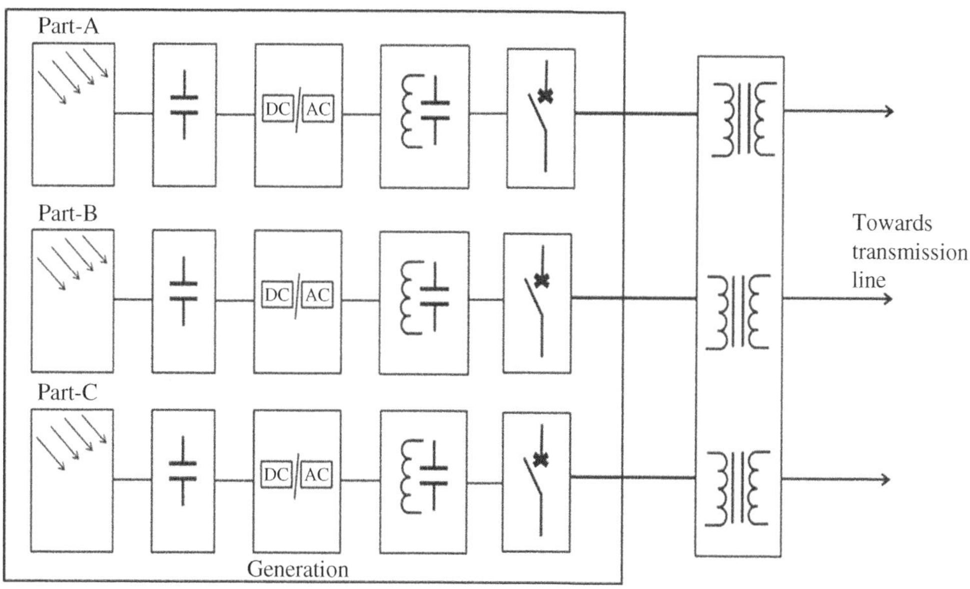

Figure 20.7 Central-inverter topology.

A two-level three-phase voltage source inverter is used in which six switches and diodes are used (Chan and Calleja 2011). The inverter failure rates are estimated with equation is:

$$\lambda_{\text{inverter}} = 6 * (\lambda_{\text{diode}} + \lambda_{\text{switch}}) \tag{20.20}$$

Diode failure rate is estimated from Eqs. (20.13)–(20.16), where operating voltage = 607 V, rated voltage = 690 V (Abdi et al. 2009), the junction temperature is 50°C (Harb and Balog 2012), n = 1, $\lambda_{\text{base}} = 0.025$, $\pi_{\text{ES}} = 1$, $\pi_{\text{QF}} = 0.7$, $\pi_{\text{CCF}} = 1$, $\pi_{\text{TF}} = 2.232$, and $\pi_{\text{ESF}} = 0.7324$. These values are considered as per the system requirement (Military Handbook, 217 F Notice 2 1995):

$$\lambda_{\text{diode}} = 1 * 0.025 * 1 * 0.7 * 1 * 2.232 * 0.7324$$

$$\lambda_{\text{diode}} = 0.0286\ \text{occur/yr}$$

Switch failure rate is estimated from Eqs. (20.17) and (20.18), where the junction temperature is 50°C (Harb and Balog 2012), $n = 1$, $\lambda_{\text{base}} = 0.012$, $\pi_{\text{ES}} = 1$, $\pi_{\text{QF}} = 0.7$, and $\pi_{\text{TF}} = 1.6487$. These values are considered as per the system requirement (Military Handbook, 217 F Notice 2 1995; Wang and Blaabjerg 2014).

$$\lambda_{\text{switch}} = 1 * 0.012 * 0.7 * 1 * 1.6487$$

$$\lambda_{\text{switch}} = 0.0138\ \text{occur/yr}$$

Now the failure rate of inverter using Eq. (20.20) (Alferidi and Karki 2017):

$$\lambda_{\text{Inverter}} = 6 * (0.0286 + 0.0138)$$

$$\lambda_{\text{Inverter}} = 0.2544\ \text{occur/yr}$$

Filter failure rate is estimated from Eq. (20.19), where $n = 1$, $\lambda_{\text{base}} = 0.0011$, $\pi_{\text{ES}} = 1$, and $\pi_{\text{QF}} = 1$. These values are considered as per the system requirement (Alferidi and Karki 2017):

$$\lambda_{\text{filter}} = 1 * 0.0011 * 1 * 1$$

$$\lambda_{\text{filter}} = 0.0011\ \text{occur/yr}$$

Reliability estimation of Block-A system using Eq. (20.8)

$$\text{Re}_{\text{A}} = \text{Re}_{\text{Capacitor}} * \text{Re}_{\text{Inverter}} * \text{Re}_{\text{Filter}}$$

$$\text{Re}_{\text{A}} = \exp^{(-0.2613*t)} * \exp^{(-0.2544*t)} * \exp^{(-0.0011*t)}$$

$$\text{Re}_{\text{A}} = \exp^{(-0.5168*t)}$$

Similarly, the Block-B and Block-C reliability values are:

$$\text{Re}_{\text{B}} = \text{Re}_{\text{C}} = \exp^{(-0.5168*t)}$$

After that, these blocks are parallel connected so their parallel equivalent reliability using Eq. (20.9):

$$\text{Re}_{\text{Parallel eq.}} = 1 - (1 - \text{Re}_{\text{A}}) * (1 - \text{Re}_{\text{B}}) * (1 - \text{Re}_{\text{C}})$$

$$\text{Re}_{\text{Parallel eq.}} = 1 - (1 - \exp^{(-0.5168*t)})^3$$

The overall reliability of central topology using Eq. (20.8):

$$\text{Re}_{\text{Cen}}(t) = \text{Re}_{\text{Parallel eq.}} * \text{Re}_{\text{Ckt. Breaker}} * \text{Re}_{\text{Trans}}$$

$$\text{Re}_{\text{Cen}}(t) = 1 - (1 - \exp^{(-0.5168*t)})^3 * \exp^{(-0.009*t)}$$

The cumulative distribution function of central topology is evaluated from Eq. (20.2):

$$\mathrm{Fa}_{\mathrm{Cen}}(t) = 1 - \mathrm{Re}_{\mathrm{Cen}}(t)$$

$$\mathrm{Fa}_{\mathrm{Cen}}(t) = 1 - (1 - (1 - \exp^{(-0.5168*t)})^3 * \exp^{(-0.009*t)})$$

The probability density function of central topology is evaluated from Eq. (20.3):

$$\mathrm{fp}_{\mathrm{Cen}}(t) = -\frac{\mathrm{dRe}_{\mathrm{Cen}}(t)}{\mathrm{d}t}$$

$$\mathrm{fp}_{\mathrm{Cen}}(t) = -\left[\frac{1 - (1 - \exp^{(-0.5168*t)})^3 * \exp^{(-0.009*t)}}{\mathrm{d}t}\right]$$

$$\begin{aligned}\mathrm{fp}_{\mathrm{Cen}}(t) = {} & 1.5504 * (1 - \exp^{(-0.5168*t)})^2 * \exp^{(-0.5168*t)} \\ & + 0.009 * 1 - (1 - \exp^{(-0.5168*t)})^3 * \exp^{(-0.009*t)}\end{aligned}$$

Figures 20.8–20.10 represent the reliability function, Cumulative distribution function and probability density function, respectively, of the Central-Inverter topology.

20.4.2 String Topology

Reliability estimation of Block-A in Figure 20.5 of string-inverter PV system consists of Two parallel DC–DC converters, Bulk DC-Link capacitance and inverter. Use Eqs. (20.10)–(20.21) to get their failure rates. Similarly, the second and third blocks are arranged. After that, all these blocks are parallel connected to the series arrangement of circuit breaker and transformer (Obeidat and

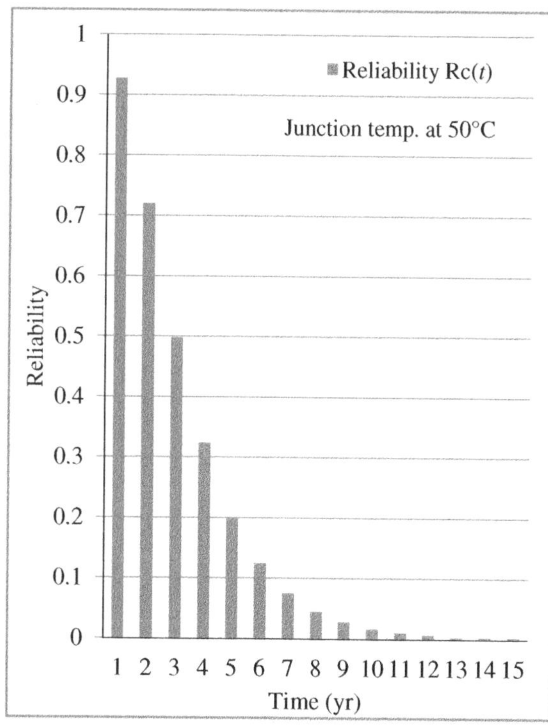

Figure 20.8 Reliability plot of central PV system.

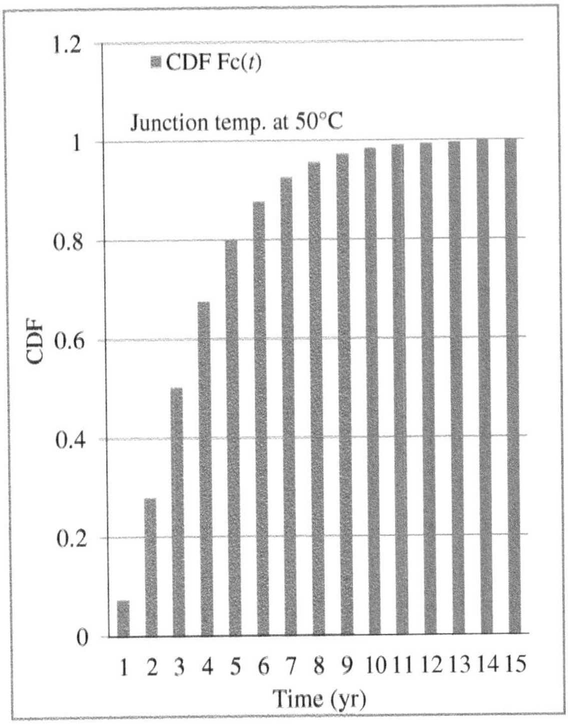

Figure 20.9 Cumulative distribution plot of central PV system.

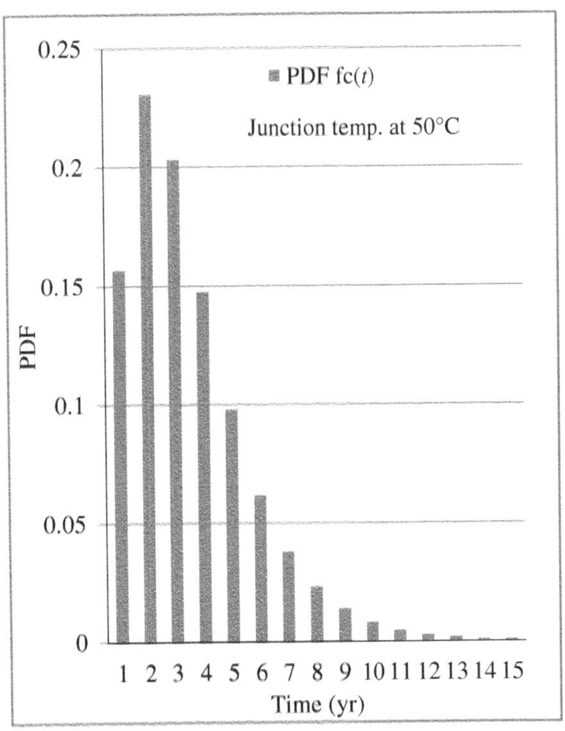

Figure 20.10 Probability density function plot of central PV system.

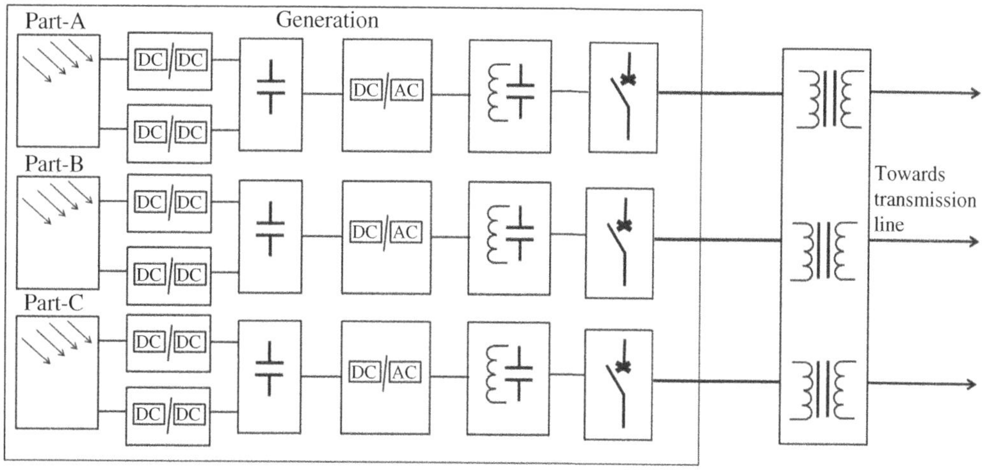

Figure 20.11 String-inverter topology.

Shuttleworth 2017; Lindquist et al. 2008; Kothari et al. 2022). The whole arrangement of the string topology is shown in Figure 20.11

DC-Link capacitor failure rate is estimated from Eqs. (20.10)–(20.12), where the capacitor value is 330 μF (Harb and Balog 2012), Junction temperature is 50°C (Harb and Balog 2012), $n = 1$, $\lambda_{base} = 0.0314$, $\pi_{ES} = 1$, $\pi_{QF} = 1$, $\pi_{CF} = 0.9656$, and $\pi_{TF} = 2.8718$. These values are considered as per the system requirement (Military Handbook, 217 F Notice 2 1995):

$$\lambda_{cap} = 1 * 0.0314 * 0.9656 * 1 * 1 * 2.8718$$
$$\lambda_{cap} = 0.0871\ \text{occur/yr}$$

Diode failure rate is estimated from Eqs. (20.13)–(20.16), where operating voltage = 607 V, rated voltage = 690 V (Abdi et al. 2009), the junction temperature is 50°C (Harb and Balog 2012), $n = 1$, $\lambda_{base} = 0.025$, $\pi_{ES} = 1$, $\pi_{QF} = 1$, $\pi_{CCF} = 1$, $\pi_{TF} = 2.232$, and $\pi_{ESF} = 0.7324$. These values are considered as per the system requirement (Military Handbook, 217 F Notice 2 1995):

$$\lambda_{diode} = 1 * 0.025 * 1 * 1 * 1 * 2.232 * 0.7324$$
$$\lambda_{diode} = 0.0409\ \text{occur/yr}$$

Switch failure rate is estimated from Eqs. (20.17) and (20.18), where the junction temperature is 50°C (Harb and Balog 2012), $n = 1$, $\lambda_{base} = 0.012$, $\pi_{ES} = 1$, $\pi_{QF} = 1$, and $\pi_{TF} = 1.6487$. These values are considered as per the system requirement (Military Handbook, 217 F Notice 2 1995; Wang and Blaabjerg 2014):

$$\lambda_{switch} = 1 * 0.012 * 1 * 1 * 1.6487$$
$$\lambda_{switch} = 0.0198\ \text{occur/yr}$$

Now the failure rate of inverter using Eq. (20.20) (Alferidi and Karki 2017):

$$\lambda_{Inverter} = 6 * (0.0409 + 0.0198)$$
$$\lambda_{Inverter} = 0.3642\ \text{occur/yr}$$

DC–DC converters are not treated as a single block; rather, it is a combination of components, and their failure rates are estimated with the given equation (Alferidi and Karki 2017):

$$\lambda_{\text{DC–DC Converter}} = (\lambda_{\text{diode}} + \lambda_{\text{Switch}})$$
$$\lambda_{\text{DC–DC Converter}} = (0.0409 + 0.0198) \tag{20.21}$$
$$\lambda_{\text{DC–DC Converter}} = 0.0607\,\text{occur/yr}$$

Filter failure rate is estimated from Eq. (20.19), where $n = 1$, $\lambda_{\text{base}} = 0.0011$, $\pi_{\text{ES}} = 1$, and $\pi_{\text{QF}} = 1$. These values are considered as per the system requirement (Alferidi and Karki 2017):

$$\lambda_{\text{filter}} = 1 * 0.0011 * 1 * 1$$
$$\lambda_{\text{filter}} = 0.0011\,\text{occur/yr}$$

Reliability estimation of Block-A system using Eqs. (20.8) and (20.9):

$$\text{Re}_{\text{A}} = (1 - (1 - \text{Re}_{\text{DC–DC Con.}})^2) * \text{Re}_{\text{Cap.}} * \text{Re}_{\text{Inv.}} * \text{Re}_{\text{Fil.}}$$
$$\text{Re}_{\text{A}} = (1 - (1 - \exp^{(-0.0607*t)})^2) * \exp^{(-0.0871*t)} * \exp^{(-0.3642*t)} * \exp^{(-0.0011*t)}$$
$$\text{Re}_{\text{A}} = (1 - (1 - \exp^{(-0.0607*t)})^2) * \exp^{(-0.4524*t)}$$

Similarly, the Block-B and Block-C and their reliability values are:

$$\text{Re}_{\text{B}} = \text{Re}_{\text{C}} = (1 - (1 - \exp^{(-0.0607*t)})^2) * \exp^{(-0.4524*t)}$$

After that, these blocks are parallel connected so their parallel equivalent reliability using Eq. (20.9):

$$\text{Re}_{\text{Par.}} = 1 - (1 - \text{Re}_{\text{A}}) * (1 - \text{Re}_{\text{B}}) * (1 - \text{Re}_{\text{C}})$$
$$\text{Re}_{\text{Par.}} = 1 - [(1 - (1 - (1 - \exp^{(-0.0607*t)})^2) * \exp^{(-0.4524*t)})^3]$$

The overall reliability of string topology using Eq. (20.8):

$$\text{Re}_{\text{Str}}(t) = \text{Re}_{\text{Parallel eq.}} * \text{Re}_{\text{Ckt. Breaker}} * \text{Re}_{\text{Trans}}$$
$$\text{Re}_{\text{Str}}(t) = (1 - [(1 - (1 - (1 - \exp^{(-0.0607*t)})^2) * \exp^{(-0.4524*t)})^3]) * \exp^{(-0.009*t)}$$

The cumulative distribution function of central topology is evaluated from Eq. (20.2):

$$\text{Fa}_{\text{Str}}(t) = 1 - \text{Re}_{\text{Str}}(t)$$
$$\text{Fa}_{\text{Str}}(t) = 1 - (1 - [(1 - (1 - (1 - \exp^{(-0.0607*t)})^2) * \exp^{(-0.4524*t)})^3] * \exp^{(-0.009*t)})$$

The probability density function of central topology is evaluated from Eq. (20.3):

$$\text{fp}_{\text{Str}}(t) = -\frac{d\text{Re}_{\text{Str}}(t)}{dt}$$
$$\begin{aligned}\text{fp}_{\text{Str}}(t) = (0.009 * (1 - (1 - [1 - (1 - \exp^{(-0.0607*t)})^2 * \exp^{(-0.4524*t)}]^3 \\ * 3 * (1 - [1 - (1 - \exp^{(-0.0607*t)})^2 * \exp^{(-0.4524*t)}]^2) * (0.4524 * \exp^{(-0.4524*t)}) \\ * 1 - (1 - \exp^{(-0.0607*t)})^2 + 0.1214 \exp^{(-0.5131*t)} - 0.1214 * \exp^{(-0.5738*t)}))) \\ * \exp^{(-0.009*t)}\end{aligned}$$

Figures 20.12–20.14 represent the reliability function, Cumulative distribution function and probability density function, respectively, of the string-inverter topology.

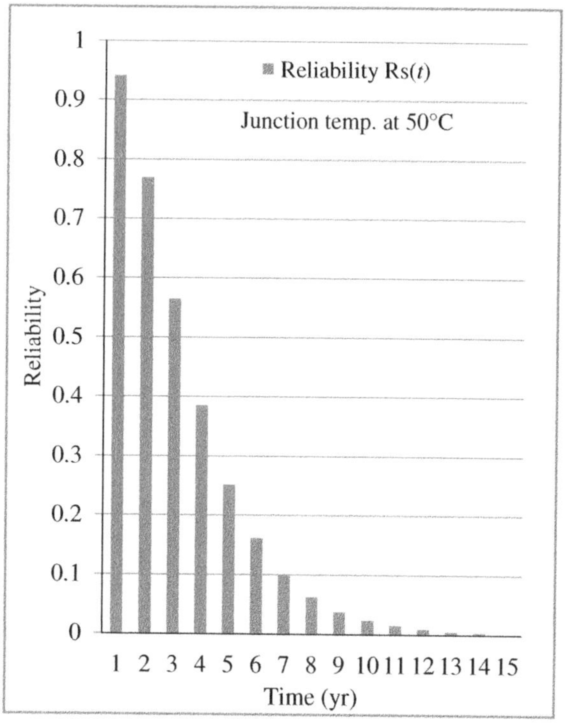

Figure 20.12 Reliability plot of string PV system.

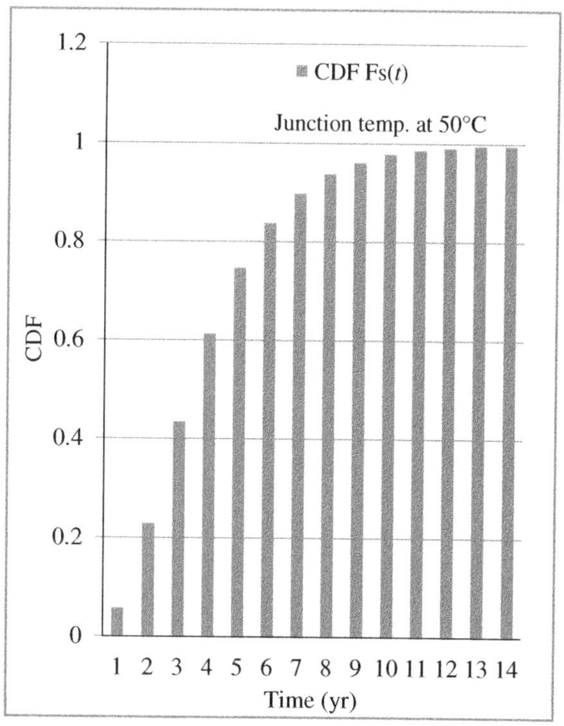

Figure 20.13 Cumulative distribution plot of string PV system.

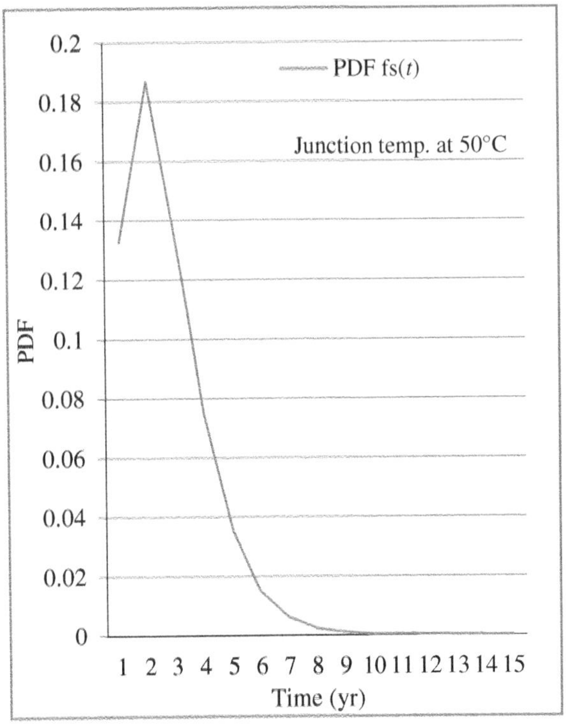

Figure 20.14 Probability density function plot of string PV system.

20.4.3 Micro Topology

Reliability estimation of Block-A in Figure 20.6 of micro-inverter solar system is DC–DC converters, Bulk DC-Link capacitance, inverter, and filter. Use Eqs. (20.10)–(20.21) to get failure rates. Similarly, the second and third blocks are arranged. After that, all phases are parallel connected to the series arrangement of circuit breaker and transformer (Obeidat and Shuttleworth 2017; Lindquist et al. 2008). The whole arrangement of the Micro topology is shown in Figure 20.15.

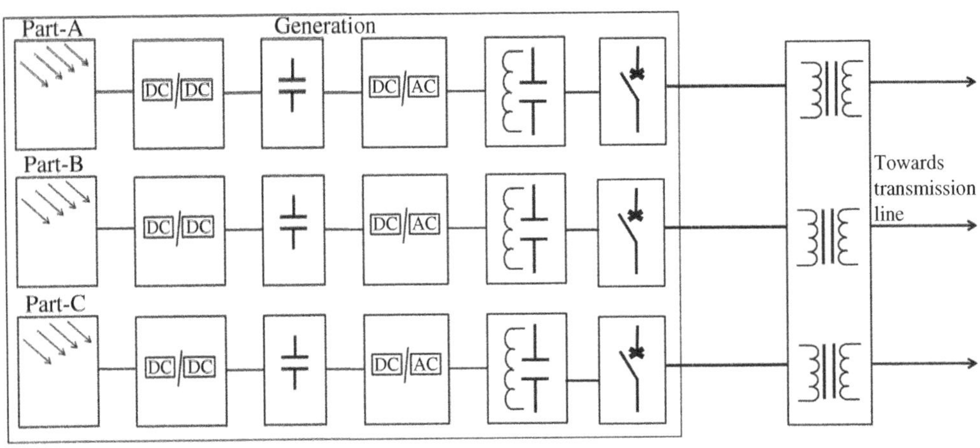

Figure 20.15 Micro-inverter topology.

DC-Link capacitor failure rate is estimated from Eqs. (20.10)–(20.12), where the capacitor value is 330 μF (Harb and Balog 2012), the junction temperature is 50°C (Harb and Balog 2012), $n = 1$, $\lambda_{base} = 0.0314$, $\pi_{ES} = 1$, $\pi_{QF} = 1$, $\pi_{CF} = 0.9656$, and $\pi_{TF} = 2.8718$. These values are considered as per the system requirement (Military Handbook, 217 F Notice 2 1995).

$$\lambda_{cap} = 1 * 0.0314 * 0.9656 * 1 * 1 * 2.8718$$

$$\lambda_{cap} = 0.0871 \text{ occur/yr}$$

Diode failure rate is estimated from Eqs. (20.13)–(20.16), where operating voltage = 607 V, rated voltage = 690 V (Abdi et al. 2009; Kothari et al. 2022; Verma et al. 2023), the junction temperature is 50°C (Harb and Balog 2012), $n = 1$, $\lambda_{base} = 0.025$, $\pi_{ES} = 1$, $\pi_{QF} = 0.7$, $\pi_{CCF} = 1$, $\pi_{TF} = 2.232$, and $\pi_{ESF} = 0.7324$. These values are considered as per the system requirement (Military Handbook, 217 F Notice 2 1995):

$$\lambda_{diode} = 1 * 0.025 * 1 * 0.7 * 1 * 2.232 * 0.7324$$

$$\lambda_{diode} = 0.0286 \text{ occur/yr}$$

Switch failure rate is estimated from Eqs. (20.17) and (20.18), where the junction temperature is 50°C (Harb and Balog 2012), $n = 1$, $\lambda_{base} = 0.012$, $\pi_{ES} = 1$, $\pi_{QF} = 0.7$, and $\pi_{TF} = 1.6487$. These values are considered as per the system requirement (Military Handbook, 217 F Notice 2 1995; Wang and Blaabjerg 2014):

$$\lambda_{switch} = 1 * 0.012 * 0.7 * 1 * 1.6487$$

$$\lambda_{switch} = 0.0138 \text{ occur/yr}$$

Now the failure rate of inverter using Eq. (20.20) (Alferidi and Karki 2017):

$$\lambda_{Inverter} = 6 * (0.0286 + 0.0138)$$

$$\lambda_{Inverter} = 0.2544 \text{ occur/yr}$$

Now the failure rate of DC–DC converter using Eq. (20.21) (Alferidi and Karki 2017):

$$\lambda_{DC\text{–}DC\ Converter} = (0.0286 + 0.0138)$$

$$\lambda_{DC\text{–}DC\ Converter} = 0.0424 \text{ occur/yr}$$

Filter failure rate is estimated from Eq. (20.19), where $n = 1$, $\lambda_{base} = 0.0011$, $\pi_{ES} = 1$, and $\pi_{QF} = 1$. These values are considered as per the system requirement (Alferidi and Karki 2017):

$$\lambda_{filter} = 1 * 0.0011 * 1 * 1$$

$$\lambda_{filter} = 0.0011 \text{ occur/yr}$$

Reliability estimation of Block-A system using Eq. (20.8):

$$Re_A = Re_{DC\text{–}DC\ Con.} * Re_{Cap} * Re_{Inv.} * Re_{Fil.}$$

$$Re_A = \exp^{(-0.0424*t)} * \exp^{(-0.0871*t)} * \exp^{(-0.2544*t)} * \exp^{(-0.0011*t)}$$

$$Re_A = \exp^{(-0.385*t)}$$

Similarly, the Block-B and Block-C and their reliability values are:

$$Re_B = Re_C = \exp^{(-0.385*t)}$$

After that, these blocks are parallel connected so their parallel equivalent reliability using Eq. (20.9):

$$\mathrm{Re}_{\text{Parallel eq.}} = 1 - (1 - \mathrm{Re}_A) * (1 - \mathrm{Re}_B) * (1 - \mathrm{Re}_C)$$
$$\mathrm{Re}_{\text{Parallel eq.}} = 1 - (1 - \exp^{(-0.385*t)})^3$$

The overall reliability of central topology using Eq. (20.8):

$$\mathrm{Re}_{\text{Micro}}(t) = \mathrm{Re}_{\text{Parallel eq.}} * \mathrm{Re}_{\text{Ckt. Breaker}} * \mathrm{Re}_{\text{Trans}}$$
$$\mathrm{Re}_{\text{Micro}}(t) = 1 - (1 - \exp^{(-0.385*t)})^3 * \exp^{(-0.009*t)}$$

The cumulative distribution function of central topology is evaluated from Eq. (20.2):

$$\mathrm{Fa}_{\text{Micro}}(t) = 1 - \mathrm{Re}_{\text{Micro}}(t)$$
$$\mathrm{Fa}_{\text{Micro}}(t) = 1 - (1 - (1 - \exp^{(-0.385*t)})^3 * \exp^{(-0.009*t)})$$

The probability density function of central topology is evaluated from Eq. (20.3):

$$\mathrm{fp}_{\text{Micro}}(t) = -\frac{d\mathrm{Re}_{\text{Micro}}(t)}{dt}$$
$$\mathrm{fp}_{\text{Micro}}(t) = -\left[\frac{1 - (1 - \exp^{(-0.385*t)})^3 * \exp^{(-0.009*t)}}{dt}\right]$$
$$\mathrm{fp}_{\text{Micro}}(t) = 1.155 * (1 - \exp^{(-0.385*t)})^2 * \exp^{(-0.385*t)} + 0.009 * 1 - (1 - \exp^{(-0.385*t)})^3 * \exp^{(-0.009*t)}$$

Figures 20.16–20.18 represent the reliability function, cumulative distribution function and probability density function, respectively, of the micro-inverter topology.

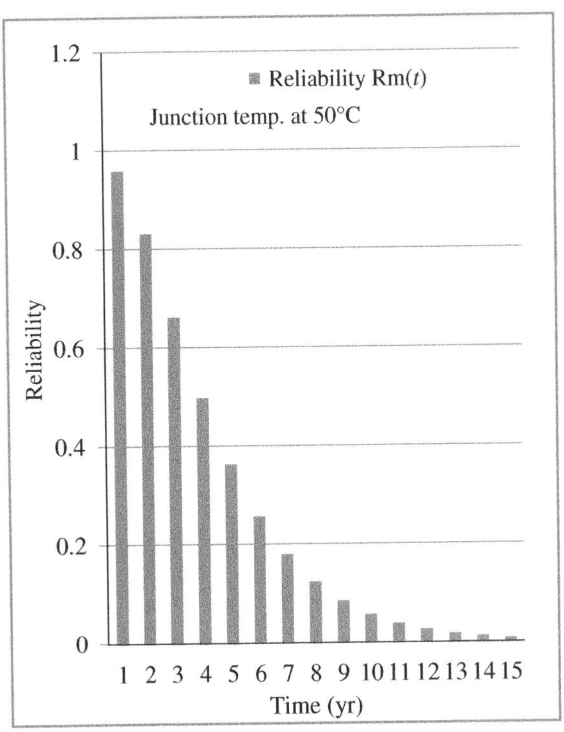

Figure 20.16 Reliability plot of micro PV system.

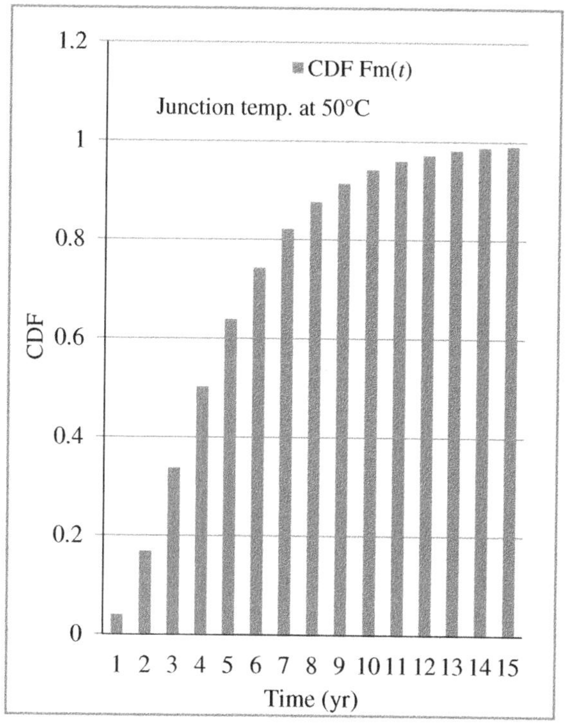

Figure 20.17 Cumulative distribution plot of micro PV system.

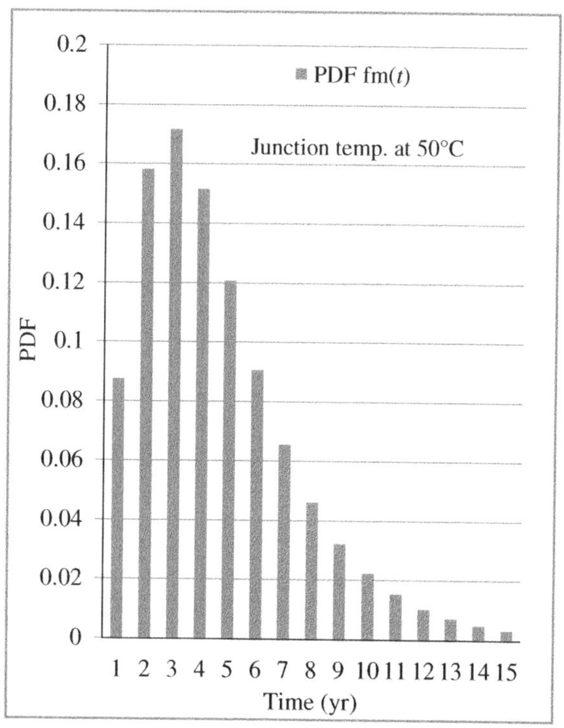

Figure 20.18 Probability density function plot of micro PV system.

20.5 Results

In Section 20.5, plot the graphs of the reliability function, cumulative distribution function, and probability density function for various topologies in PV system at a junction temperature of 50°C. Similar, evaluation at junction temperature of 25 and 75°C of these topologies. This section now shows a comparison plot of these junction temperatures for various topologies in PV system.

The reliability function plots of 25, 50, and 75°C temperatures are shown in Figures 20.19–20.21, respectively. The reliability value decreases at a slow rate at low temperature compared to high temperature so the system is more reliable at low temperature.

The cumulative distribution function plots of 25, 50, and 75°C temperatures as shown in Figures 20.22–20.24, respectively. The failure probability value increasing slow rate at low temperature compare to high temperature so the systems get failure early at high temperature.

The probability density function plots of 25, 50, and 75°C temperatures as shown in Figures 20.25–20.27, respectively. At a particular time "*t*" before that time, it represents failure probability and, after that time, it represents reliability function nature. In all these probability density function plot represents micro has less failure before that time "*t*" and more reliable after that time compared to all topologies.

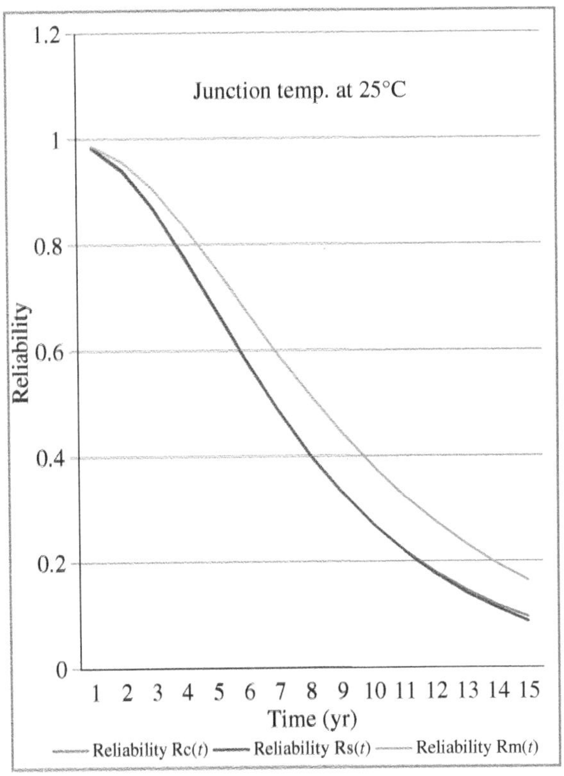

Figure 20.19 Reliability plots at 25°C of different topologies.

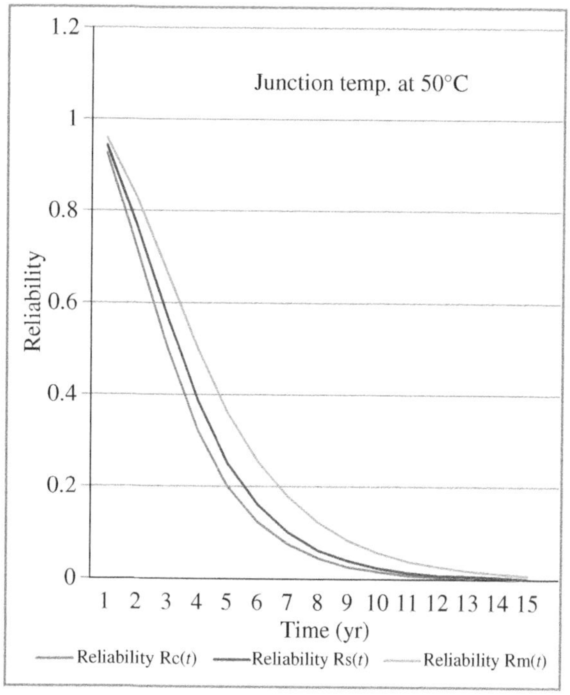

Figure 20.20 Reliability plots at 50°C of different topologies.

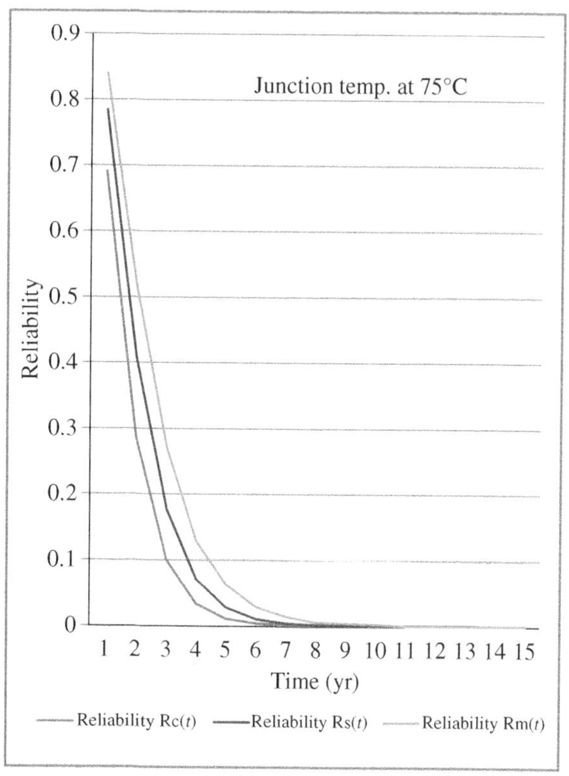

Figure 20.21 Reliability plots at 75°C of different topologies.

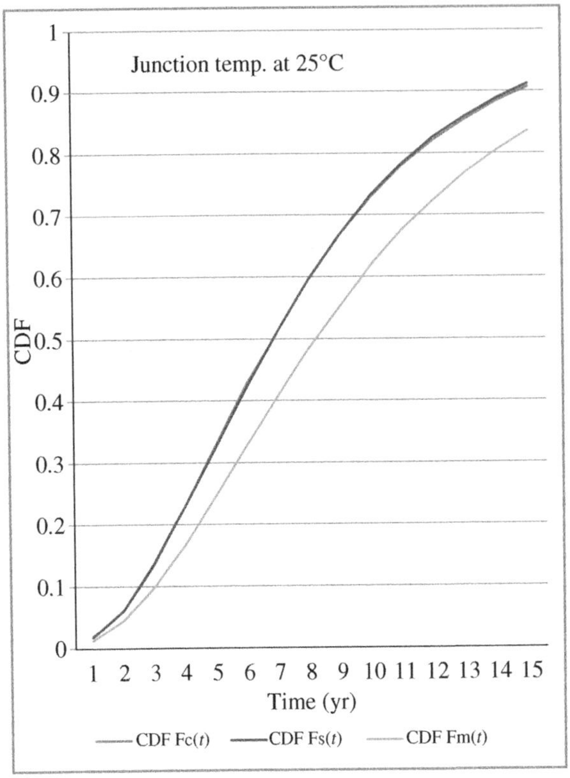

Figure 20.22 Cumulative distribution plots at 25°C of different topologies.

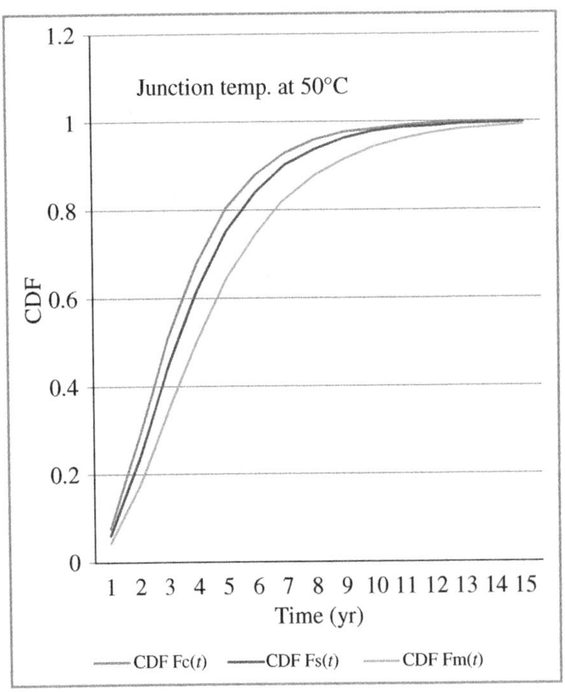

Figure 20.23 Cumulative distribution plots at 50°C of different topologies.

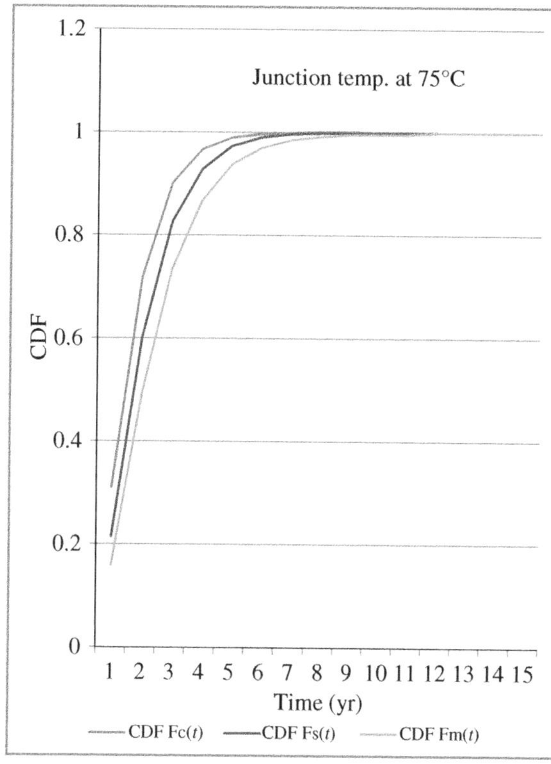

Figure 20.24 Cumulative distribution plots at 75°C of different topologies.

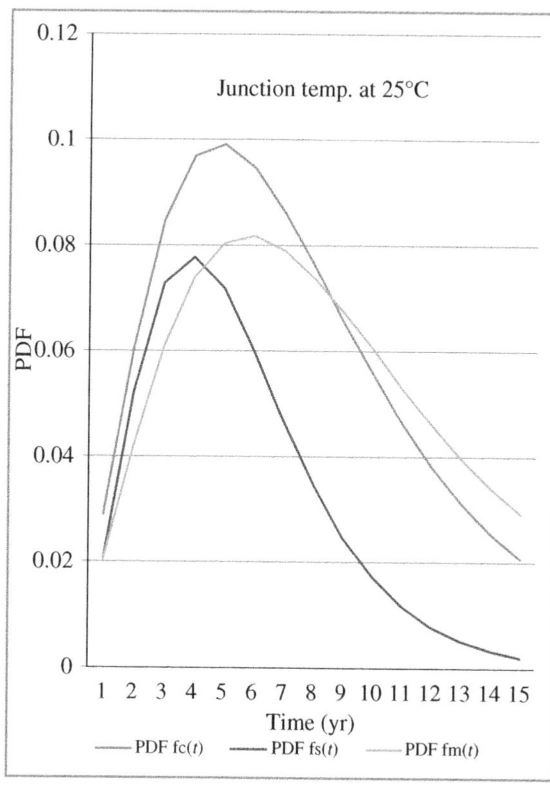

Figure 20.25 Probability density plots at 25°C of different topologies.

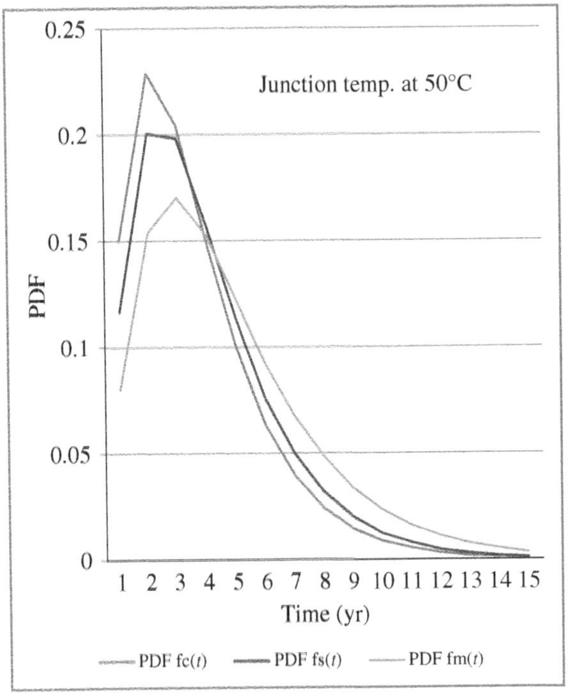

Figure 20.26 Probability density plots at 50°C of different topologies.

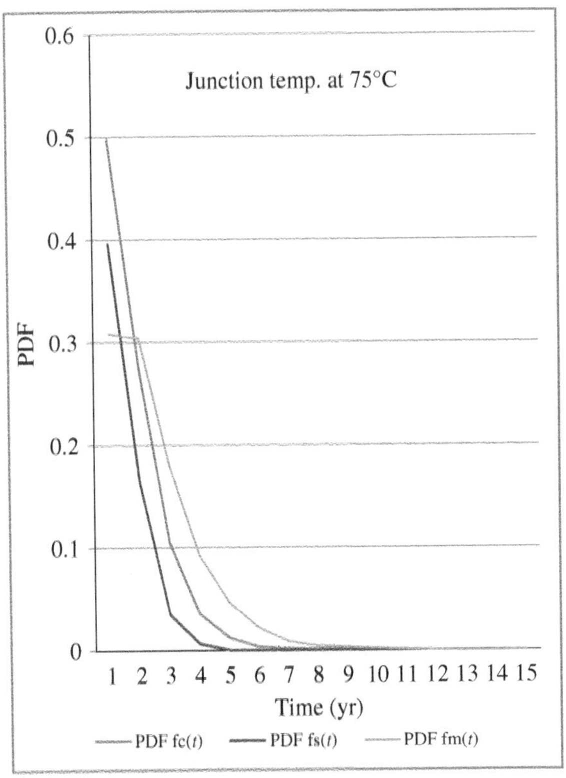

Figure 20.27 Probability density plots at 75°C of different topologies.

20.6 Conclusions

From the reliability assessment of central-inverter, string-inverter, and micro-inverter on 25, 50, and 75°C it is found that micro-inverter topology shows the better result in comparison with central-inverter and String-inverter.

The most crucial component in the most reliable solar topology, according to this study, is an inverter, which has a failure rate of 0.2544 per year, which is high when compared to all other components utilized in the micro-inverter solar topology. The author proposed a decrement in the failure rate of an inverter will significantly change system reliability by optimizing the inverter architecture.

References

Abdi, B., Ranjbar, A.H., Gharehpetian, G.B., and Milimonfared, J. (2009). Reliability considerations for parallel performance of semiconductor switches in high-power switching power supplies. *IEEE Transactions on Industrial Electronics* 56: 2133–2139.

Alferidi, A. and Karki, R. (2017). Development of probabilistic reliability models of photovoltaic system topologies for system adequacy evaluation. *Applied Sciences* 7(2): 176.

Allan, R.N., Billinton, R., Breipohl, A.M., and Grigg, C.H. (1994). Bibliography on the application of probability methods in power system reliability evaluation: 1987–1991. *IEEE Transactions on Power Systems* 9(1): 41–49.

Anand, S., Gundlapalli, S.K., and Fernandes, B.G. (2014). Transformer-less grid feeding current source inverter for solar photovoltaic system. *IEEE Transactions on Industrial Electronics* 61: 5334–5344.

Atasoy, B. and Tezcan, S.S. (2021). Comparison of string and central inverter for 10MW PV plant. *2021 5th International Symposium on Multidisciplinary Studies and Innovative Technologies (ISMSIT)*, 182–186. https://doi.org/10.1109/ISMSIT52890.2021.9604588.

Baker Home Energy (2022). https://bakerhomeenergy.com/blog/2016-06-14/advantages-and-disadvantages-of-microinverters/ (accessed 06 April 2022).

Billinton, R. and Allan, R.N. (1984). Power-system reliability in perspective. *Electronics and Power* 30(3): 231–236.

Billinton, R. and Allan, R.N. (1992). *Reliability Evaluation of Engineering Systems: Concepts and Techniques*. New York: Plenum Press.

Billinton, R. and Allan, R.N. (1996). *Reliability Evaluation of Power Systems*, 2e. New York and London: Plenum Publishing.

Billinton, R., Fotuhi-Firuzabad, M., and Bertling, L. (2001). Bibliography on the application of probability methods in power system reliability evaluation 1996–1999. *IEEE Transactions on Power Systems* 16(4): 595–602.

Bose, B.K., Szczesny, P.M., and Steigerwald, R.L. (1985). Microcomputer control of a residential photovoltaic power conditioning system. *IEEE Transactions on Industry Applications* 5: 1182–1191.

Catelani, M., Ciani, L., and Venzi, M. (2019). RBD model-based approach for reliability assessment in complex systems. *IEEE Systems Journal* 13(3): 2089–2097. https://doi.org/10.1109/JSYST.2018.2840220.

Chan, F. and Calleja, H. (2011). Reliability estimation of three single-phase topologies in grid-connected PV systems. *IEEE Transactions on Industrial Electronics* 58: 2683–2689.

Denson, W. (1998). The history of reliability prediction. *IEEE Transactions on Reliability* 47(3): SP321–SP328. https://doi.org/10.1109/24.740547.

Dogga, R. and Pathak, M.K. (2019). Recent trends in solar PV inverter topologies. *Solar Energy* 183: 57–73.

Ebeling, C.E. (1997). *An Introduction to Reliability and Maintainability Engineering*, 1e. Tata McGraw-Hill Education.

Harb, S. and Balog, R.S. (2012). Reliability of candidate photovoltaic module-integrated-inverter (PV-MII) topologies–a usage model approach. *IEEE Transactions on Power Electronics* 28(6): 3019–3027.

Jana, J., Saha, H., and Bhattacharya, K.D. (2017). A review of inverter topologies for single-phase grid-connected photovoltaic systems. *Renewable and Sustainable Energy Reviews* 72: 1256–1270.

Kjaer, S.B., Pedersen, J.K., and Blaabjerg, F. (2005). A review of single-phase grid-connected inverters for photovoltaic modules. *IEEE Transactions on Industry Applications* 41(5): 1292–1306.

Kothari, D.P., Nagrath, I.J., and Saket, R.K. (2022). *Modern Power System Analysis*. Tata McGraw-Hill Publishing Company.

Li, W. and Billinton, R. (1994). *Reliability Assessment of Electrical Power Systems Using Monte Carlo Methods*. New York: Plenum Press.

Li, Q. and Wolfs, P. (2008). A review of the single phase photovoltaic module integrated converter topologies with three different DC link configurations. *IEEE Transactions on Power Electronics* 23(3): 1320–1333.

Lindquist, T.M., Bertling, L., and Eriksson, R. (2008). Circuit breaker failure data and reliability modelling. *IET Generation, Transmission & Distribution* 2(6): 813–820.

Military Handbook, 217 F Notice 2 (1995). Reliability Prediction of Electronic Equipment.

Obeidat, F. and Shuttleworth, R. (2017). PV inverters reliability prediction. *World Applied Sciences Journal* 35(2): 275–287.

Patel, M.R. (2005). *Wind and Solar Power Systems: Design, Analysis, and Operation*. CRC Press.

Peyghami, S., Blaabjerg, F., and Palensky, P. (2021). Incorporating power electronic converters reliability into modern power system reliability analysis. *IEEE Journal of Emerging and Selected Topics in Power Electronics* 9(2): 1668–1681.

Ristow, A., Begovic, M., Pregelj, A., and Rohatgi, A. (2008). Development of a methodology for improving photovoltaic inverter reliability. *IEEE Transactions on Industrial Electronics* 55: 2581–2592.

Shimizu, T., Wada, K., and Nakamura, N. (2006). Flyback-type single-phase utility interactive inverter with power pulsation decoupling on the DC input for an AC photovoltaic module system. *IEEE Transactions on Power Electronics* 21(5): 1264–1272.

Sino Voltaics (2022). https://sinovoltaics.com/learing-center/inverters/string-inverters/ (accessed 05 April 2022).

Solar Mango (2022). https://www.solarmango.com/ask/2015/10/23/what-are-central-inverters/ (accessed 04 April 2022).

Solar Power Europe (2021). Global Market Outlook for Solar Power 2021–2025.

Song, Y. and Wang, B. (2013). Survey on reliability of power electronic systems. *IEEE Transactions on Power Electronics* 28(1): 591–604. https://doi.org/10.1109/TPEL.2012.2192503.

Tariq, M.S., Butt, S.A., and Khan, H.A. (2018). Impact of module and inverter failures on the performance of central-, string-, and micro-inverter PV systems. *Microelectronics Reliability* 88: 1042–1046.

Verma, A., Singh, A., Anand Kumar, K. et al. (2023). Reliability analysis of multilevel and matrix converters used in more electric aircraft. *IET Electrical Systems in Transportation* 13(2): e12078.

Villanueva, E., Correa, P., Rodriguez, J.É., and Pacas, M. (2009). Control of a single-phase cascaded H-bridge multilevel inverter for grid-connected photovoltaic systems. *IEEE Transactions on Industrial Electronics* 56(11): 4399–4406.

Wang, H. and Blaabjerg, F. (2014). Reliability of capacitors for DC-link applications in power electronic converters–an overview. *IEEE Transactions on Industry Applications* 50(5): 3569–3578. https://doi.org/10.1109/TIA.2014.2308357.

Wikstrom, P., Terens, L.A., and Kobi, H. (2000). Reliability, availability, and maintainability of high-power variable-speed drive systems. *IEEE Transactions on Industry Applications* 36(1): 231–241.

Xiao, B., Hang, L., Mei, J. et al. (2014). Modular cascaded H-bridge multilevel PV inverter with distributed MPPT for grid-connected applications. *IEEE Transactions on Industry Applications* 51(2): 1722–1731.

Xue, Y., Chang, L., Kjær, S. et al. (2004). Topologies of single-phase inverters for small distributed power generators: an overview. *IEEE Transactions on Power Electronics* 19: 1305–1314. https://doi.org/10.1109/TPEL.2004.833460.

Zhang, P., Li, W., Li, S. et al. (2013). Reliability assessment of photovoltaic power systems: review of current status and future perspectives. *Applied Energy* 104: 822–833.

Section 6

Reliability Analysis of Power Electronics Components and Systems for Modern Power System Applications

21

Reliability Evaluation of Power Electronics Converters for Modern Power System Applications

Amit Kumar[1], *Sachin Kumar*[2], *Sunil K. Singh*[3], *R. K. Saket*[4], *and P. Sanjeevikumar*[5]

[1] *Flameproof and Equipment Safety Department, CSIR-Central Institute of Mining and Fuel Research, Dhanbad, Jharkhand, India*
[2] *Department of Electrical Engineering, G.B. Pant Institute of Engineering and Technology, Pauri-Garhwal, Uttarakhand, India*
[3] *Department of Electrical Engineering, Shri Ramswaroop Memorial University, Lucknow, Uttar Pradesh, India*
[4] *Department of Electrical Engineering, Indian Institute of Technology (BHU), Varanasi, Uttar Pradesh, India*
[5] *Department of Electrical Engineering, IT and Cybernetics, University of South-Eastern Norway, Porsgrunn, Norway*

21.1 Introduction

Power electronics systems are widely used in various applications such as variable speed drives, electric vehicles, renewable energy systems, and power quality control stations (Kumar et al. 2020; Singh et al. 2023). Reliability is essential for power electronics systems due to increased dependability in many crucial areas like energy and transportation. Reliability is the ability of any instrument to perform as desired for a certain period. Reliability is the maintainability represented by the probability of an item working in a healthy condition after a period. The primary purpose of reliability engineering is to prevent the occurrence of faults (Kothari et al. 2022). Power electronics system components are individually responsible for the reliable and safe operation of the complete system. A single component failure may cause a total shutdown, downtime, and a huge maintenance cost. Therefore, throughout the lifetime, reliable operation is essential. The chapter covers the reliability evaluation and enhancement methods of the power electronics system and their fragile components like semiconductor switches, capacitors, and sensors. According to a survey, power electronics semiconductors are the most vulnerable among the other components (Sarita et al. 2023). In almost 31% of cases, the failure is due to some fault in semiconductors. In contrast, 60% of the failure in the semiconductors is due to thermal stress. The chance of failure of power semiconductor devices increases twice for every 10°C rise in working temperature.

Following the explanation above, semiconductor switches provide a significant role in the reliability of the electrical converter system. Various methodologies have been developed to address the challenges related to semiconductor reliability, such as remaining useful life estimation, active thermal control, and condition monitoring.

In a PV system, PV inverters efficiently convert DC voltage to AC voltage suitable for the load or grid. These PV inverters have the highest failure rate among all the other constituents of the system. Inverter failure for five years in the PV production plant was estimated to account for 37% of all unscheduled maintenance and 59% of the associated expenditures. Throughout the PV system's life, inverters must be replaced approximately three to five times.

Reliability Analysis of Modern Power Systems, First Edition. Edited by R. K. Saket and P. Sanjeevikumar.

In a wind generation system, generated power has high fluctuations. Power electronic converters regulate the power and maximize the generation from wind. Since even a single wind turbine failure significantly affects the grid. Generators have a lesser failure rate but higher downtime; on the other hand, power converters are prone to failure, and repair time is less than a wind generator (Kumar et al. 2021).

21.2 Failures in Power Electronics Converters

Any power electronics converter consists mainly of semiconductor switches, gate drivers, electrolytic capacitors, sensors, and microcontrollers. Semiconductor switches have the highest failure rate. The reason behind the higher failure rate is thermal degradation. Since semiconductor switches are made up of layers of different materials with different temperature coefficients. Two types of failures are observed in the converter: random and systematic. Random failures, such as hardware degradation and human error. Manufacturing processes, design modifications, or operational procedures are the only ways to eliminate systematic failures. Figure 21.1 provides the failure mechanism of a semiconductor switch.

21.2.1 Catastrophic Failure

Failure at a consistent rate is referred to as catastrophic failure. Long-term historical failure information collected from the use of a product in a similar operating environment can be used to estimate the failure rate. The failure rate for similar items can be taken from the handbook whenever historical data is unavailable. For electronic components, the military handbook MIL-HDBK-217F provides a generic failure rate. The handbook method does not consider the physics of failure;

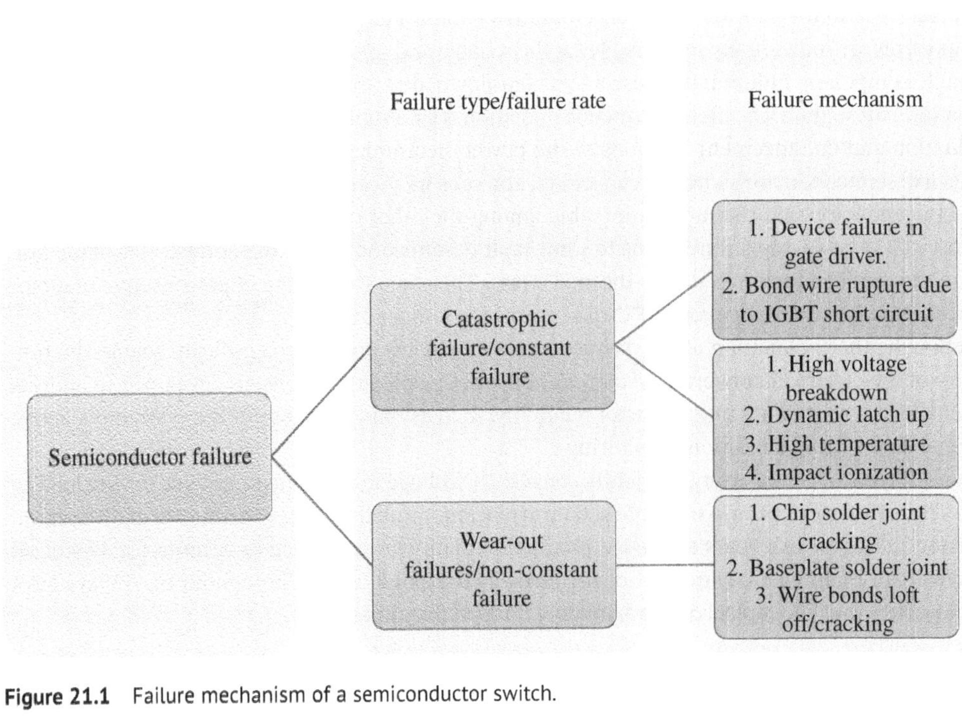

Figure 21.1 Failure mechanism of a semiconductor switch.

therefore, reliability may sometimes need to be revised. The FIDES approach was used, which considered the factor of physics of failure. Equations (21.1)–(21.5) can be used to forecast the failure rate (λ) using the FIDES technique:

$$\lambda = \Pi_{\text{pm}} \Pi_{\text{process}} \lambda_{\text{phy}} \tag{21.1}$$

$$\lambda_{\text{phy}} = \sum_{i=1}^{\text{phase}} \frac{t_{\text{annual}}}{8760} \Pi_i \lambda_i \tag{21.2}$$

where

Π_{pm} = factor for manufacturing quality impact
Π_{process} = factor for all processes such as operation and maintenance
λ_{phy} = Failure rate due to physical contributions
t_{annual} = the ith phase's duration in a year
Π_i = the ith phase's factor for induced electrical, mechanical, and thermal overstress
λ_i = Failure rate of ith phase

$$\Pi_i = \left(\Pi_{\text{placement}} \Pi_{\text{App}} \Pi_{\text{Rugg}}\right)^{0.511 \ln(c_g)} \tag{21.3}$$

$$\lambda_i = \sum_{1}^{k} \lambda_{0k} \Pi_k \tag{21.4}$$

where

$\Pi_{\text{placement}}$, Π_{App}, and Π_{Rugg} = constituent factors of Π_i
λ_{0k} = base failure rate
Π_k = factor physical constraints experienced during operation.

Using the above equations, we can predict the constant failure rate of the two most vulnerable components, semiconductor ($\lambda_{\text{phy\_scd}}$) and capacitor ($\lambda_{\text{phy\_cap}}$). Therefore, the total constant failure rate ($\lambda_{\text{C\_useful}}$)

$$\lambda_{\text{C\_useful}} = \sum \left(\lambda_{\text{phy\_scd}} + \lambda_{\text{phy\_cap}}\right) \tag{21.5}$$

21.2.2 Wear-Out Failure

Stress strength analysis is used to estimate wear-out failure. A solid grasp of the failure mechanics is required to estimate the strength or lifetime of the device connected to the failure mechanism. Based on the likelihood that the applied stress will exceed the device's strength, the probability of failure is estimated (Rani and Kumar 2022). Equation (21.6) has previously analyzed the lifespan model of the electrolytic capacitor (L_{cap}):

$$L_{\text{cap}} = L_0 2^{\frac{T_0 - T_{\text{cap}}}{n_1}} \left(\frac{V_{\text{cap}}}{V_0}\right)^{-n_2} \tag{21.6}$$

where L_0 denotes lifetime under voltage V_0, V_0 is nominal voltage, T_0 is nominal temperature, V_{cap} and T_{cap} are the operating voltage and temperature for the Capacitor lifetime L_{cap}, and n_1 and n_2 are constants. The Number cycle to failure (N_f) may be estimated as described in Eq. (21.7):

$$N_f = A \cdot \Delta T_j^{\alpha} e^{\left(\frac{\beta}{T_{jn}}\right)} \left(\frac{t_{\text{on}}}{1.5}\right)^{-0.3} \tag{21.7}$$

where the constants for curve fitting are A, α, and β, respectively, T_{jn} is junction temperature, ΔT_j is the deviation in junction temperature, and t_{on} is the rise time of the temperature cycle.

The wear-out failure rate estimation process evaluation method using a mission profile is prevalent (Miveh et al. 2015). The mission profile is translated into stress on the component. Dividing by the applied stress, the device's damage is calculated. This can be interpreted as how many operations must be completed before failure. Therefore, the damage distribution function for a mission profile may be predicted using Monte Carlo simulation. The failure rate and wear-out failure density function can be calculated using this technique.

The failure rate for wear-out ($\lambda_{C\_wearout}$) is provided by Eq. (21.8):

$$\lambda_{C\_wearout} = \sum(\lambda_{wearout_{scd}} + \lambda_{wearout_{cap}}) \tag{21.8}$$

Therefore, the total failure rate in a converter ($\lambda_{C\_converter}$) can be calculated and depicted as Eq. (21.9):

$$\lambda_{C\_converter} = \lambda_{C\_useful} + \lambda_{C\_wearout} \tag{21.9}$$

So, the converter's Total reliability $R(t)$ is calculated using Eq. (21.10):

$$R(t) = e^{-\int_0^t \lambda_{C\_converter}(\tau)d\tau} \tag{21.10}$$

21.3 Estimation and Monitoring of Junction Temperature

The wear-out failure mechanism is the most common in power semiconductor devices due to high thermo-mechanical stress during thermal and power cycles. Apart from this change in junction temperature, high-frequency operation of semiconductor switches leads to high temperatures in the nearby areas. Therefore, a better understanding of junction temperature helps design the converter to ensure reliable operation. The junction temperature information helps predict the remaining useful life and assists in developing thermal management algorithms for optimal loss management of IGBT modules during degraded operation (Choi et al. 2016). Temperature-dependent electrical parameters (TDEPs), such as infrared or thermocouples, can indirectly be used to determine junction temperature. The junction temperature measurement by the direct approach is only possible with some changes in package modification.

21.3.1 Direct Method

Modern-day semiconductor packaging is done so that direct contact with the thermocouple, thermistor, and sensors is impossible. Direct methods can be used if direct contact with the switches is possible. The coating material determines the spatial resolution of a thermocouple probe, and the thermal capacitance of the probes regulates the response time. In laboratory testing-based converters having IGBT ICs with exposed legs, a direct method can be used. However, indirect electrical methods for junction temperature estimation in commercial compact converters with modules are preferred.

21.3.2 Indirect Method

Electrical methods are the most widely used methods to determine junction temperature. The monitoring of electrical parameters is more accessible than the physical direct methods due to practicability and fast response time. Continuous monitoring of electrical parameters on which junction temperature depends is an excellent method to estimate junction temperature.

21.3.2.1 Estimation by $V_{ceon(Ihigh)}$

The $V_{\text{ceon(Ihigh)}}$ is nonlinearly related to the junction temperature and collector current. For error within an acceptable bound limit, the relationship based on electro-thermal modeling is approximated by Eq. (21.11):

$$V_{\text{ceon(Ihigh)}} = V_{\text{ceo}} + \alpha_{\text{vceo}}\left(T_j - T_{\text{jo}}\right) + I_{\text{c}}\left[r_{\text{o}} + \beta_{\text{ro}}\left(T_j - T_{\text{jo}}\right)\right] - \gamma_{\text{ge}}\Delta V_{\text{ge}} \tag{21.11}$$

where

$V_{\text{ceon(Ihigh)}}$ = collector-emitter voltage under high current
V_{ceo} = collector-emitter voltage at T_{jo}
T_{jo} = Base junction temperature
α_{vceo} = Temperature coefficient of V_{ce}
I_{C} = Collector current
r_0 = Series resistance at T_{jo}
β_{ro} = Temperature coefficient of *ro*
γ_{ge} = Temperature coefficient of ΔV_{ge}
ΔV_{ge} = Gate emitter voltage variation

During normal operation, V_{ge} barely varies. Therefore, ΔV_{ge} may be ignored in the equation. Considering the series module resistance, T_j is represented in Eq. (21.12):

$$T_j = \frac{V_{\text{ceon(Ihigh)}} - V_{\text{ceo}} - I_{\text{C}}\left[r_{\text{o}} + \beta_{\text{ro}}\left(T_j - T_{\text{jo}}\right)\right]}{\alpha_{\text{vceo}}} + T_{\text{jo}} \tag{21.12}$$

Assuming temperature distribution is homogeneous, the estimated junction temperature (T_{jest}) can be expressed as Eq. (21.13):

$$T_{\text{jest}} = \frac{V_{\text{ceon(Ihigh)}} - V_{\text{ceo}} - I_{\text{C}}r_{\text{o}}}{\alpha_{\text{vceo}}} + T_{\text{jo}} \tag{21.13}$$

Therefore, the estimated junction temperature using $V_{\text{ceon(Ihigh)}}$ is highly sensitive to the temperature coefficients α_{vceo}, β_{ro}, and γ_{ge}. In one of the studies, the sensitivity range is from 1 to 5 mV/°C for 300–1000 A, respectively.

This measurement method provides an option to measure junction temperature in each switching period. Many researchers came up with different measurement configurations to estimate the $V_{\text{ceon(Ihigh)}}$ of semiconductor devices (Falck et al. 2018; Lelis et al. 2014). However, continuous high ohmic drop across modules at high currents is a big disadvantage and leads to significant errors in temperature measurement. Moreover, V_{ceon} is heavily affected by the aging of the power module. As the device ages, correlation factors need to be updated occasionally. A circuit to measure V_{ceon} is shown in Figure 21.2.

21.3.2.2 Estimation by Threshold Voltage (V_{thr})

It is suggested by the researchers if the TDEP parameters have a linear dependency on the temperature, it becomes a favorable condition for temperature estimation. When the device is turned on, the gate emitter voltage is equivalent to the threshold voltage. The authors explained that the threshold voltage highly depends upon the junction temperature because parameters like carrier concentration and oxide layer charges depend on the manufacturing process and remain fixed. It is described in Eq. (21.14):

$$V_{\text{thr}}\left(T_j\right) = \frac{k}{q}\ln\left\{\frac{N_{\text{A}}}{N_{\text{D}}}\right\}\cdot T_j - \frac{Q_{\text{f}} + Q_{\text{m}} + Q_{\text{ot}}}{C_{\text{ox}}} + \frac{\sqrt{4\cdot\varepsilon_{\text{S}}N_{\text{A}}\cdot k}}{C_{\text{ox}}}\cdot\sqrt{\ln\left\{\frac{N_{\text{A}}}{n_i T_j}\right\}\cdot T_j} \tag{21.14}$$

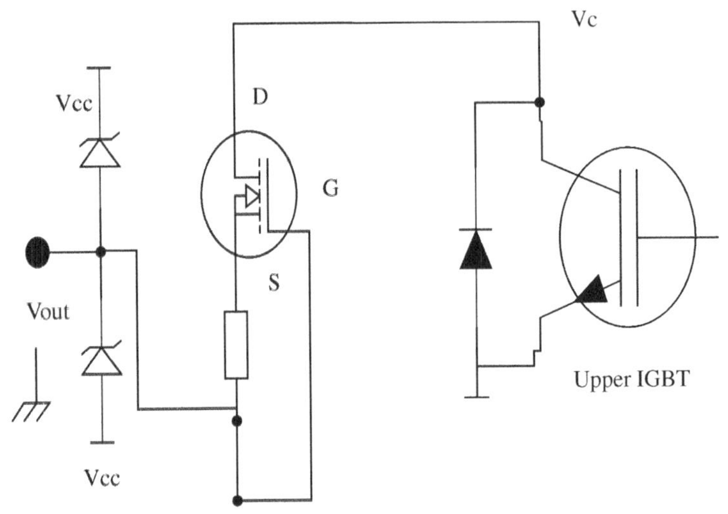

Figure 21.2 Circuit for measurement of V_{ceon}.

where k = Boltzmann constant, q = elementary charge, N_A = acceptors concentration, N_D denotes donors concentration, C_{ox} = oxide layer capacitance, and Q_f, Q_m, and Q_{ot} are the fixed oxide charge, mobile ion charge, and intrinsic charge of oxide, respectively. ε_s represents silicon permittivity. For continuous monitoring of V_{thr}, the voltage developed in the parasitic inductance due to the turn-on current is compared with the reference voltage. The choice of the reference voltage is crucial and should be done carefully, as the transient present in the turn-on process is affected by the junction temperature. The pulse generated by the comparator is used to create the measuring pulse with the help of the NAND gate and driver output voltage. The measuring pulse is applied to held V_{thr} by sample and hold circuit, as shown in Figure 21.3. Based on temperature sensitivity, the V_{thr} monitoring method is better than the V_{ceon} because the external gate resistance is less than the internal gate resistance. Also, lesser resistance from the internal gate side may lead to the induction of synchronization jitter, which may give a significant error.

The V_{thr} calculation can be incorporated in the gate-driver circuit, as the reference of the ground for both circuits is standard. From the graph, it can be concluded that temperature-dependent voltage variation is independent of the collector current for a limited range.

21.3.2.3 Estimation by Short Circuit Current (SCC)

A relation between temperature and short circuit current (SCC) must be established to measure the junction temperature by the SCC method, independent of the gate voltage. A short circuit and an additional bypass IGBT switch are developed to achieve this. This switch is turned on for short intervals when the switch under test (SUT) is in conducting mode. This arrangement requires an additional protection circuit, which increases the complexity of the circuit. The SCC of IGBT is calibrated in terms of temperature, and a curve is plotted between SCC and temperature. This curve gives the junction temperature for a particular SCC. The IGBT switch will experience high voltage and high current stress concurrently throughout the test, so assessing its short-circuit handling capacity is crucial before the test. The curve in Figure 21.4 (SCC decreases with temperature rise) shows that the SCC exhibits a negative temperature coefficient.

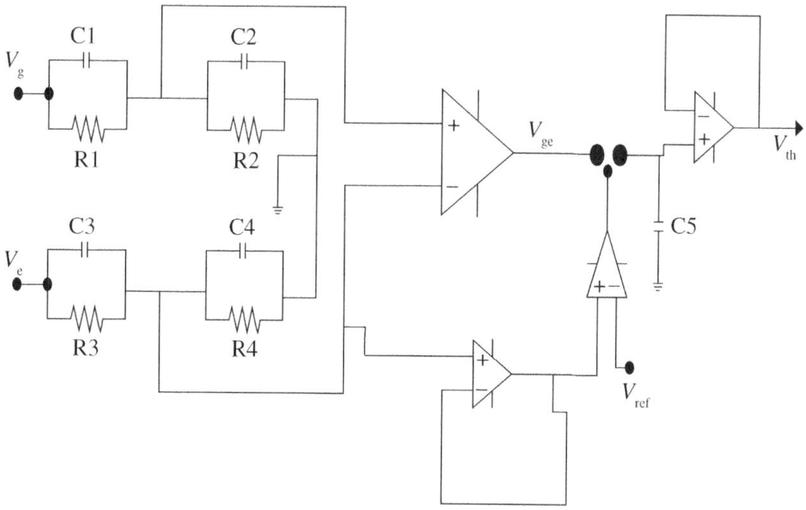

Figure 21.3 Circuit to measure V_{thr}.

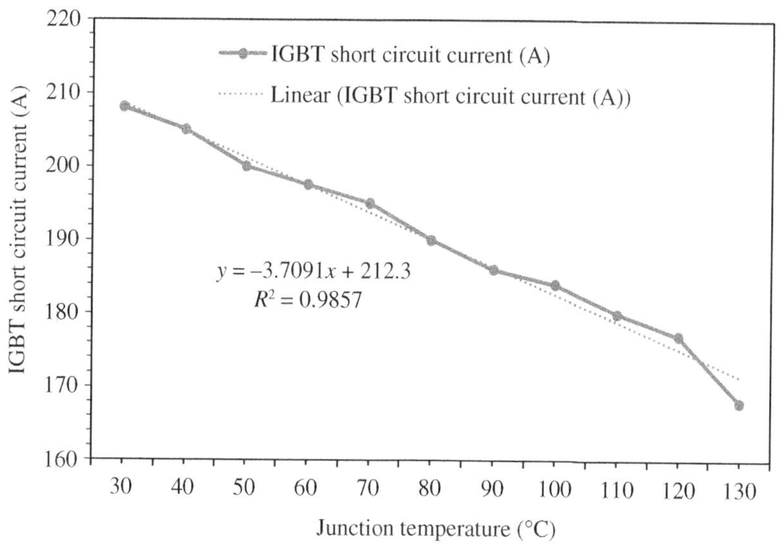

Figure 21.4 Short circuit current versus junction temperature.

21.3.2.4 Estimation by Turn-On/Off Delay Time

T_j can be estimated by turn-on delay time t_{don} also. The switch's t_{don} depends upon the external and internal gate resistance, gate capacitance, and threshold voltage, as Eq. (21.15) describes. The capacitance (C) includes depletion capacitance and oxide capacitance and R_{Gtotal} the external and internal gate resistance

$$t_{don} = \left(R_{Gtotal}\right) \cdot C \cdot \ln\left(1 - \frac{V_{th}}{\Delta U}\right) \tag{21.15}$$

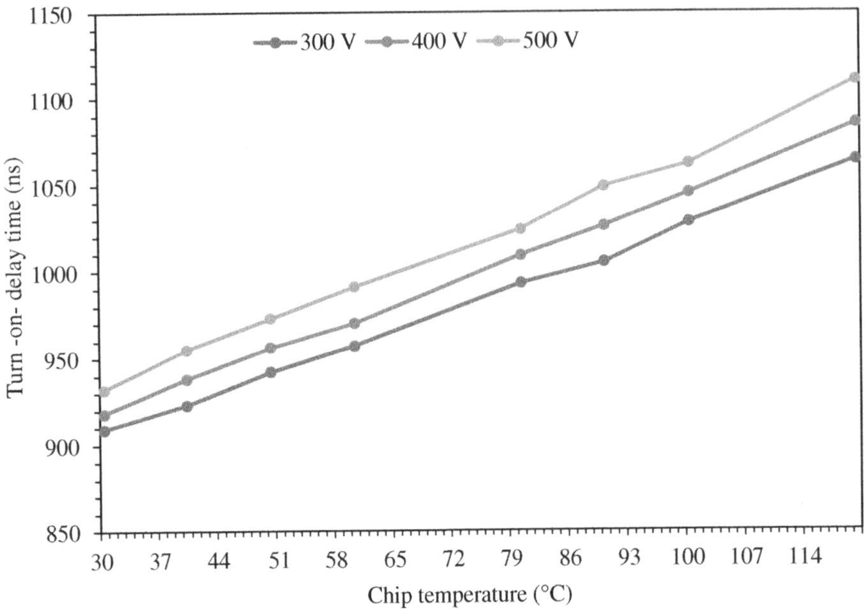

Figure 21.5 t_{don} versus chip temperature.

Depletion capacitance is highly governed by the carrier concentrations, which are the function of temperature. As the temperature increases, carrier concentration increases, which leads to an increase in the depletion capacitance. A turn-on delay time has a linear relationship with the overall gate capacitance, and as temperature increases, t_{don} will also increase. To estimate t_{don}, an additional timing-counter circuit is used, and a plot between the t_{don} and T_j is depicted. In this method, t_{don} is measured when the current in the switch is zero. The plot between t_{don} and T_j is given in Figure 21.5, which also explains that the t_{don} decreases as the V_{dc} increases and is independent of almost independent of the current in the switch (Luo et al. 2015). The method is more suitable compared to other current dependent methods, and the sensitivity of this method follows in the range of 1–2 ns per degree centigrade. For high current injection, temperature measurement of the SUT can be done by $V_{ceon(Ihigh)}$, which has good temperature sensitivity in the range of 0.5–3.5 mV/°C. This temperature sensitivity varies according to the load current. But for a lower range of current, this method is not suitable. However, the turn-off delay t_{doff} approach has a relatively constant temperature sensitivity, approximately 4 ns/ °C. Therefore, the t_{doff} method is appropriate for the lower current range but cannot be employed for the power diodes (Luo et al. 2015).

The plot between junction temperature and turn-off delay is depicted in Figure 21.6. It can be concluded from the plot that turn-off delay time monotonically rises with a rise in junction temperature. Also, the junction temperature for a particular DC link voltage increases for the specific turn-off time if the load current increases. An increase in DC link voltage also shifts the curve upwards, which means that for particular junction temperatures, turn-off delay time increases with an increase in DC link voltage.

21.3.2.5 Estimation by R_{gint} with the Help of I_{gpeak}

In this method, the temperature measurement of the internal thermo-sensitive gate internal resistance (R_{gint}) is positioned in the module's center. In this method, the Junction temperature can be

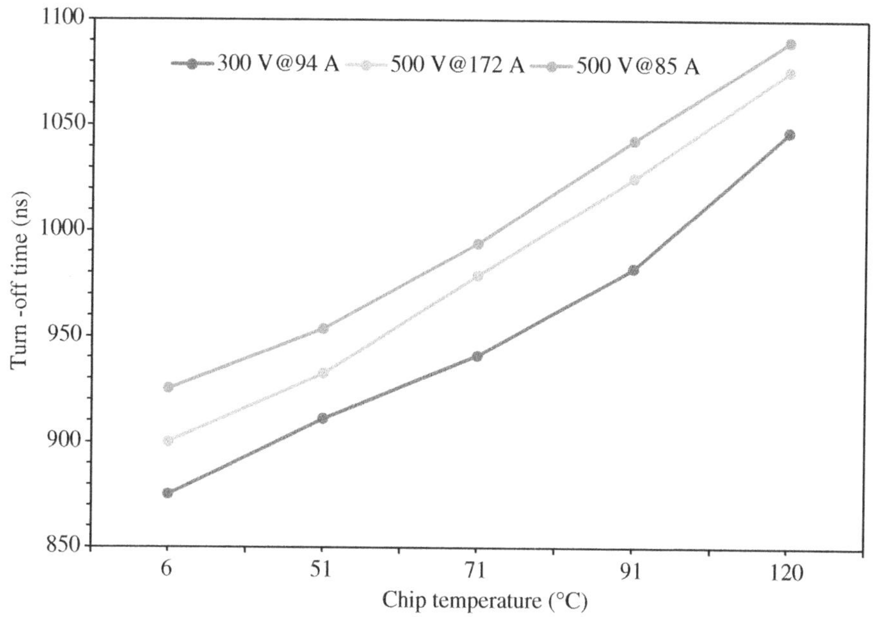

Figure 21.6 t_{doff} versus chip temperature.

measured by measuring the temperature of the internal gate resistor in the middle of the module. This R_{gint} is thermos-sensitive. During the running conditions, it is nearly impossible to open the power module and measure the temperature. To eliminate this problem, a researcher proposed a method to measure the gate current variation because of a thermos-sensitive internal gate resistance change by measuring the voltage drop across external gate resistance (R_{gext}). R-L-C network for gate drive circuit is used (Baker et al. 2015), which operates in overdamped condition. Therefore, gate current (I_g) is estimated by Eq. (21.16).

$$I_g = \frac{V}{R_{Total}} e^{\frac{-t}{R_{total}C}} \tag{21.16}$$

where $R_{total} = R_{gext} + R_{gint}$. This study assumes that the changing temperature does not affect external resistance; therefore, R_{gint} is calculated using Eq. (21.17).

$$R_{gint} = \frac{V_{gneg} - V_{gpos}}{\frac{V_{peak}}{R_{gext}}} - R_{gext} \tag{21.17}$$

T_j can be estimated based on calibration. The experiment conducted by the researcher obtained a temperature sensitivity of +3 mΩ/°C, which gives a linear relation between the external temperature and internal resistance, as shown in Figure 21.7a,b. The main advantage of this method is that there is no loading effect on the measurement circuit, which is an integral part of the gate driver. However, errors may appear due to different assumptions, and the aging of the module correction factor needs to be calibrated from time to time. Other than these above-discussed methods, there are various parameters like rate of change of collector current (dI_C/dt), Forward voltage drop, Voltage between Kelvin emitter and power emitter, Miller plateau voltage, Saturation current, etc., which can be monitored continuously to estimate junction temperature.

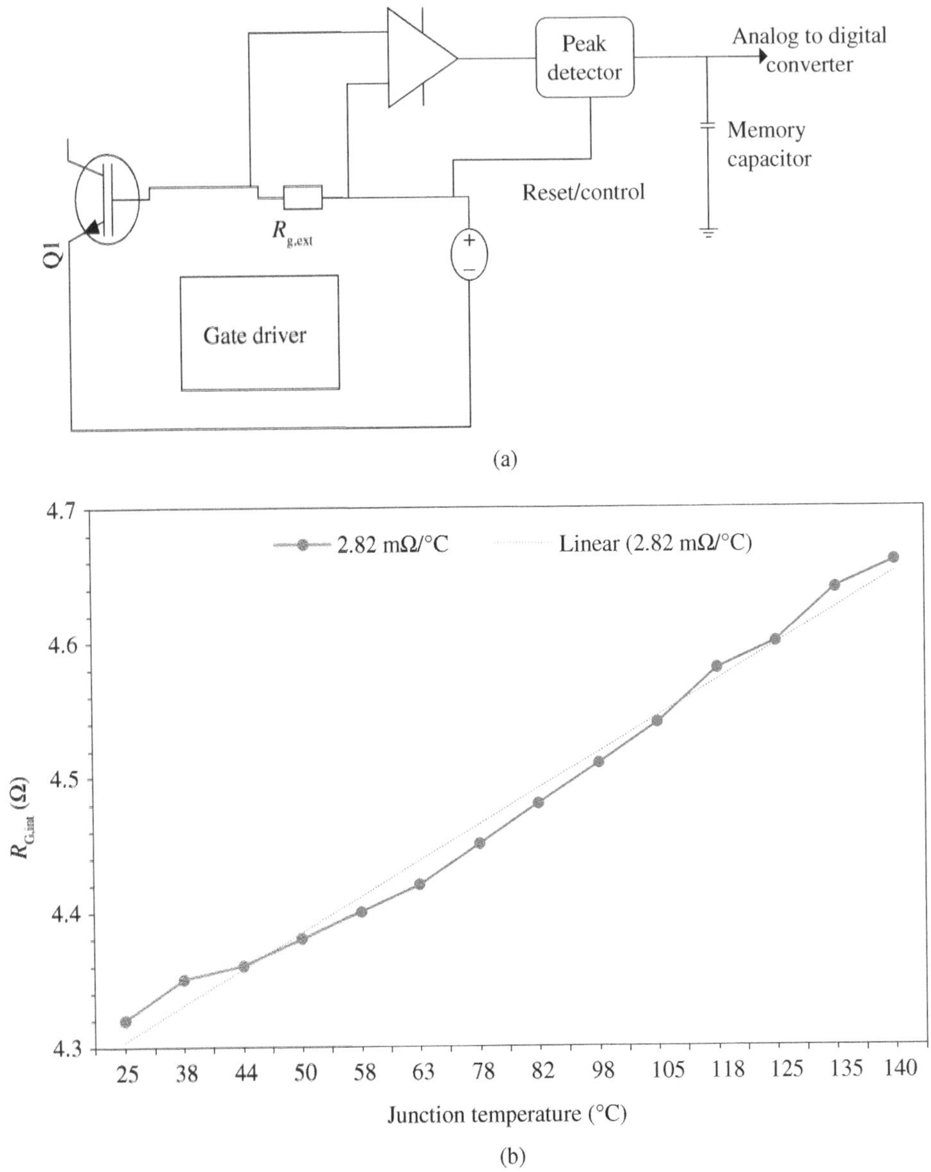

Figure 21.7 (a) Circuit for detecting the gate peak voltage and (b) R_g versus T_j plot.

21.4 Reliability of a Modern Power System

The uninterrupted ability of the MPS to satisfy its electrical power demand is a measure of its reliability. The availability of the component is considered an important parameter to measure reliability. Availability is the probability that any item operates properly at any instant of time "t" after starting at $t = 0$. Failure rates may be constant or time-varying. Markov's principle is used to obtain the availability of the exponentially distributed system. Availability is obtained as stated in Eq. (21.18):

$$A = 1 - U = 1 - \frac{\mu}{\lambda + \mu} \tag{21.18}$$

where
A = Availability
U = Unavailability
λ = Failure rate
μ = Repair rate.

According to Markov's principle, availability (A) is the probability of being in the operating state "1." "2" is downstate in the state space model. Markov's principle cannot be used for a system with nonexponential failures. In such cases, the failure rate is separated into two failure rates, which are constant and variable ($\lambda(t) = \lambda_{\text{constant}} + \lambda_{\text{wear-out}}(t)$) as presented in Figure 21.8a,b. An inconsistent operation or impulse overstressing might lead to the random failure of converter components like switches and capacitors. In contrast, wear-out failures happen as a result of long-term deterioration. Therefore, these two failures can be split into two different failure rates. As Eq. (21.19) describes, total availability can be calculated:

$$A_{\text{total}}(t) = A_{\text{constant}} \cdot A_{\text{wearout}}(t) \tag{21.19}$$

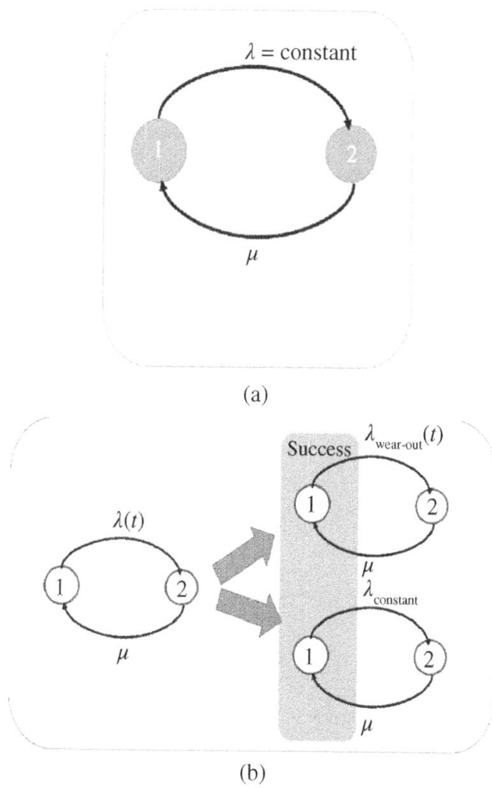

Figure 21.8 Markov's models. (a) For constant failure rate. (b) For time-varying failure rate.

where
$A_{total}(t)$ = Availability of total system
$A_{constant}(t)$ = Availability associated with time-constant failure rate
$A_{wearout}(t)$ = Availability associated with time-varying failure rate

Semi-Markov's process can be used to calculate the availability of time-varying failure systems.

21.4.1 Power Converter Availability Model

The converter's reliability can be modeled using its component's reliability. The failure rate of capacitor and switches are very high. Series relation and Markov's model of the converter are shown in Figure 21.9. When all the components are available, the availability can be obtained using Eq. (21.20) (Falck et al. 2018):

$$A_{converter}(t) = A_{switch}(t) \cdot A_{capacitor}(t) \cdot A_{oc} \tag{21.20}$$

where
$A_{converter}(t)$ = Availability of the converter
$A_{switch}(t)$ = Availability of semiconductor switches
A_{OC} = Availability of other components of converter.

21.4.2 HVDC Availability Model

A sending end converter, a DC transmission line, and a receiving end resistor are the three major components of the HVDC system. Modeling the system as a series network is possible, as seen in Figure 21.10. Equation (21.21) can be used to determine the availability of the HVDC system $A_{HVDC}(t)$:

$$A_{HVDC}(t) = A_{SEC}(t) \cdot A_{REC}(t) \cdot A_{DC} \tag{21.21}$$

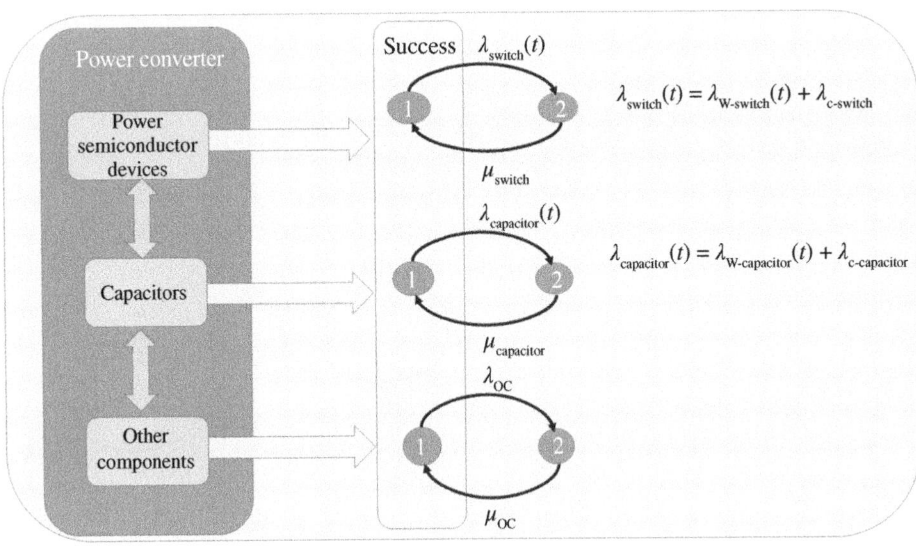

Figure 21.9 Power converter availability block diagram with Markov's model.

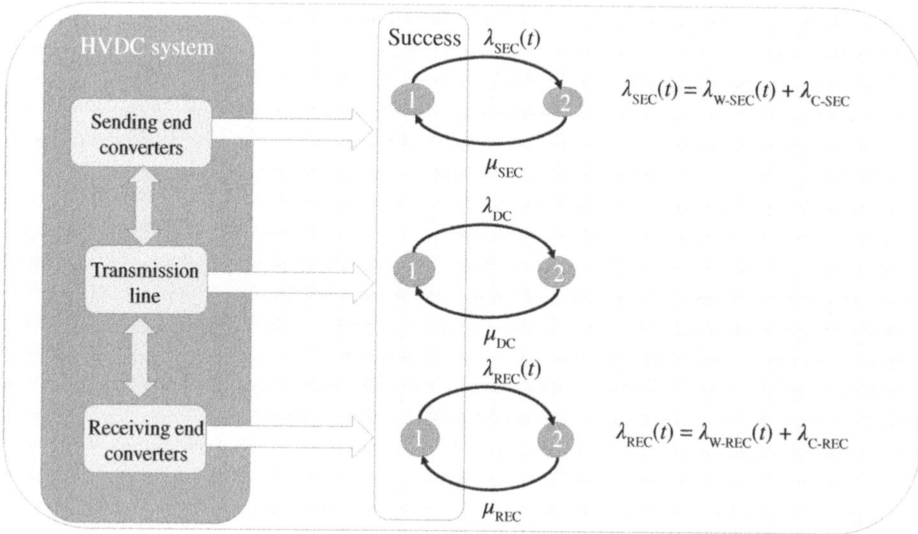

Figure 21.10 HVDC system availability block diagram with Markov's model.

where
$A_{SEC}(t)$ = Availability of sending end converter
$A_{REC}(t)$ = Availability of receiving end converter
A_{DC} = Availability of DC transmission line.

21.4.3 Reliability Model of Modern Power System

MPS reliability is measured with the help of probabilistic indices. LOLE is the number of days per hours in a specified time when power cannot be supplied due to energy shortage, which is represented as Eq. (21.22):

$$\text{LOLE} = \sum_{x=1}^{n} P_x \cdot (C_x - L_x) \tag{21.22}$$

where
C_x = Available capacity
L_x = Peak load forecast
P_x = Probability of loss of load

EENS = curtailed energy (E_x) because of generation shortage, calculated as Eq. (21.23):

$$\text{EENS} = \sum_{x=1}^{n} P_x \cdot E_x \tag{21.23}$$

A wear-out failure rate of the wind turbine converter is estimated for the wind farm power system reliability model under a specific mission profile. Then, the wind turbine converter's availability is evaluated using random and wear-out failure rates. Availability of the wind turbine system is evaluated using components and power availability. The availability of wind farm HVDC can be estimated using historical data and failure rates. Combining the models for the transmission lines and the wind farm's turbine system yields the wind farm reliability model. Similarly, the availability of the HVDC transmission line and conventional generators can also be estimated using converter chance failure and individual availability, respectively. The combined reliability is shown

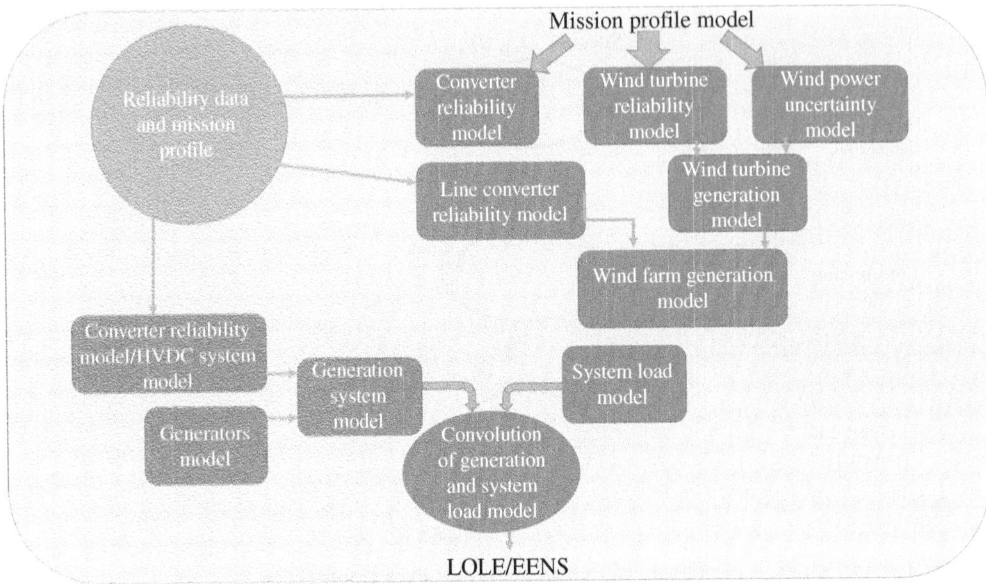

Figure 21.11 Reliability evaluation in MPS.

in Figure 21.11. The generation system reliability and load can be convolved to produce power system reliability indices like LOLE and EENS.

21.5 Challenges and Future Directions

The chapter has discussed the types of failures (constant and time-varying) common in a power electronics-based system. Temperature effect on the performance of switches is crucial. Therefore, continuously monitoring temperature-dependent parameters is an excellent option to increase system reliability. The reliability modeling of a MPS with the help of Markov's model has been discussed.

Condition monitoring of TDEPs is a low-cost and affordable solution for reliability enhancement. Still, it is necessary to distinguish temperature-dependent from damage-dependent electrical parameters. A more acute study of the physics of failure is needed to accomplish this goal. Further advancement of condition monitoring methods at the converter level is required to identify more failure indices based on power system output parameters. The model-based lifespan estimation approach is preferred over the data-driven based model due to its simplicity and accuracy. Real-time monitoring for power electronics converter systems will be effective for better mission profile data, ultimately enhancing the system's availability and reliability.

References

Baker, N., Munk-Nielsen, S., Iannuzzo, F., and Liserre, M. (2015). IGBT junction temperature measurement via peak gate current. *IEEE Transactions on Power Electronics* 31 (5): 3784–3793.

Choi, U.-M., Blaabjerg, F., Jørgensen, S. et al. (2016). Reliability improvement of power converters by means of condition monitoring of IGBT modules. *IEEE Transactions on Power Electronics* 32 (10): 7990–7997.

Falck, J., Felgemacher, C., Rojko, A. et al. (2018). Reliability of power electronic systems: an industry perspective. *IEEE Industrial Electronics Magazine* 12 (2): 24–35.

Kothari, D.P., Nagrath, I.J., and Saket, R.K. (2022). *Modern Power System Analysis*. Tata McGraw-Hill.

Kumar, S., Saket, R.K., Dheer, D.K. et al. (2020). Reliability enhancement of electrical power system including impacts of renewable energy sources: a comprehensive review. *IET Generation, Transmission & Distribution* 14 (10): 1799–1815.

Kumar, S., Sarita, K., Saket, R.K. et al. (2021). Reliability assessment for DFIG-based WECS considering the impact of 3-phase fault and lightning impulse voltage. *International Transactions on Electrical Energy Systems* 31 (8): e12952.

Lelis, A.J., Green, R., Habersat, D.B., and El, M. (2014). Basic mechanisms of threshold-voltage instability and implications for reliability testing of SiC MOSFETs. *IEEE Transactions on Electron Devices* 62 (2): 316–323.

Luo, H., Chen, Y., Sun, P. et al. (2015). Junction temperature extraction approach with turn-off delay time for high-voltage high-power IGBT modules. *IEEE Transactions on Power Electronics* 31 (7): 5122–5132.

Miveh, M.R., Rahmat, M.F., Ghadimi, A.A., and Mustafa, M.W. (2015). Power quality improvement in autonomous microgrids using multi-functional voltage source inverters: a comprehensive review. *Journal of Power Electronics* 15 (4): 1054–1065.

Rani, P. and Kumar, S. (2022). Probabilistic analysis of wind integrated power system considering 3-phase line fault. In: *2022 IEEE 10th Power India International Conference (PIICON)*, 1–6. IEEE.

Sarita, K., Saket, R.K., and Khan, B. (2023). Reliability, availability, and condition monitoring of inverters of grid-connected solar photovoltaic systems. *IET Renewable Power Generation* 17 (7): 1635–1653.

Singh, S., Saket, R.K., and Khan, B. (2023). A comprehensive state-of-the-art review on reliability assessment and charging methodologies of grid-integrated electric vehicles. *IET Electrical Systems in Transportation* 13 (1): e12073.

22

Reliability Assessment of Sub-components of Electric Vehicle for Performance Enhancement Grid Integrated Power System

Saumya Singh[1], *Dhawal Dwivedi*[1], *Sandeep K. Soni*[2], *R. K. Saket*[1], *and Dwarkadas P. Kothari*[3]

[1] *Department of Electrical Engineering, Indian Institute of Technology (BHU), Varanasi, Uttar Pradesh, India*
[2] *Department of Electrical Engineering, Shiv Nadar Institution of Eminence, Delhi, NCR, India*
[3] *Department of Electrical Engineering, Indian Institute of Technology, Delhi, India*

22.1 Introduction

EVs and the demand for sustainable mobility are transforming the global automotive sector. The broad adoption of EVs is essential to reducing greenhouse gas emissions and combating climate change. EV zero-tailpipe pollution and energy efficiency will change urban and long-distance travel. The success of EVs is not solely dependent on their electrification but also on the seamless integration of these vehicles into existing power systems. Grid integration represents a crucial milestone in the EV revolution, as it allows for efficient charging infrastructure, vehicle-to-grid (V2G) capabilities, and optimal power management to improve overall system performance. Achieving a harmonious coexistence between EVs and the electric grid requires a comprehensive understanding of the reliability of their sub-components. This chapter serves as a comprehensive guide for researchers, engineers, and policymakers seeking to understand the multifaceted aspects of reliability assessment and performance enhancement in the context of EVs and their integration into grid-connected power systems. Through synthesizing theoretical insights and practical applications, we aim to contribute to the evolving landscape of sustainable transportation and its pivotal role in shaping our future. When individuals are in the process of acquiring a vehicle, their primary concerns often revolve around the vehicle's reliability, safety, cost, and ease of maintenance. Plug-in electric vehicles (PEVs) are engineered with many electrical components and systems, including batteries, motor drives, controllers, and energy management systems. These intricate electrical systems inherently possess a higher susceptibility to failure. Consequently, PEVs typically exhibit lower reliability than mechanically driven internal combustion engine-based vehicle systems.

The subsequent sections of this chapter are organized as follows: Section 22.2 presents a comprehensive discussion on exploring and analyzing various sub-components within EVs. Section 22.3 includes charging infrastructure, various types of drive motors used in EVs, reliability studies of sub-components in EVs, methodologies of reliability studies, load management, and the benefits and challenges of grid integration. Section 22.4 exhibits reliability assessment techniques in EVs and battery management systems (BMS). Section 22.5 explains reliability parameters, distribution network (DN) reliability assessment, reliability evaluation using a V2G approach and high EV penetration. Finally, Section 22.6 brings the chapter to a conclusion.

Reliability Analysis of Modern Power Systems, First Edition. Edited by R. K. Saket and P. Sanjeevikumar.

22.2 Electric Vehicles and Grid Integration

Grid integration for electric cars integrates them into their existing electrical systems. To maintain an uninterrupted and steady power supply, regulating EV's impact on the grid is crucial as more people embrace them. EVs are becoming integral components of modern power systems, pivotal in the transition toward more sustainable transportation and energy consumption. This chapter explores the intricate interplay between EVs and power systems, focusing on the grid integration aspects from a power system perspective. The following are some EV grid integration aspects described in terms of the power system (Kothari et al. 2022).

Future Solutions for Grid Integration

a. **Vehicle-to-Grid (V2G) Technology**: Two-way power transmission between EVs and the grid is made possible by V2G technology. EVs have the ability to release electricity back to the grid during peak demand, aiding in the reduction of strain on the grid. V2G can also provide a source of distributed energy storage to enhance grid resilience. EVs can use V2G to draw electricity from the grid and feed surplus energy back into it, serving as distributed energy resources to improve grid stability, as discussed in Singh et al. (2023b).
b. **Grid Modernization**: The ongoing modernization of electrical grids with advanced sensors, automation, and communication technologies will enable more efficient load management and better grid stability.
c. **Renewable Energy Integration**: By synchronizing the charging of EVs with the availability of renewable energy sources (RES), such as solar and wind power, we can achieve two significant advantages:
   (i) **Reduced Environmental Impact**: When EVs are charged using clean, renewable energy (RE), their carbon emissions are greatly diminished. This means that the electricity powering these vehicles comes from eco-friendly sources, resulting in a lower overall environmental footprint for EV usage.
   (ii) **Enhanced Grid Efficiency**: Coordinating the charging of EVs with periods of high RE production helps maintain a well-balanced electrical grid. During times when wind or solar power generation is abundant, we can encourage more EV charging, thus relieving strain on the grid and ensuring efficient utilization of RES.
   Coordinating EV charging with RE generation, such as wind and solar, can reduce the carbon footprint of EV charging while optimizing grid usage. Integrating EVs with RES, such as solar and wind, can enhance the sustainability of the transportation sector. EV charging can be encouraged during high RE generation periods, promoting a greener and more balanced power system.
d. **Interconnected Networks and Distribution System Planning**: Grid operators are exploring ways to share load and resources between neighboring grids to mitigate the localized impact of EV charging.
   EV adoption estimates must be considered by utilities and grid operators when developing their long-term distribution systems. As a result of the aforementioned, the grid is guaranteed that it will be able to support both user's overall battery-powered cars and regular power users with the same level of efficiency.
e. **Data and Communication Systems**: Successful management of EV grid integration depends on maintaining reliable data and communication systems. Grid operators can make informed judgments and optimize charging schedules using real-time data on EV charging routine, grid load, and RE availability.

Integrating EVs into the electrical grid is a complex task that requires careful planning, investment, and innovative solutions. Load management and grid stability are critical aspects that must be addressed to ensure a smooth transition to a cleaner and more sustainable transportation system. Collaboration among government agencies, utilities, technology providers, and EV manufacturers is essential to successfully manage the challenges and opportunities presented by the electrification of transportation (Das et al. 2020).

22.2.1 Charging Infrastructure

EV grid integration involves establishing charging infrastructure in several locations to meet EV charging demands. This infrastructure can include public charging stations, fast chargers, home chargers, and workplace chargers. Charging infrastructure plays a pivotal role in successfully integrating EVs into our transportation ecosystem. As EVs continue to gain popularity and become a significant part of the global vehicle fleet, developing an extensive and reliable charging network is crucial for promoting widespread adoption and addressing concerns related to range anxiety. In this discussion, we will delve into the various aspects of charging infrastructure, its importance, and the challenges as discussed in Yilmaz and Krein (2012).

Types of Charging Infrastructure: Charging infrastructure encompasses a range of technologies and charging speeds to cater to the diverse needs of EV users. The following levels of charging comprise the main categories of charging infrastructure:

- **Level 1**: This level is the slowest charging option, using a standard 120-V household outlet. Level 1 chargers are suitable for overnight charging at home and provide an average of 2–5 miles of range/h.
- **Level 2**: This charging level uses a 240-V power source and is commonly found in residential and commercial settings. They offer faster charging speeds, providing around 10–30 miles of range per hour of charging.
- **Level 3 (DC Fast Charging)**: DC fast chargers are high-power chargers that can deliver a substantial amount of energy quickly, providing up to 80% charge in 30–60 minutes. These are typically located along highways and major travel routes.

  Importance of Charging Infrastructure
  - **Range Confidence**: An extensive charging network helps alleviate range anxiety, a common concern among potential EV buyers. Knowing that charging stations (CS) are readily available enhances driver confidence in the viability of EVs for long trips.
  - **Accessibility**: Charging infrastructure is essential to ensure that EVs can be conveniently charged at home, work, or on the road. This accessibility is essential for EV owners who may not have access to home charging.
  - **Urban Mobility**: In urban areas, public charging infrastructure is critical for residents who may not have private garages or dedicated parking spots. It encourages the use of EVs in densely populated regions.
  - **Reduced Carbon Emissions**: Promoting EV adoption through charging infrastructure reduces greenhouse gas emissions from the transportation sector, thereby addressing environmental concerns.

22.2.2 Load Management and Grid Stability

EVs increase grid demand, especially during peak charging. Load management must balance this additional load without overloading the grid. Intelligent charging, time-of-use (TOU) rates, and demand response programs may encourage off-peak charging. EV adoption is on the rise worldwide, and while this transition to cleaner transportation is essential for reducing greenhouse gas emissions, it also challenges the stability and management of electrical grids. The variable and sometimes unpredictable nature of EV charging can impact the grid's stability. Localized voltage fluctuations and power quality concerns can result from rapidly charging several EVs in a crowded location. Grid operators must regulate voltage and power factors to preserve grid stability.

Load management and grid stability are critical to integrating EVs into the existing power infrastructure. This discussion will explore the implications, strategies, and technologies involved in managing EV loads and ensuring grid stability.

Impact of EV Charging on Grids

- **Increased Demand**: Electricity demand will increase as more EVs are put on the road, particularly during peak charging hours. This can strain local grids and lead to overloads.
- **Load Distribution**: The distribution of EVs across regions and neighborhoods can vary significantly, creating localized demand spikes that challenge grid operators.
- **Charging Speeds**: The speed at which EVs charge, from slow Level 1 charging to fast DC charging, affects the grid differently. High-speed chargers draw a substantial amount of power quickly, requiring careful management.

Load Management Strategies

- **Time-of-Use (TOU) Pricing**: TOU pricing serves as an incentive for EV owners to choose off-peak hours for charging when electricity rates are more affordable, ultimately helping to relieve grid congestion during peak demand periods.
- **Demand Response Programs**: Grid operators can communicate with EVs to temporarily reduce or delay charging during grid stress events, ensuring grid stability.
- **Smart Charging**: The use of smart charging systems empowers EVs to communicate with the grid, making real-time adjustments to their charging rates based on grid conditions, the availability of RES, and user-defined preferences.
- **Fleet Charging Optimization**: For commercial fleets of EVs, advanced software can optimize charging schedules to minimize operational disruptions and reduce peak load impact.

Grid Stability Considerations

1. **Voltage and Frequency Control**: Rapid and uncontrolled EV charging can disrupt grid voltage and frequency. Grid operators need advanced monitoring and control systems to maintain stability.
2. **Distribution System Upgrades**: In some cases, distribution systems may require upgrades to handle the additional load from EVs. This includes transformers, substations, and grid infrastructure improvements.
3. **Energy Storage Integration**: The deployment of energy storage systems (ESS), such as large-scale batteries, can absorb excess energy during low-demand periods and discharge it when needed, helping stabilize the grid.

22.2.3 Benefits and Challenges of Grid Integration

The development of a reliable and consistent charging infrastructure for EVs is made possible through grid integration. The important features that will enhance grid integration are efficient charging infrastructure, V2G capabilities, enhanced RE integration, demand response and load flexibility, grid support during emergencies, etc.

Some of the common challenges of grid integration are:

- Grid Capacity and Infrastructure Upgrades,
- Peak Demand Management,
- Battery Degradation and Lifetime,
- Interoperability and Communication Standards,
- Regulatory and Policy Frameworks, and
- Cybersecurity Concerns.

Addressing the above challenges and reaping the benefits of grid integration will be critical in promoting the effective integration of EVs into the larger energy ecosystem, thereby contributing to a more sustainable and electrified transportation future.

Challenges in Charging Infrastructure Development

Cost: Establishing a comprehensive charging network requires significant investment in infrastructure development, including installation, maintenance, and grid upgrades.

Grid Capacity: Rapid adoption of EVs can strain the electrical grid, necessitating upgrades to accommodate the increased demand for electricity during peak charging times.

Standardization: Ensuring compatibility between different EV models and charging stations is essential. Establishing universal charging standards, such as CCS (Combined Charging System) and CHAdeMO, helps in this regard.

Balancing Demand: Managing peak demand for electricity during charging times is crucial to avoid grid instability and high energy costs (Giri et al. 2019).

22.3 Sub-components of EVs

The powertrain subsystem of a battery electric vehicle (BEV) is primarily composed of key components, including a battery system, a power distribution unit (PDU), a motor controller, and a drive motor, as illustrated in Figure 22.1. When BEVs usually function, the electric energy kept in the battery system is fed into the PDU and then, via the PDU, to the motor controller. Powering the motor system allows for converting electric energy into mechanical energy, which is necessary for the BEV's operation.

The battery system, which comprises many battery cells connected in series and parallel, mostly stores electrical energy. In addition to battery cells, the battery system includes components such as a battery management system (BMS) controller, power electronics elements, and other related components. The BMS controller is in charge of managing and monitoring the battery modules.

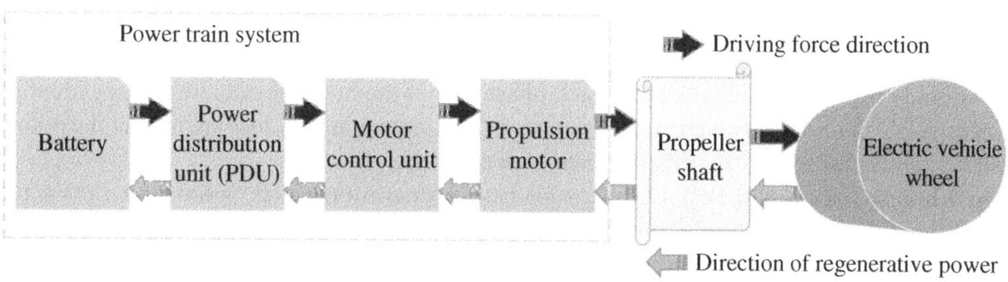

Figure 22.1 Sub-components of EVs.

Individual battery cells' voltage, current, and temperature can all be measured. By analyzing the collected data, we devise a suitable control method to safeguard against irregular conditions in the battery compartment, which includes challenges like overcharging, excessive discharging, and temperature spikes. The components of power electronics act as a shield for battery cells. They avert harm from high currents, handle the initiation and termination of the electrical system, and sever the power link to the BEV motor during urgent scenarios.

- The PDU channels electric power from the battery system and provides interfaces to various components or systems within the BEV.
- The motor controller primarily supervises the drive motor. It ensures steady and dependable performance and delivers instant feedback about the drive system, including the motor and its controller, to the main vehicle controller.
- The drive motor in EVs converts electrical energy from the battery into mechanical energy to propel the vehicle.
- The drive motor's efficiency, power output, and control systems significantly influence an EV's performance, range, and overall driving experience.

Concept of drive motors utilized in EVs can be described as follows:

The drive motor in EVs converts electrical energy from the battery into mechanical energy to propel the vehicle. It is an essential part of the electric drivetrain and serves as the primary power source for the wheels.

The drive motor operates based on the principles of electromagnetism, where electrical current flowing through coils generates a magnetic field. This magnetic field interacts with permanent magnets or other magnetic components, resulting in a rotational motion that drives the wheels.

There are various types of drive motors used in EVs.

- **Permanent Magnet Synchronous Motor (PMSM)**: PMSMs have permanent magnets in the rotor, and the flow of direct current generates the stator's magnetic field. They are efficient, have a high torque-to-weight ratio and less acoustic noise, offer high power density and are widely used in EVs (Krishnan 2017).
- **Induction Motor (IM)**: IMs have rotor windings that induce a current when exposed to the stator's rotating magnetic field. They are robust, reliable, and cost-effective but may have slightly lower efficiency than PMSMs.
- **Switched Reluctance Motor (SRM)**: SRM uses the principle of magnetic reluctance, where the rotor aligns itself with the minimum magnetic reluctance path. SRMs are relatively simple in construction and can offer high torque at low speeds. PMSM and Synchronous Reluctance Machines (SynRel) typically require fewer conductor connections than traditional three-phase IMs. Additionally, external control for PMSM and SynRel motors is more intricate when compared to induction and DC motors of similar types. SynRel machines excel in high-speed operations and have low inertia, making them ideal actuators and traction drive choices. Furthermore, SRM is naturally resilient to faults because each phase within the machine operates independently electrically. Therefore, if one phase experiences a short-circuit fault, it does not impact the other phases (Jayasawal et al. 2022; Krishnan and Materu 1990).
- **Brushless DC Motor (BLDC)**: BLDC motors have permanent magnets on the rotor, similar to PMSM, but they use electronic commutation instead of brushes. They provide good efficiency and control but may require more complex electronics. The utilization of BLDC motors featuring trapezoidal back-EMF has become prevalent across industrial, consumer, and military sectors due to their superior torque capabilities when compared to PMSM with sinusoidal

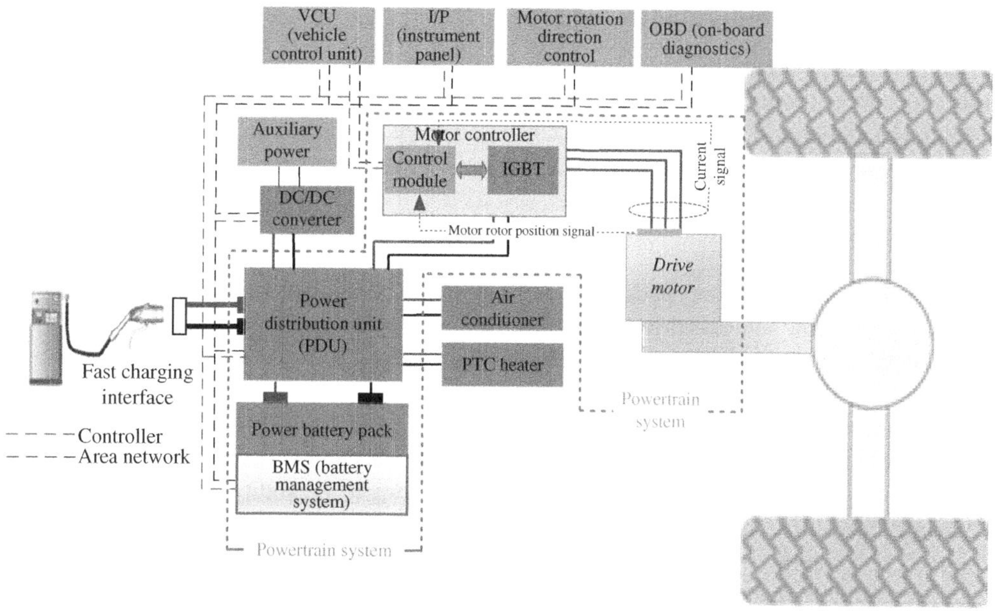

Figure 22.2 Structural diagram of power-train system in a BEV.

back-EMF. This popularity is attributed to the BLDC motor's significant advantages, including a high torque-to-weight ratio, exceptional efficiency, and reduced acoustic noise (Dwivedi et al. 2022).

As outlined previously and illustrated in Figure 22.2, the BEV's power-train system is intricate with many components, and a malfunction in any one of them could compromise the whole propulsion mechanism.

22.3.1 Reliability Study of Sub-components in EVs

Reliability refers to the capacity of an item or system to perform its intended functions within predetermined conditions and over a specified period. The adoption of EVs has seen a significant rise due to their environmental benefits and increasing interest in sustainable transportation. However, ensuring the reliability of key sub-components in EVs is crucial for their widespread acceptance and long-term success. EV usage has increased significantly due to its environmental benefits and growing interest in sustainable mobility. However, assuring the dependability of essential subsystems in EVs is critical for their general acceptance and long-term success. This research aims to thoroughly evaluate the dependability of critical EV subsystems such as battery packs, electric motors, power electronics, charging systems, and thermal management. The current investigation attempts to identify possible deficiencies, failure patterns, and improvement possibilities in EV dependability by utilizing real-world data, models, and accelerated testing. As the automotive industry transitions toward more sustainable and eco-friendly transportation solutions, EVs have gained significant attention due to their potential to reduce carbon emissions and dependence on fossil fuels. One critical aspect of ensuring EVs' success and widespread adoption is their reliability. Reliability studies of sub-components within EVs play a crucial role in establishing the overall dependability of these vehicles (Gandoman et al. 2019).

Importance of Reliability Studies

Reliability refers to the ability of a component, system, or vehicle to perform its intended function consistently and without failure over a specified period (Sarita et al. 2023). In the context of EVs, reliability is paramount because any unexpected breakdown or failure can lead to inconvenience for the owner, increase maintenance costs, and potentially negatively impact the public perception of electric cars.

Scope of Reliability Studies

1. **Battery Systems**: The battery pack is one of the most critical components of an EV. Its reliability directly impacts the vehicle's driving range, performance, and overall user experience. Studying the battery's degradation rate, thermal management system effectiveness, and safety measures against thermal runaways are essential to its reliability assessment.
2. **Electric Motors**: The electric motor is another key component that drives the vehicle's movement. Studying the motor's durability, efficiency, cooling mechanisms, and potential wear and tear under various driving conditions is crucial for ensuring its reliability.
3. **Power Electronics**: Power electronics control the flow of electricity between the battery and the motor. The reliability of inverters, converters, and other power electronic components is critical to maintaining a smooth and efficient power transfer. Numerous scenarios, such as operating electric motors or feeding power into the electrical grid, there is a frequent need for a consistently high and stable voltage level (Soni et al. 2023).
4. **Charging Infrastructure**: The reliability of charging stations (CS) is equally important, as it directly affects the convenience of EV ownership. Studying factors such as charging speed, connectivity, and the ability to withstand different environmental conditions is vital.
5. **Connectivity and Software**: Many modern EVs have advanced software and connectivity features. Ensuring the reliability of infotainment systems, over-the-air updates, and autonomous driving capabilities involves rigorous testing and validation.
6. **Auxiliary Systems**: Heating ventilation and air conditioning (HVAC), lighting, and other auxiliary systems contribute to the overall comfort and functionality of the vehicle. Reliability studies in these areas address performance, longevity, and energy efficiency (Kamruzzaman and Benidris 2019).

22.3.1.1 Methodologies of Reliability Studies

Reliability studies involve various methodologies, including:

1. **Accelerated Testing**: Components are subjected to extreme conditions and stresses to simulate years of use within a shorter time frame. This helps identify potential failure points and weaknesses.
2. **Field Testing**: Real-world usage data is collected from a fleet of vehicles to analyze component performance over extended periods and under diverse conditions.
3. **Failure Analysis**: Investigating the root causes of component failures provides insights into design improvements and preventive measures.
4. **Statistical Analysis**: Data from tests and field studies are analyzed statistically to assess failure rates, mean time between failures (MTBF), and other reliability metrics.

Benefits of Reliable Sub-components in EVs

- **Lower Maintenance Costs**: Reliable components result in fewer breakdowns and repairs, reducing maintenance expenses for EV owners.

- **Sustainability**: Longer-lasting components lead to a longer vehicle lifespan, contributing to the overall environmental benefits of EVs.
- **Brand Image**: Automakers prioritize reliability to establish a positive brand image and gain a competitive edge in the market.

In conclusion, conducting reliability studies on sub-components within EVs is essential for ensuring EVs' long-term success and acceptance in the automotive landscape. These studies enhance the consumer experience and contribute to a more sustainable transportation future. Manufacturers and researchers continue to refine their approaches to reliability assessment to provide EV owners with vehicles that are not only environmentally friendly but also dependable and trustworthy.

22.4 Reliability Assessment Techniques in EVs

Reliability assessment uses investigative approaches based on traditional methods and computational simulations. Several novel techniques have been recently suggested for reliability analysis. To gain deeper insight into the reliability of the BEV power train, a review of the reliability metrics was carried out for both the complete system and its separate elements. These techniques ensure that EVs meet safety, durability, and performance standards. Here are some common reliability assessment techniques used in the EV industry. Figure 22.3 presents the categorization of reliability evaluation methods.

22.4.1 Markov Model

Markov models are frequently employed to analyze the reliability of engineering and systems that are resilient to faults. The Markov model outperforms alternative approaches when assessing

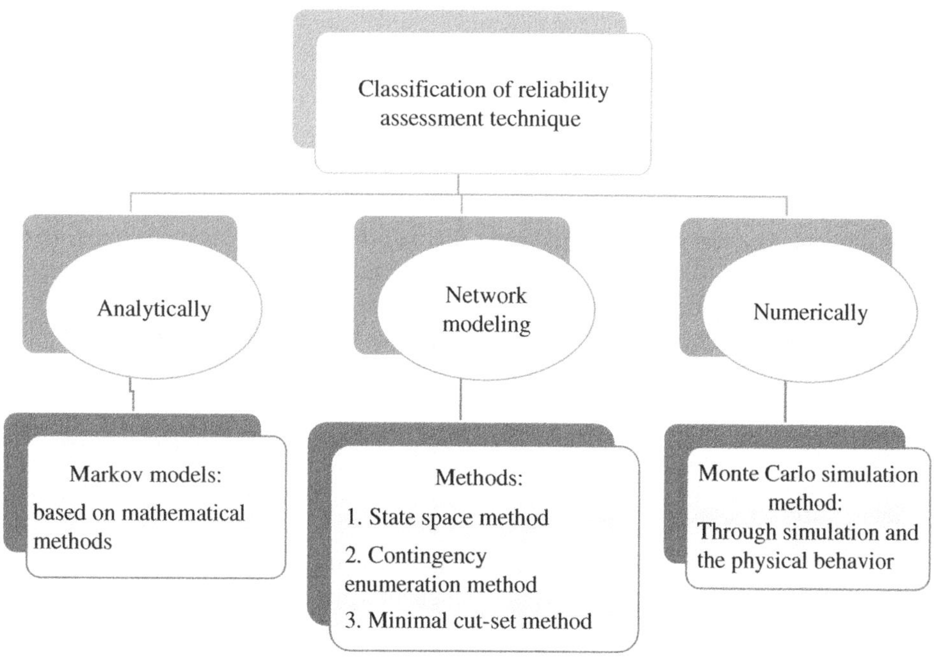

Figure 22.3 Classification of reliability assessment technique.

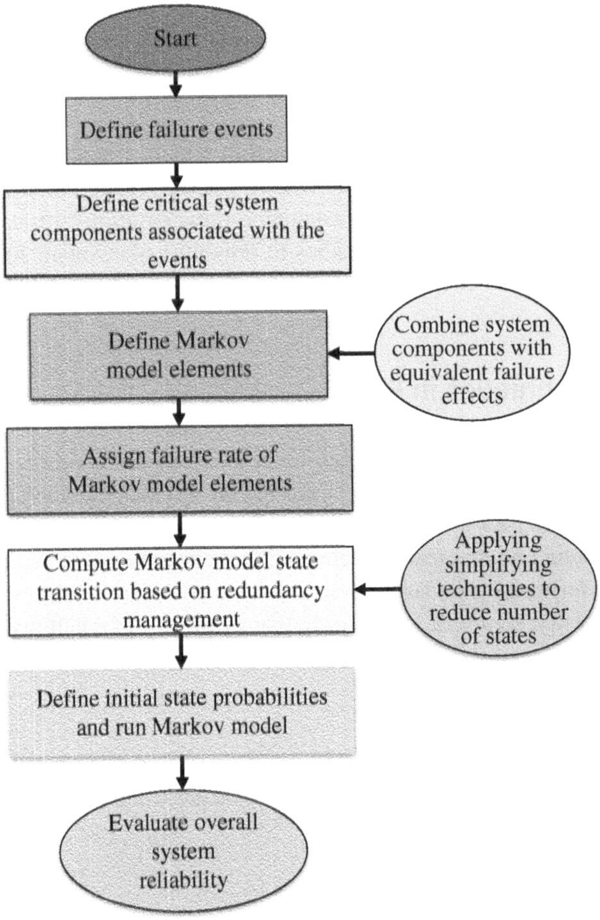

Figure 22.4 Markov model design process.

the reliability of a complex system where components are meticulously modeled. This modeling technique is particularly effective for components that can be accurately characterized using mathematical methods. The Markov model can be depicted as discrete, encompassing temporal and spatial aspects. In evaluating reliability for electrical systems, using stationary Markov models is fundamental. These models exhibit discreteness in spatial representation while maintaining continuous temporal. Figure 22.4 presents an overview of the design process for the Markov model analysis.

22.4.2 Monte Carlo Simulation (MCS)

The utilization of Monte Carlo Simulation (MCS) provides the option to incorporate more advanced component models, encompassing the effects of component aging. The application of the MCS method in managing EV charging within the electric power system (EPS) operation has a notable impact on the planning of the electric market. The EV charging load mainly drives the influence. Such an effect emerges because of the irregular distribution of EVs both in space and time.

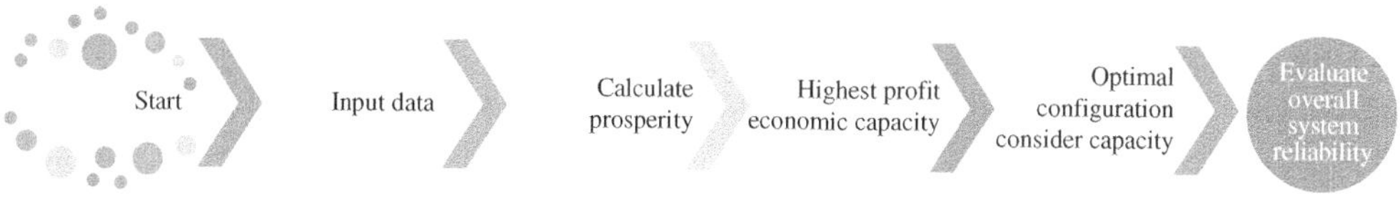

Figure 22.5 Monte Carlo simulation design process.

The MCS is the premier method for predicting EV charging duration and power. It is rooted in the traditional probability distribution of vehicular data. In forecasting the EV charging load using the MCS method, a Probability Distribution Function (PDF) is crafted considering starting charging moments and conditions, factoring in aspects like the count of EVs, battery size, and other pertinent factors. Following this, MCS utilizes mathematical simulations to sum up the charging demands of each separate EV to compute the overall charging demand. Numerous statistical tests are conducted through the MCS methodology to tackle both computational and tangible issues.

This specific technique simulates EVs' driving and charging actions to determine charging time and power based on the PDF of EV attributes. The starting moment for charging is typically depicted, as illustrated in Eq. (22.1):

$$f_x(y) = \begin{cases} \dfrac{1}{\sigma_x\sqrt{2\pi}}\left[\exp\left(-\dfrac{(y-\mu_x)^2}{2\sigma_x^2}\right)\right], & \mu_x - 12 < y \le 24 \\ \dfrac{1}{\sigma_x\sqrt{2\pi}}\left[\exp\left(-\dfrac{(y+24-\mu_x)^2}{2\sigma_x^2}\right)\right], & 0 < y \le \mu_x - 12 \end{cases} \tag{22.1}$$

Furthermore, the PDF for the daily mileage is established, as depicted in Eq. (22.2):

$$f_z(y) = \frac{1}{y\sigma_z\sqrt{2\pi}}\left[\exp\left(-\frac{(\ln y-\mu_z)^2}{2\sigma_z^2}\right)\right] \tag{22.2}$$

in which the pertinent shape parameters for the PDFs are denoted as σ_x, μ_x, σ_z, and μ_z, enhancing reliability. When applying the MCS technique to address challenges, the outcomes frequently manifest as mathematical forecasts for a designated random factor. This factor is produced via a conceptual test using exact technical information. The issue is fundamentally tackled by using the average value of this particular factor. The significant need for random numbers in MCS cannot be overstated. Figure 22.5 illustrates the design process of the MCS.

22.4.3 Contingency Enumeration Method

Another analysis method is the contingency enumeration technique, which gauges reliability by scrutinizing a set number of contingencies. This method is particularly useful for assessing the trustworthiness of electrical systems. The contingency enumeration procedure can be broken down into four consecutive phases. Figure 22.6 illustrates the contingency enumeration method. The initial phase outlines the analysis framework, encompassing system boundaries and the chosen load flow technique. Moving on to the subsequent phase, the focus shifts to handling contingencies involving various outage combinations. Here, it is essential to account for all conceivable contingencies, even in cases where this leads to prolonged computational efforts. These calculations are crucial in identifying potential electrical system issues arising from these contingencies. Ultimately, the reliability metrics are computed, and the cumulative annualized indices are derived by aggregating results across all relevant operational scenarios and potentialities.

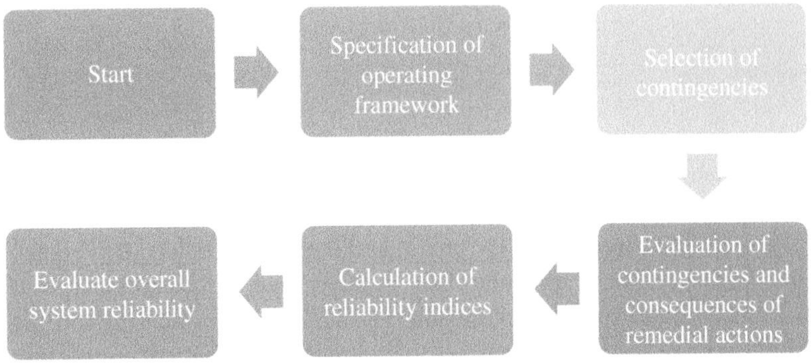

Figure 22.6 Contingency enumeration method process.

22.4.4 State Space Method

The state space approach generally hinges on the ideas of "upstate" and "downstate." Fundamentally, this approach is anchored in three core parameters, as depicted in Figure 22.7.

In Figure 22.8, m represents the standard operational time of the system, while r signifies the duration of failure, and T stands for the average time between instances of failure.

The diagram in Figure 22.8 illustrates the state space of an individual component within a larger system. As in this context, t, λ represents the rate of failures, while μ pertains to the rate of repairs (Ghosh et al. 2013).

In the context of reliability analysis using state space methods, MTTF (mean time to failure [uptime]), MTBF (mean time between failures-cycle time: [sum of the uptime and downtime]), and MTTR (mean time to repair-downtime) are important reliability metrics. These metrics help assess and understand the performance and availability of systems or components, as already discussed in Chapters 20 and 21.

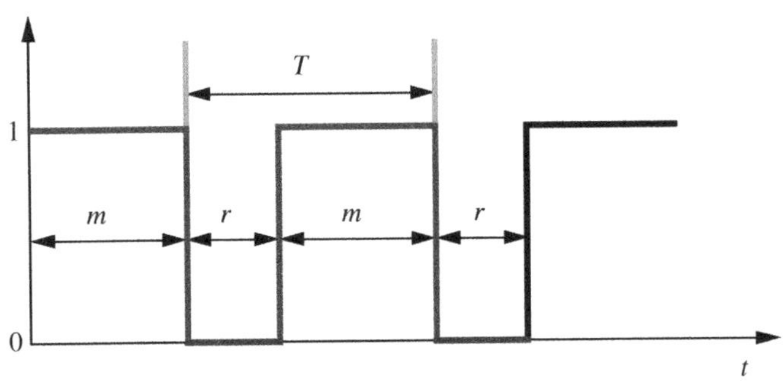

Figure 22.7 State space method process.

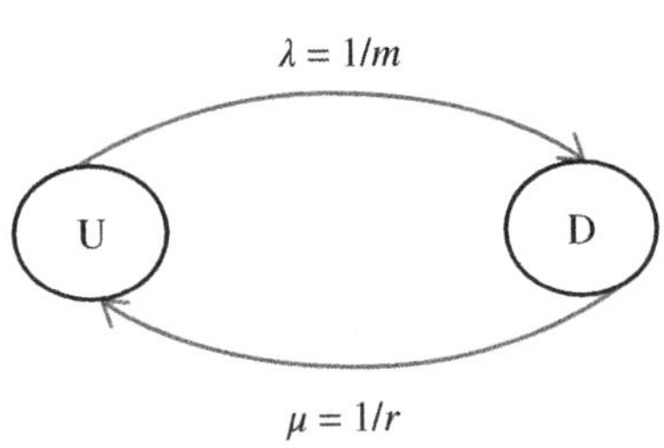

Figure 22.8 Diagram depicting the state space method for an individual element within a system.

22.4.5 Electric Vehicles Trend

Based on data from the transport ministry, India's car sector presently holds the fifth position worldwide and is forecasted to climb to the third spot by 2030. In the quest to reduce greenhouse gas releases, EVs emerge as vital solutions. Utilizing RESs for recharging or generating energy onboard can eradicate fuel expenses. Based on background information, categorization, and adoption rates, EVs can be divided into hybrid and battery EVs. The literature has addressed challenges, including inadequate battery charging infrastructure, extended charging durations, limited driving ranges, and performance shortcomings (Singh et al. 2023a).

Numerous obstacles must be addressed, encompassing battery expenses, efficient charging techniques, and enhancing charging station effectiveness and efficiency. Ensuring dependable charging speeds stands out as a critical factor. For bridging the link between the electric grid and EV battery setups, affordable and dependable power converters play a crucial role in enhancing EV charging techniques. Recent research predicts that the EV market was valued at 170 billion in 2021 and is projected to exceed 1103 billion by 2030, as illustrated in Figure 22.9. With such substantial global sales growth expected in a decade, it would be reasonable to anticipate that EVs will become commonplace in everyday life.

In India, statistics show that roughly 1 in every 125 vehicle is electric, predominantly two-wheelers and three-wheelers. The main elements impacting EV efficiency have been steady for the last 20 years. These elements encompass limited product advancements, cost issues, and an inadequate density of charging facilities, collectively hindering the broad acceptance of EVs in the country (Rivera et al. 2021).

22.4.6 Fundamental Considerations in Assessing the Reliability of EVs

Car manufacturers have undergone substantial policy and strategy adjustments with the increasing global demand for EVs. Within this context, a critical emphasis lies on assessing reliability, which

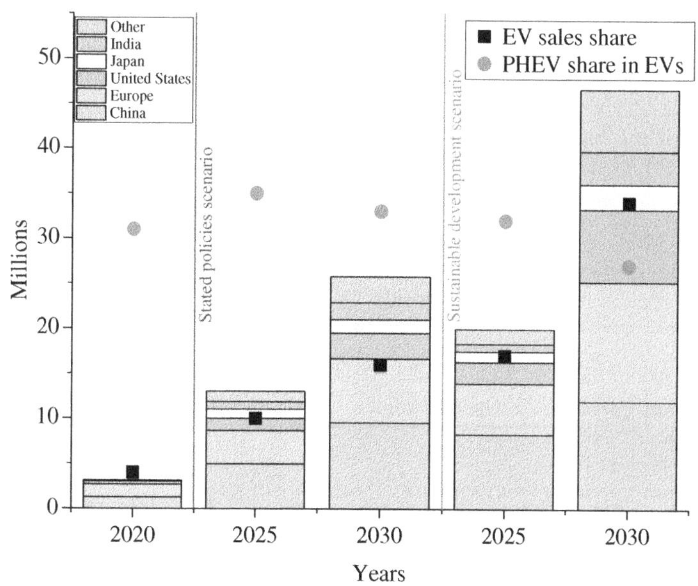

Figure 22.9 Predictive data for global EV sales from 2020 to 2030.

can be divided into two principal domains:

1. The evaluation of reliability in contemporary electrical grids incorporating EVs.
2. The analysis of reliability concerning the internal systems of EVs.

When assessing the reliability of EVs, various viewpoints need to be considered, such as:

- Understanding reliability from the perspective of the customer.
- Interpreting reliability from the standpoint of the manufacturer.
- Grasping reliability from the viewpoint of the seller.

The evaluation of reliability from the aforementioned perspectives holds substantial importance in the strategic planning and overall operational lifespan of various components within EVs, encompassing aspects like power electronic converters, battery packs, and electric motors. Within the operational phase of EVs, three distinct areas (production, sales, and customers) carry relevance from a reliability standpoint. Figure 22.10 visually represents these three zones, demonstrating the implementation of each element in assessing EV reliability. The utmost significance of the reliability concept for the components of EVs.

Due to the presence of multiple sub-systems within EVs, conducting reliability and safety analyses becomes notably challenging. Moreover, the intricate interdependence among components adds uncertainty, prompting the utilization of a logical framework to explore the uncertain relationships of failures in EV components. To evaluate the reliability of electric vehicles (EVs), it is essential to consider the failures that arise in crucial electrical components. Figure 22.11 depicts the categorization of primary failures in the electrical components of EVs. The crucial factors to be considered in the reliability assessment of EVs are outlined as follows:

- Understanding the functioning of power components within EVs.
- Pinpointing component malfunctions in EVs.
- Analyzing the sequence of failures.
- Proposing a framework to depict these malfunctions.
- Choosing an approach to gauge the reliability of EVs.
- Evaluating reliability from the manufacturing angle is complex, given the myriad of procedures in battery cell production.

To investigate the reliability of the battery pack, it is imperative to address four significant concerns.

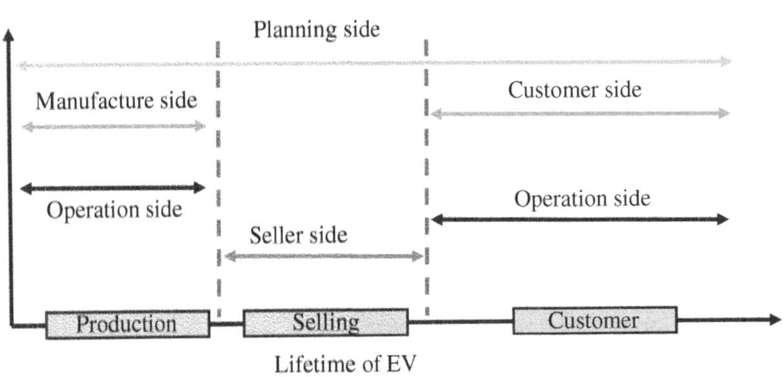

Figure 22.10 The utmost significance of the reliability concept for the components of EVs.

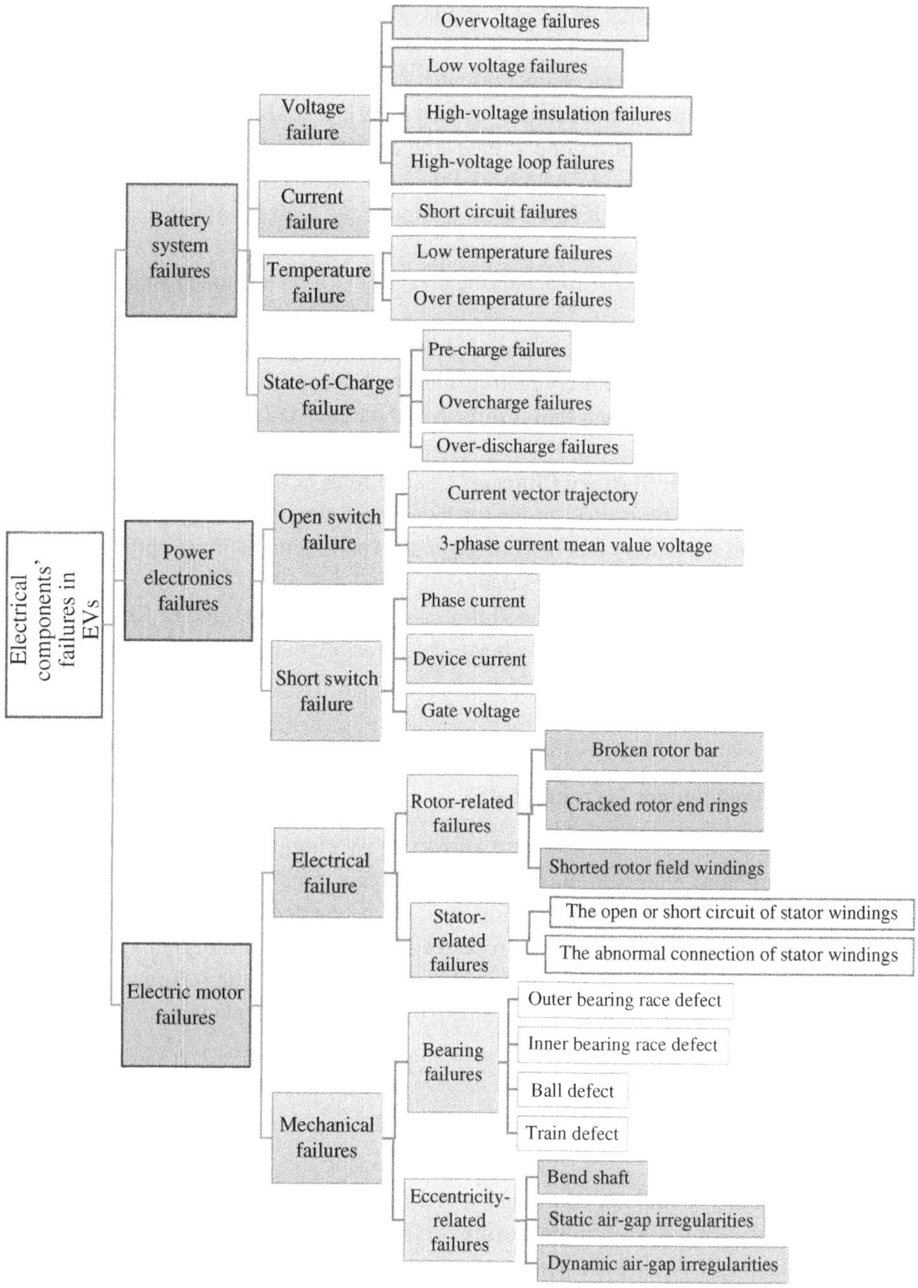

Figure 22.11 Categorization of primary electrical component failures in EVs.

1. **Chemical Issues**: During the battery manufacturing process, it is essential to account for the electrochemical factors contributing to battery failures.
2. **Thermal Issues**: An additional crucial aspect in evaluating the reliability of the battery pack pertains to maintaining safe conditions through thermal management. The thermal analysis for battery pack heating is subdivided into two key components: battery cell thermal analysis and

comprehensive battery pack thermal analysis. Various cooling systems have been employed to regulate the temperature of the cells.

3. **Mechanical Issues**: The primary mechanical-related failure that can influence the reliability of EV batteries is the result of mechanical stresses experienced during different modes of operation and collisions. These failures encompass stress-based and strain-based mechanisms. Notably, advancements are being made in both testing and modeling the mechanical characteristics of the battery pack, along with its individual components, such as cathodes, anodes, and separators.
4. **Electrical Issues**: Several critical electrical parameters within the battery are pivotal in addressing reliability concerns.
   (i) **Over-Voltage**: The condition where the battery charging voltage surpasses the established standard, leading to issues like battery plating and overheating, is called over-voltage.
   (ii) **Under-Voltage**: Another problem that results in battery pack failure is under-voltage during the charging process.
5. **Power Electronic Converter/Battery Charger**: Central roles of the battery charger encompass channeling energy from the electric grid to the battery, refining the charging sequence, and halting the charge when the battery reaches full capacity. Any functional interruptions directly influence the dependability of the battery charger.
   One approach to mitigate the aforementioned issues is to restrict the State of Charge (SoC) within a specific range. Operating the battery beyond this range can harm battery lifespan, reliability, and safety.

22.4.7 EVs Battery Management Systems (BMS)

The BMS recharges the batteries of the EVs, also known as EV charging points or EV charging systems. The critical functions of an automotive BMS encompass important aspects such as monitoring the battery's state of charge (SoC), gauging its state of health (SoH), overseeing voltage, temperature, and current levels, and ensuring proper cell balancing for lithium-ion (Li-ion) batteries. The BMS incorporates a range of essential indicators to guarantee the secure and efficient functioning of EV batteries over their intended lifespan, including the following:

(i) SoC (State of Charge)
(ii) SoH (State of Health)
(iii) SoT (State of Temperature)

1. **State of Charge (SoC)**: SoC indicates the amount of energy left in the battery as a percentage of its total capacity. It is essential to accurately estimate the remaining driving range and prevent over-discharge, which can damage the battery.
2. **State of Health (SoH)**: SoH reflects the battery's overall health and degradation level. It helps assess the battery's capacity reduction over time and ensures that it is replaced or serviced when its capacity falls below a certain threshold.
3. **State of Temperature (SoT)**: Temperature significantly influences and is highly responsive to the dynamic attributes of a battery. Additionally, due to the requirement for high-energy ESS, there is a consistent demand for battery packs with elevated energy density. Consequently, high-energy-density battery packs inherently encounter challenges related to thermal management due to the heat produced during charge and discharge operations.

22.5 Evaluation of Distribution Systems Reliability with Integrated EVs

The evaluation of distribution systems' reliability with the integration of EVs is a critical aspect in the field of power system planning and management. As the adoption of EVs continues to rise, their interaction with the electrical grid introduces both opportunities and challenges, particularly concerning the overall reliability of the DN. This evaluation aims to assess how the presence of EVs influences the reliability of the distribution system and to develop strategies for ensuring a robust and resilient power supply

Importance of Distribution System: Reliable electricity distribution is essential for the smooth functioning of various societal and economic activities. Any disruptions in power supply can lead to inconveniences, financial losses, and even safety hazards. Effective power system planning, control, and maintenance hinge on the synchronized management of power flow within the power grid and are fundamental aspects of power system planning and modeling (Giri et al. 2019).

- The integration of EVs into the distribution system adds a dynamic load and introduces bidirectional power flow due to charging and discharging, thereby necessitating a comprehensive reliability assessment.
- **Impact of EVs on Reliability**: The presence of EVs impacts the distribution system in several ways.

(i) **Load Variability**: EV charging introduces additional load variations, increasing stress on the distribution infrastructure during peak charging times.
(ii) **Bidirectional Power Flow**: V2G and V2H functionalities allow EVs to inject power back into the grid. While this can provide grid support, it also introduces complexities in managing power flows.
(iii) **Infrastructure Overload**: Rapid adoption of EVs in specific regions can potentially overload local distribution infrastructure if not properly managed.

Reliability Assessment Techniques

Reliability assessment involves analyzing the potential failure points, evaluating the impact of these failures on the system, and devising strategies to enhance the system's reliability. Key assessment techniques include:

(i) **Load Flow Analysis**: Analyzing power flows to identify potential bottlenecks and overloads
(ii) **Fault Analysis**: Simulating fault conditions to determine the system's response and the impact on reliability
(iii) **Monte Carlo Simulation**: Incorporating uncertainty in load and EV charging patterns to assess the system's performance under varying conditions.
(iv) **Sensitivity Analysis**: Evaluating how changes in EV penetration levels and charging behaviors affect reliability metrics.

Strategies for Enhancing Reliability: To ensure the reliability of distribution systems integrated with EVs, various strategies can be employed:

- **Advanced Metering Infrastructure (AMI)**: Smart metering allows real-time monitoring of EV charging and grid conditions, enabling better load management.
- **Demand Response**: Implementing demand response programs to shift EV charging away from peak load times, reducing stress on the grid.

- **Energy Storage**: Integrating ESS with EV charging infrastructure can help balance supply and demand and mitigate load fluctuations.
- **Grid Upgrades**: Proactive grid upgrades, including transformer replacements and infrastructure reinforcement, can accommodate the increased load from EVs.

Future Directions: As EV technology advances and their adoption becomes more widespread, continued research is necessary to understand the evolving dynamics between EVs and distribution system reliability. This includes exploring optimized charging strategies, incorporating RES, and developing advanced control algorithms to manage EV-grid interactions effectively.

22.5.1 Distribution Network (DN) Reliability Assessment

The reliability of a DN is a measure of its capability to provide an uninterrupted power supply. A DN comprises interconnected elements such as busbars, distributors, service mains, cables, isolators, circuit breakers, protecting devices, and transformers (Manoj et al. 2015). From a reliability perspective, these components are arranged in a series configuration, meaning that each element must remain operational to ensure continuous power supply. The expected failure rates influence the overall reliability of the DN λ and interruption times τ of these individual components. Additionally, a range of customer interruption metrics has been developed in the literature to assess the impact and severity of system failures and to conduct reliability forecasting assessments. This study utilizes several metrics, including ASAI, SAIFI, SAIDI, CAIDI, and ENS. The following analytical probabilistic models, chronological probabilistic models and their scientific interpretation, mathematical description, and importance are illustrated in Chapters 20 and 21. The above description of these indices is also discussed in Billinton et al. (1996).

22.5.2 Reliability Evaluation Using a V2G Approach and High EV Penetration

Reliability assessments can indeed be conducted using analytical or chronological probabilistic models, depending on the specific context and the nature of the system or process being analyzed. These models help evaluate the likelihood of a system or component functioning correctly over a given period or under certain conditions.

1. **Analytical Probabilistic Models**
   - **Probability Density Functions (PDFs)**: Analytical models often use probability density functions to describe the distribution of failure or reliability over time. For example, the exponential distribution might be used to model the failure rate of a component.
   - **Weibull Distribution**: The Weibull distribution is frequently used in reliability engineering to model the failure rates of components, with its shape parameter allowing for modeling various failure patterns.
   - **Monte Carlo Simulation**: Analytical models can also incorporate MCSs to account for complex, interconnected systems or situations where exact mathematical models are not available. This involves generating random inputs and analyzing the output distribution to assess reliability.
2. **Chronological Probabilistic Models**
   - **Markov Models**: These models are used when analyzing systems with discrete states and transitions. Markov models consider the probability of transitioning from one state to another over time, allowing for the assessment of system reliability and availability.

- **Fault Tree Analysis (FTA)**: FTA is a top-down deductive approach used to identify and evaluate the probability of failure events in a system. It uses a graphical tree structure to represent how different events or failures can lead to system failure.
- **Event Tree Analysis (ETA)**: ETA is complementary to FTA and is used to assess the consequences of various events or failures, considering different possible outcomes and their probabilities.
- **Reliability Block Diagrams (RBD)**: RBDs are graphical representations that break down a system into individual components and their interconnections, allowing for the analysis of system reliability by combining component reliability data.

To address the requirement for chronological modeling of EV consumption, the study will employ a method known as "chronology." This method involves generating virtual scenarios for both electricity generation and demand, which will then be utilized to compute reliability indices.

Traditional generating units are characterized by two conditions: By applying the exponential CDF to create a virtual scenario for the time-to-failure (TTF), time-to-repair (TTR), the mean-time-to-failure (MTTF), mean-time-to-repair (MTTR) are stated, respectively, in years. Uf is consistently between 0 and 1, corresponding to interruption f, dispersed random number (Negarestani et al. 2016). There may be problems with the generation units that cause load interruptions. The annual system and interruption indices can be used to separate the various reliability indices (Talukdar and Deka 2021). There are many available annual system indices:

(i) Loss of load expectation (LOLE) in h/yr
(ii) Loss of energy expectation (LOEE) in megawatt hours/yr,
(iii) Loss of load frequency (LOLF) in interruptions/yr (Sandelic et al. 2022).

Numerous Interruption indices are available

(i) Energy Not Served per Interruption (ENSPI) in megawatt hours/interruption
(ii) System Average Interruption Frequency Index (SAIFI) in interruption/customer (int./cu.)
(iii) System Average Interruption Duration Index (SAIDI) in h/cu.

Various reliability indices for NI interruptions over N years can be determined in Eqs. (22.3)–(22.8):

$$\text{LOLE} = \sum_{i=1}^{\text{NI}} \frac{h_i}{N} \tag{22.3}$$

$$\text{LOEE} = \sum_{i=1}^{\text{NI}} \frac{\text{ENS}_i}{N} \tag{22.4}$$

$$\text{LOLF} = \frac{\text{NI}}{N} \tag{22.5}$$

$$\text{ENSPI} = \sum_{i=1}^{\text{NI}} \frac{\text{ENS}_i}{N} = \frac{\text{LOEE}}{\text{LOLF}} \tag{22.6}$$

$$\text{SAIFI} = \frac{\text{The number of customers interruptions}}{\text{Total customers served}} \tag{22.7}$$

$$\text{SAIDI} = \frac{\text{Total durations of customers interruptions}}{\text{Total customers served}} \tag{22.8}$$

22.5.3 Reliability Parameters

EVs possess the dual capability of serving as an energy provider and an electricity user, attributed to their engagement in V2G operations (Xu and Chung 2015). The extensive interplay between their V2G and G2V functionalities and their time-space characteristics necessitates recognizing this unique aspect and formulating fresh criteria for EV reliability. In alignment with this goal, the reliability metrics for EVs predominantly bifurcate into two categories, illustrated in Figure 22.12. A more detailed categorization and explanation can be derived from recent research findings (Hou et al. 2016).

22.5.3.1 Reliability Parameters of EVs

1. **Average Interruption Frequency of EV Charging Index (E_{AIFI}, t/yr):**

$$E_{AIFI} = \frac{\sum_{i=1}^{N} \lambda_i}{N \cdot Y} \tag{22.9}$$

The symbol λ_i denotes the count of charge interruptions over the period of EV simulation. Here, N signifies the total number of EVs, while Y indicates the simulation duration in years.

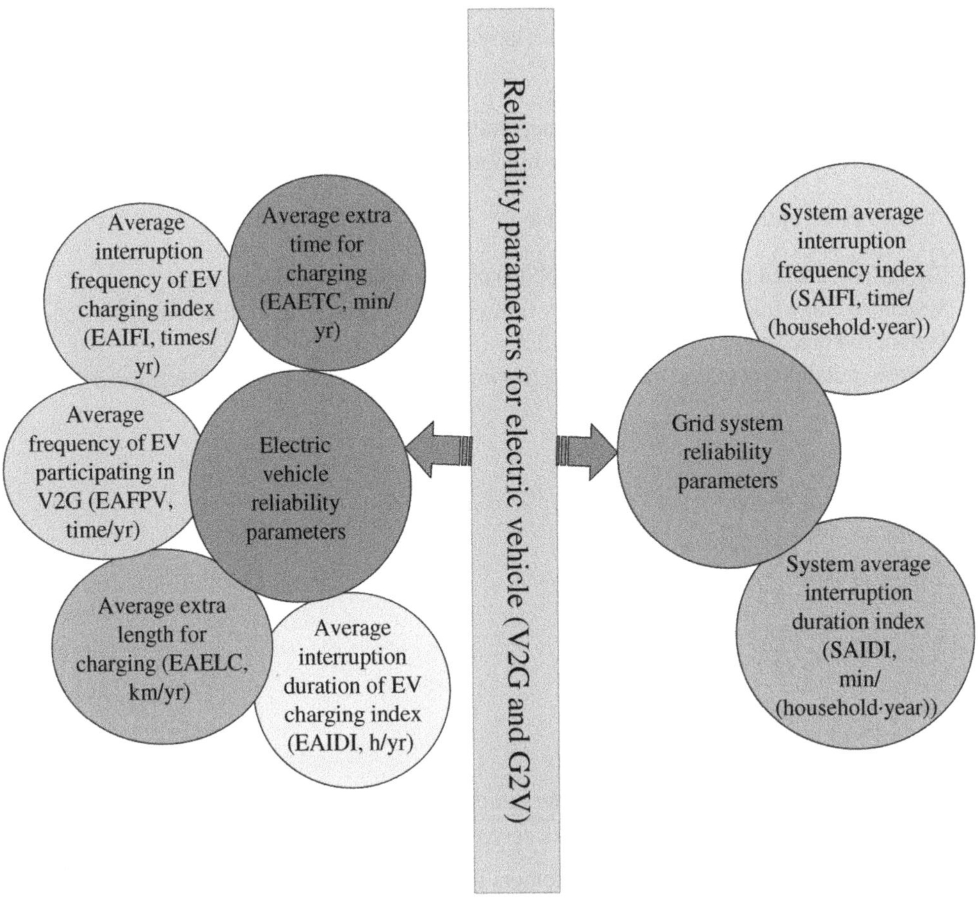

Figure 22.12 Reliability parameters for EVs.

2. **Average Interruption Duration of EV Charging Index (E_{AIDI}, h/yr):**

$$E_{\text{AIDI}} = \frac{\sum_{i=1}^{N} t_i}{N \cdot Y} \tag{22.10}$$

In this context, t_i stands for the cumulative charging disruption time during the simulation period for the ith EV.

3. **Average Extra Length for Charging (E_{AELC}, km/yr):**

$$E_{\text{AELC}} = \frac{\sum_{i=1}^{N} \Delta l_i}{N \cdot Y} \tag{22.11}$$

In this scenario, l_i denotes the distance covered by the ith EV due to a power outage when $\text{SOC}_i < S_i$, as it searches for an alternative charging station throughout the simulation duration.

4. **Average Extra Time for Charging (E_{AETC}, min/yr):**

$$E_{\text{AETC}} = \sum_{i=1}^{N} \frac{t_i^E}{N \cdot Y} \tag{22.12}$$

Here, (t_i^E) represents the time taken by the ith EV due to a power failure when $\text{SOC}_i < S_i$ as it seeks a new charging station over the course of the simulation interval.

5. **Average Frequency of EV Participating in V2G (E_{AFPV}, t/yr):**

$$E_{\text{AFPV}} = \frac{\sum_{i=1}^{N} c_i}{N \cdot Y} \tag{22.13}$$

Here, c_i indicates the cumulative count of instances the ith EV engaged in V2G activities during the simulation duration.

22.5.3.2 Grid System Reliability Parameters

Although EVs are transportable and interrupt load for the distribution system, the impact of power loss on them is relatively minimal. In order to remove an electric car, the existing DN described below is referred to by system reliability indices.

1. **System Average Interruption Frequency (S_{AIFI}, t/(household·yr)):**

$$S_{\text{AIFI}} = \frac{\sum_{i=1}^{M} f_i}{M \cdot Y} \tag{22.14}$$

In this context, M refers to the total number of households, and fi_i denotes the count of power interruptions experienced by the ith user throughout the simulation period.

2. **System Average Interruption Duration Index (S_{AIDI}, min/(household·yr)):**

$$S_{\text{AIDI}} = \frac{\sum_{i=1}^{N} t_i^S}{M \cdot Y} \tag{22.15}$$

Here, t_s^i signifies the total duration of the power outage experienced by the user.

3. **System Expected Energy Not Supplied (S_{EENS}, MWh/(household·yr)):**

$$S_{\text{EENS}} = \frac{\int_{i=0}^{Y} P_c(t)\text{d}t}{Y} \tag{22.16}$$

Here, $P_c(t)$ indicates the system's load decrease at time t and is set to zero when the network operates normally.

4. **Average Extra Length for Charging (E_{AELC}, km/yr):**

$$E_{AELC} = \frac{\sum_{i=1}^{N} \Delta l_i}{N \cdot Y} \tag{22.17}$$

The methodology and Reliability parameters outlined above necessitate a profound understanding of the mechanical, electrothermal, and lifespan characteristics of the vehicle's drive-train and its sub-components (Blaabjerg et al. 2021).

22.6 Conclusion

In conclusion, evaluating the reliability of distribution systems in the context of integrated EVs is a multifaceted task that requires a holistic approach. It involves considering the impact of EV charging behaviors, bidirectional power flow, and load variations on the DN. In the context of the advancements in EVs, it is crucial to emphasize the critical importance of evaluating the reliability and safety aspects concerning the longevity of EVs' electrical components, especially when it comes to planning, servicing, and maintenance. Among the various electrical components in EVs, the battery pack, power electronic components, and the electric motor stand out as the most pivotal elements requiring a thorough assessment from both reliability and safety perspectives. By utilizing advanced assessment techniques and implementing strategies to manage the challenges posed by EV integration, distribution system operators can ensure a reliable and efficient power supply while accommodating the growing demand for electric mobility.

References

Billinton, R., Chen, H., and Ghajar, R. (1996). Time-series models for reliability evaluation of power systems including wind energy. *Microelectronics Reliability* 36 (9): 1253–1261.

Blaabjerg, F., Wang, H., Vernica, I. et al. (2021). Reliability of power electronic systems for EV/HEV applications. *Proceedings of the IEEE* 109 (6): 1060–1076. https://doi.org/10.1109/JPROC.2020.3031041.

Das, H.S., Rahman, M.M., Li, S., and Tan, C.W. (2020). Electric vehicles standards, charging infrastructure, and impact on grid integration: a technological review. *Renewable and Sustainable Energy Reviews* 120: 109618.

Dwivedi, D., Singh, S., and Kalaiselvi, J. (2022). Impact of PWM and duty ratio control on voltage for SiC fed three phase BLDC motor drive. *2022 IEEE International Conference on Power Electronics, Smart Grid, and Renewable Energy (PESGRE)*, 1–5. https://doi.org/10.1109/PESGRE52268.2022.9715861.

Gandoman, F.H., Ahmadi, A., Van den Bossche, P. et al. (2019). Status and future perspectives of reliability assessment for electric vehicles. *Reliability Engineering & System Safety* 183: 1–16.

Ghosh, R., Kim, D.S., and Trivedi, K.S. (2013). System resiliency quantification using non-state-space and state-space analytic models. *Reliability Engineering & System Safety* 116: 109–125.

Giri, A.K., Singh, S., Dash, R. et al. (2019). A review on application of ANN and machine learning algorithm for the optimal placement of STATCOM. *2019 Innovations in Power and Advanced Computing Technologies (i-PACT)*, volume 1, 1–5. https://doi.org/10.1109/i-PACT44901.2019.8960138.

Hou, K., Xu, X., Jia, H. et al. (2016). A reliability assessment approach for integrated transportation and electrical power systems incorporating electric vehicles. *IEEE Transactions on Smart Grid* 9 (1): 88–100.

Jayasawal, K., Rana, A.K., and Teja, A.V.R. (2022). Computationally efficient model predictive torque control of switched reluctance motor drives. *IECON 2022 - 48th Annual Conference of the IEEE Industrial Electronics Society*, 1–6. https://doi.org/10.1109/IECON49645.2022.9968991.

Kamruzzaman, M.D. and Benidris, M. (2019). Reliability-based metrics to quantify the maximum permissible load demand of electric vehicles. *IEEE Transactions on Industry Applications* 55 (4): 3365–3375.

Kothari, D.P., Nagrath, I.J., and Saket, R.K. (2022). *Modern Power System Analysis*. Tata McGraw-Hill Publishing Company.

Krishnan, R. (2017). *Permanent Magnet Synchronous and Brushless DC Motor Drives*. CRC Press.

Krishnan, R. and Materu, P.N. (1990). Design of a single-switch-per-phase converter for switched reluctance motor drives. *IEEE Transactions on Industrial Electronics* 37 (6): 469–476. https://doi.org/10.1109/41.103450.

Manoj, K.N., Sethi, V.K., and Singh, S. (2015). Impact of distributed generation in electrical power system. *International Journal of Electrical and Electronics Research* 3 (3): 146–152.

Negarestani, S., Fotuhi-Firuzabad, M., Rastegar, M., and Rajabi-Ghahnavieh, A. (2016). Optimal sizing of storage system in a fast charging station for plug-in hybrid electric vehicles. *IEEE Transactions on Transportation Electrification* 2 (4): 443–453.

Rivera, S., Kouro, S., Vazquez, S. et al. (2021). Electric vehicle charging infrastructure: from grid to battery. *IEEE Industrial Electronics Magazine* 15 (2): 37–51. https://doi.org/10.1109/MIE.2020.3039039.

Sandelic, M., Peyghami, S., Sangwongwanich, A., and Blaabjerg, F. (2022). Reliability aspects in microgrid design and planning: status and power electronics-induced challenges. *Renewable and Sustainable Energy Reviews* 159: 112127.

Sarita, K., Saket, R.K., and Khan, B. (2023). Reliability, availability, and condition monitoring of inverters of grid-connected solar photovoltaic systems. *IET Renewable Power Generation* 17 (7): 1635–1653.

Singh, S., Saket, R.K., and Khan, B. (2023a). A comprehensive state-of-the-art review on reliability assessment and charging methodologies of grid-integrated electric vehicles. *IET Electrical Systems in Transportation* 13 (1): e12073.

Singh, S., Soni, S.K., Singh, K.A., and Saket, R.K. (2023b). Sliding mode control for bidirectional DC-DC power converter in electric vehicle charger for G2V and V2G applications. *2023 IEEE 3rd International Conference on Sustainable Energy and Future Electric Transportation (SEFET)*, 1–6. https://doi.org/10.1109/SeFeT57834.2023.10245726.

Soni, S.K., Singh, S., Singh, K.A. et al. (2023). Event-triggered control for LPV modeling of DC-DC boost converter. *IEEE Transactions on Circuits and Systems II: Express Briefs* 70 (6): 2062–2066. https://doi.org/10.1109/TCSII.2022.3230418.

Talukdar, B.K. and Deka, B.C. (2021). An approach to reliability, availability and maintainability analysis of a plug-in electric vehicle. *World Electric Vehicle Journal* 12 (1): 34.

Xu, N.Z. and Chung, C.Y. (2015). Reliability evaluation of distribution systems including vehicle-to-home and vehicle-to-grid. *IEEE Transactions on Power Systems* 31 (1): 759–768.

Yilmaz, M. and Krein, P.T. (2012). Review of battery charger topologies, charging power levels, and infrastructure for plug-in electric and hybrid vehicles. *IEEE Transactions on Power Electronics* 28 (5): 2151–2169.

23

Reliability Assessment of Multilevel Inverter for Modern Power System Applications

Saumya Singh[1], Dhawal Dwivedi[1], Kumari Sarita[2], R. K. Saket[1], and P. Sanjeevikumar[3]

[1] *Electrical Engineering Department, Indian Institute of Technology (BHU), Varanasi, Uttar Pradesh, India*
[2] *Department of Electrical Engineering, Government Engineering College, Aurangabad, Aurangabad, Bihar, India*
[3] *Department of Electrical Engineering, IT and Cybernetics, University of South-Eastern Norway, Porsgrunn, Norway*

23.1 Introduction

Around the world, fossil fuels are the primary source of electricity. But using fossil fuels in power plants creates greenhouse gases like carbon dioxide that cause climate change. Many nations use renewable energy as an alternative energy source due to the problems associated with burning fossil fuels. Several renewable energy sources (RES) have been created to generate power, including wind, tidal, wave, and photovoltaic (PV). Shortly, it is anticipated that these RES, particularly photovoltaic, will overtake all other forms of renewable energy (Surana and Jordaan 2019; Kothari et al. 2022). The three most important performance parameters for grid-connected solar systems are conversion efficiency, maximum power point tracking (MPPT), and total harmonic distortion (THD). The inverter's control method and topology selection significantly impact these performance metrics. MLI produces a terminal voltage using multiple DC voltage sources at different levels, typically in steps, allowing for reduced THD and better output waveform quality than traditional two-level inverters. The common configurations include:

- Capacitor-Clamped Flying Configuration
- Neutral-Point Clamped Diode Configuration
- H-Bridge Cascade Configuration

Consequently, MLIs primarily serve as a bridge connecting different electrical equipment inverters (DC/AC) and (DC/DC) converters. This is necessary because the loads or system cannot directly generate electricity from RES. The grid frequency has to match the frequency of the injected signal and can be fed by a voltage or current with less harmonic content, or a pure sinusoidal wave can power it (Li et al. 2014). The MLI with a reduced LC filter size produces sinusoidal three-phase voltage and current with lower harmonics. Voltage sources for DC input to the MLIs can be created using RES, such as PV arrays with DC/DC converters, induction generator/synchronized generator-fed wind turbines using rectifier configurations, and fuel cells as converters (DC/DC) (Mahalakshmi and Thampatty 2015). MLIs deliver enhanced power quality and more significant levels in their terminal voltage waveforms than two-level inverters.

Reliability Analysis of Modern Power Systems, First Edition. Edited by R. K. Saket and P. Sanjeevikumar.

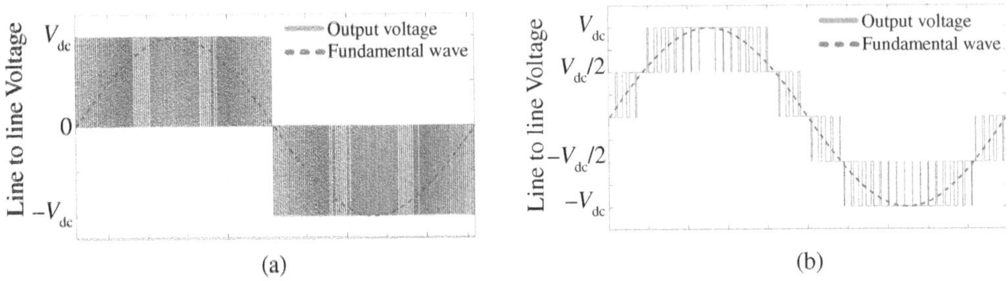

Figure 23.1 Output line voltage, and fundamental waveform (a) two-level, and (b) three-level inverters.

MLIs offer numerous benefits:

- Enhanced output waveforms with minimized THD.
- Lower stress on power electronic devices due to reduced voltage steps.
- Modularity and scalability.
- Improved efficiency in certain configurations.

The output voltage waveforms of MLIs have higher levels than those of two-level inverters, as illustrated in Figure 23.1. This means that the power quality is better. Commencing with three-level converters and progressing through various MLI topologies for several years, the first MLI application existed in 1975. MLI switches are capable of blocking lower voltages due to the distribution of DC-link stress voltage across various switching units. Consequently, creating power semiconductor devices with reduced voltage specifications enables their utilization in a broader spectrum of products, thereby enhancing their accessibility and adaptability (Wang et al. 2023). The increased switching voltages will also increase power losses on high-voltage switches. Higher efficiency, lower dV/dt, and less electromagnetic interference (EMI) are further benefits of MLIs. Higher dv/dt values result in issues with insulation and EMI. The conventional MLI, includes neutral-point clamped (NPC) (Nabae et al. 1981), flying capacitor (FC) (Kim et al. 2019), cascaded H-bridge (CHB) (Malinowski et al. 2010). The two other most popular MLIs are modular multilevel converter (MMC) and T-type. Due to its modular design, lack of energy-storing components, and capacity to isolate faults, the CHB is the most prevalent subclass (Franquelo et al. 2008; Villanueva et al. 2009). In the case of conventional MLIs, more components are needed to achieve higher output voltage levels. External controllers and switches, clamped diodes, are needed for balancing the voltage across the capacitors in the case of NPC, the no. of levels in terminal voltage increases. The presence of clamped capacitors in the FC design imposes limitations and necessitates the use of controllers to maintain the capacitor voltage balance. However, as the number of levels increases in the output voltage for CHB, need for switches and DC sources increases. Because CHB lacks a voltage balancing circuit, it is better suited for substantial-power applications (Malinowski et al. 2010). The inclusion of sub-module capacitor voltage balancing in MMC introduces complexity to its control, yet its modular design renders it well-suited for a range of high-power applications, particularly in high-voltage direct current (HVDC) transmission (Debnath et al. 2015).

The researchers have developed a switched-capacitor-based MLI, which can generate boosted multilevel output voltage because the traditional MLI topologies cannot increase voltage, which can raise the terminal voltage to appropriate grid voltage by utilizing switched-capacitor voltages found in the circuits with minimize components. In the matter of conventional MLI, extra high-gain DC/DC converter DC/DC converter is required to raise the voltage of the dispersed power sources to required level for grid integration purposes. As the no. of voltage levels rises, the circuit's count of the switched capacitors also increases, contributing to an elevation in the output voltage. Suitable switching must be given to make self-voltage balance less voltage ripple since this switched capacitor draws an enormous magnitude of charging current as the ripple voltage across the capacitor voltage increases (Tayyab et al. 2022). Various techniques, such as hysteresis current control and CHB inverters, have been employed to optimize the harmonic characteristics of voltage and current waveforms for effective grid integration (Yao and Xiao 2012; Rashid 2011). In addition to its fault-tolerant and versatile traits, it produced a higher number of voltage levels and a well-matched sinusoidal voltage waveform.

Additionally, has examined new suggested MLIs with fewer components in light of comparative variables, component selection, and the acceptability and availability of the components (Agrawal and Jain 2017). Finally, research studies have suggested the design of a modular CHB MLI in single or three-phase grid-connected applications to increase the flexibility and efficiency of power systems and maximize power extraction and efficiency (Xiao and Tolbert 2014; Bana et al. 2019). The number of power semiconductor devices employed in DC/AC converters increased to obtain high-quality voltage waveforms, making them more prone to failure. Additionally, MLIs provide economical ways for electric aircraft to improve reliability, as some MLI topologies may function in faulty conditions without additional hardware components. According to the research presented above, it is evident that additional work is needed to build a fault-tolerant MLI topology with the express purpose of implementing fault-tolerance characteristics by adopting the lowest device count possible. As a result, reliability is one of the main obstacles to the widespread use of MLIs in industrial applications.

The factors influencing the reliability:

- **Component Failures**: Every electronic component within the inverter has its failure rate. Factors like operating temperature, switching frequency, and stress can accelerate these failures (Sarita et al. 2023).
- **Control Complexity**: Higher-level multilevel inverters (MLI) require more complex control strategies, potentially increasing the likelihood of control-related issues.
- **Thermal Management**: Thermal stress can accelerate component degradation.Exploring the application of artificial neural networks for optimizing battery temperature control to improve system efficiency and minimize energy consumption. Effective cooling systems and designs that distribute thermal load evenly can significantly influence reliability (Saket 2011; Singh et al. 2023).

The following Sections 23.2–23.4 in this chapter are structured as follows: Section 23.2 emphasizes the use of reliability assessment techniques (RAT) for the mathematical evaluation of reliability within the framework of MLIs, Section 23.3 presents the different types of MLIs popular in grid-connected inverters, and their configurations are discussed in depth, Section 23.4 covers the comparative reliability assessment of MLIs, and Section 23.5 concludes the chapter.

23.2 Reliability Assessment Techniques

Reliability engineering is an intricate discipline that employs a variety of mathematical approaches to evaluate the probability of a system performing without failure over a specified time. In this section, some basic mathematical formulas and metrics related to reliability are discussed briefly that can be applied to the MLIs. Previous chapters provide a detailed description of mathematical formulas, related computations, and the applicability of these techniques.

Reliability $R(t)$:

$$R(t) = e^{-\lambda t} \tag{23.1}$$

where λ = failure rate. Failure rate λ: The rate at which failures occur in a system. If $f(t)$ represents probability density function (PDF) of failures, then:

$$\lambda(t) = \frac{f(t)}{R(t)} \tag{23.2}$$

- **Series Systems**: For components connected in series (all components must work for the system to work):

$$R_{\text{system}}(t) = R_1(t) \times R_2(t) \times \cdots \times R_n(t) \tag{23.3}$$

where $R_i(t)$ = Reliability of the ith component.

- **Parallel Systems (Redundancy)**: For components connected in parallel (only one component must work for the system to work):

$$R_{\text{system}}(t) = 1 - [(1 - R_1(t)) \times (1 - R_2(t)) \times \cdots \times (1 - R_n(t))] \tag{23.4}$$

- **Mean time to failure (MTTF) and mean time between failures (MTBF)**:
  - For nonrepairable systems:

$$\text{MTTF} = \int_0^{\infty} R(t)\text{d}t \tag{23.5}$$

  - For repairable systems:

$$\text{MTBF} = 1/\lambda \tag{23.6}$$

- **Availability**: For repairable systems, Availability A is the proportion of time a system is operational:

$$A = \frac{\text{MTBF}}{\text{MTBF} + \text{MTTR}} \tag{23.7}$$

where MTTR represents as mean time to repair. The above following parameters have been discussed in detail in previous chapters.

- **Reliability Application to MLIs**:
  - **Diode-Clamped**: Given the nature of diode-clamped inverters, components are typically in series. Thus, the reliability of the inverter would be a product of the individual reliabilities.
  - **Flying Capacitor**: Capacitors can be thought of as parallel components (redundant paths). The reliability would therefore be evaluated using the formula for parallel systems.
  - **Cascaded H-Bridge**: Modules can act in parallel for redundancy purposes, but components are typically in series within each module.
  - **T-type**: Components within a T-type inverter are often in series.

- **Failure Mode and Effects Analysis (FMEA)**: A systematic approach that evaluates potential failure modes of each component and estimates the impact on the overall system (Kara-Zaitri et al. 1991).

  FMEA is a step-by-step methodology used to identify potential failure modes at the earliest stages and evaluate their potential impact. While FMEA is primarily a qualitative technique, it uses quantitative metrics such as Risk Priority Number (RPN), which is calculated as follows:

$$\text{RPN} = \text{Severity} \times \text{Occurrence} \times \text{Detection} \tag{23.8}$$

where Severity (S) rates the seriousness of the failure mode effect on a scale of 1 (not severe) to 10 (very severe), Occurrence (O) estimates the likelihood of the failure mode occurring, rated from 1 (unlikely) to 10 (very likely), and Detection (D) assesses the probability of the failure mode being detected before it affects the customer, rated from 1 (highly likely) to 10 (very unlikely).

- **Markov Chain Analysis**: Useful in modeling the reliability of systems where different states (like healthy, degraded, or failed) are present and transition probabilities can be determined (Hou et al. 2015).
  Markov chain analysis models systems as they transition between states based on certain probabilities. In the context of MLI, the states might represent different operational conditions such as healthy, degraded, and failed. Let's assume there are n states for the inverter's health. The transition probability matrix (P) represents the likelihood of transitioning from one state to another. For reliability assessment, if one state represents "failed," the probability of being in this state after a certain number of transitions (time steps) can be determined. This can be calculated by raising the matrix (P) to the power of the number of transitions and examining the appropriate matrix entry. The reliability is assessed by evaluating the long-term state probabilities, which are the stable probabilities of being in each state as time tends to infinity.
  For inverters, factors affecting the transition probabilities might include:
  - Component degradation,
  - Operational stresses (e.g. temperature and load variations),
  - Control strategies implemented.
- **Life Cycle Cost Analysis (LCCA)**: By considering the total cost of ownership, including repair and maintenance, this analysis provides insight into the long-term reliability and economic feasibility of different inverters (Dhillon 1989).
  LCCA evaluates the total cost of ownership of a product, considering not just the initial cost but all costs over its lifecycle. The total Life Cycle Cost (LCC) is given by:

$$\text{LCC} = \text{CC} + \text{OC} + \text{MC} + \text{FC} - \text{RV} \tag{23.9}$$

  where
  - **CC (Capital Costs)**: Initial costs, including purchase, transportation, and installation.
  - **OC (Operating Costs)**: Costs associated with energy, water, and other utilities required for the system's operation.
  - **MC (Maintenance Costs)**: Predicted maintenance, repairs, and replacement costs.
  - **Failure Costs**: Costs related to system downtime, reactive maintenance, and potential penalties for failing to meet energy delivery commitments.
  - **RV (Residual Value)**: Salvage value or end-of-life disposal costs.

  For reliability, the Failure Costs term is crucial. If a multilevel inverter has a higher probability of failure due to its design or operational conditions, the Failure Costs will be higher, thus increasing the LCC. By comparing the LCCs of different inverter designs or operational strategies, one can determine which option is most cost-effective over the long run, taking reliability into account.

Integrating Markov Chain Analysis with LCCA makes it possible to estimate the Failure Costs more accurately. For instance, knowing the probability of failure from the Markov model can estimate the expected downtime and associated costs for the inverter over its lifecycle. This holistic approach comprehensively views the reliability and economic implications of different multilevel inverter designs and operational strategies.

23.3 Types of Multilevel Inverters (MLIs)

The ability of MLIs to meet specific requirements for power rating and electric power quality has led to an increase in their use in systems with low levels of electronic radiation and harmonic distortion. The commercial viability of MLIs for high-power quality applications and dynamic performance needs is now being studied. For these applications, the power ranges from 1 to 30 MW. MLIs are excellent for medium to high-voltage applications because they provide higher voltages and may produce output voltage waveforms with lower THD even with constrained device ratings (Rodriguez et al. 2007). The flying capacitor (FC), NPC, CHB, T-type, and MMC topologies are now the most popular and commonly used topologies in the commercial sector. As shown in Figure 23.3, there are two ways that MLIs can be connected to the grid.

For MLI, there are various types of modulation strategies. The different modulation methods are space vector modulation (SVM), selective harmonic elimination (SHE), third harmonic injection, and sinusoidal pulse-width modulation (SPWM) (Kouro et al. 2010; Ahmed and Borghate 2014) (Figure 23.2). The most well-known of these traditional techniques is SPWM, which is further divided into level-shifted pulse-width modulation (LSPWM) and phase-shifted pulse-width modulation (PSPWM) (Carrara et al. 1992). In this, the LSPWM technique is used for MLIs.

23.3.1 Cascaded H-Bridge Multilevel Inverters

Baker and Bannister introduced the CHB inverter in 1975. A generalized CHB MLI is demonstrated using a single-phase full-bridge inverter that is series-linked and has a separate DC supply. A series connection aims to improve the output voltage levels while lowering harmonic distortions. Figure 23.3 shows the five-level CHB MLI architecture, which consists of two full-bridge inverters with a distinct DC supply whose values can be the same or different. Each H-bridge output voltage is added together to create the output voltage waveform. An MLI arrangement is considered to be symmetrical if the magnitude of the input side DC supply is same for every H-bridge, otherwise, it is said to be asymmetrical. The generation of output voltage for a five-level CHB is shown in Figure 23.4. Table 23.1 displays switching states to achieve appropriate voltage values for five-level CHB.

There are different forms of modulation techniques for MLI. Generally, two signals are used in the pulse width modulation technique: the reference (modulating)signal and the carrier signal, but in MLI, multiple carriers are used. For five-level CHB (5-1), i.e. four carriers are used with one modulating signal. To generate a PWM, the modulating signal is compared with carriers. If the triangle carrier waveform has a peak amplitude of A_c with a frequency of f_c and the sinusoidal reference waveform has a peak amplitude of A_m with a frequency of f_m, respectively. Then, the

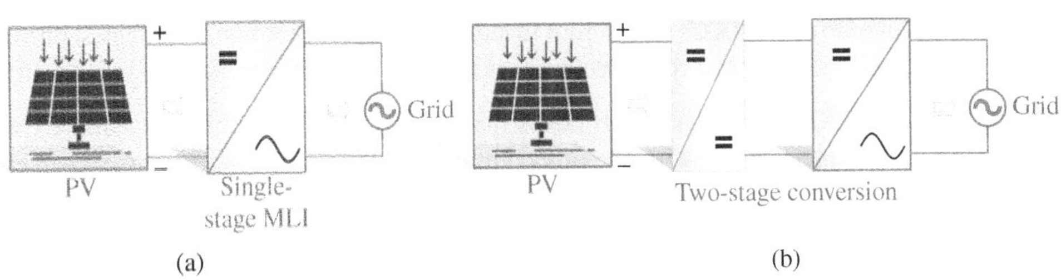

Figure 23.2 Types of inverter in accordance with stage of conversion (a) one-stage or direct conversion and (b) dual-stage conversion.

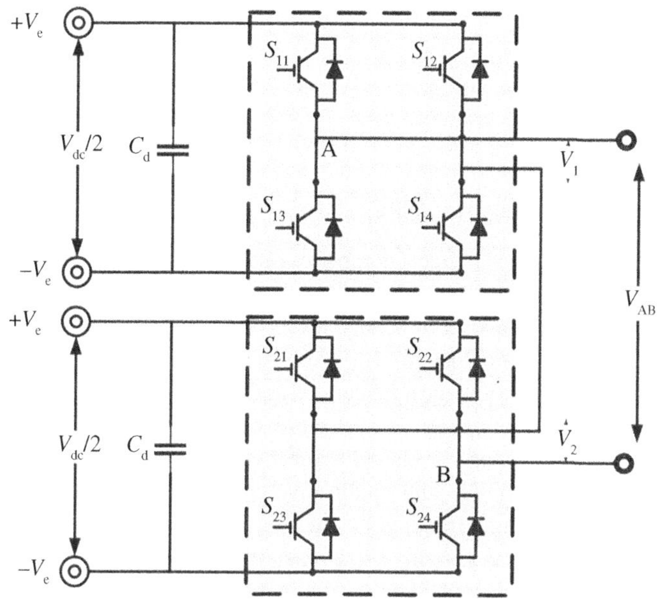

Figure 23.3 Single phase five-level CHB inverter.

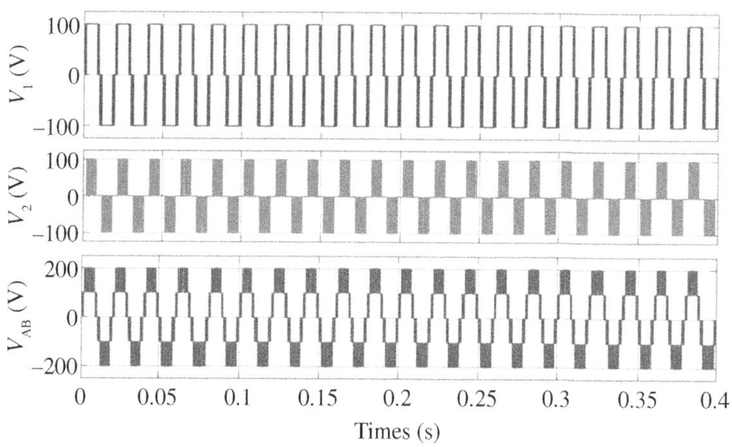

Figure 23.4 The generation of output voltage for five-level CHB.

Table 23.1 Switching states for five-level CHB.

	Switching signals								Output voltage
States	S_{11}	S_{12}	S_{13}	S_{14}	S_{21}	S_{22}	S_{23}	S_{24}	V_{AB}
1	OFF	ON	ON	OFF	OFF	OFF	OFF	OFF	$-V_{dc}/2$
2	OFF	ON	ON	OFF	OFF	ON	ON	OFF	$-V_{dc}$
3	OFF	OFF	OFF	OFF	OFF	OFF	OFF	OFF	0
4	ON	OFF	OFF	ON	OFF	OFF	OFF	OFF	$+V_{dc}/2$
5	ON	OFF	OFF	ON	ON	OFF	OFF	ON	$+V_{dc}$

amplitude modulation index and frequency modulation ratio are given as:

$$M_a = \frac{A_m}{A_c} \tag{23.10}$$

$$m_f = \frac{m_m}{m_c} \tag{23.11}$$

23.3.2 Neutral-Point Clamped Three-Level Inverter (NPC)

A three-level converter, known as an NPC or diode-clamped, comprises two capacitor voltages connected in series with the middle tap as the neutral. Baker released another diode-clamped inverter in 1980. This inverter splits the DC bus voltage in a set of voltage levels using a series-connected semiconductor device. Then, the most effective topology in MLI is the NPC introduced by Nabae et al. (1981). The excellent capabilities in medium-voltage applications and the relatively high efficiency of NPC inverters have led to widespread adoption. Furthermore, the neutral point gives a zero-voltage level, resulting in three separate voltage levels. NPC inverters are typically used in electrical drives, uninterruptible power supply (UPS), and medium-voltage grid-connected converters because they are efficient in the fundamental frequency switching range. Entire frequency switching increases voltage and current THD, necessitating extra reactors to minimize the problem, which might raise the implementation cost. On the other hand, increasing the voltage levels necessitates expanding the number of clamping diodes and capacitors, which may compromise the overall structure's durability. As opposed to that, this topology has the advantage of lowering the harmonic content of the produced voltage by adding the zero-voltage level. Higher levels can be achieved by increasing switching devices, clamping diodes, and capacitors. The output phase voltage of an n-level diode-clamped inverter is n-level, while the output line voltage is $(2n - 1)$. Figure 23.5 shows the diagram of a three-level NPC. The output voltage of the NPC three-level inverter varies between $(+V_{dc}/2)$ and (0) or $(-V_{dc}/2)$ and (0), while a conventional inverter varies between $(+V_{dc}/2)$ and $(-V_{dc}/2)$. Table 23.2 lists the NPC converter's switching states and the output voltages. Figure 23.6 presents the outcomes for phase and line voltage, with Figure 23.7 showing the capacitor balancing of NPC topology.

23.3.3 Flying Capacitors Three-Level Inverter

Meynard and Foch (1992) created a new topology, the FC; this design uses the floating capacitor to clamp the voltage. The topologies of the diode-clamped multilevel inverter (DCMLI) or NPC and

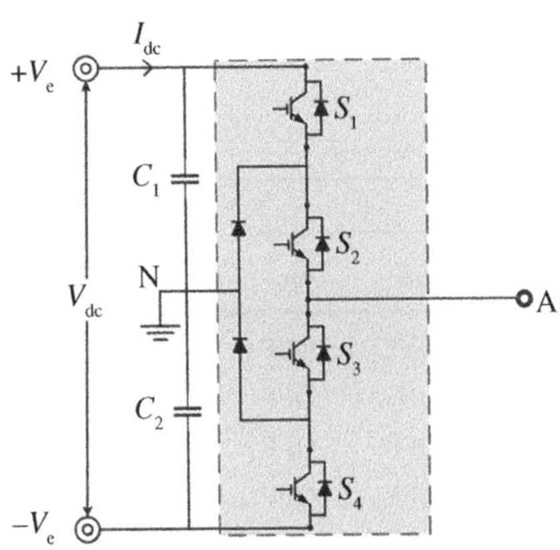

Figure 23.5 Three-level NPC inverter.

Table 23.2 Switching states for three-level NPC.

	Device switching status				Terminal voltage
States	S_1	S_2	S_3	S_4	V_{AN}
1	OFF	OFF	ON	ON	$-V_{dc}/2$
2	ON	OFF	ON	OFF	0
3	ON	ON	OFF	OFF	$+V_{dc}/2$

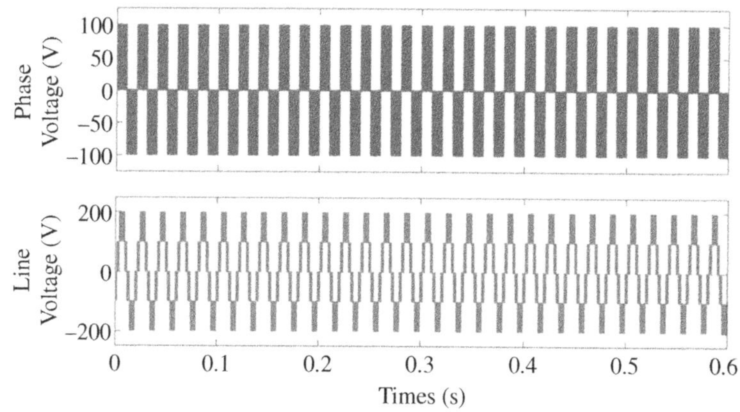

Figure 23.6 Phase and line voltages waveform of three-level NPC inverter.

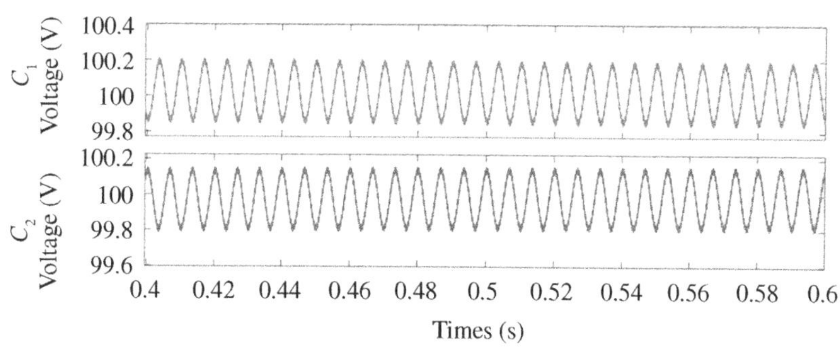

Figure 23.7 Balancing of capacitor voltage for three-level NPC inverter.

the flying capacitor multilevel inverter (FCMLI) are very similar, with the only difference that the DCMLI employs clamping diodes while the FCMLI uses floating capacitors. Figure 23.8 depicts the inverter with three capacitor clamping levels. One DC source can power all the FCs because of switching state redundancy inside the phase of the FC inverter. The capacitor clamping requires many bulk capacitors to clamp the voltage, just like the diode-clamped inverter. An N-level converter will need $(N - 1) * (N - 2)/2$ clamping capacitors per phase. Additionally, the $N - 1$ main DC bus capacitors, provided that each capacitor used has the same voltage rating as that of the power switch. The rating of the capacitors, which have significant amounts of the DC bus voltage across them, is a downside of the topology. The switches S_1 and S_4 are complementary. Thus, if S_1

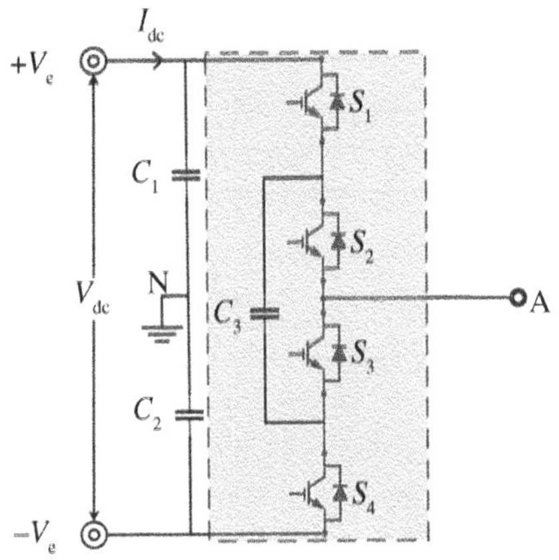

Figure 23.8 The three-level FCMLI.

Table 23.3 Switching states for three-level FCMLI.

	Switching status				Terminal voltages	
States	S_1	S_2	S_3	S_4	V_{AN}	C_3
1	OFF	OFF	ON	ON	$-V_{dc}/2$	NC
2	OFF	ON	OFF	ON	0	↓
3	ON	OFF	ON	OFF	0	↑
4	ON	ON	OFF	OFF	$+V_{dc}/2$	NC

is ON, S_4 is OFF, and vice-versa. Similarly, the switches S_2 and S_3 are complementary. The state of the FC corresponding to the selected switching combination is also shown in Table 23.3. The NC indicates the capacitor is not charged or discharged in this mode. ↑ and ↓ signifies the charging and discharging of the capacitors. The switching states are specified for the positive half of the current waveform, i.e. the load current is going out. The capacitor's state (↑ and ↓) will get reversed for the negative half cycle of the current. The equation of output phase voltage for FCMLI is given below:

$$V_{AN} = S_1(V_C - V_{C_3}) + S_2 V_{C_3} - \frac{V_C}{2} \tag{23.12}$$

where, V_{C_1}=V_{C_2}=V_C. Figure 23.9 presents the outcomes for phase and line voltage, with Figures 23.10 and 23.11 showing the capacitor balancing of FCMLI topology.

23.3.4 Three-Level T-Type Inverter

Depending on the multilevel inverter topology's structure, they can have issues with voltage balancing, isolated power supply requirements (for CHB inverters), large capacitor requirements (for FC inverters), and massive capacitor requirements (for NPC inverters). In addition to these topologies, a unique variety of NPC inverter called the T-type inverter is gaining popularity. T-type inverters, which have advantages like simple operation and fewer switches, can replace complex three-level inverter topologies since they do not require an isolated source, a large capacitor, or clamping diodes. This architecture is also more effective at medium frequencies than the two- and three-level NPC inverters (Schweizer and Kolar 2011, 2013). The clamping diodes are eliminated despite the

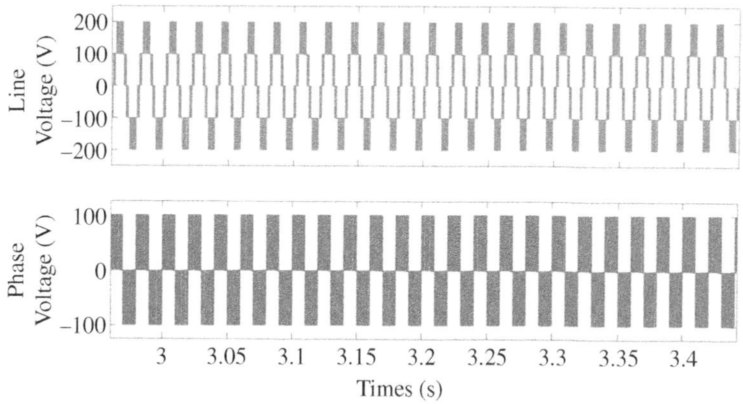

Figure 23.9 Line and phase voltages of three-level FCMLI.

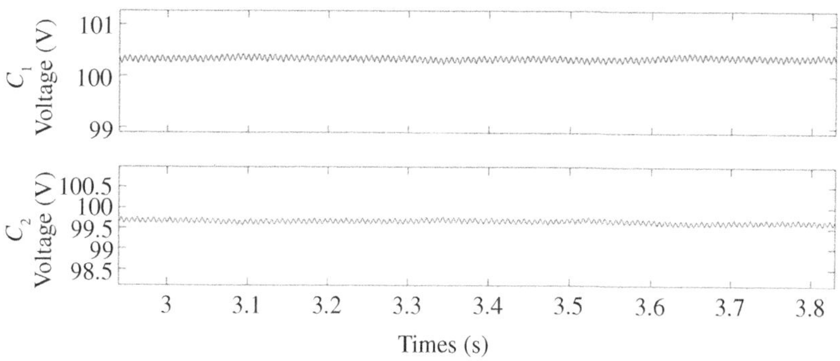

Figure 23.10 Balancing of DC-link capacitor voltage for three-level FCMLI.

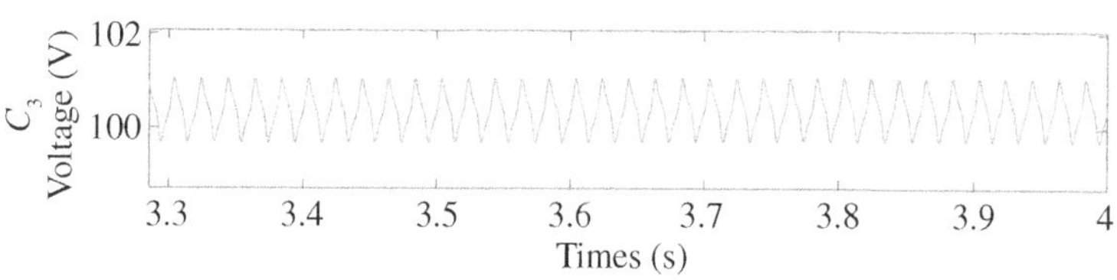

Figure 23.11 Balancing of flying capacitor voltage for three-level FCMLI.

inverter having the same active switches as an NPC inverter. Figure 23.12 shows the three-level T-type converter ($3LT^2C$) fundamental topology. A dynamic, bidirectional switch is added to the two-level, traditional VSC architecture to extend it to the DC-link midpoint (Schweizer and Kolar 2013). A less isolated gate driver is required than the other three-level topologies and the neutral point is also connected to the common emitter switches, which operate at low frequency. The bidirectional switches used in T-type are of common emitter or collector configuration. The $3LT^2C$ can use the same modulation techniques as the three-level NPC converter. When the bidirectional switch (combination of S_4 and S_3) is ON, the voltage from the load terminal to the neutral point is zero, $-V_{dc}/2$ when S_4 is ON, and $V_{dc}/2$ when S_1 is ON. The switching states and output voltages for the $3LT^2C$ are shown in Table 23.4. Figure 23.13 presents the outcomes for phase and line voltage, with Figure 23.14 showing the capacitor balancing of $3LT^2C$ topology.

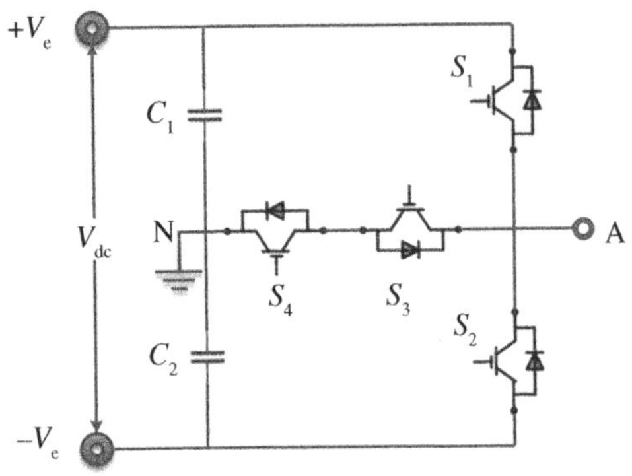

Figure 23.12 Three-level T-type topology with common emitter configuration.

Table 23.4 Switching states for three-level T-type MLI.

	Switching signals				Output voltage
States	S_1	S_2	S_3	S_4	V_{AN}
1	OFF	OFF	ON	ON	$-V_{dc}/2$
2	OFF	ON	ON	OFF	0
3	ON	ON	OFF	OFF	$+V_{dc}/2$

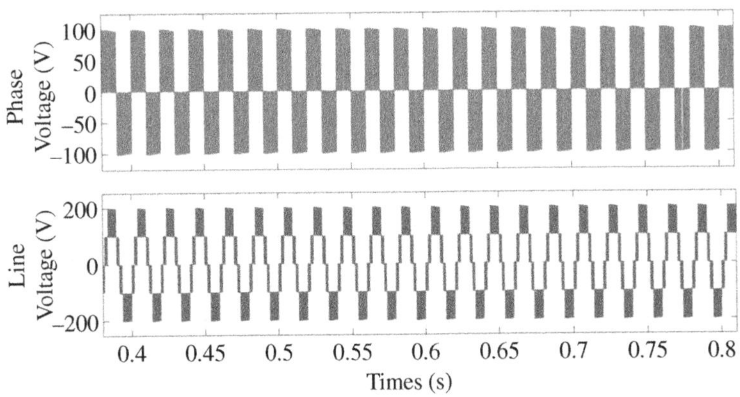

Figure 23.13 Phase and line voltages of three-level T-type converter.

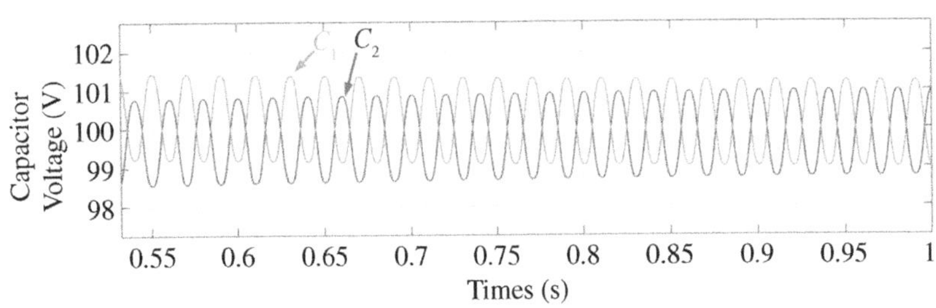

Figure 23.14 Balancing of DC-link capacitor voltage for three-level T-type converter.

23.4 Comparative Reliability Assessment of MLIs

MLI have become increasingly popular in contemporary power systems because they can generate high-quality output voltages, Minimize harmonic interference, and function effectively at elevated voltage levels. This section delves into the reliability of four prominent configurations: CHB, diode-clamped or NPC, flying capacitor, and T-type.

- **Diode-Clamped (Neutral-Point Clamped)**
  - **Advantages**:
    * Simple structure and widely used in medium-voltage applications.
    * Reduced voltage stress across each switch due to multiple clamping diodes.
  - **Reliability Concerns**:
    * Voltage balancing issues for the clamping diodes, especially at higher levels.
    * A higher number of clamping diodes increases component count, potentially affecting reliability.
- **Flying Capacitor (Capacitor-Clamped)**
  - **Advantages**:
    * Offers the possibility of redundant paths, which can be a boon for reliability.
    * Natural voltage balancing under certain conditions.
  - **Reliability Concerns**:
    * Requires voltage balancing control, which adds complexity.
    * Capacitors are prone to aging and require maintenance and periodic replacement.
- **Cascaded H-Bridge (CHB)**
  - **Advantages**:
    * Modular structure allows for increased system reliability; a failure in one module doesn't incapacitate the entire system.
    * Uses conventional two-level switching devices, simplifying design and control.
  - **Reliability Concerns**:
    * Requires separate DC sources for each H-bridge, which might be a limitation in some applications.
    * Increase in the number of semiconductor devices can raise concerns for maintenance and potential points of failure.
- **T-type (Three-level inverter)**
  - **Advantages**:
    * Fewer components compared to other three-level topologies.
    * Efficient utilization of the DC supply voltage, allowing for higher output voltages.
  - **Reliability Concerns**:
    * Neutral-point voltage balancing can be challenging, similar to the diode-clamped topology.
    * Potential for shoot-through (short-circuit) if both switches in a leg turn on simultaneously. Requires meticulous control.

A comparative summary of the MLIs based on different important parameters is given:

- **Component Count**: Generally, as the component count increases, potential failure points also increase. CHB has the highest component count, while T-type tends to have the least among three-level inverters.
- **Voltage Balancing**: Both diode-clamped and T-type configurations face challenges with voltage balancing, while FC inverters require intricate voltage balancing control.

- **Redundancy**: FC inverters offer redundant switching paths, providing an inherent reliability advantage. The modularity of CHB also offers redundancy benefits.
- **Control Complexity**: T-type inverters have simpler control strategies than Diode-Clamped or FC configurations, making them less prone to control-induced failures.
- **Maintenance**: FC inverters might need more frequent maintenance due to capacitor aging, while CHB designs, due to their high component count, can have more potential maintenance points.

23.5 Conclusion

In conclusion of this book chapter, it becomes evident that the increasing need for efficient and reliable power conversion in contemporary power systems magnifies the importance of MLI. A comprehensive understanding of their reliability, influenced by design, control, and environmental factors, ensures that these systems can effectively serve the evolving needs of the energy landscape. No single inverter configuration is universally the most reliable. The choice depends on the specific application, operational environment, and reliability challenges that need addressing. For instance, a T-type or CHB inverter might be preferable in climates where maintenance is complex due to their modularity and lower component count. Meanwhile, applications that require redundancy might benefit more from the FC design. As always, careful design, control, and operational practices will significantly influence the reliability of any chosen topology.

References

Agrawal, R. and Jain, S. (2017). Comparison of reduced part count multilevel inverters (RPC-MLIs) for integration to the grid. *International Journal of Electrical Power & Energy Systems* 84: 214–224. https://doi.org/10.1016/j.ijepes.2016.05.011.

Ahmed, I. and Borghate, V.B. (2014). Simplified space vector modulation technique for seven-level cascaded H-bridge inverter. *IET Power Electronics* 7: 604–613. https://doi.org/10.1049/iet-pel.2013.0135.

Bana, P.R., Panda, K.P., Naayagi, R.T. et al. (2019). Recently developed reduced switch multilevel inverter for renewable energy integration and drives application: topologies, comprehensive analysis and comparative evaluation. *IEEE Access* 7: 54888–54909. https://doi.org/10.1109/ACCESS.2019.2913447.

Carrara, G., Gardella, S., Marchesoni, M. et al. (1992). A new multilevel PWM method: a theoretical analysis. *IEEE Transactions on Power Electronics* 7 (3): 497–505. https://doi.org/10.1109/63.145137.

Debnath, S., Qin, J., Bahrani, B. et al. (2015). Operation, control, and applications of the modular multilevel converter: a review. *IEEE Transactions on Power Electronics* 30 (1): 37–53. https://doi.org/10.1109/TPEL.2014.2309937.

Dhillon, B. (1989). *Life Cycle Costing: Techniques, Models and Applications*. Routledge.

Franquelo, L.G., Rodriguez, J., Leon, J.I. et al. (2008). The age of multilevel converters arrives. *IEEE Industrial Electronics Magazine* 2 (2): 28–39. https://doi.org/10.1109/MIE.2008.923519.

Hou, K., Jia, H., Xu, X. et al. (2015). A continuous time Markov chain based sequential analytical approach for composite power system reliability assessment. *IEEE Transactions on Power Systems* 31 (1): 738–748.

Kara-Zaitri, C., Keller, A.Z., Barody, I., and Fleming, P.V. (1991). An improved FMEA methodology. In: *Annual Reliability and Maintainability Symposium. 1991 Proceedings*, 248–252. IEEE.

Kim, S.-H., Kim, R.-Y., and Kim, S.-I. (2019). Generalized model predictive control method for single-phase *N*-level flying capacitor multilevel rectifiers for solid state transformer. *IEEE Transactions on Industry Applications* 55 (6): 7505–7514. https://doi.org/10.1109/TIA.2019.2917869.

Kothari, D.P., Nagrath, I.J., and Saket, R.K. (2022). *Modern Power System Analysis*. Tata McGraw-Hill Publishing Company.

Kouro, S., Malinowski, M., Gopakumar, K. et al. (2010). Recent advances and industrial applications of multilevel converters. *IEEE Transactions on Industrial Electronics* 57 (8): 2553–2580. https://doi.org/10.1109/TIE.2010.2049719.

Li, W., Ruan, X., Bao, C. et al. (2014). Grid synchronization systems of three-phase grid-connected power converters: a complex-vector-filter perspective. *IEEE Transactions on Industrial Electronics* 61 (4): 1855–1870. https://doi.org/10.1109/TIE.2013.2262762.

Mahalakshmi, R. and Thampatty, K.C.S. (2015). Grid connected multilevel inverter for renewable energy applications. *Procedia Technology* 21: 636–642. https://doi.org/10.1016/j.protcy.2015.10.076.

Malinowski, M., Gopakumar, K., Rodriguez, J., and Pérez, M.A. (2010). A survey on cascaded multilevel inverters. *IEEE Transactions on Industrial Electronics* 57 (7): 2197–2206. https://doi.org/10.1109/TIE.2009.2030767.

Meynard, T.A. and Foch, H. (1992). Multi-level conversion: high voltage choppers and voltage-source inverters. In: *PESC '92 Record. 23rd Annual IEEE Power Electronics Specialists Conference*, vol. 1, 397–403. IEEE https://doi.org/10.1109/PESC.1992.254717.

Nabae, A., Takahashi, I., and Akagi, H. (1981). A new neutral-point-clamped PWM inverter. *IEEE Transactions on Industry Applications* IA-17 (5): 518–523. https://doi.org/10.1109/TIA.1981.4503992.

Rashid, M.H. (2011). *Power Electronics: Circuits, Devices, and Application (for Anna University)*. Pearson Education India.

Rodriguez, J.É., Bernet, S., Wu, B. et al. (2007). Multilevel voltage-source-converter topologies for industrial medium-voltage drives. *IEEE Transactions on Industrial Electronics* 54 (6): 2930–2945. https://doi.org/10.1109/TIE.2007.907044.

Saket, R.K. (2011). *Reliability Evaluation of Defence Support Systems*, 241–286. Berlin, Heidelberg: Springer-Verlag. ISBN 978-3-642-17764-4. https://doi.org/10.1007/978-3-642-17764-4_8.

Sarita, K., Saket, R.K., and Khan, B. (2023). Reliability, availability, and condition monitoring of inverters of grid-connected solar photovoltaic systems. *IET Renewable Power Generation* 17 (7): 1635–1653.

Schweizer, M. and Kolar, J.W. (2011). High efficiency drive system with 3-level T-type inverter. In: *Proceedings of the 2011 14th European Conference on Power Electronics and Applications*, 1–10. IEEE.

Schweizer, M. and Kolar, J.W. (2013). Design and implementation of a highly efficient three-level T-type converter for low-voltage applications. *IEEE Transactions on Power Electronics* 28 (2): 899–907. https://doi.org/10.1109/TPEL.2012.2203151.

Singh, S., Saket, R.K., and Khan, B. (2023). A comprehensive state-of-the-art review on reliability assessment and charging methodologies of grid-integrated electric vehicles. *IET Electrical Systems in Transportation* 13 (1): e12073.

Surana, K. and Jordaan, S. (2019). The climate mitigation opportunity behind global power transmission and distribution. *Nature Climate Change* 9: 1–6. https://doi.org/10.1038/s41558-019-0544-3.

Tayyab, M., Sarwar, A., Murshid, S. et al. (2022). Grid-connected operation and control of single-phase asymmetrical multilevel inverter for distributed power generation. *IET Renewable Power Generation* 16 (16): 3629–3642. https://doi.org/10.1049/rpg2.12581.

Villanueva, E., Correa, P., Rodriguez, J.É., and Pacas, M. (2009). Control of a single-phase cascaded H-bridge multilevel inverter for grid-connected photovoltaic systems. *IEEE Transactions on Industrial Electronics* 56 (11): 4399–4406. https://doi.org/10.1109/TIE.2009.2029579.

Wang, D., Hemming, S., Yang, Y. et al. (2023). Multilevel inverters for electric aircraft applications: current status and future trends. *IEEE Transactions on Transportation Electrification* 1. https://doi.org/10.1109/TTE.2023.3296284.

Xiao, B. and Tolbert, L.M. (2014). Efficiency improved and current balanced three-phase modular cascaded H-bridge multilevel PV inverter for grid-connected applications. In: *2014 IEEE Energy Conversion Congress and Exposition (ECCE)*, 4661–4669. IEEE https://doi.org/10.1109/ECCE.2014.6954039.

Yao, Z. and Xiao, L. (2012). Two-switch dual-buck grid-connected inverter with hysteresis current control. *IEEE Transactions on Power Electronics* 27 (7): 3310–3318. https://doi.org/10.1109/TPEL.2011.2179318.

24

Reliability Aspects in Snubber Circuit for Industrial Power Applications

Dhawal Dwivedi[1], Saumya Singh[1], Kumari Sarita[2], R. K. Saket[1], and P. Sanjeevikumar[3]

[1] *Department of Electrical Engineering, Indian Institute of Technology (BHU), Varanasi, Uttar Pradesh, India*
[2] *Department of Electrical Engineering, Government Engineering College, Aurangabad, Aurangabad, Bihar, India*
[3] *Department of Electrical Engineering, IT and Cybernetic, University of South-Eastern Norway, Notodden, Norway*

24.1 Introduction

The early workers quickly noticed the inherent issue of triggering a switch with an inductive load due to their repetitive switching. When the switch opened at first, an arc was seen across the main switch contacts. Two things happened as a result of this arc. The connections were quickly corroded. Second, the primary switch, which frequently heated quickly, used a portion or possibly the majority of the energy available intended for the secondary spark. There was an excessive switching loss! The connection between switch behavior and loss in an inductive circuit was discovered more than 160 years ago (Kothari et al. 2022; Sarita et al. 2023). Armand Fizeau devised a solution in 1853 by placing a capacitor across the contacts. When the switch is turned off, the current of switch is commutated to the capacitor (Severns 2008). Still, because the switch has discharged it, the voltage across the capacitor is very low and only slowly increases as the integral of the current. The primary arc is reduced. As a result, he is allowing the switch contacts to open with relatively little voltage applied across them. The action is the same as that found in modern semiconductor capacitive turn-off snubbers. Fizeau's paper shows that he saw the issue clearly and devised a solution. Of course, the ringing voltage waveform that frequently associated with modern snubbers was caused by a capacitor was placed across the primary while the secondary was being discharged. Snubbers can serve various extremely valuable purposes, but they almost always come with "side effects" that must be dealt with. Research efforts were redirected toward the development of resonant converters in the 1980s. The idea was to use resonant tanks in the converters to produce oscillating voltage and/or current waveforms, which are typically sinusoidal, so that the power switches may operate in a zero voltage switching (ZVS) or zero current switching (ZCS) condition. The switching frequency of resonant converters can now reach hundreds of kilo-Hertz (usually 100 to 500 kHz) thanks to the reduction of switching loss and ongoing advancements in power switches (Rashid et al. 2018). Power electronics would only be complete with snubbers. The purpose of snubbers, compact network components in power-switching circuits, is to regulate the effects of circuit reactances. Snubbers improve switching circuit performance and produce smaller size, lighter size, more efficient, and more reliable switching circuits with greater switching frequencies. A snubber's main purpose is to remove energy from the circuit's reactive components. Circuit damping, regulating the voltage or current change rate, or clamping voltage overshoot are possible benefits. A snubber lowers the stress and hence increases the reliability (Todd 1993). Snubbers are

Reliability Analysis of Modern Power Systems, First Edition. Edited by R. K. Saket and P. Sanjeevikumar.

often divided into two categories: passive networks and active networks. The only passive snubbers covered in this chapter are resistors, capacitors, inductors, and diodes. Transistors and other active switches are examples of active snubbers, frequently suggesting additional circuitry. The way of energy is transferred to and from snubbers may influence their classification. The snubber is categorized as dissipative if its energy goes to a resistor for dissipation but as nondissipative if the energy is returned to the input or output (Todd 1993; García-Caraveo et al. 2010). Another way to categorize snubber circuits is based on their mode of operation, which allows us to modify the load line to maintain it inside the safe operating area (SOA) for overload snubbers and minimize or eliminate voltage or current spikes (García-Caraveo et al. 2010). The structure of this chapter is organized as follows. Section 24.2 deals with the theory of passive snubber circuits and the conventional design methods, Section 24.3 discusses the selection of turn-OFF snubber, Section 24.4 discusses the design of a discharge-suppressing resistor-capacitor-diode (RCD) snubber, the simulation results of RCD snubber are given in Sections 24.5 and 24.6 discusses reliability aspects in snubber design for industrial power applications, and Section 24.7 concludes the chapter.

24.2 Passive Snubber Circuit

During turn-off, transient voltages are protected by snubber circuits. A series combination of resistor R with a capacitor C is typically used in a simple RC snubber. A power MOSFET and an RC snubber are connected in parallel. A circuit's voltage rises significantly when the current source is eliminated because of stray inductances. The snubber dampens this surge voltage to save the power of MOSFET and other components nearby. Snubbers are capable of a wide range of actions, including (Severns and EMI Reduce 2006):

1. Reduce or eliminate voltage or current spikes.
2. Limit dI/dt or dV/dt.
3. Shape the load line to keep it within the SOA.
4. Transfer power dissipation from the switch to a resistor or a useful load.
5. Reduce total losses due to switching.
6. Reduce EMI by damping voltage and current ringing.

The RCD turn-off snubber and the resistor-capacitor (RC) damping network are the two most used types of snubbers.

24.2.1 Surge Across Switch

The stray inductance of the wire on the PCB layout stores energy when a MOSFET turns on. Surge current occurs when the stored energy resonates with the MOSFET's parasitic capacitance. Surge voltage develops at the drain-source across the switch when it is turn-off due to a resonant phenomenon between Parasitic/stray inductance L_{stray} and the parasitic capacitance of the MOSFET C_{oss} (C_{ds}+C_{gd}). The path of the "ringing current" in a half-bridge topology with high-side and low-side switches is shown in Figure 24.1. Current I_L flows through the stray inductance L_{stray} when either the low-side or high-side is turned on. From ROHM application note (Rohm Semiconductor 2020), the frequency of surge (f_{surge}) is given as:

$$f_{\text{surge}} = \frac{1}{\sqrt{L_s C_{\text{OSS}}}} \sqrt{1 - \left\{ \frac{\sqrt{L_S/C_{\text{OSS}}}}{2R_{\text{OFF}}} \right\}^2} \tag{24.1}$$

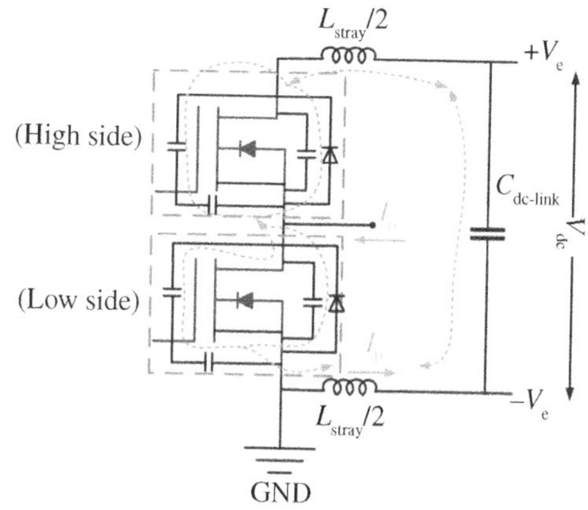

Figure 24.1 Surge current path when turn-off occurs.

24.3 Selection of Turn-OFF Snubber

A circuit's stray inductances generate surge voltage. To absorb the surge voltage, a turn-OFF snubber should be connected in parallel with a switching device. Snubbers can be connected across individual switching devices or collectively across a power bus for all switching devices. The snubber should ideally be positioned as close to the MOSFET as feasible to enhance its effectiveness.

24.3.1 Snubbers Across Each Switching Device

24.3.1.1 RC Snubber Circuit

It can be put close to the MOSFET, but every time the MOSFET switches on, R_S (snubber resistance) must release the energy held in C_S (snubber capacitance). The size of C_S is constrained by the significant amount of energy (several watts) that the R_S would dissipate if the switching frequency were high enough so, it is not suitable for high-switching applications. As a result, the snubber's ability to inhibit surges is diminished. A small-value resistor is required for an RC snubber to prevent the drain current from increasing during turn-on in a high-capacitance switching device. It is ideal for chopper circuits. The snubber resistor's power dissipation is calculated as follows:

$$P = C_S V_{dc}^2 f_{sw} \tag{24.2}$$

where V_{dc} = DC supply voltage and f_{sw} = Switching frequency.

24.3.1.2 RCD Charge-Discharge Snubber

The RC snubber's resistance can be increased by including a diode. This makes it possible to eliminate a current that switching devices have to share while turning on. During switch ON, R_S loses energy at a similar rate to (a), but C_S's capacity for surge absorption is superior to (a) because surge current passes through the diode. High di/dt in the snubber circuit can occur during the switching transient. Hence, it is essential to take the diode's recovery characteristics into account. To prevent overvoltages, stray inductances should be reduced to a minimum. The outcome is the same if R_S and C_S are connected in parallel. An RDC snubber is inappropriate for high-frequency applications due to the significant power lost through the snubber resistor (TOSHIBA 2020). The snubber resistor's power dissipation is calculated as follows:

$$P = \frac{L_{stray} I_0 f_{sw}}{2} + \frac{C_S V_{dc}^2 f_{sw}}{2} \tag{24.3}$$

where L_{stray} = Stray inductance of the main circuit and I_0 = Drain current during device turn-off.

24.3.1.3 Discharge-Suppressing RCD Snubber

In this turn-off surge voltage is suppressed, and it does not discharge all of the energy held in C_S during each switching transient because R_S only dissipates the energy absorbed by C_S produced during the overvoltage and hence the dissipation of power by a snubber is small. Therefore, even at high switching frequencies, the energy consumption at R_S does not significantly rise and makes it ideal for high-frequency switching applications As a result, a huge C_S can be used, realizing a very efficient snubber circuit. However, it should be emphasized that this method demands an extremely complicated wire layout that can be accomplished by a PCB with more than four layers. The snubber resistor's power dissipation is calculated as follows:

$$P = \frac{L_{\text{stray}} I_0^2 f_{\text{sw}}}{2} \tag{24.4}$$

Figure 24.2 shows different snubbers across each switching device examples: (a) RC snubber, (b) RCD charge-discharge snubber, and (c) Discharge-suppressing RCD snubber.

24.3.2 Lumped Snubbers Between Power Buses

24.3.2.1 C Snubber

The simplest is C snubbers, but the main circuit's stray inductances and the snubber capacitor's LC resonance make them vulnerable to voltage oscillation.

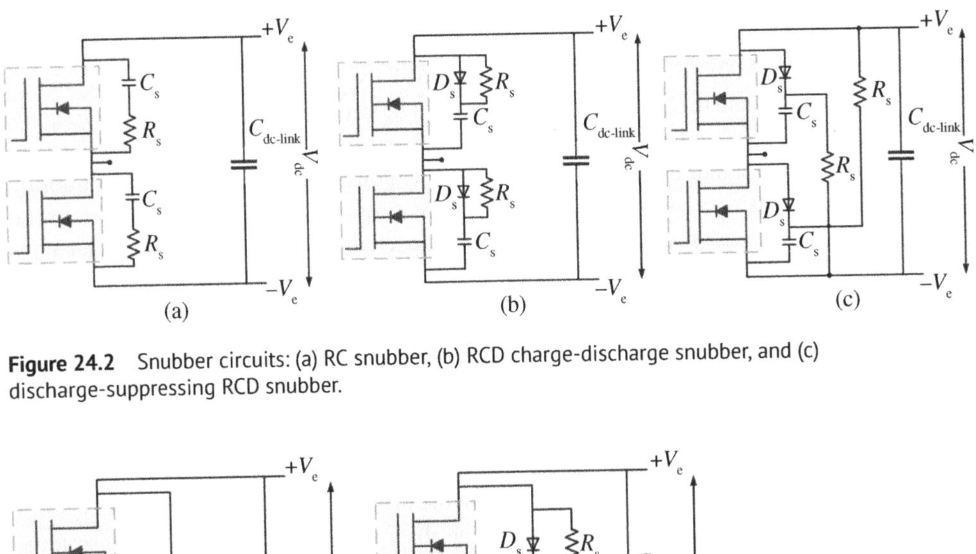

Figure 24.2 Snubber circuits: (a) RC snubber, (b) RCD charge-discharge snubber, and (c) discharge-suppressing RCD snubber.

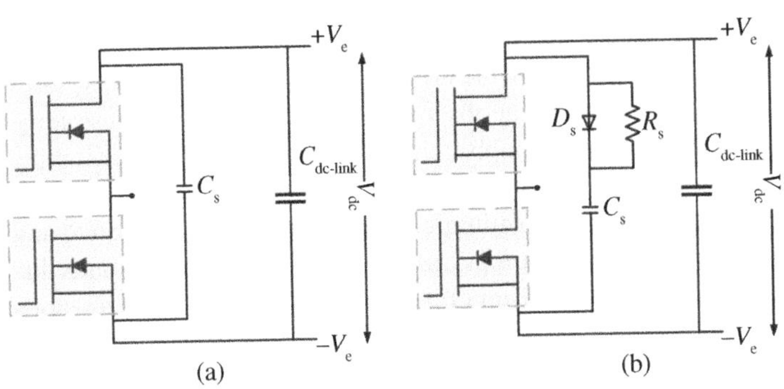

Figure 24.3 Lumped snubbers between power buses: (a) C snubber and (b) RCD snubber.

24.3.2.2 RCD Snubber

Snubber diodes should be chosen carefully since they could result in voltage oscillation and a significant voltage spike during reverse recovery. Figure 24.3 shows the lumped snubber (a) C snubber circuit and (b) RCD snubber.

24.4 Design of a Discharge-Suppressing RCD Snubber

Figure 24.4 shows a detailed circuit of RCD. The following equations can be used to compute the voltages and current constants shown in the waveform.

(a) Figure 24.4a V_{L_s} is the voltage generated by the snubber's inductance L_s, and it can be calculated as follows:

$$V_{L_s} = V_{\text{dc}} + V_{\text{fr}} + L_s \times \frac{\text{d}i}{\text{d}t} \tag{24.5}$$

The forward voltage V_{fr} of the diode D_s should be kept as low as possible and decrease L_s that can result in a voltage surge in order to achieve this.

(b) $V_{L_{s2}}$ and C_s When the snubber capacitor C_s is overcharged with energy from the main circuit's stray inductance L_{stray}, a peak voltage known as $V_{L_{s2}}$ is observed across the capacitor C_s. Their energies are equal since the L_{stray} energy is transferred to C_s. As a result, the following equation is valid:

$$\frac{1}{2} \times L_{\text{stray}} \times I_{\text{OFF}}^2 = \frac{1}{2} \times C_s \times (V_{L_{s2}} - V_{\text{dc}})^2 \tag{24.6}$$

This equation yields the following result for the value of C_s:

$$C_s = \frac{L_{\text{stray}} \times I_{\text{OFF}}^2}{(V_{L_{s2}} - V_{\text{dc}})^2} \tag{24.7}$$

where L_s = inductance of the snubber circuit, V_{fr} = forward voltage of D_s, C_s = capacitor of the snubber circuit, $V_{L_{s1}}$ = peak surge voltage of L_s, $V_{L_{s2}}$ = peak forward voltage of C_s and I_{OFF} = turn OFF current. It is necessary to determine $V_{L_{s2}}$, considering the MOSFET's sustained voltage.

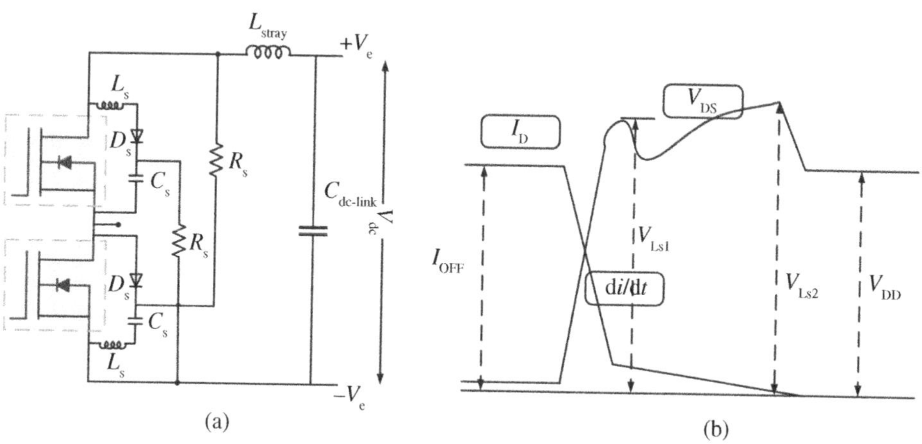

Figure 24.4 Discharge-suppressing RCD (a) snubber circuit and (b) waveform.

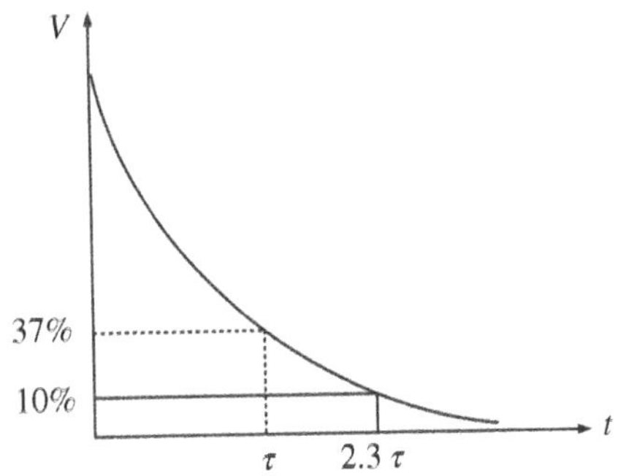

Figure 24.5 Time constant versus the amount of discharge.

(c) Selection of the value of snubber resistor R_S: The snubber resistor R_s aims to release the electric charge stored in the snubber capacitor before a MOSFET starts its further turn-on action. Suppose that is the discharge time constant τ, then:

$$\tau = R \cdot C \tag{24.8}$$

Where, τ represents the time it takes for the voltage to drop to 37% of the stored voltage, as illustrated in Figure 24.5, the voltage drops to 10% in 2.3τ, or 90% of the capacitor's stored charge is discharged.

Before performing the next turn-off, the capacitor needs to be discharged. As a result, the following equation needs to be satisfied. The following formula can be used to compute R_S:

$$2.3\tau \leq \frac{1}{f_{\text{sw}}}$$

$$2.3 \times R_S \cdot C_S \leq \frac{1}{f_{\text{sw}}}$$

$$R_S \leq \frac{1}{2.3 \cdot C_S \cdot f_{\text{sw}}} \tag{24.9}$$

The snubber may display current oscillation if the resistor's value is too low. Use a resistor with a resistance that is as high as possible.

24.5 Simulation Results of RCD Snubber

Consider a two-level half-bridge inverter with a stray or parasitic induction of around 2.9 nH. The input V_{dc} voltage is 100 V, the load resistance is 20 Ω, and the switching frequency is 10 kHz. The waveform of surge/spike is shown in Figure 24.6. Figure 24.6a For snubber across each switching device: Figure 24.7 displays the waveform of the drain to source voltage after inserting a C snubber of 560 pf across the switching devices. Typically, the range of snubber capacitors varies from a few pf to 1 μf depending upon the application and is made up of MLCC (multi-layer ceramic capacitor), film capacitor, etc. But for a higher current rating, the snubber capacitor goes up to 10 μf. Then analyze the waveform of one cycle, T_r (ringing time period)=80 μs. For optimum snubber design:

$$f_r = \frac{1}{2\pi\sqrt{L_{\text{stray}} C_s}} \tag{24.10}$$

For the underdamped condition:

$$R_s < \sqrt{\frac{L}{C}} \tag{24.11}$$

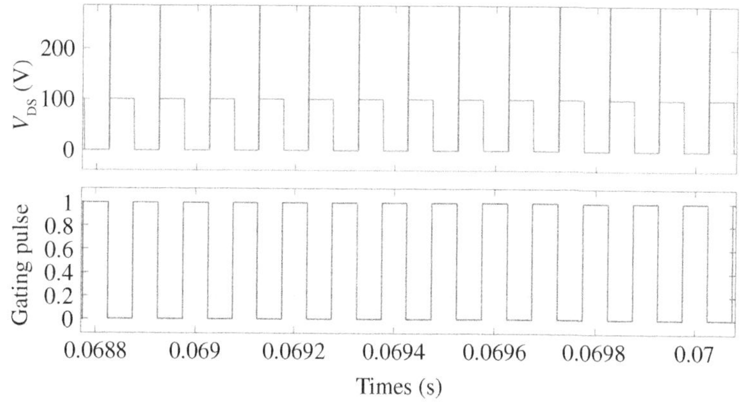

Figure 24.6 Surge across the drain to source voltage (V_{DS}) with gating pulse of MOSFET.

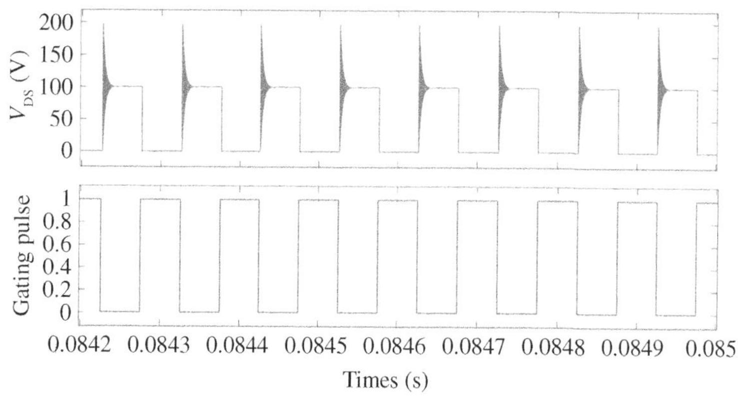

Figure 24.7 The drain to source voltage (V_{DS}) with the gating pulse of MOSFET after inserting C snubber.

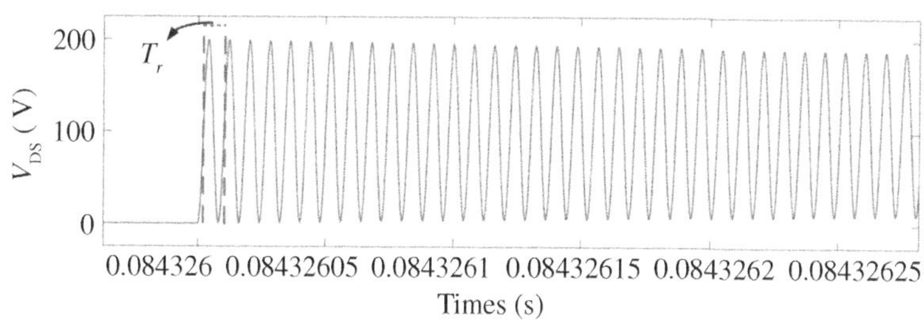

Figure 24.8 The oscillatory view of drain to source voltage (V_{DS}) after inserting C snubber.

R_s lesser than 10 Ω. This R_s is for RC or RCD snubber circuits. Figure 24.9 shows the drain to source voltage with gating pulse of MOSFET after inserting RC snubber.

Example 24.1

Calculate the required parameters for a snubber circuit to provide protection to a MOSFET used in a single-phase bridge converter. The MOSFET has a maximum capability of 60 V/μs. The input line to line voltage has a peak value of 425 V and the stray inductance is 0.2 mH.

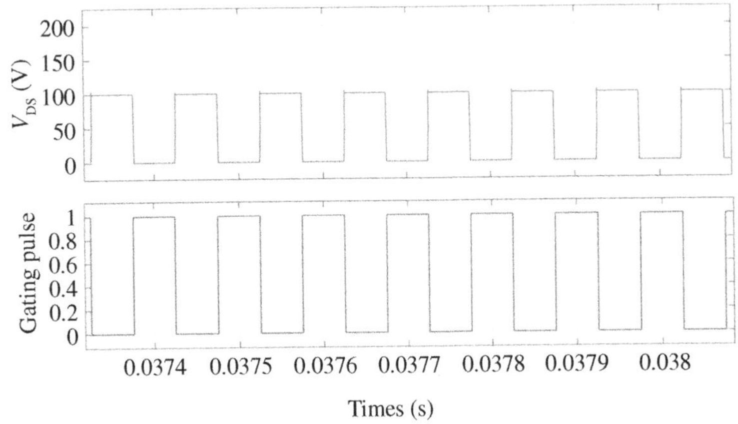

Figure 24.9 The drain to source voltage (V_{DS}) with the gating pulse of MOSFET after inserting RC snubber.

Solution: Given:

$$\frac{dv}{dt} = 60\,\text{V}/\mu\text{s}, L = 0.2\text{mH}, V_m = 425\text{V}$$

Then,

$$C = \frac{1}{2L}\left(\frac{0.564V_m}{\frac{dv}{dt}}\right)^2 = \frac{1}{2\times 0.2\times 10^{-3}}\left(\frac{0.564\times 425}{60}\times 10^{-6}\right)^2$$

$$C = 0.04\mu\text{F}$$

In the above equation observe that 10^{-6} have been multiplied in numerator inside the brackets. It comes from 60 V/μs, i.e. $\frac{60}{1\times10^{-6}}$ to be substituted in equation for C. Let the damping factor $\zeta = 0.65$. Then,

$$\zeta = \frac{R}{2}\sqrt{\frac{C}{L}}$$

$$R = 2\zeta\sqrt{\frac{L}{C}} = 2\times\sqrt{\frac{0.2\times 10^{-3}}{0.04\times 10^{-6}}}$$

$$R = 92\Omega.$$

Example 24.2

For the circuit shown in Figure 24.10, the IGBT is operated at 2 kHz. The $\frac{dv}{dt}$ required is 100 V/μs. The discharge current is to be limited to 100 A. Determine

i) Values of R_s, and C_s
ii) Snubber loss
iii) Power rating of R_s

Load and stray inductances are negligible.

Solution: Given:

Load resistance, $R = 5\Omega$, frequency= 2 kHz, $V_s = 200$ V, $\frac{dv}{dt} = 100$ V/μs, $I_{TD} = 100$ A, $L = 0$

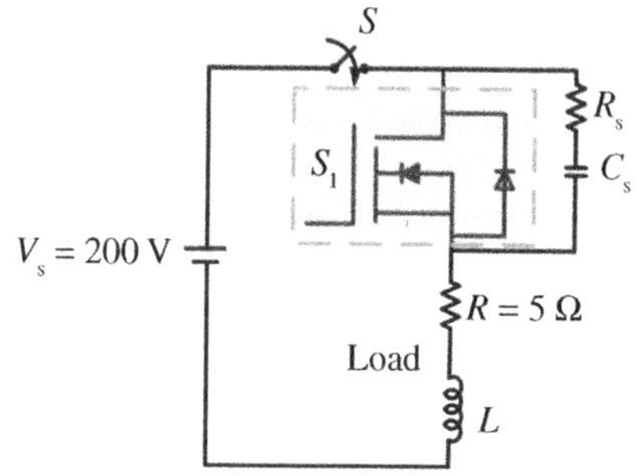

Figure 24.10 Circuit for Example 24.2.

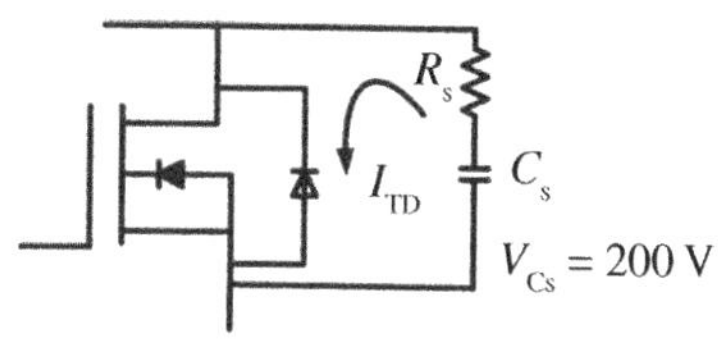

Figure 24.11 Path of I_{TD}.

(i) **To obtain the value of R_s and C_s**

R_s limits the discharging current through MOSFET S_1. $S_1 - R_s - C_s$ forms a loop when S_1 turns ON. Prior to turning ON of S_1, C_s charges to 200 V. From this Figure 24.11,

$$V_{C_s} = R_s I_{TD}$$

$$R_s = \frac{V_{C_s}}{I_{TD}} = \frac{V_s}{I_{TD}} = \frac{200}{100} = 2\Omega$$

KVL can express the charging current of C_s to $V_s - R_s - C_s - R - L$ loop as,

$$V_s = R_s i_C(t) + \frac{1}{C}\int i_C(t)\,dt + V_C(t-0) + Ri_C(t)$$

on solving the above equation,

$$i_C(t) = \frac{V_s}{R_s + R}\exp^{-\frac{t}{\tau}}, \text{Here } \tau = R_s + R \quad C_s \text{ and } V_c(t=0) = 0.$$

Hence, the voltage across the MOSFET can be expressed as,

$$i_C(t) = \frac{V_s}{R_s + R}\exp -\frac{t}{\tau}, \text{Here } \tau = (R_s + R)C_s \text{ and } C_s \text{ and } v_C(t=0) = 0. \qquad (24.12)$$

Hence, the voltage across the MOSFET can be expressed as,

$$v_{S_1}(t) = V_s - Ri_c(t) = V_s - \frac{RV_s}{R + R_s}e^{-\frac{t}{\tau}}$$

$$At\ t = 0,\ v_{S_1}(0) = V_s - \frac{RV_s}{R_s + R}\frac{R_s V_s}{R_s + R}$$

$$\text{and at } t = \tau,\ v_{S_1}(0) = V_s - \frac{RV_s}{R_s + R}\exp^{-1} = V_s - \frac{0.9678RV_s}{R_s + R}$$

$$= \frac{R_s V_s + 0.6321RV_s}{R_s + R}$$

Now $\frac{dv}{dt}$ can be expressed as,

$$\frac{dv}{dt} = \frac{v_{T_1}(\tau) - v_{T_1}(0)}{\tau} = \frac{\frac{R_sV_s + 0.6321RV_s}{R_s + R} - \frac{R_sV_s}{R_s + R}}{(R_s + R)C_s}$$

$$= \frac{0.632RV_s}{(R_s + R)^2C_s}$$

$$C_s = \frac{0.632RV_s}{\frac{dv}{dt}(R_s + R)^2} = \frac{0.632 \times 5 \times 200}{\frac{100}{1 \times 10^{-6}}(2 + 5)^2} = 0.1289\mu\text{F} \quad (24.13)$$

(ii) **To obtain snubber loss**

The power stored in C_s is dissipated in R_s. Hence, it is snubber loss

$$P_s = \frac{1}{2}C_sV_s^2f_s$$

$$= \frac{1}{2} \times 0.1289 \times 10^{-6} \times 200^2 \times 2 \times 10^3$$

$$= 5.2\text{W}$$

(iii) **To obtain power rating of R_s**

The power stored in C_s is dissipated in R_s. It is 5.2 W. Hence power rating of R_s will be 5.2 W.

24.6 Reliability Aspects in Snubber Design for Industrial Power Applications

Before diving into the reliability aspects, it was essential to understand the basic structure and functionality of snubber circuits that were covered in Sections 24.3–24.5. This section will discuss the reliability aspects of the snubber circuit.

In the realm of industrial power applications, snubber circuits play a pivotal role. Their primary function is to suppress voltage transients, limit voltage spikes, and protect power semiconductor devices. Ensuring the reliability of snubber circuits is of paramount importance to maintain the overall efficiency, safety, and longevity of power systems. In industrial applications, these are the main reasons behind the need for reliable operation of snubber circuit.

- **Protection of Expensive Equipment**: Voltage spikes can cause damage to semiconductor devices, leading to costly replacements.
- **Operational Continuity**: An unreliable snubber can cause outages, affecting production timelines.
- **Safety**: High-voltage transients can cause accidents if not managed appropriately.

The following are the factors affecting system reliability in industrial power electronics applications.

- **Component Selection**: The quality and specifications of resistors, capacitors, and diodes determine their performance under stress.
- **Thermal Management**: Excessive heat can reduce the life of snubber components.
- **Circuit Design**: Proper placement and interconnections can play a role in enhancing or undermining reliability.
- **Environmental Factors**: Moisture, dust, and corrosive agents can degrade components over time.

System reliability is typically expressed in terms of the MTTF, representing the average lifespan of a system until it encounters its first failure as discussed by Singh et al. (2023). This parameter can be determined by examining the system's failure rate, especially when components showcase exponential behaviors during their useful operational periods, as discussed by the authors Sarita et al. (2023).

For streamlining reliability analysis, certain assumptions are taken into account:

- The snubber's failure would result in an open circuit.
- Each individual component's failure within the snubber circuit would render the entire circuit nonfunctional.
- The combined failure rate of the snubber is the aggregate of its individual components' failure rates:

$$\lambda_S = \lambda_R + \lambda_C + \lambda_D \tag{24.14}$$

This supposition suggests that a malfunction in the snubber circuit doesn't lead to the entire system's breakdown. Put differently, even if the snubber circuit encounters an issue, the primary circuit remains operational. However, the entire system would only collapse if the switch malfunctions.

The switch's likelihood of breaking down is affected by the condition of the snubber circuit. As long as the snubber circuit is working, the switch's failure chance, represented by λ_W, depends on the snubber's capacitor size. But if the snubber breaks, the switch uses more power and has a greater risk of failure, represented by λ_{W1}.

The system's reliability can be determined by combining the reliability of separate, nonoverlapping events that result in the system working correctly, as illustrated in Figure 24.12. These successful outcomes are: (i) both the switch and snubber remain functional for a time period from 0 to t, or (ii) the snubber breaks between 0 to t_1, but the switch continues to work from 0 to t. We can represent the reliability of these events with functions R_1 and R_2, which are detailed in the the following equations:

$$R_1(t) = \exp(-\lambda_W(t))\exp(-\lambda_S(t)) \tag{24.15}$$

$$R_2(t) = \int_0^t (\exp(-\lambda_W t_1))(\lambda_S \exp(-\lambda_S t_1 dt_1))(\exp(-\lambda_{W1}(t - t_1))) \tag{24.16}$$

The terms on the right side of $R_2(t)$ represent the following: the first term denotes the switch working successfully up to time t_1 with a failure rate of λ_W; the second term indicates the likelihood of the snubber failing between times t_1 and $t_1 + dt_1$; and the third term describes the switch's

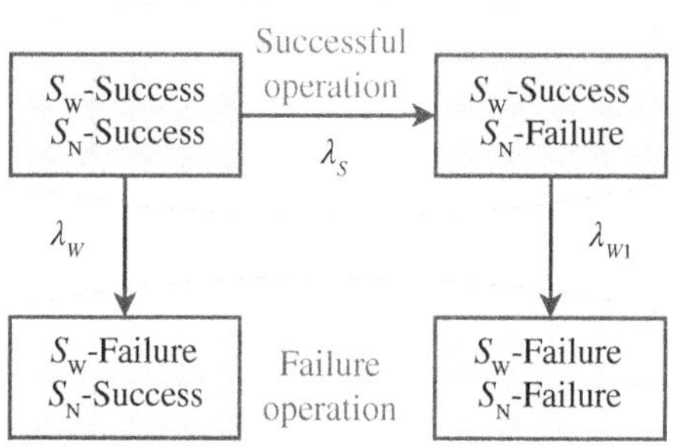

Figure 24.12 Reliability of switch and snubber represented in state space diagram.

successful operation from time t_1 to t with a failure rate of λ_{W_1}, assuming the snubber isn't functioning. By combining R_1 and R_2, we can determine the overall system reliability and then compute the MTTF:

$$\text{MTTF} = \frac{\lambda_{W1} + \lambda_S}{(\lambda_W + \lambda_S)\lambda_{W1}} \tag{24.17}$$

The average failure time for a circuit without a snubber is given by $1/\lambda_{W1}$. Thus, the MTTF ratio comparing a system with a snubber to one without it can be defined as:

$$\text{MTTF}_{\text{ratio}} = \frac{\lambda_{W1} + \lambda_S}{(\lambda_W + \lambda_S)} \tag{24.18}$$

A compelling observation emerges when examining MTTF as a function of snubber capacitance, as explained by Parvari et al. (2018). The Authors concluded that the initial increments in capacitance dramatically enhance MTTF due to a sharp decline in the switch's failure rate. However, as capacitance continues to rise, the benefits plateau. This behavior stems from the balance between reducing switch power loss and increasing snubber-set energy storage. This trade-off suggests an optimum capacitance value that strikes a balance between efficiency and reliability. To achieve reliable operation of the system, the following methods can be implemented.

- **Component Derating**: Using components at levels below their maximum rated values ensures longer lifespans.
- **Heat Sinks and Cooling**: Prevents thermal overload and ensures steady performance.
- **Quality Checks**: Regular inspection and testing can identify issues before they escalate.
- **Protective Coatings**: Can protect against environmental hazards.

24.7 Conclusion

In the vast landscape of industrial power applications, snubber circuits might appear to be minute details. Yet, their reliability ensures the smooth operation, safety, and efficiency of much larger and more intricate systems. By understanding and prioritizing their reliability aspects, industries can ensure operational excellence and reduce downtimes. This chapter dealt with different aspects of the reliability assessment of snubber circuits. The choice of optimal capacitor values does not ensure reliable operation. The chosen values of R, C, and D must be such that reliability is enhanced as discussed in this chapter.

References

García-Caraveo, A., Soto, Á., González, R., and Bañuelos-Sánchez, P. (2010). Brief review on snubber circuits. In: *2010 20th International Conference on Electronics Communications and Computers (CONIELECOMP)*, 271–275. https://doi.org/10.1109/CONIELECOMP.2010.5440754.

Kothari, D.P., Nagrath, I.J., and Saket, R.K. (2022). *Modern Power System Analysis*. Tata McGraw-Hill Publishing Company.

Parvari, R., Zarghani, M., and Kaboli, S. (2018). RCD snubber design based on reliability consideration: a case study for thermal balancing in power electronic converters. *Microelectronics Reliability* 88: 1311–1315.

Rashid, M.H., Hui, S.Y.R., and Chung, H.S.-H. (2018). Chapter 12 - Resonant and soft-switching converters. In: *Power Electronics Handbook*, 4e (ed. M.H. Rashid), 339–383. Butterworth-Heinemann. ISBN 978-0-12-811407-0. https://doi.org/10.1016/B978-0-12-811407-0.00012-X. https://www.sciencedirect.com/science/article/pii/B978012811407000012X.

Rohm Semiconductor (2020). Snubber circuit design methods. *Application Note*. https://fscdn.rohm.com/en/products/databook/applinote/discrete/sic/mosfet/sic-mos_snubber_circuit_design_an-e.pdf (accessed 9 February 2024).

Sarita, K., Saket, R.K., and Khan, B. (2023). Reliability, availability, and condition monitoring of inverters of grid-connected solar photovoltaic systems. *IET Renewable Power Generation* 17 (7): 1635–1653.

Severns, R.P. (2008). *Snubber Circuits for Power Electronics*. Rudy Severns. https://books.google.co.in/books?id=SmVFPwAACAAJ.

Severns, R. and EMI Reduce (2006). *Design of Snubbers for Power Circuits*. International Rectifier Corporation.

Singh, S., Saket, R.K., and Khan, B. (2023). A comprehensive state-of-the-art review on reliability assessment and charging methodologies of grid-integrated electric vehicles. *IET Electrical Systems in Transportation* 13 (1): e12073. https://doi.org/10.1049/els2.12073.

Todd, P.C. (1993). Snubber circuits: theory, design and application. In: *Unitrode-Power Supply Design Seminar*. Texas Instruments Incorporated.

TOSHIBA (2020). Power MOSFET selecting MOSFFETs and consideration for circuit design. *Application Note*. https://toshiba.semicon-storage.com/info/application_note_en_20180726_AKX00064.pdf?did=13416 (accessed 9 February 2024).

25

Reliability Assessment of Power Electronics Devices and Systems for Modern Power Applications

Jyoti Maurya[1], Saumya Singh[1], Sachin Kumar[2], P. Sanjeevikumar[3], and R. K. Saket[1]

[1] *Department of Electrical Engineering, Indian Institute of Technology (BHU), Varanasi, Uttar Pradesh, India*
[2] *Department of Electrical Engineering, G.B. Pant Institute of Engineering and Technology, Pauri-Garhwal, Uttarakhand, India*
[3] *Department of Electrical Engineering, IT and Cybernetics, University of South-Eastern Norway, Porsgrunn, Norway*

25.1 Introduction

Power electronic applications involve the management and control of power flow. In power electronics (PE), the predominant focus is efficiently controlling power flow by employing various switching techniques (Peyghami et al. 2020). Different power semiconductor devices, highlighting their usage as switches in power electronic applications. Some commonly used power semiconductor switches include bipolar junction transistors (BJTs), diodes, thyristors, gate turn-off thyristors (GTOs), insulated gate bipolar transistors (IGBTs), metal oxide semiconductor field effect transistors (MOSFETs), and metal oxide semiconductor-controlled thyristor switches (MCTs). Power MOSFETs and IGBTs significantly impact the control of modern power equipment, so their study is emphasized.

The stability of these systems is today a primary concern and most challenging difficulty in modern power applications, where these semiconductor switches have been integrated with power converters for grid-connected power systems. Power system modernization is required for reliable and secure power delivery with a zero carbon footprint. It necessitates the deployment of new technology and infrastructure and the deregulation of the electrical sector.

Electronic transmission, renewable energy resources (RER), distribution system storage, and e-mobility are well-established technologies essential to power system modernization (Kothari et al. 2022). Additionally, PE are a vital component in the energy conversion processes of the above-mentioned technologies. Power systems are heading for 100% renewable energies in the future, which increases the necessity of PE. However, as power converters are a common source of failure in many applications (Peyghami et al. 2021), substantial downtime and maintenance expenses are required. Furthermore, industrial experience and field data show that power converters are susceptible to wear-out and growing defeats resting on operation conditions (Peyghami et al. 2019).

Power converters consist of various elements, which include gate drivers, power modules, capacitors, cooling systems, and control units. Among these, electrolytic power modules and capacitors are the most delicate parts, susceptible to wear-out failures. Various elements, such as the durability of the device's structure, the amount of electrical load it handles, the surrounding environment, and the efficiency of its control and switching systems, collectively impact the reliability of these

Reliability Analysis of Modern Power Systems, First Edition. Edited by R. K. Saket and P. Sanjeevikumar.

parts. These factors collectively contribute to the gradual deterioration of component materials over the converter's extended operation, ultimately activating potential failure modes.

Utilizing principles of failure physics within stress-strength analysis (SSA) allows for anticipation of component reliability and improvement of wear-out performance across different failure modes. Electrothermal modeling is crucial in SSA and operates on three levels: device, converter, and system.

System modeling identifies the load on the converter by analyzing its utilization within the power system. Methods like power-sharing are employed to gauge the load on converters that are connected in parallel. Additionally, converter-level examinations utilize electrical models and simulations to evaluate strain on individual components due to control and switching systems in place. Electrothermal device modeling becomes indispensable for capturing crucial thermal parameters such as hot-spot temperature, which, in turn, help predict failure modes within a designated mission profile. Consequently, SSA analysis encompasses both fast dynamics at the device level in the switching frequency range and slow dynamics at a system level in the hourly load fluctuation zone. As a result, forecasting wear-out failures over a yearly mission profile takes time. It will place a substantial computational load on system-level analysis, especially in enormous-scale power systems.

In the context of power electronic devices and systems (PEDS), the technique of SSA has been harnessed to model and extend the lifespan of both capacitors and semiconductors. Focusing the converter level, active thermal management strategies, including adaptive switching frequency, suitable modulation methods, and controlling reactive power, are applied to bolster the converter's overall lifespan.

Converter topologies and solar array parameters are two further factors that influence converter lifespan. Furthermore, capacitors' lifespan expansion is examined by interleaving converters. Again, by correctly altering converter shifting and loading device damages from a high-stress converter to a low-stress converter, system level dependability in multi-converter systems is increased (Singh et al. 2019).

The reliability assessments mentioned earlier focus solely on forecasting and improving the lifespan of converters, even within broader system-level investigations. Nevertheless, making the best decisions when designing, planning, operating, and maintaining power system converters requires comprehensively exploring their impact on the reliability of the power system. This calls for the fusion of concepts from power electronic reliability with evaluation methods for reliability in power systems. Power system reliability refers to the capacity of a power system to fulfill consumer demands (Fischer et al. 2019).

The measurement of this capacity is primarily assessed using adequacy indicators like loss of load expectation (LOLE) and anticipated energy deficit Expected Energy Not Supplied. Moreover, apart from conventional evaluation of power system dependability, there has been significant research into the reliability of power systems based on PE, such as wind farms (WFs) and high-voltage direct current (HVDC) transmission systems.

Furthermore, Carroll et al. (2014) investigate microgrid systems' dependability by considering conditions affecting power electronic converters. In power system analysis, historical data from advanced research on analogous situations is used to establish the failure rate of converters. Additionally, the assessment of power system reliability has yet to incorporate wear-out failures of converters.

In real-world applications, converter and component wear-out may occur sooner than expected (Enslin et al. 2019). Consequently, an operational failure rate of converter will rise, and its effective lifespan will become restricted. This will lead to the deterioration of converter components over time, undermining the reliability of the entire system and elevating the potential for downtime and maintenance expenses in Maurya and Saket (2023). This scenario is particularly pertinent in extensive power electronic-based power systems (PEPS). To circumvent these challenges, implementing appropriate protocols for achieving the best decisions is crucial for designing, running, and up-keeping contemporary systems reliant on PE.

At system level, this requires an in-depth analysis of reliability, involving the modeling of converter reliability within the evaluation of power system reliability. However, this method is highly time-intensive and almost unfeasible for extensive, large-scale power plants. This is because when utilizing electrothermal modeling relying on SSA, it requires a time-domain examination covering time spans from microseconds, which are associated with the converter switching frequency, all the way to hourly adjustments in load. Consequently, there is a need for streamlined approaches to electrothermal modeling across various time scales to effectively model the reliability of power electronic systems at a system level. This will easily tackle the mentioned challenges stemming from converting converters into power systems.

(1) Due to necessity of making decisions about converter operation, planning, and maintenance system-wide basis, significance of system-level dependability modeling in power electronic-based systems cannot be overlooked. This chapter strives to connect models of converter reliability with power system reliability principles, intending to assess the reliability of PE-based power systems.
(2) Due to growing utilization of converters within power systems, the likelihood of their malfunctions becomes intertwined with stochastic factors, while deteriorative breakdowns influence the power system's effectiveness. This chapter aims to demonstrate interplay between converter malfunctions, aging, and their repercussions on power system dependability.
(3) Due to intricate nature of reliability modeling in SSA-based converters, it is imperative to introduce more straightforward approaches for analyzing systems at macro level. This piece elaborates on the required precision when it comes to modeling and reliability of converters, particularly in the context of power system analysis across different applications.
(4) Due to effects of component aging, there will be a rise in the failure likelihood of converters. Consequently, it becomes necessary to replace them in accordance with an appropriate maintenance strategy.

    As followed by an introduction, this chapter is structured in the following manner. An introduction, considering the recent literature available, is discussed in Section 25.1. Section 25.2 includes concept of PEDS reliability in modern power systems, Section 25.3 delves into V-shape model-based reliability assessment in PEDS, and converter reliability modeling has been elaborated in Section 25.4. At last, Section 25.5 serves as the chapter's conclusion and future challenges.

25.2 Concept of PEDS Reliability in Modern Power System

The schematic of evaluating the modern power system reliability is depicted in Figure 25.1. Reliability plays a vital role in PEDS utilized within contemporary modern power systems. Reliability pertains to the capacity of an entity or system to execute its designated operations under predefined circumstances and across a specified duration. As defined, system's effectiveness must remain consistent within a defined range over a specific time frame, as depicted in Figure 25.2.

The specific system application dictates evaluation of performance metrics. The measure of dependability performance in scenarios involving missions, such as spacecraft, relates to the

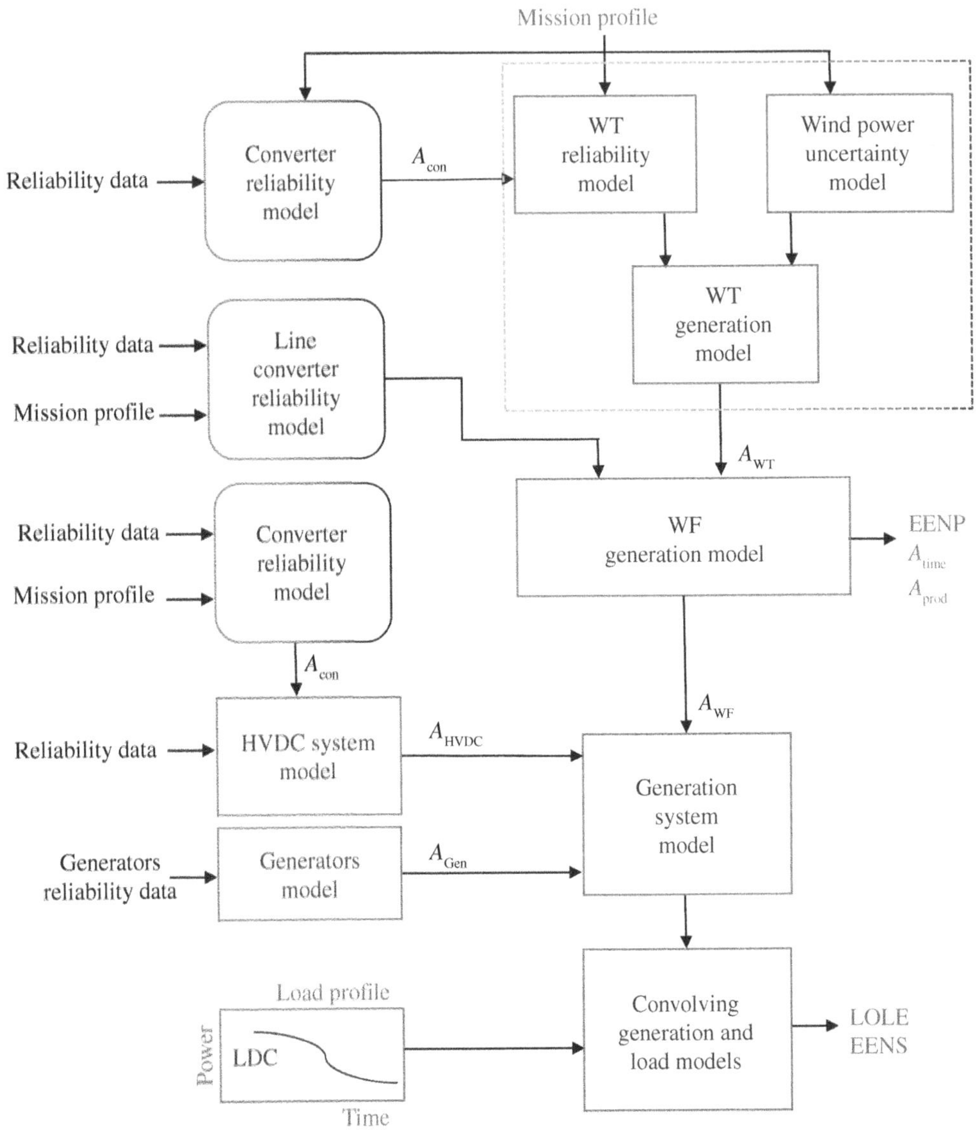

Figure 25.1 Modern power systems reliability evaluation.

possibility of lasting for the entire specified period. Similarly, in systems that can be maintained or repaired, like automobiles, ensuring continuous functionality irrespective of past malfunctions is paramount. This is gauged through availability metrics, which indicate the probability of the system is operational at any point in time (Alhmoud 2018). This description asserts that the system's up-time should be upheld at a desirable level by employing either proactive maintenance methods before potential failures or reactive maintenance approaches following malfunctions available in Figure 25.3.

Moreover, performance metrics could represent a tangible characteristic of an item or system, like capacitance of a capacitor, which falls below 80% of its nominal value, indicating a malfunction. A power system constitutes an intricate amalgamation of numerous subsystems and parts that must

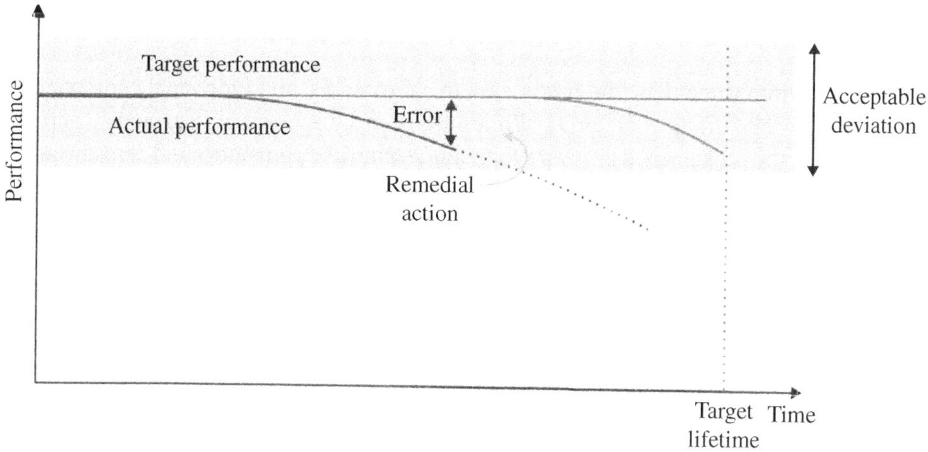

Figure 25.2 Characterization of reliability as a metric for evaluating performance within a system.

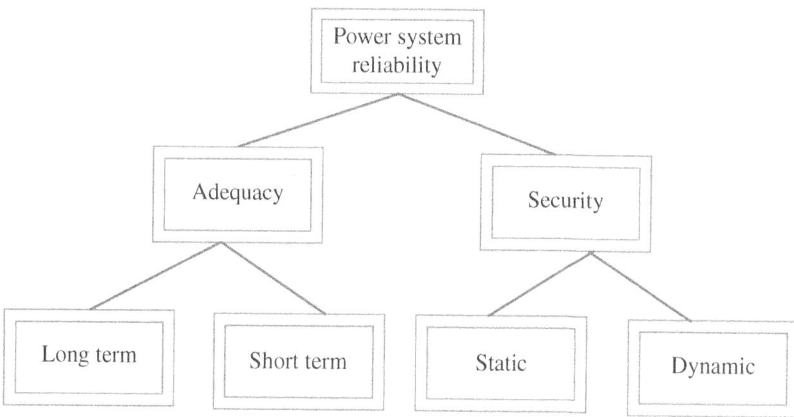

Figure 25.3 Notion of reliability within electric power systems encompasses both adequacy and security considerations.

function uninterrupted over a prolonged period of time. Following this declaration, it is imperative to guarantee the power system's capability to satisfy demand consistently. The operational phase of power systems is intertwined with their immediate effectiveness, delineated as their ability to endure unforeseen circumstances and disruptions. This capacity is characterized as a security connection related to stability of a power system (Jiang et al. 2021). Moreover, sustained effectiveness of a power system is connected to its capacity to provide power to demands despite fluctuations in generator accessibility and load requirements. This requires appropriate infrastructure for generation, transmission, and distribution. This capacity is referred to as power system adequacy, a metric for both design and planning (facility and operational planning) aspects.

Consequently, power system reliability is characterized by power system's capability to address sudden events in short run by reacting to unexpected situations and, in long run, by establishing and utilizing suitable infrastructure. To guarantee the system's effectiveness, these infrastructures must exhibit a fair degree of accessibility as per these benchmarks.

25.3 V-Shape Model-Based Reliability Assessment in PEDS

PEPS are structured into three tiers: the power system, subsystems, and individual components, as depicted in Figure 25.4 (Yang et al. 2011). While subsystems discussed here primarily pertain to power converters, the scope can also cover other subsystems like generators and transformers. Within PEPS, each converter has a distinct function in facilitating power and energy distribution, contingent on its designated purpose. These converters interlink renewable energy setups, electronic transmission systems, e-transportation charging points, and energy storage units. Consequently, their influence on PEPS performance varies. Moreover, each converter's reliability level is dictated by its specific capabilities within the PEPS framework.

Moreover, various elements exist within each converter, including power devices, capacitors, cooling mechanisms, and control and protection modules. The specifications for converter's reliability within the PEPS dictate design and dimensions of these elements. Hence, performance of the converter in PEPS is determined by its level of dependability, and its constituent parts are crafted according to this measure of reliability. Consequently, modeling, evaluating dependability, and enhancing the PEPS must be approached hierarchically.

25.3.1 Hierarchical Reliability Modeling and Assessment

Power converters represent one of the most precarious components within contemporary power grids, frequently susceptible to deterioration-related malfunctions. This segment will specifically delve into power converter breakdowns while maintaining relevance to other subsystems. Failures in power converters often stem from capacitors and power modules. Operational specifics, surrounding conditions, and task requirements dictate the duration of their operational efficiency

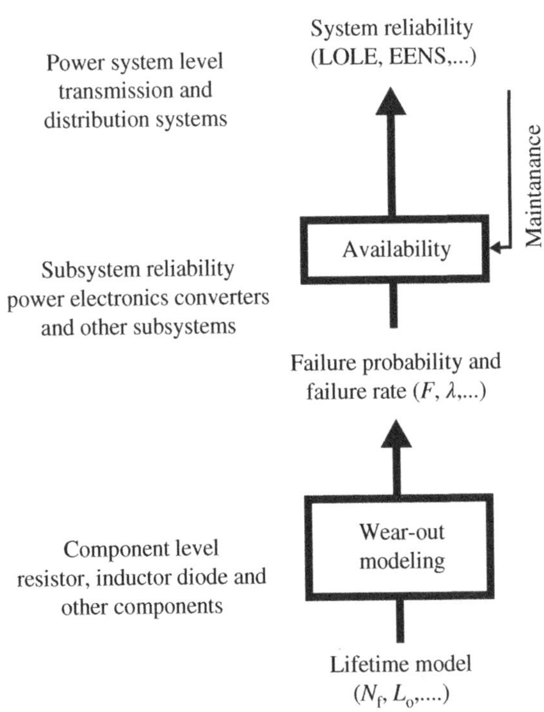

Figure 25.4 Correlation between PE and power system reliability concept. Source: Adapted from Peyghami et al. (2021).

and reliability. The longevity of power modules is directly linked to their susceptibility to failure across numerous operational cycles (N_f), an aspect that is notably influenced by fluctuations in junction operating temperature. These temperature variations, in turn, are governed by variables such as converter workloads and external environmental factors, including ambient temperature and fluctuations.

Different manufacturers and technologies offer a range of lifespan models for power equipment. The choice of an appropriate model for assessing reliability depends on specific device and availability of relevant data. For instance, in realm of power devices, one can calculate the cycles until failure (N_f) using subsequent equation:

$$N_f = A \cdot \Delta T_j^{\alpha} \cdot \exp\left(\frac{\beta}{T_j}\right) \tag{25.1}$$

where ΔT and T are the junction temperature swing and its average value. The constants of A, α, and β are curve-fitting constants obtained from aging tests.

$$L_o = L_r \cdot 2^{\frac{T_r - T_o}{n_f}} \left(\frac{V_o}{V_r}\right)^{-n_2} \tag{25.2}$$

L_r signifies rated lifespan under rated voltage V_r and upper category temperature T_r, while L_o denotes operational life of capacitor under operating voltage V_o, and temperature T_o. As stated in Eq. (25.2), longevity and dependability of a capacitor are influenced by its voltage and temperature. These factors are intertwined with operational conditions, encompassing capacitor's load and ambient temperature. It's worth highlighting that more sophisticated models for a lifespan of capacitors and power devices will encompass additional factors n_2 such as vibration, humidity, and other stresses, particularly pertinent in specific applications.

Due to the influence of converter loading on lifespan of capacitors and power devices, it becomes essential to comprehensively examine the converter's operational circumstances at the system level. Hence, the hierarchical approach's initial phase to ensure reliability involves analyzing power distribution. This analysis is grounded in energy management strategy employed within broader power system framework.

Once operational profiles for individual converters are established, next task involves translating these profiles into thermal and mechanical domains. This is essential for gathering information about thermal factors that impact the longevity of converter components, including metrics like junction temperature (i) and capacitor voltage (ii). Consequently, subsequent stage entails executing an electrothermal modeling exercise at the subsystem level. This modeling effort aims to gauge thermal stress experienced by each device, thereby contributing to a comprehensive understanding of their performance and durability.

Thermal stresses are linked to distinctive and predominant ways individual devices fail at the device level. It's worth noting that each device exhibits distinct failure modes and mechanisms, necessitating an appropriate lifetime model for accurate modeling. In this research, a general lifetime model is considered for each device. Nevertheless, this approach can be expanded to encompass diverse failure scenarios, enhancing precision of the dependability model. The thermal variables collected at sub-system level can compute the expended lifetime based on a specified mission profile and failure mode. With a particular lifetime model in a component, an uncertainty analysis becomes necessary to determine time until failure for a specific failure mode associated with it.

One way to achieve this is by conducting a SSA, which assesses the probability of failure associated with a specific failure mechanism. This analysis also helps quantify the manufacturing variations and modeling uncertainties affecting the reliability of a component when it operates according to a specified mission profile. Consequently, analysis at device level will establish a connection between thermal stresses and functions that describe reliability and failure rates for each specific failure mechanism.

After estimating device's reliability, converter's likelihood of failure and operational availability is established using failure rates of its parts. The last step in evaluating PEPS reliability involves merging converter's availability with the energy accessibility of the primary generation mechanisms and load patterns. This calculation determines extent of energy or load loss due to the absence of different components. This can be achieved by concurrently refining load pattern and generation system model. A pertinent dependability measure like LOLE, EENS, and others can represent resulting outcome.

The initial phase entails the dissection of PEPS into its subsystems, constituents, and potential points of failure. This process results in segmentation of overall system into sub-systems. Within this phase, a power flow analysis is conducted to establish load characteristics of each converter. Subsequently, converter is deconstructed down to its component levels. During this stage, an electrothermal mapping technique is utilized to assess the stress experienced by components. Lastly, analysis progresses to categorize failure mechanisms of components and various failure scenarios to ascertain their respective lifetimes under applied stress. The second stage of proposed V-shape model involves amalgamating reliability models derived from distinct failure causes for each component, extending up to PEPS.

Consequently, a cumulative failure rate of various failure modes is aggregated to calculate the overall failure rate for each device. This device failure rate is subsequently employed in modeling the availability of converters. Ultimately, converter's availability is integrated with load model to assess the performance of PEPS using metrics like LOLE, EENS, and similar indices.

25.3.2 Model-Based Reliability Enhancement

Reliability evaluations usually occur two times: before system construction to facilitate planning and design goals and during operation to improve reliability, plan for expansion, and schedule maintenance. The primary objective in both instances is to configure and run PEPS with optimal cost efficiency. After assessing system performance in these dual scenarios, the subsequent task involves ascertaining:

(1) Acceptability of reliability levels
(2) Economic viability and appeal of achieved reliability

If these criteria aren't met, appropriate actions must be taken to tackle these challenges, which will be further elaborated below. In outlining and creating the PEPS, it's vital to ensure system's effective functioning, such as maintaining a Low Occurrence of Loss of Energy (LOE) that is below a specified threshold, like 7.5 hours per year. This assessment of LOE will be conducted by considering the chosen equipment and their individual reliability and failure rates. If LOE surpasses the target level, two approaches are available to maintain it.

- The primary objective is to enhance vulnerabilities within the system. The proposed assessment tool, which relies on a model-based approach incorporating the principles of failure mechanisms, assists in identifying the least resilient subsystems, their most fragile components, and corresponding failure modes. Consequently, planners of power systems can opt for subsystems with superior dependability, such as converters, to guarantee the reliability of PEPS. Moreover,

developers of power converters can create converters tailored to specific applications, while upholding the reliability standards sought by PEPS planners. In this scenario, manufacturers can personalize their products, enabling them to formulate more economical and efficient converters. Collaboration between power system planners and power converter manufacturers has potential to enhance overall efficiency of the supply chain.
- Next approach involves maintaining system according to the initial plan and getting ready to replace converters at suitable intervals. Once more, since this recommended technique relies on models, PEPS planner can identify least resilient components in the system and their impact on power system changes over time. Consequently, planner can strategize for having an adequate number of backup units during appropriate time frames, ensuring swift replacement of these units to uphold overall reliability of the PEPS in alignment with intended performance targets.

Significantly, initial choice is notably more appropriate for systems with a restricted period of operation, like a vessel, perhaps around 20 years. Nevertheless, all elements must be regularly upgraded since the electrical grid is anticipated to run indefinitely. Consequently, second alternative is better suited for distribution power systems like the electric network. Establishing cost-effective replacement schedules and designing converters for that duration is imperative.

Moreover, during the operation, it is imperative to assess the system's reliability due to many uncertainties that can impact its performance, such as unexpected breakdowns and aging of individual components. Consequently, an annual evaluation of LOLE will be conducted based on the reliability model and past system performance. If LOLE exceeds the acceptable limit, a suitable maintenance approach must be implemented to maintain the system's efficiency. Maintenance plans can be categorized into two primary types: corrective and preventive maintenance. Corrective maintenance will be applied in instances where impact of a converter on system performance is minimal. In such cases, converter will be replaced as soon as it malfunctions, following this approach. Conversely, if failure of the converter significantly diminishes overall system performance; specifically the LOLE, preventive maintenance techniques must be employed.

Various preventive maintenance methods are available, including calendar-based, age-based, and condition-based approaches. These methods may entail differing maintenance costs, contingent on factors like the system's size, the characteristics of failures, the role of the converter in Electronic Power Systems (PEPS), and more. Depending on the extent of system impairment resulting from unit failure, one of these maintenance protocols will be chosen to replace or repair converter before any failure occurs.

25.4 Converter Reliability Modeling

The converter's failure patterns align with failure zone depicted by bathtub curve, similar to other systems. These patterns encompass three distinct phases: an initial stage of early failures, a period of operational effectiveness, and a later phase marked by wear-related issues. Issues during the initial phase often stem from debugging and production processes. Consequently, converter is susceptible to random failures and wear-related deterioration during its operational lifespan. External factors like overcurrent and overvoltage commonly trigger these random failure events (Yang et al. 2011). Consequently, within bathtub curve, failures follow an exponential distribution throughout the device's usable lifespan. Usually, the suitable rate of loss is estimated by considering historical reliability information and operational knowledge. Moreover, wear-out failure rate is connected to deterioration of device materials as they age, and this is depicted by Stress-Strength-Aging (SSA) model for components prone to aging:

$$N_{\mathrm{f}} = A \cdot \Delta T_j^{\alpha} \cdot \exp\left(\frac{\beta}{T_j}\right) \cdot \left(\frac{t_{\mathrm{on}}}{1.5}\right)^{-0.3} \tag{25.3}$$

where ΔT and T denote the swing in junction temperature and its mean value, while ton represents the rise time of the temperature cycle. A, α, and β are constants obtained through aging experiments. It's important to note that the junction temperature is influenced by the thermal dissipation of the device, its thermal impedance, and the surrounding temperature. Moreover, thermal dissipation is split into conduction loss and switching loss. The conduction loss is linked to the device's current, voltage, and ON-state resistance. Additionally, the switching loss is impacted by the switching frequency and ON-OFF energy loss. To assess the device's lifespan, the yearly mission profile needs to be converted into thermal parameters in Eqs. (25.1) and (25.4), specifically temperature and voltage, using electrothermal analysis.

Consequently, a comprehensive electrical and thermal representation of converter elements must be employed, requiring time-domain examination. The approach for predicting reliability is bifurcated into two phases. The initial phase is accountable for electrothermal simulation, while the subsequent step pertains to reliability simulation. The primary stage is further segmented into two domains: electrical and thermal. Within these domains, yearly mission profile (potentially encompassing factors like solar irradiance, wind speed, and ambient temperature) is transformed into electrical parameters such as device voltage and current patterns and subsequently into thermal parameters like junction and hot-spot temperatures.

This procedure necessitates electrical modeling that takes into account evolving frequency characteristics of switching, which can vary from a few kilohertz to several tens of kilohertz, depending on the converter's capacity, across the entire mission profile that covers various power levels. Furthermore, the durability of power components is affected by changes in junction temperature, which is controlled by alterations in the converter's power load and the thermal impedance of the device.

Thermal impedance is commonly represented using a higher-order transfer function with three to five orders. To accurately translate variations in the mission profile into fluctuations in junction temperature, which notably impact operational lifespan of switch as outlined in Eq. (25.2), it becomes essential to carry out a comprehensive electrothermal mapping. The alterations in temperature will be determined by solving differential equation of a thermal model designed for extended mission profiles. This endeavor will introduce greater intricacy and computational load. Evaluating the thermo-mechanical degradation of devices necessitates converting the mission profile into thermal parameters to ascertain longevity of converter. The accrued damage to capacitors, referred to as accumulated capacitor damage (ADCap), for a specific mission profile is computed subsequently:

$$\mathrm{AD}_{\mathrm{Cap}} = \sum_{k} \frac{t_k}{L_{\text{o-}k}} \tag{25.4}$$

In a context where t_k represents the operating period of a capacitor within parameters (V_{o}, T_{o}), and $L_{\text{o-}k}$ is the resultant lifetime derived; process continues by categorizing the junction temperature profile to calculate accumulated damage in semiconductor devices (ADS). This involves defining different classes for temperature profile, with class "h" characterized by a specific set of variables (T_{h}, t_{onh}, $\mathrm{n}_{\text{cycle-h}}$). Subsequently, the accumulated damages (ADs) are computed according to following procedure:

$$\mathrm{AD}_{\mathrm{S}} = \sum_{\mathrm{h}} \frac{n_{\text{cycle-h}}}{N_{\text{f-h}}} \tag{25.5}$$

In Eqs. (25.4) and (25.5), device's longevity corresponds to the inverse of AD, where $n_{\text{cycle-h}}$ represents the cycle count in category h, and $N_{\text{f-h}}$ signifies the cycle count in category h. These AD values, derived from individual devices, are combined with distinct values from the lifespan model alongside temperature models specific to each device. Practical implementation involves accounting for diverse fluctuations in lifetime models and device thermal characteristics. Consequently, it becomes essential to establish the distribution of AD considering uncertainties stemming from the model and manufacturing variations. Monte Carlo simulations offer a means to ascertain the distribution function of the annual downtime (AD). It should be recognized that employing Monte Carlo simulations for dependability modeling can be an uphill task. The function representing the unreliability of device, denoted as $F(t)$, is essentially the inverse of the AD distribution function. Once unreliability function has been derived, it becomes possible to estimate failure rate due to wear-out of components through the following procedure:

$$\lambda_{\text{w-X}}(t) = \frac{1}{1 - F(t)} \frac{\mathrm{d}}{\mathrm{d}t} F(t) \tag{25.6}$$

If we denote the differential operator as $\mathrm{d}/\mathrm{d}t$ and represent the wear-out failure of device X as $\lambda_{\text{w-X}}$, it indicates the decline in device X's functionality over time. Degradation in performance is often described by a Weibull distribution function, which incorporates a hazard function denoted as $h(t)$. This structure is illustrated below:

$$h(t) = \lambda_{\text{w-X}}(t) + \lambda_{\text{c-X}} \tag{25.7}$$

$\lambda_{\text{c-X}}$ denotes a consistent failure rate that can be expected based on historical failure data and operational observations over the valid lifetime.

This chapter has examined reliability within traditional power systems by elucidating its core components. Subsequently, it has depicted the configuration of contemporary power systems, encompassing emerging technologies like intelligent microgrids, renewable sources, and energy reservoir systems. The difficulties confronting modern power systems have been acknowledged and discussed. Lastly, a novel structure for evaluating reliability in modern power systems has been introduced. Effective and secure management of modern power systems necessitates novel regulations for intelligent microgrids and the establishment of consistent standards for gauging reliability in cyber-physical distributed power systems stated in Figure 25.5.

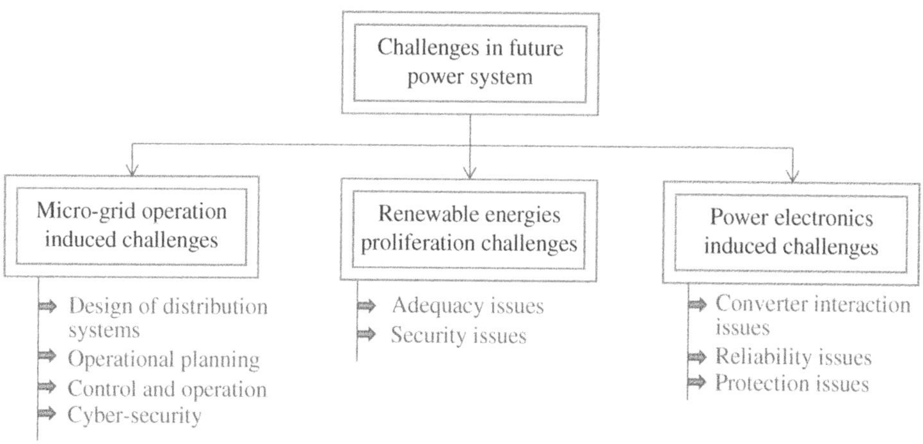

Figure 25.5 Challenges in upcoming power systems reliant on power electronics.

25.5 Conclusion and Future Challenges

Furthermore, conducting reliability analyses across various tiers demands efficient tools and strategies for the planning and operation of cyber-physical power systems containing a multitude of components and elements, particularly in light of the variable failure rates associated with novel technologies. Consequently, decentralized, self-organizing methods can facilitate optimal and dependable planning and operation of forthcoming intelligent grids. In the future, research will be directed toward appraising the reliability of contemporary power systems using the proposed framework while accounting for diverse aspects of reliability.

References

Alhmoud, L. (2018). Reliability improvement for a high-power IGBT in wind energy applications. *IEEE Transactions on Industrial Electronics* 65(9): 7129–7137.

Carroll, J., McDonald, A., and McMillan, D. (2014). Reliability comparison of DFIG drive train configuration with PMG drive train configuration in the first 5 years of operation. In: *3rd Renewable Power Generation Conference (RPG 2014)*, 1–6. IET.

Enslin, J.H., Whisenant, S.G., and Hadidi, R. (2019). Third eGrid workshop maps the grid of the future: attendees engage to examine the role of power electronic applications in modern electric power systems. *IEEE Power Electronics Magazine* 6(1): 48–55.

Fischer, K., Pelka, K., Bartschat, A. et al. (2019). Reliability of power converters in wind turbines: exploratory analysis of failure and operating data from a worldwide turbine fleet. *IEEE Transactions on Power Electronics* 34(7): 6332–6344.

Jiang, J., Peyghami, S., Coates, C., and Blaabjerg, F. (2021). A decentralized reliability-enhanced power sharing strategy for PV-based microgrids. *IEEE Transactions on Power Electronics* 36(6): 7281–7293.

Kothari, D.P., Nagrath, I.J., and Saket, R.K. (2022). *Modern Power System Analysis*. New Delhi: Tata McGraw-Hill Publishing Company.

Maurya, J. and Saket, R.K. (2023). Performance analysis of single-stage and two-stage VSI-fed IM drive for solar pump irrigation systems. In: *2023 IEEE IAS Global Conference on Renewable Energy and Hydrogen Technologies (GlobConHT)*, 1–6. IEEE.

Peyghami, S., Davari, P., and Blaabjerg, F. (2019). System-level reliability-oriented power sharing strategy for DC power systems. *IEEE Transactions on Industry Applications* 55(5): 4865–4875.

Peyghami, S., Wang, Z., and Blaabjerg, F. (2020). A guideline for reliability prediction in power electronic converters. *IEEE Transactions on Power Electronics* 35(10): 10958–10968.

Peyghami, S., Blaabjerg, F., and Palensky, P. (2021). Incorporating power electronic converters reliability into modern power system reliability analysis. *IEEE Journal of Emerging and Selected Topics in Power Electronics* 9(2): 1668–1681.

Singh, C., Jirutitijaroen, P., and Mitra, J. (2019). *Introduction to Reliability*, 1–9. Wiley-IEEE Press.

Yang, S., Bryant, A., Mawby, P. et al. (2011). An industry-based survey of reliability in power electronic converters. *IEEE Transactions on Industry Applications* 47(3): 1441–1451.

26

Reliability Aspects in the Design and Development of Microgrids

Amit Kumar[1], *Sachin Kumar*[2], *Almoataz Y. Abdelaziz*[3], *R. K. Saket*[4], *and D. P. Kothari*[5]

[1] *Flameproof and Equipment Safety Department, CSIR-Central Institute of Mining and Fuel Research, Dhanbad, Jharkhand, India*
[2] *Department of Electrical Engineering, G.B. Pant Institute of Engineering and Technology, Pauri-Garhwal, Uttarakhand, India*
[3] *Faculty of Engineering & Technology, Future University in Egypt, Cairo, Egypt*
[4] *Department of Electrical Engineering, Indian Institute of Technology (BHU), Varanasi, Uttar Pradesh, India*
[5] *Department of Electrical Engineering, Indian Institute of Technology, Delhi, India*

26.1 Introduction

The electric power system is the most extensively manufactured system that significantly improves humankind's quality of life. A power system consists of power generation, distribution, and utilization. The high-voltage AC transmission system is famous for transmitting power from the generation end to the distribution area because of low transmission losses (Kothari et al. 2022). Due to the increasing energy demand, modern grids have certain constraints, like continuous power supply, quality of power, and reliability. Due to specific reasons like energy shortages and environmental effects, fossil fuels are becoming unpopular in the modern power system. Renewable energy sources are being integrated into the grid swiftly (Shahgholian 2020).

Microgrids are becoming popular because they operate independently in an islanded mode without external power grid interference. The microgrid concept provides a solution for demand and location problems. Local generation, utilization, and renewable energy integration make the microgrid popular. Using integrated renewable energy generation and storage technologies, microgrids are small-scale energy systems that can deliver enough electricity to meet local demand (Kumar et al. 2021).

Microgrids may be designed as AC or DC according to their distribution system. Power electronics converters are an intrinsic part of microgrid technology since efficient power conversion and control are necessary at various points. Due to the increased use of power electronics converters, control and utilization flexibility have increased, but reliability has become a more significant concern. Components of the converters, like semiconductor switches, are prone to failure, which makes the whole system highly vulnerable. The chapter covers the microgrid's reliability, considering its technology in detail.

An introduction that includes a literature survey has already been described in Section 26.1. The rest of the chapter has been organized as follows: A detailed description of the architecture and operation of the microgrid is explained in Section 26.2. Section 26.3 discusses microgrid control, which includes multiple modes to ensure stable and secure operations. Reliability Aspects of microgrid planning and design, which include reliability indices, optimization, layer-wise reliability, etc., are discussed in Section 26.4. Finally, the conclusion, challenges, and future directions are discussed in Section 26.5.

Reliability Analysis of Modern Power Systems, First Edition. Edited by R. K. Saket and P. Sanjeevikumar.

26.2 Architecture and Operation of Microgrid

Microgrid systems have several distributed energy resources and use a low-voltage distribution. Microgrids may or may not be connected to any external grid. A microgrid mainly consists of distributed generators, variable loads, energy storage devices, control systems, and points of standard coupling (PCC). A typical microgrid is shown in Figure 26.1, and most of the components are denoted.

26.2.1 Types of Microgrid

Microgrid classification is based on various design parameters such as size, application, operational distribution system, architecture, distribution configuration, scenario, and type of sources. Microgrids are broadly classified into three categories based on their operation mode.

1. grid connected,
2. transited, and
3. standalone.

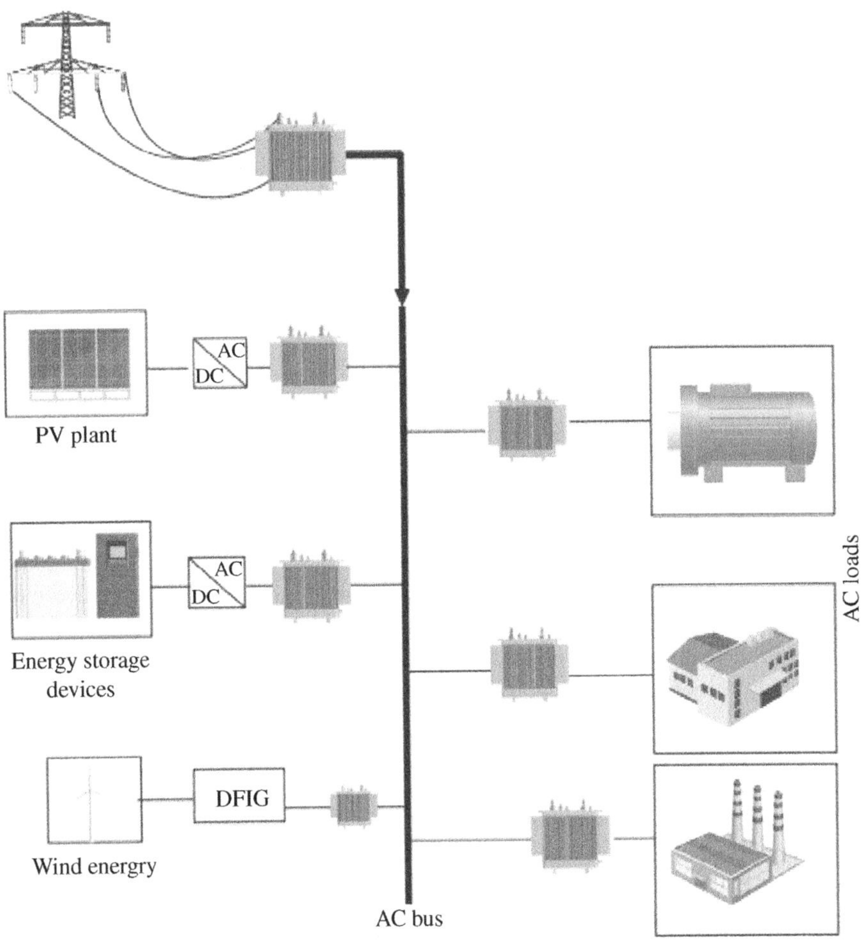

Figure 26.1 Architecture of a microgrid.

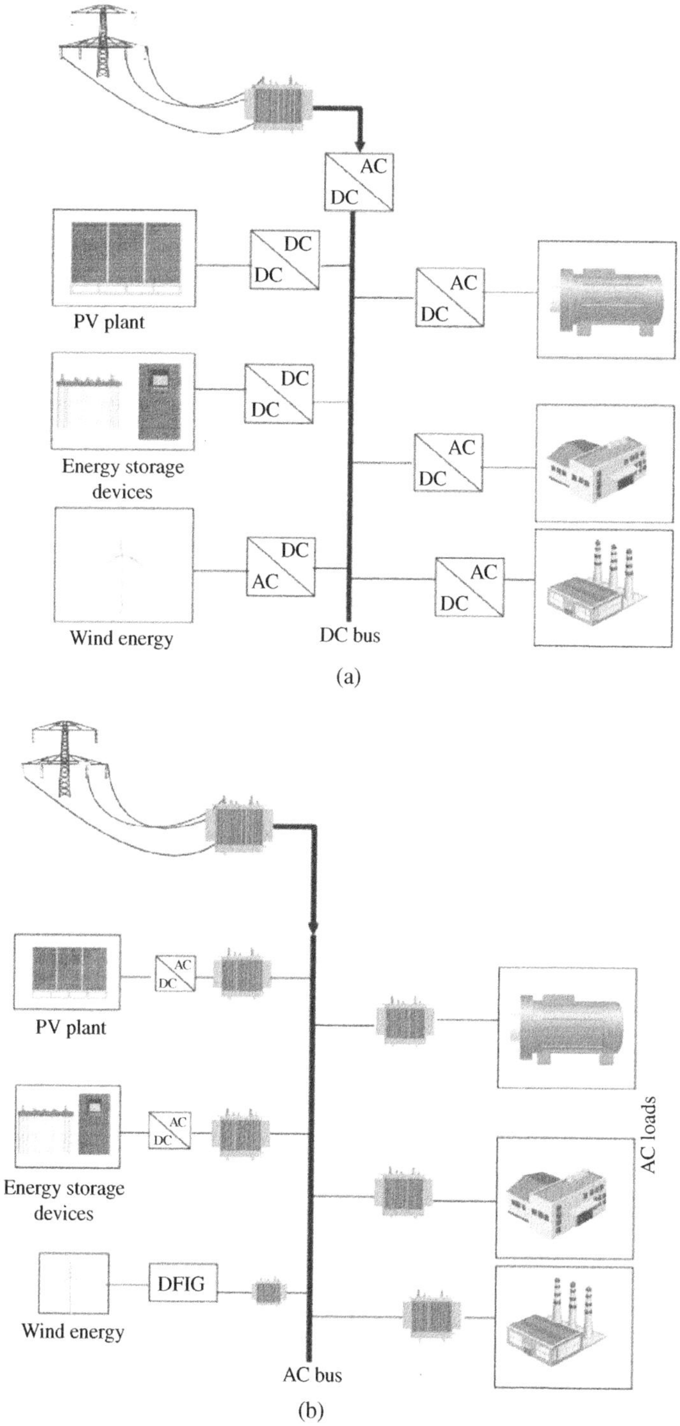

Figure 26.2 Different classifications of microgrid. (a) DC microgrid and (b) AC microgrid.

In the standalone microgrid, the reactive power balance shall be maintained independently due to the absence of an infinite bus (Kaushal and Basak 2019). The challenges for standalone microgrids are as follows.

1. Stable magnitude and frequency of the voltage
2. Stable power balance

In grid-connected microgrids, the control of microgrids becomes much more complex since the grid is connected to the microgrid through a static transfer switch. Based on the size of the microgrid, it is classified as small scale (<10 kW), medium scale (10 kW–1 MW), and large scale (>1 MW). Microgrids are also classified as AC and DC microgrids based on their distribution systems. AC and DC microgrids are based on their distribution systems (Shahparasti et al. 2015). Both AC and DC have their own merits and demerits. These microgrids are designed to fulfill local demands according to their size. Similarly, suitability, stability, and reliability are considered on every basis; different classifications are done, as shown in Figure 26.2a, and 26.2b.

26.2.2 Microgrid Technology

As previously discussed, the microgrid system comprises generation units, converters, energy storage systems, control stations, and interconnected switches. Technology supports all of these microgrid components. As day-by-day research is going on in this area, the latest technologies in every section are used to improve the grid's efficiency, reliability, and safety. The distributed generation unit technology utilizes solar, wind, and fuel cells. In modern-day microgrids, fuel gas with combined heat and power technology as a microturbine helps improve efficiency significantly.

Along with batteries, supercapacitors and flywheels are also used in the microgrid. Due to this new energy storage system, several benefits are observed, such as power stabilization (less fluctuations), reduced load spikes, and enhanced reliability. Similarly, microgrid switching technology uses a digital signal processor to improve the response time. Due to advancements in switches, high-frequency switches are being used. Microgrid control technologies can be grouped into islanding mode and network mode. Control technologies introduced in power converters regulate active and reactive power supply and frequency control.

26.3 Microgrid Control Strategies

From the consumer's perspective, a microgrid is a grid system that provides reliable, independent, and high-quality electric power (Hosseini and Shahgholian 2017). According to Hu et al. (2015), it is challenging to coordinate various micropower grid types to create a stable frequency and voltage-controlling microgrid system. Its primary goals are to control active and reactive power separately, fix voltage sag and system imbalances, and meet the grid's load dynamics needs. Microgrids have a control system unlike traditional power systems for the following reasons:

1. Frequent unbalanced loading due to one phase.
2. Energy storage plays a huge role, for which a control mechanism is required.
3. Different static and dynamic characteristics than other conventional grid.

Power systems need the right control strategies to ensure proper operation. Microsource and load controllers, a central controller for the microgrid system, and a distribution management system comprise the microgrid control.

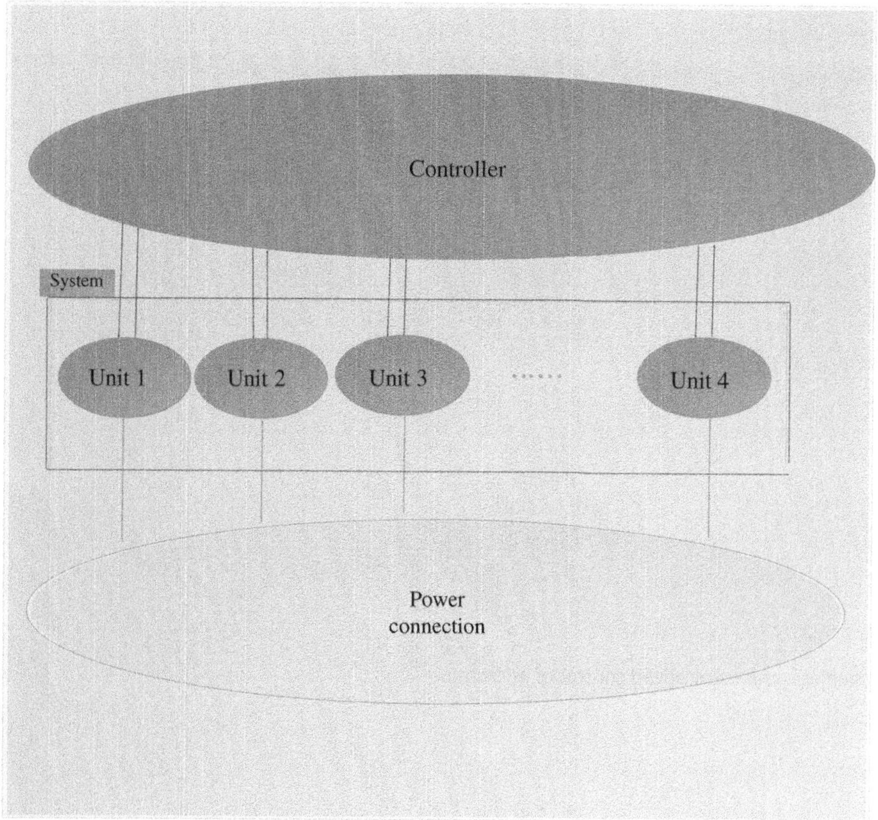

Figure 26.3 Schematic of centralized control of microgrid.

1. **Centralized Control**: As shown in Figure 26.3, in this type of control strategy, the central control unit determines the output of each local control unit based on the input production capacity of each unit. Various parameters, such as energy demand, market price, and production, determine the control strategy.
2. **Decentralized Control**: In a decentralized control strategy, each local control determines the operation command to fulfill the energy production demand, as shown in Figure 26.4. This type of control strategy focuses on overall performance rather than financial income.
3. **Multi-agent System**: Multi-agent system (MAS) is a development of the traditional distribution control strategy that can be used to manage very large and complex entities. The primary characteristic of MAS is its capacity to integrate intelligence components into each local control.

Both centralized and decentralized systems need trustworthy data communication infrastructure. A data communication network carrier can be a power line, radio, or telephone. As shown in Figure 26.5, these communication networks connect the main control system or applications for the energy management system to each distributed energy unit. They also run the microgrid.

26.3.1 Advantages and Disadvantages of Microgrid

Despite the advancements in microgrid technology, its implementation is a complicated process. Poor understanding and unsuitable government policies are the main obstacles to applying

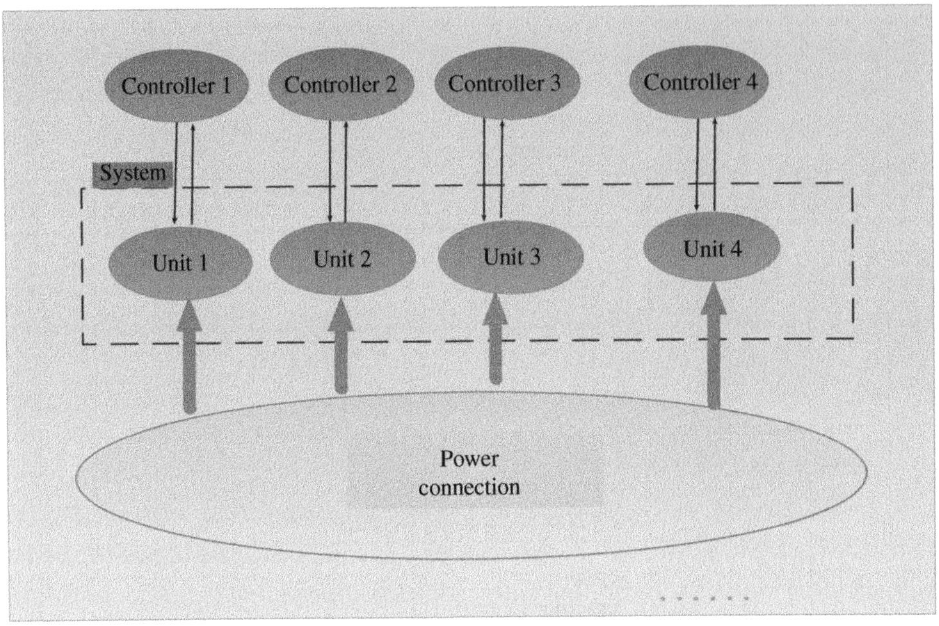

Figure 26.4 Schematic of decentralized control of microgrid.

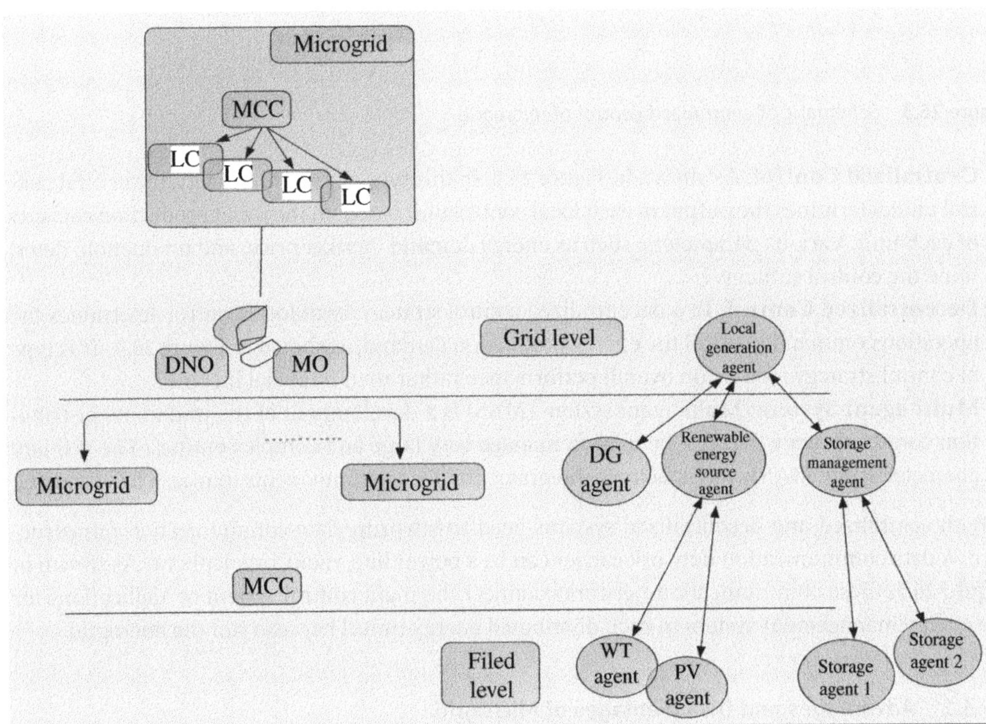

Figure 26.5 Schematic of multi-agent control of microgrid.

Table 26.1 Advantages and disadvantages of microgrid

Microgrid advantages	Microgrid disadvantages
1. Microgrids benefit from operating separately or with the outside grid 2. During peak load periods, energy may be supplied using the generation or energy storage units to provide uninterrupted power 3. Microgrids with more renewable generation units provide low emissions	1. Energy storage units require significant space and regular maintenance 2. Resynchronization with the outer grid is a difficult job 3. Maintaining the power quality up to the applicable standard requires complete control over the grid 4. The problem of safety in microgrids

microgrids. Microgrid has drawbacks in addition to its many benefits, and the only way to overcome them is through ongoing research into the problems. Table 26.1 covers the advantages and disadvantages of microgrids.

26.4 Reliability Aspects in Microgrid Planning and Design

Reliability is a critical aspect in the planning and design of microgrids. Microgrids are localized energy systems operating independently or in conjunction with the main grid. Ensuring their reliability is essential to providing uninterrupted power to the connected loads. Here are some key reliability aspects to consider in microgrid planning and design:

1. **Redundancy and Resilience**: Microgrids should be designed with redundancy to ensure they can continue operating despite equipment failures or external disruptions. Redundant components such as generators, energy storage systems, and inverters can help maintain reliability. The microgrid should also be designed to be resilient against extreme weather events, cyberattacks, and other unforeseen circumstances.
2. **Component Selection**: The choice of components, such as generators, energy storage systems, and renewable energy sources, plays a crucial role in microgrid reliability. High-quality and reliable equipment should be selected to minimize the chances of failures (Kumar et al. 2020).
3. **Control and Management Systems**: Robust control and management systems are essential for reliability. These systems should monitor the state of the microgrid, predict potential issues, and take corrective actions. Automation and advanced control algorithms can optimize microgrid performance and enhance reliability.
4. **Black Start Capability**: A microgrid should be able to "black start," meaning it can restart and re-establish its operations independently in case of a complete grid outage. This often involves having dedicated generators or energy storage systems that can initiate the microgrid's recovery.
5. **Islanding Capability**: Microgrids should be designed to transition between grid-connected and islanded modes seamlessly. The islanding capability allows the microgrid to operate autonomously when the main grid is down, ensuring power continuity.
6. **Predictive Maintenance**: Implementing predictive maintenance strategies can help identify potential issues in advance and schedule maintenance activities to prevent unplanned downtime.

7. **Cybersecurity**: Microgrids are vulnerable to cyberattacks, which can compromise their reliability. Robust cybersecurity measures, including firewalls, intrusion detection systems, and regular security assessments, should be part of the microgrid design.
8. **Energy Storage**: Energy storage systems can enhance reliability by providing backup power during grid outages and smoothing the integration of intermittent renewable resources.
9. **Load Management**: Efficient load management strategies can help balance supply and demand within the microgrid, ensuring a reliable power supply to critical loads.
10. **Grid Interaction**: Microgrids can be designed to interact with the main grid to import or export power as needed. This interaction can enhance reliability by accessing additional resources during peak demand or emergencies.
11. **Testing and Simulation**: Before deployment, microgrids should undergo rigorous testing and simulation to validate their reliability in various scenarios.
12. **Training and Maintenance**: Ensuring that personnel are adequately trained in operating and maintaining the microgrid is essential for long-term reliability.

Reliability in microgrid planning and design is a multifaceted challenge that requires a holistic approach. By addressing these aspects, microgrid operators can provide a stable and continuous power supply to meet critical infrastructure needs and enhance energy resilience.

26.4.1 Reliability Evaluation

While designing the microgrid, it is necessary to include economic generation, utilization, and adequate reliability. Reliability evaluation of the designed microgrid helps determine the factors responsible for the power interruption or low power quality. In IEEE Standard 2030.9-2019 (Hosseini and Shahgholian 2019), reliability indices are elaborated, as shown in the flow chart in Figure 26.6. Load point and system indices are discussed in IEEE Standard 2030.9-2019. Load point indices indicate the service continuity of an individual load. On the other hand, system indices indicate the service continuity of a group of loads (Allan 2013).

1. **Load Point Indices**
   The average failure rate denotes the interruption rate (number of interruptions per year)

$$\lambda_s = \sum_{i=1}^{n} \lambda_i \tag{26.1}$$

   λ_i = average failure rate at ith load point.
   Average Annual Outage Time (U_s): The average annual outage time denotes the overall downtime in one year. It is denoted in hours. This index represents the unavailability of the supply.

$$U_s = \sum_{i=1}^{n} \lambda_i * r_i \tag{26.2}$$

   r_i = duration of interruption per failure at ith load point.
   Average Outage Time per Failure r_s: The average outage time per failure denotes the average duration of interruption. It is generally denoted in hour per interruption.

$$r_s = \frac{U_s}{\lambda_s} = \frac{\sum_{i=1}^{n} \lambda_i * r_i}{\sum_{i=1}^{n} \lambda_i} \tag{26.3}$$

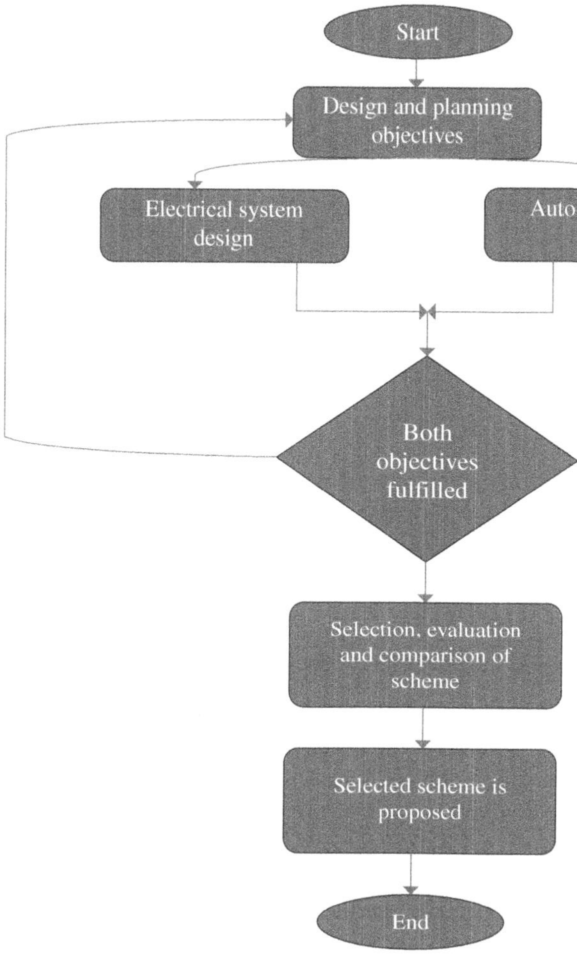

Figure 26.6 Microgrid planning procedure as per IEEE 2030.9-2019. Source: Adapted from (Hosseini and Shahgholian 2019).

2. **System Indices**

System Average Interruption Frequency Index (SAIFI): SAIFI is the ratio of the total number of interrupted customers to the total number of served customers

$$\text{SAIFI} = \frac{\sum_{i=1}^{n} \lambda_i * N_i}{N_\text{T}} \tag{26.4}$$

where

N_T = Number of customers served

N_i = number of customers interrupted for ith interruption.

Average Service Availability Index (ASAI): ASAI is the ratio of actual served hours (hours for which service was available) to the total demanded customer service hours.

$$\text{ASAI} = \frac{N - \sum_{i=1}^{n} C_i * L_i}{N} \tag{26.5}$$

where
C_i = Available generation capacity
L_i = Maximum load of the ith day
N = Number of days.

System Average Interruption Duration Index (SAIDI): SAIDI is the ratio of the total duration of customer interruption to the total number of customers for which service was available.

$$\text{SAIDI} = \frac{\sum_{i=1}^{n} r_i * N_i}{N_\text{T}} \tag{26.6}$$

where
r_i = Restoration time for ith interruption.

Loss of Load Probability (LOLP): LOLP represents the probability that the load demand exceeds the generating capacity.

$$\text{LOLP} = \frac{\sum_{i=1}^{n} P * (C_i * L_i)}{N} \tag{26.7}$$

Analytical methods like Markov's and Monte Carlo simulations are used to calculate the system's reliability using these indices. The flowchart shown in Figure 26.6 describes the procedure of microgrid planning and design (IEEE 2019).

26.4.2 Energy Scheduling, Forecasting, and Optimization Techniques

A microgrid's short-term planning and energy management are easily implemented due to distributed energy resources. Since it is easier to predict loads in large conventional power systems than in small-load microgrids, this unpredictability is the main reason for implementing a proper energy management and scheduling scheme to supply uninterrupted power. This scheduling time scale may vary from minutes to hours on several-day scales. Optimization techniques are used in scheduling with the help of parameters like energy unit capacity, minimized cost, and environmental effects. High reliability can be achieved with these parameters and a proper optimization method. For good optimization, adequate sizing of distributed energy units is required (Yuan et al. 2017; Swaminathan et al. 2020).

Corrective scheduling is a crucial method of energy management in microgrids. The concept mitigates errors due to uncertainty by adjusting the scheduling based on new information about load demand and environmental conditions. Optimization techniques are selected based on the complexity of the grid. Various optimization techniques used are linear programming, convex optimization, Genetic algorithms, particle swarm optimization techniques, etc. A similar study provided a unified framework for optimal and reliable microgrid design with the help of two different optimization techniques, integer linear programming, and particle swarm technique (Machado et al. 2018). A practical planning procedure is provided for microgrids on the robust optimization method within the operational constraints, and the total cost of planning is minimized (Khodaei 2014).

26.4.3 Reliability Aspects of Power Electronics in Microgrids

Determination of failure rate plays an essential role in evaluating the reliability of the power electronics converters. Failure rate follows a bathtub curve with time, which includes three stages: infant mortality, helpful life, and wear-out phase. Since the installed power electronics converter passes its infant mortality stage, the failure rate during the useful life and wear-out phase is considered for reliability evaluation.

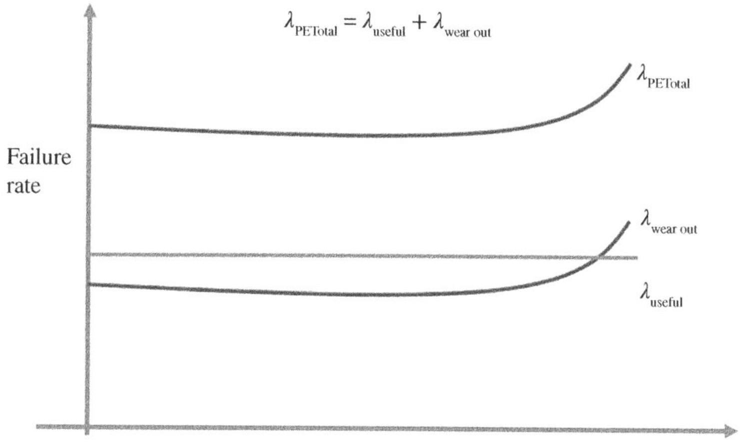

Figure 26.7 Failure rate versus time curve.

The failure rate during useful life is nearly constant and significantly increases during the wear-out period due to the aging effect of components. Both the failure rates are considered for accurate and precise conversion reliability. However, due to the uncertainty of wear-out, most of the study ignores the wear-out failure rate and relies on the constant failure rate only, as shown in Figure 26.7. Analysis of field failure data of similar components provided by supply gives an accurate failure rate.

Reliability handbooks provide failure rates based on historical field data. Some popular reliability handbooks are military handbook (MIL-HDBK-217), Handbook of Reliability Data (HRD5), International Electro-technical Commission standards IEC-61709, and IEC-62380-TR. These handbooks provide an empirical model. The main disadvantage of these handbooks is that they are based on historical data on the technologies that have yet to be practiced.

To determine the Constant failure rate (constant), FIDES reliability prediction method is used due to its higher accuracy. This method uses the component's sensitivity concerning stress for a longer period. The input mission profiles are disintegrated into different phases. Since these phases represent different stress parameters. Physical failure rates are determined after this decomposition of mission profiles into phases. With the help of this physical failure rate in each phase, constant failure rates are determined (Carton et al. 2017).

Modeling based on the physics of failure and stress analysis is used to obtain the wear-out failure rate. For Power electronics converters, electrolytic capacitors and switches are considered critical components. Both components are prone to failure due to thermal and electrical stress. Electro-thermal modeling is developed to obtain the stress profile of these too-critical components from the mission profile. With the help of loading decomposition, the impact of stress level can be evaluated. Lifetime consumption estimation is done by plotting conception with time. This lifetime consumption concerning time gives a corresponding distribution function. A probabilistic assessment uses this probability distribution function to determine the wear-out failure rate.

26.4.4 Component to System: Layer-Wise Reliability

Reliability is calculated at different levels, such as system, subsystem, and component levels, as shown in Figure 26.8. The component level reliability helps determine the reliability of the system. In the same way, while evaluating the reliability of microgrids' reliability, the smallest components'

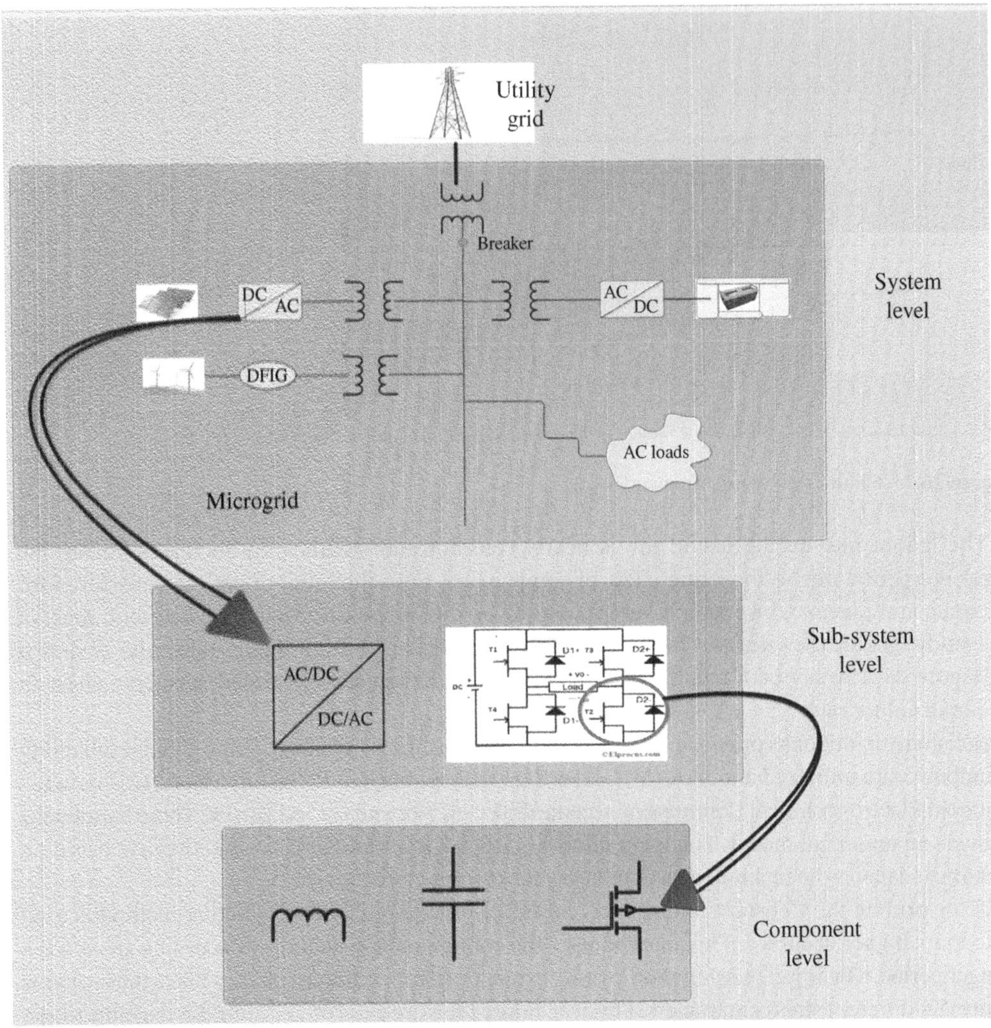

Figure 26.8 Layer-wise reliability evaluation of microgrid.

reliability must be known. Microgrid reliability can be evaluated in three levels: component level, converter level, and microgrid level (Peyghami et al. 2020). The first level includes switches, capacitors, resistors, inductors, etc. The physics of failure is studied in this layer, and possible failure reasons are found. The expected lifetime determination of components in this layer helps in evaluating reliability. In the second level, the reliability of converters is evaluated with the help of component reliability. The healthiness of this layer shows that healthy inverters are available to use. In the third level, overall microgrid reliability is evaluated with the help of the other two layers and their corresponding components' connection. The topology and control methodology become essential in this layer since their interconnection differs.

26.5 Conclusion and Future Challenges

In the long-term planning of microgrids, adopting new methods for forecasting and scheduling energy demand is necessary. Distributed energy resource sizing with suitable optimization techniques can be used to develop new standards for future microgrid design. Incorporating different stressors in reliability studies will enhance the reliability-oriented design of microgrids. A better optimization technique can help to meet the cost minimization and climate requirements. Transformation of existing microgrids into intelligent, robust, reliable, and safe microgrids is a big challenge.

Microgrid is seen as an alternative to conventional grids, which is becoming increasingly popular. In this chapter, microgrid type, control techniques, and the importance of reliability are discussed, and some popular reliability evaluation methods are provided. Since uninterrupted and long duration operation of grid is a demand of the time. It is vital to develop more reliable methods and techniques for energy scheduling. Proper inclusion of external stress in the planning and study is required.

References

Allan, R.N. (2013). *Reliability Evaluation of Power Systems*. Springer Science & Business Media.

Carton, P., Giraudeau, M., and Davenel, F. (2017). New fides models for emerging technologies. In: *2017 Annual Reliability and Maintainability Symposium (RAMS)*, 1–6. IEEE.

Hosseini, E. and Shahgholian, G. (2017). Partial-or full-power production in WECS: a survey of control and structural strategies. *EPE Journal* 27 (3): 125–142.

Hosseini, E. and Shahgholian, G. (2019). Recommended practice for the planning and design of the microgrid. In: *IEEE Std 2030. 9-2019. BOG/SC21 - Distributed Generation, Energy Storage and Interoperability Standards Committee*, 1–6. IEEE.

Hu, J., Zhang, T., Du, S., and Zhao, Y. (2015). An overview on analysis and control of micro-grid system. *International Journal of Control and Automation* 8 (6): 65–76.

Kaushal, J. and Basak, P. (2019). A decision making methodology to assess power quality monitoring index of an AC microgrid using fuzzy inference systems. *Electric Power Components and Systems* 47 (14-15): 1349–1361.

Khodaei, A. (2014). Provisional microgrids. *IEEE Transactions on Smart Grid* 6 (3): 1107–1115.

Kothari, D.P., Nagrath, I.J., and Saket, R.K. (2022). *Modern Power System Analysis*. Tata McGraw-Hill.

Kumar, S., Saket, R.K., Dheer, D.K. et al. (2020). Reliability enhancement of electrical power system including impacts of renewable energy sources: a comprehensive review. *IET Generation, Transmission & Distribution* 14 (10): 1799–1815.

Kumar, S., Sarita, K., Saket, R.K. et al. (2021). Reliability assessment for DFIG-based WECS considering the impact of 3-phase fault and lightning impulse voltage. *International Transactions on Electrical Energy Systems* 31 (8): e12952.

Machado, P., de Souza, L.E., and Netto, R.S. (2018). Framework to support grid-connected microgrid optimal planning. In: *2018 IEEE International Conference on Environment and Electrical Engineering and 2018 IEEE Industrial and Commercial Power Systems Europe (EEEIC/I&CPS Europe)*, 1–6. IEEE.

Peyghami, S., Blaabjerg, F., and Palensky, P. (2020). Incorporating power electronic converters reliability into modern power system reliability analysis. *IEEE Journal of Emerging and Selected Topics in Power Electronics* 9 (2): 1668–1681.

Shahgholian, G. (2020). Coordinated design of power system stabilizer and variable impedance devices to increase damping of inter-area modes using genetic algorithm. *Nashriyyah-i Muhandisi-i Barq va Muhandisi-i Kampyutar-i Iran* 75 (4): 271.

Shahparasti, M., Mohamadian, M., Baboli, P.T., and Yazdianp, A. (2015). Toward power quality management in hybrid AC–DC microgrid using LTC-L utility interactive inverter: load voltage–grid current tradeoff. *IEEE Transactions on Smart Grid* 8 (2): 857–867.

Swaminathan, S., Pavlak, G.S., and Freihaut, J. (2020). Sizing and dispatch of an islanded microgrid with energy flexible buildings. *Applied Energy* 276: 115355.

Yuan, C., Liu, G., Wang, Z. et al. (2017). Economic power capacity design of distributed energy resources for reliable community microgrids. *Energy Procedia* 142: 2561–2567.

Abbreviations

2TBN	Two-slice temporal Bayesian network
$3LT^2C$	Three-level T-type converter
ABAO	As-bad-as-old
AC	Alternating current
ac	Air cooling with extended fin
ADS	Accumulated damage in semiconductor devices
AELC	Average extra length for charging
AENS	Average energy not supplied
AETC	Average extra time for charging
AGAN	As-good-as-new
AIDI	Average interruption duration of EV charging index
AIFI	Average interruption frequency of EV charging index
ALT	Accelerated life testing
AM	Asset management
AMI	Advanced metering infrastructure
ANN	Artificial neural network
APM	Application of probability methods
ARA	Arithmetic reduction of age
ARI	Arithmetic reduction of intensity
ASAI	Average service availability index
ASAI	Average system availability index
ASUI	Average system unavailability index
ATHEANA	A technique for human event analysis
BITE	Built-in test equipment
BJT	Bipolar junction transistors
BLDC	Brushless DC motor
BMS	Battery management systems
BN	Bayesian network
BTO	Break-through technologies
CAIDI	Customer average interruption duration index
CBM	Condition based maintenance
CC	Capital costs
CCS	Combined charging system
CDF	Cumulative distribution function
CDS	Composite distribution system

Reliability Analysis of Modern Power Systems, First Edition. Edited by R. K. Saket and P. Sanjeevikumar.

CFR	Constant failure rate
CHB	Cascaded H-bridge
CHP	Combined heat and power
CIC	Customer interruption cost
CIL	Critical items list
CLOE	Complete loss of excitation
CM	Corrective maintenance
cm	Centimeter
CMF	Common mode failure
COS	Change-over switch
CPT	Conditional probability table
CREAM	Cognitive reliability and error analysis method
CRT	Cathode ray tube
CS	Charging stations
dB	Decibel
DBN	Dynamic Bayesian network
DC	Direct current
DE	Differential evolution
DFIG	Doubly fed induction generator
DfR	Design for reliability
DG	Diesel generator
DG	Distributed generation
DIgSILENT	Digital simulation and electrical network
DN	Distribution network
DOE	Design of experiment
DPLVC	Daily peak load variation curve
DSP	Digital signal processor
DSR	Distribution system reliability
E	Erying
EENS	Expected energy not supplied
EENS	Expected energy not served
EM	Electromagnetic
EMI	Electromagnetic interference
EMS	Energy management systems
ENSPI	Energy not served per interruption
EPDN	Electrical power distribution network
EPNS	Expected power not supplied
EPS	Electric power system
ESS	Energy storage systems
ETA	Event tree analysis
EV	Electric vehicle
FC	Flying capacitor
FC	Failure costs
FFTA	Fuzzy fault tree analysis
FLISR	Fault location, isolation, and service restoration
FMEA	Failure modes and effects analysis
FMECA	Failure modes effects and criticality analysis

FORM	First-order reliability method
FP	Flower pollination
FTA	Fault tree analysis
G2V	Grid to vehicle
GA(s)	Genetic algorithm(s)
GDA	Gaussian distribution approach
GRP	Generalized renewal process
GTO	Gate turn-off thyristors
H, M, and R	Three-state model (healthy state, marginal state, and risky state) for reliability analysis
HAWT	Horizontal-axis wind turbines
HL	Hierarchical level
HL-1	Hierarchical level 1
HL-2	Hierarchical level 2
HL-3	Hierarchical level 3
HMM	Horizon mission methodology
HMM	Hidden Markov models
HPP	Homogeneous Poisson process
HRD	Handbook of reliability data
HRES	Hybrid renewable energy systems
HVAC	Heating ventilation and air conditioning
HVDC	High-voltage direct current
I, O	Denotes input and output nodes
ICA	International Commission on Accreditation
ICs	Integrated circuits
IEC	International Electrotechnical Commission
IEEE	Institute of Electrical and Electronics Engineers (USA)
IGBT	Insulated gate bipolar transistor
IIoT	Industrial Internet of Things
IM	Induction motor
LCCA	Life cycle cost analysis
LCC	Life cycle cost
LCM	Life cycle management
LDC	Load duration curve
LHS	Latin hypercube sampling
LOE	Loss of energy
LOEE	Loss of energy expectation
LOLE	Loss of load expectation
LOLF	Loss of load frequency
LOLP	Loss of load probability
LRU	Line replaceable unit
LSAR	Logistics support analysis record
LSPWM	Level-shifted pulse-width modulation
MAS	Multi-agent system
MC	Monte Carlo
MC	Maintenance costs
MCS	Monte Carlo simulation

MDT	Mean down time
MHT	Mean healthy time
MIL-HDBK	Military-handbook
MK	Markov models
ML	Machine learning
MLEs	Maximum likelihood estimators
MLI	Multilevel inverter
MMC	Modular multilevel converter
MMF	Magneto-motive force
MMT	Mean marginal time
MOSFET	Metal oxide semiconductor field effect transistors
MPP	Most probable point
MPP	Maximum power point
MPPT	Maximum power point tracking
MRO	Maintain, repair and overhaul
MRT	Mean risky time
MTBF	Mean time between failures
MTTF	Mean time to failure
MTTR	Mean time to repair
MUT	Mean up time
MVar	Mega volt ampere reactive
MW	Mega watt
mW	Milli watt
NASA	National Aeronautics and Space Administration (USA)
NCPI	Novel collapse prediction index
NHPP	Nonhomogeneous Poisson process
NOCT	Normal operating conditions temperature
NPC	Neutral-point clamped
OC	Operating costs
OLS	Ordinary least squares
ORU	Orbital replaceable unit
P	Peck
pc	Passive cooling with heat pipe
PCB	Printed circuit board
PDF	Probability density function
PDF	Probability distribution function
PDU	Power distribution unit
PE	Power electronic
PEDS	Power electronic devices and systems
PEPS	Power electronic-based power system
PEVs	Plug-in electric vehicles
PF	Probability of failure
PI	Performance indices
PI	Performance indicators
PII	Probabilistic insecurity index
PLC	Peak load consideration
PLOE	Partial loss of excitation

PM	Preventive maintenance
PMSM	Permanent magnet synchronous motor
PoF	Physics of failure
Pof	Probability of failure
PRE	Power reliability engineer
PSCAD	Power systems computer-aided design
PSF	Probabilistic sufficiency factor
PSPWM	Phase-shifted pulse-width modulation
PSR	Power system reliability
PSS/E	Power system simulation for engineering
PV	Photovoltaic
RAT	Reliability assessment techniques
RBD	Reliability block diagram
RBDO	Reliability-based design optimization
RBTS	Roy Billinton test system
RC	Resistor–capacitor
RCD	Resistor–capacitor–diode
RCM	Reliability-centered maintenance
RDB	Reliability block diagrams
RE	Renewable energy
RER	Renewable energy resources
RES	Renewable energy sources
RHPVES	Renewable hybrid PV energy system
RNN	Recurrent neural network
RP	Renewal process
RPN	Risk priority number
RPS	Reward and penalty system
RSM	Response surface method
RTS	Reliability test system
RUL	Remaining useful life
RV	Residual value
SAIDI	System average interruption duration index
SAIFI	System average interruption frequency index
SBX	Simulated binary crossover
SCC	Short circuit current
SCR	Silicon-controlled rectifiers
SE	Shielding effectiveness
SEIG	Self-exited induction generator
SFC	Safety factor concept
SHE	Selective harmonic elimination
SLDC	Stepped load duration curve
SOA	Safe operating area
SoC	State of Charge
SoH	State of Health
SORM	Second-order reliability method
SoT	State of temperature
SPAR-H	Standardized plant analysis risk human reliability

SPE	Solar power Europe
SPWM	Sinusoidal pulse-width modulation
SRM	Switched reluctance motor
SSA	Stress–strength analysis
STC	Standard temperature condition
SUT	Switch under test
SVM	Space vector modulation
TBF	Time between failures
TDEP	Temperature-dependent electrical parameters
TE	Transverse electric
THD	Total harmonic distortion
TL	Transmission line
TLBO	Teaching-learning-based optimization
TM	Transverse magnetic
TOU	Time-of-use
TTL	Total technical life
TVSM	Thevenin-based voltage stability margin
UPS	Uninterruptible power supply
V2D	Vehicle to distribution
V2G	Vehicle-to-grid
V2H	Vehicle to Home
VAWT	Vertical-axis wind turbines
VVO	Volt-VAR optimization
"W" and "F"	Working (up) and failed (down) state, respectively
W and F	Success set of event and fail set of event
WAMC	Wide-area monitoring and control
WC	Water cooling
WECS	Wind energy conversion system
WF	wind farms
WS	Wick structure
WT	Wind turbine
YoY	Year-on-Year
ZCS	Zero current switching
ZVS	Zero voltage switching

Notations

α	Failure mode ratio (for an item, α = 1)
α	Re-installation rate of component
α	Distribution parameter
αI	Relative frequency of occurrences
α_{K}, β_{K}	Coefficients of cost for rate of failure and time of repair, respectively
α_{vceo}	Temperature coefficient of V_{ce}
β	Distribution parameter
β	Ratio of common cause failure to total failure rate
β	The likelihood of function or mission loss given specific conditions
β	Safety index
β_{ro}	Temperature coefficient
$\Lambda(t)$	Expected number of failures through time
δ_j	Slope of error
ξ_i	The right eigenvector corresponding to λ_i
γ_{ge}	Temperature coefficient of ΔV_{ge}
$\eta_{battery}$	Battery efficiency
η_c	Distribution index
$\eta_{inverter}$	Inverter efficiency
λ	Failure rate
λ and μ	Single component failure and repair rate, respectively
λ_{base}	Basic failure rate of components (capacitors/diodes/switches)
λ_c	Failure rate due to common cause
λ_C	Capacitance failure rates
$\lambda_{c\text{-}X}$	Consistent failure rate
λ_{cap}	Capacitor failure rate
$\lambda_{constant}$	Content failure rate
$\lambda_{C\_converter}$	Failure rate in a converter
λ_{C_useful}	Useful life failure rate of converter
$\lambda_{C\_wearout}$	Wear-out failure rate of converter
λ_D	Diode failure rates
$\lambda_{DC-DC\ Converter}$	DC–DC converters failure rates
λ_{diode}	Diode failure rate

Reliability Analysis of Modern Power Systems, First Edition. Edited by R. K. Saket and P. Sanjeevikumar.

λ_e and μ_e	Equivalent failure and repair rate, respectively
λ_f	Expected failure rate
λ_{Filter}	Filter failure rate
λ_i	The *i*th eigenvalue of the Jacobian for load flow
λ_i	Failure rate of *i*th phase
λ_i	Average failure rate at *i*th load point
$\lambda_{inverter}$	Inverter failure rates
λ_j	Transmission line failure
λ_k, r_k	The *k*th distributor segment's rate of failure and average repair time, correspondingly
$\lambda_{k,\min}$ and $r_{k,\min}$	Attainable minimums for the *k*th distributor segment's rate of failure and repair time
$\lambda_{k,\max}$ and $r_{k,\max}$	The *k*th distributor segment's maximum permitted rate of failure and time to repair
λ_{max}	Maximum load flow eigenvalue Jacobian
λ_{min}	Minimum eigenvalue of [*H*] or [*J*]
λ_p	Item failure or hazard rate
$\lambda_{PETotal}$	Power electronics converters failure rate
λ_{phy_cap}	Failure rate of capacitor
λ_{phy_scd}	Failure rate of semiconductor
λ_R	Resistance failure rates
λ_s	Failure rate of series system
λ_s	Switch failure rate
λ_s	Snubber failure rates
λ_{Switch}	Switch failure rate
λ_{sys}	Failure rate of the system
$\lambda_{sys,i}$	Failure rate of the system at the *i*th load point
λ_r	Failure rates of *r*th component
λ_{useful}	Failure rate during useful
λ_W	Switch's failure chance
$\lambda_{w\text{-}X}$	Wear-out failure of device X
$\lambda_{wearout}$	Wear-out failure rate
λ_{XY}	Total transition rate from state X to Y
λ_1^+ *and* λ_2^+	Failure rates of components while only one of them is working for load sharing system
λ_1, λ_2	Failure rates of each component while both02 are in operation
$\lambda_1, \lambda_2, \ldots, \lambda_n$	Component failure rates
λ_{0k}	Base failure rate
μ	Repair rate
μ_{B_j}	Mean of vector B, the *j*th constraint
μ_{C_l}	Mean of vector C, the *l*th constraint
μ_i and λ_i	Repair and failure rates of *i*th DG unit
μ_j	Repair rates
μ_r	Expected repair rate
ω_r	Rotor speed
ω_s	Synchronous speed

πCF	Capacitance factor
πCCF	Contact construction factor
πES	Effect of environment stress
πQF	Quality factor
πTF	Temperature factor
Π_i	Factor for electrical mechanical and thermal overstress in ith phase
Π_{pm}	Factor for manufacturing quality impact
Π_{process}	Factor for all processes such as operation and maintenance
σ_{c}	Standard deviation of capacity
σ_d	Standard deviation of demand
σ_{X}, μ_{X}, σ_{Z}, and μ_{Z}	Pertinent shape parameters for the PDFs
τ	Discharge time constant
A	Availability
A	Hypothetical event A
A	Transition intensity matrix
A	Area of cylinder bore
A, α, and β	Curve-fitting constants
A_{c}	Peak amplitude of carrier waveform
$A_{\text{converter}}(t)$	Availability of the converter
$A_{\text{constant}}(t)$	Availability associated with time-constant failure rate
AD_{Cap}	Accumulated capacitor damage
ADCOST	Extra money put toward distributed generations (DGs) to buy power
A_{H}, A_{M}, and A_{R}	Availability in H, M, and R states
a_i	The network's output for the ith training dataset
a_{ij}	The ijth load flow element $[J']$ for Jacobian
A_i	Series of exhaustive
A_i and $\overline{A}_i$	Availability and unavailability of ith DG unit
A_j	Long-term availability
A_j	Transmission line's unavailability
A_{OC}	Availability of other components of converter
A_{DC}	Availability of DC transmission line
A_{m}	Peak amplitude of reference waveform
$A_{\text{REC}}(t)$	Availability of receiving end converter
$A_{\text{SEC}}(t)$	Availability of sending end converter
$A_{\text{switch}}(t)$	Availability of semiconductor switches
A_{sys}	System availability
$\overline{A}_{\text{sys}}$	System unavailability
$A_{\text{total}}(t)$	Availability of total system
$\overline{A}_{\text{Ti}}$	Unavailability of the transmission line
A_{Ti}	Availability of the transmission line
$A_{\text{wearout}}(t)$	Availability associated with time-varying failure rate
B	Hypothetical event B
B	Event
B_{c}	Susceptance to lines charging
$B_{\text{c}_{\text{T}}}$	Total susceptance of the line charging

b_{h} and b_{y}	Bias vectors that let every node to learn and offset
C	The total number of training data samples
C	Generation capacity
C	Capacitor value in micro Farad
C	Capacitor
$\overline{C}$	Mean generation capacity
CC	Capital costs
C_{dgUS}	Capacity available due to distribution substation DG units
c_i	Indicates the cumulative count of instances the ith EV engaged in V2G activities during the simulation duration
C_i	ith component
C_i	L_{ith} load level mean generation capacity
C_i	Available capacity in interval i
C_i	Available generation capacity
C_{m}	Criticality number
C_{network}	Network's overall reliability costs
$C_{\mathrm{oss}} = C_{\mathrm{ds}} + C_{\mathrm{gd}}$	Parasitic capacitance of the MOSFET
C_{ox}	Oxide layer capacitance
C_{RP}	The utility's cost of the reward or penalty
C_{pk}	Interruption costs in rupees per kW
C_{s}	Snubber capacitance
$C_{\mathrm{S,S}}$	Capacity available due to distribution substation
$\mathrm{Cs}_1, \mathrm{Cs}_2, \ldots, \mathrm{Cs}_n$, and Cs_m	Minimal cut-sets
C_{T}	Capacity of a composite distribution system
C_T	Tth minimal cut-set
$C_{\mathrm{total}}^{\mathrm{society}}$	Total reliability cost of society
$C_{\mathrm{total}}^{\mathrm{utility}}$	The utility's overall reliability costs
d	Duty cycle
D	Detection
DA	Battery autonomy
$\mathrm{d}I_{\mathrm{C}}/\mathrm{d}t$	Rate of change of collector current
D_{L}	Repair rate in hour
e	Electron charge
E_{avg}	Average demand energy
E_{act}	Thermal activation energy of the degradation process (eV)
E_{cap}	Energy capacity of the battery
$E_{\mathrm{cur}}, E_{\mathrm{cel}}$	Current energy level of the battery
EENSO	Annual estimated or expected energy supply not provided when distributed generations are not linked
EENSD	Annual estimated or expected energy supply not provided when distributed generations are linked
E_i	Curtailed energy in interval i
E_i and E_j	Events
E_{m}	Maximum energy capacity of the battery
f	Failure tests
F	Failure probability

F	Objective function to decide the optimized values of various parameters
$\mathrm{Fa}_{\mathrm{Cen}(t)}$	Cumulative distribution function of central topology
$\mathrm{Fa}(t)$	Cumulative distribution function
$\mathrm{Fa}_{\mathrm{Str}(t)}$	Cumulative distribution function of string topology
FC	Failure costs
$f_{\mathrm{cu}}(t)$	Cumulative distribution function
$f_{\mathrm{f}}(t)$	Failure density function
$f_{\mathrm{H}}, f_{\mathrm{M}}$, and f_{R}	Frequency of the system in H, M, and R states
f_{ii}	Denotes the count of power interruptions experienced by the *i*th user throughout the simulation period
f_{m}	Frequency of reference signal
$f_{\max}$	Objective function's maximum value
F_{P}	Failure probability
$\mathrm{fp}_{\mathrm{Cen}(t)}$	Probability density function of central topology
$\mathrm{fp}_{\mathrm{Str}(t)}$	Probability density function of string topology
$\mathrm{fp}(t)$	Probability density function
f_{surge}	Frequency of surge
f_{sw}	Switching frequency
$f(t)$	Failure density function
$f(t)$	Probability density function
$F(\cdot)$	Probability of failure
$F(t)$	Cumulative distribution function
$F_{(t)}$	Unreliability of device
$F(\vec{x})$	The fitness function
$f(\vec{x})$	The objective function
fi	Frequency of transitions from "W" to "F" state
$f_{\mathrm{r}}(t_{\mathrm{r}})$	Repair density function
f_{pos}	Posterior distribution
f_{por}	Prior distribution
$f_1(t), f_2(t), \ldots, f_n(t)$	Failure density function of components
$f_{\mu_{\mathrm{B}},\mu_{\mathrm{C}}}$	The joint *pdf* of all random design variables and parameters
$\mathrm{fz}(y)$	PDF for the daily mileage
$g(t)$	Distribution function
$g(t)$	Failure density function of parallel system
$G(t)$	Cumulative distribution function
$G_{\max}$	Maximum generation number
g_{R_i}	The nominal resistance, i.e. capacity or strength
g_{S_i}	The nominal load, i.e. demand or stress
h	Characterized by a specific set of variables (T_{h}, t_{onh}, $n_{\text{cycle-h}}$)
$h(t)$	Hazard function of the system or device
$h^{(t)}$	The hidden layer's neuronal output
$h_{(t)}$	Hazard function
$\mathscr{H}_t$	History of failure process through time *t*
$\mathrm{Hz}(t)$	Hazard rate function

I	Solar cell output current
I_C	Collector current
I_c	EV charging current
I_D	Diode current
I_{dis}	Discharging power
$I_{g\ peak}$	Peak gate current
I_L	Light-generated current
I_L	Current through stray inductance
I_{OFF}	Turn off current
IR_k	Reliability importance of kth component
I_0	Drain current during device turn-off
J	An objective function to decide the locations of DGs
J	jth cut-set or tie-set
J	Possible outcomes
J	Load flow Jacobian
$J' \lambda_{min}$	The load flow Jacobian's minimum eigenvalue
k	Boltzmanns constant
k/n	k-out of n configuration for partially redundant system
K_B	Boltzmann constant
K_f	Probability of failure
K_s	Probability of success
L	Likelihood function
L	Step in FP for L'evy flight distribution
L	Length of stroke
L_{avg}	Average load
L_{cap}	Capacitor lifetime
LCC	Life cycle cost
LCCA	Life cycle cost analysis
$\overline{\text{ld}}$ and σ_{ld}	Mean of normally distributed load and standard deviation
L_i	Mean of L_{ith} load level
L_i	Connected average load at the ith load point
L_i	Peak load forecast in interval i
L_i	Maximum load of the ith day
l_i	Denotes the distance covered by the ith EV
Lik	Likelihood function
L_o	Operational life of capacitor
$L_{o\text{-}k}$	Resultant lifetime derived
L_r	Rated lifespan
L_s	Snubber's inductance
L_s/L_{stray}	Parasitic/stray inductance
L_0	Lifetime under voltage V_0
m	Standard operational time of the system
M	Total number of households
M_a	Amplitude modulating index
MC	Maintenance costs
m_f	Frequency modulating ratio
MWh	Megawatt hours

n	Total study duration
n	Number of components (capacitors/diodes/switches/filters)
N	Combined count of simulation trials
N	Cut-set quantity
N	Connection matrix
N	Argument an interval N $(a, b]$
N_{A}	Acceptor concentration
NB	Set of buses
N_{c}	A total of all distributor segments
$n_{\text{cycle-h}}$	Cycle count in category h
N_{D}	Donor concentration
N_{f}	Failure across numerous operational cycles
NG	Set of PV buses
N_i	Time duration for occurrence of $L_{i\text{th}}$ load level
N_i	Customer count at load location i
N_i	Number of customers interrupted for i_{th} interruption
N_{f}	Number cycle to failure
NL	Set of transmission lines
N_L	Number of customers at L_{th} load
NS	Number of samples
N_{T}	Number of customers served
$N(t)$	Value of counting random variable, i.e. failure numbers occurred in the interval $[0, t]$
$n_r(t_r, \Delta t_r)$	Number of repairs completed between t_r to $t_r + \Delta t_r$
n_j	Quantity of blocks in the jth cut set or tie set
N_{r}	Total number of repair carried over in a specified period of study
n_{XY}	Number of transitions from state X to Y
$N(0,1)$	Normal distribution with mean = 0 and standard deviation = 1
O	Network's output value
O	Occurrence
P	Power output of a solar cell
P	Transition probability matrix
P	Operative pressure
P	Probability matrix
P	Snubber resistor's power dissipation
$\hat{p}$	Security index without including the unavailability of a transmission line
p, q	Reliability and unreliability of component
p_{c}	Crossover probability
P_{c}	EV charging power
Pc(t)	Indicates the system's load decrease at time t and is set to zero when the network operates normally
P_{d}	Load demand
P_{d}	Average load demand

P_{dis}	Discharging voltage
p_H, p_M, and p_R	Probability of H, M, and R states
p_m	Mutation probability
P_f	Probability of failure
p_{f_i}	The *pof* for the solution with respect to the *i*th constraint
p_{f_i}	The approximated *pof* for the *i*th performance criteria
P_{f_i}	Probability of failure of L_{ith} load level
p_f^T	The true probability of the failure
P_i	Probability of loss of load in interval *i*
P_{Load}	Load power
p_n, p_O, p_S	Probabilities of normal, open circuit, and short circuit for diode
$p_0(t + dt)$	Probability of system being in "0th" state at $t + dt$
P_{pvg}	PV generator power
P_r	Rotor active power
$Pr(t)$	Probability function
p_s	Probability of successful change-over for standby system
P_s	Stator active power
P_S	Success probability
p_{tf_i}	The target *pof*
$p_1(t)$	Availability
$p_2(t)$	Unavailability
p_{1ss}	Steady-state availability
p_{2ss}	Steady-state unavailability
$P_a(X_i)$	Parent nodes
$P_a(X_t^i)$	Parents of $X_t^i$and X_t^i is the *i*th node at time *t*
$P^k_{g,min}$, $P^k_{g,max}$	Lower and higher limits to the active power output at the *k*th bus
$P(A_i)$	Mutually exclusive events
PII_o	Current insecurity index obtained from the ANN
PII_{th}	Threshold value of the indicator
$P(A, B)$	Joint probability
$P(A/B)$	Conditional probability
$P(B)$	Probability of event *B*
$P(B/A)$	Conditional probability
$P(E_i)$	Probability of occurrence of *i*th event
p_{ij}	Transition probabilities from state *i* to state *j*
$p_{ij}(t, \Delta t)$	Transitional probability
$P(S/X)$	Probability of system success provided "*X*" is working.
$P\left(S/\overline{X}\right)$	Probability of system success given that "*X*" has failed $\left(\overline{X}\right)$
$P(X)$	Reliability of "*X*" component
$P(x_i/P_a(x_i))$	Conditional probability
$P(X_t \mid X_{t-1})$	Transition probability
$P(X_{1:T})$	Joint probability distribution of 2 TBN network
P_W	Wind Turbine power output
$P_{W,r}$	Rated wind turbine output power
q	Total rate of transition

Q_{f}	Fixed oxide charge
$Q_{\mathrm{f_b}}$	The cumulative number of simulations resulting in failures
$Q^k_{\mathrm{g,min}}$, $Q^k_{\mathrm{g,max}}$	Lower and higher limits to the reactive power output at the kth bus
q_{ij}	Transition rate from state i to j
q_{jk}	Transition rates from jth to kth state
Q_{L}	Reactive load
Q_{m}	Mobile ion charge
Q_{ot}	Intrinsic charge of oxide
Q_{P}	Unreliability of parallel system
"r"	Number of standby component
r	Interruption duration
r	Signifies the duration of failure
R	Resistance to lines
R	Reliability
rand	Arbitrary digit between [0, 1]
$\mathrm{Re}(t)$	Reliability function
R_{D} (%/h)	Inverse to the meantime to failure (MTTF)
$R_{\mathrm{D\ Erying}}$	Rate of degradation by Erying
$R_{\mathrm{D\ Peck}}$	Rate of degradation by Peck
$\mathrm{Re}(t)$	Reliability function
$\mathrm{Rep}(t)$	Parallel equivalent reliability function
$\mathrm{Res}(t)$	Series equivalent reliability function
$\mathrm{Re}_{\mathrm{Str}(t)}$	Reliability of string topology
R_{Gtotal}	Total gate resistance
$R_{\mathrm{g\ ext}}$	External gate resistance
rh	Relative humidity (%)
r_i	Duration of interruption per failure at ith load point
R_i	Reliability of block/component i
$R_i(t)$	Reliability of the ith component
R_{off}	Resistance when the MOSFET turns-off
R_{P}	Reliability of parallel system
R_{s}	Reliability of series system
R_{s}	Reliability of system
R_{s}	Snubber resistance
Rs.	Indian currency rupees
R_{T}	Total resistance of the line
$R(\cdot)$	Reliability of a component at the respective time
$R(t)$	Reliability
$R(t)$	Reliability function
$R(t)$	Reliability of the system
RV	Residual value
$r_1, r_2,\ldots,r_n$	Time independent reliability
S	Sample space
S	Sensitivity matrix
S	Success tests

S	Severity
SAIFI_t, SAIDI_t, CAIDI_t and AENS_t	The desired/target values of the corresponding indices
S_{d}	System load overall (real and reactive load)
SD_{B_j}	Standard deviation of vector B, the jth constraint
SD_{C_l}	Standard deviation of vector C, the lth constraint
SI	Security index with including the unavailability of a transmission line
$S\lambda_{\mathrm{max}}$	Higher limit on the highest eigenvalue of $[S]$
SPII_i	Sensitivity of PII with respect to U_i
SPII_r	Sensitivity of indicator with respect to the "r" control variable
SPIID_i	Sensitivity of indicator (PII) with respect to load shedding at ith bus
t	Sequence of time
t	Operating or at-risk time of item
t_{don}	Turn-on delay time
$t_1, t_2 \ldots$	Failure times in success
T	Network's target value
T	Tie set quantity
T	Average time between instances of failure
T	Average value of temperature
ΔT	Junction temperature swing
T_{amb}	Ambient atmospheric temperature
T_{c}	Standby system the cycle time
T_{c}	Battery charging time
T_{cel}	Cell temperature
T_{dis}	Discharging time
T_{f}	Life length of component
t_i	The target output for the ith training dataset
t_i	Stands for the cumulative charging disruption time during the simulation period for the ith EV
T_j	Junction temperature in degree Celsius
T_{jest}	Estimated junction temperature
T_{jn}	Junction temperature
T_{jo}	Base junction temperature
t_k	Operating period
T_k	kth system state duration
T_n	Tournament size
T_{o}	Operating temperature
T_{r}	Time to repair of component
T_{r}	Upper category temperature
Tr	rth minimal tie-set
$T_1, T_2, \ldots, T_n$	Life lengths of components
u	Random number [0, 1]
U	Unavailability
$U_{\mathrm{best}}^{(k)}$	The best choice discovered in FP among all solutions in the generation that currently exists

$U_{\text{best}}^{(k)}$	The best choice discovered in Rao optimization method among all solutions in the generation that currently exists
$U_{\text{worst}}^{(k)}$	The current worst solution discovered using Rao optimization techniques, out of all the solutions at this generation
U_i	Input variable at ith node
U_i	Component of the input vector $[U]$
$\mathrm{U}_i^{(k)}$	At kth generation, the resultant solution vector
$U_i^{(k+1)}$	Vector modified at the kth generation
U_{ij}^0	jth parameter of U_i vector
$U_{j,\max}$ and $U_{j,\min}$	Upper and lower limits for the variable U_j
U_{L}	Hour/year
U_{s}	Average annual outage time
$U_{\text{sys},i}$	Duration of the system's yearly outage at the ith load point
$u(t)$	Intensity function of a point process
V	Voltage across the solar cell terminals
V_{c}	EV charging voltage
V_{ceo}	Collector-emitter voltage at T_{jo}
$V_{\text{ceon(I high)}}$	Collector-emitter voltage under high current
Vc,in	Cut-in speed
Vc,out	Cut-out speed
V_{dc}	DC-Link voltage
V_{dis}	Discharging current
V_{fr}	Forward voltage of the diode
ΔV_{ge}	Gate emitter voltage variation
V_{L_s}	Voltage across snubber′s inductance
$V_{L_{s2}}$	Peak voltage across snubber capacitor
$\Delta V_{\text{m},i}$	The change in the load bus voltage vector as a result of modal variation
$V_{\min}^i, V_{\max}^i$	ith load bus voltage's lower and higher limits
V_n	Virtual age after nth repair of the system
V_{o}	Operating voltage
V_{r}	Rated wind speed
V_{r}	Rated voltage
V_{Sr}	Voltage stress factor
V_{T}	Thermal voltage
V_{thr}	Threshold voltage
V_0	Nominal voltage
W_{hh}	The weight matrix that recurs between neighboring time steps' hidden layers
W_{hx}	The matrix of weights between the i/p and the hidden layers
W_{ij}	Weight that connects the ith input node to the jth hidden neuron
ΔW_{jk}	Weight change W_{jk}
X	Key-stone component in decomposition method
X	An element of the input vector

X	Reactance to lines
X_n	The time between failures
X_r	Rotor circuit reactance
X_s	Stator circuit reactance
X_T	Total reactance of the line
Y	The network's output
$Z(t)$	Instantaneous failure rate

Index

Reliability Analysis of Modern Power Systems, First Edition. Edited by R. K. Saket and P. Sanjeevikumar.

e

m

t

Printed and bound by CPI Group (UK) Ltd, Croydon, CR0 4YY

17/09/2026

14964315-0007